D0142678

The Evolving Female

The Evolving Female

A Life-History Perspective

Mary Ellen Morbeck,
Alison Galloway, and
Adrienne L. Zihlman, EDITORS

PRINCETON UNIVERSITY PRESS
PRINCETON, N. J.

Published by Princeton University Press,
41 William Street, Princeton, New Jersey 08540
In the United Kingdom: Princeton University Press,
Chichester, West Sussex

Library of Congress Cataloging-in-Publication Data

The evolving female : a life-history perspective /
edited by Mary Ellen Morbeck, Alison Galloway, and
Adrienne L. Zihlman.
p. cm.
Includes bibliographical references and index.
ISBN 0-691-02748-X (cloth : alk. paper). —
ISBN 0-691-02747-1 (pbk. : alk. paper)
1. Human evolution. 2. Women—Evolution.
3. Females—Physiology. 4. Women's studies—
Biographical methods.
I. Morbeck, Mary Ellen, 1945- . II. Galloway,
Alison, 1953- . III. Zihlman, Adrienne L.
GN281.E93 1996
573.2—dc20 96-20402

This book has been composed in Galliard

Princeton University Press books are printed
on acid-free paper and meet the guidelines for
permanence and durability of the Committee on
Production Guidelines for Book Longevity
of the Council on Library Resources

Printed in the United States of America
by Princeton Academic Press

10 9 8 7 6 5 4 3 2 1

10 9 8 7 6 5 4 3 2 1

Contents

Contributors

Silvana M. Borgognini Tarli
Department of Behavioural and Human Sciences
Section of Anthropology
University of Pisa
Via S. Maria 55 - 56100 Pisa, Italy

Patricia Draper
Departments of Anthropology and Human Development
Pennsylvania State University
University Park, PA 16802

Linda Marie Fedigan
Department of Anthropology
University of Alberta
Edmonton, Alberta
Canada T6G 2H4

Alison Galloway
Board of Studies in Anthropology
University of California, Santa Cruz
Santa Cruz, CA 95064

Mariko Hiraiwa-Hasegawa
Institute of Natural Science
Senshu University
Tokyo, Japan

Alison Jolly
Department of Ecology and Evolutionary Biology
Princeton University
Princeton, NJ 08544

Robin McFarland
Department of Anthropology
De Anza College
21250 Stevens Creek Boulevard
Cupertino, CA 95014

Beverly McLeod
National Center for Research on Cultural Diversity and Second Language Learning
University of California, Santa Cruz
Santa Cruz, CA 95064

Mary Ellen Morbeck
Departments of Anthropology and Cell Biology & Anatomy
University of Arizona
Tucson, AZ 85721

Gilda A. Morelli
Boston College
Department of Psychology
Chestnut Hill, MA 02167

Kathryn Ono
Department of Life Sciences
University of New England
Hills Beach Road
Biddeford, ME 04005

Catherine Panter-Brick
Department of Anthropology
43 Old Elvet
University of Durham
Durham, DH1 3HN
United Kingdom

Mary S. McDonald Pavelka
Department of Anthropology
The University of Calgary
Calgary, Alberta,
Canada T2N 1N4

Caroline M. Pond
Department of Biology
The Open University
Milton Keynes, MK4 6AA
United Kingdom

Joanne Reiter
Institute of Marine Sciences
University of California,
Santa Cruz
Santa Cruz, CA 95064

Elena Repetto
Department of Behavioural and Human Sciences
Section of Anthropology
University of Pisa
Via S. Maria 55–56100 Pisa, Italy

Barbara B. Smuts
Departments of Psychology and Anthropology
Center for Human Growth and Development
University of Michigan
Ann Arbor, MI 48109-1346

Virginia J. Vitzthum
Department of Anthropology
University of California, Riverside
Riverside, CA 92521

Adrienne L. Zihlman
Board of Studies in Anthropology
University of California, Santa Cruz
Santa Cruz, CA 95064

Acknowledgments

MANY PEOPLE contributed to this project and we are grateful to them all. We each owe an intellectual debt to Sherwood Washburn, who served as a mentor at various points in our careers. Through his comparative, functional, and evolutionary perspective that integrated the anatomical, behavioral, and social dimensions of apes, early hominids and ourselves as *Homo sapiens,* we first gained a sense of "real animals" and their lives.

We are indebted to the colleagues who contributed chapters and to their subjects, both human and nonhuman, from whom we acquired the information to formulate the discussions presented here. They allowed us to share their lives for a time, and we thank them.

Our appreciation goes to Emily Wilkinson of Princeton University Press and to her assistant, Kevin J. Downing. Both were there to answer the long stream of questions, provide practical advice on editing, and they showed confidence in the worth of this project.

We owe a huge debt to those who enabled each of us to communicate among ourselves and transmit manuscripts, revisions, and comments. Debbie Neal, Penny Stinson, Bill Hyder, and Jane Nyberg (UCSC) assisted in the translation of computer disks. Doris Sample's (UA) good humor in the face of repeated changes in word processing was also a great blessing. The late Walt Allen (UA), with infinite patience and wit, persevered in adapting computer resources to meet the needs of a changing volume. The Word Processing Center at UCSC word-scanned hard copy, under sponsorship of the Social Science Division. Jacky Leighton and the UCSC Social Science Steno Pool facilitated the printing of this manuscript. Michelle Bezanson (UA) and Kim Nichols (UCSC) toiled in checking references and quotes.

As editors, we each experienced some very personal debts. We now recognize that (1) you can't write a book without hands, and (2) you can't write a book with a two-year-old, now a three-year old. Yet despite these revelations, we managed to complete the bulk of the work during a time when one of us was disabled by repetitive strain injury and another was balancing work and child care. Although these factors substantially slowed our pace of work, they also brought new insights into "what it means to be human." For those who helped maintain our sanity and our survival during this period, we would like to express the following personal thanks.

From M. E. Morbeck: As I continue to learn how to build a new way of life, thanks to John Hoffman who kept me together and took me away at all the right times. Many other friends also continue to provide much-appreciated intellectual as well as logistical "life support" and I thank them: Virginia Morbeck and Michelle Bezanson, Alison Galloway, Virginia Landau, Ken Mowbray, Lita Osmundsen, and Jane Underwood. My Tumamoc Hill friends, especially Paul Martin, deserve special thanks for creating and maintaining an intellectually exciting and productive place to work. The natural history of "The Hill" itself has revealed its healing powers for which I am very grateful. In addition, Jane Church, Nancy and Bill Cook, and Nancy and Jim Hoffman provided much-needed retreats for quiet thinking, reading, and writing. Thanks also to a myriad of highly skilled physicians, other health-care specialists and, especially, my ever-cheery hand therapists, who nurtured my positive outlook as part of their medical practice. I am grateful to Jane Hill, John Olsen, and Holly Smith for their use of administrative powers in the best possible way—to facilitate faculty to achieve research and teaching goals no matter how their lives are constrained.

From A. Galloway: I would like to express gratitude to my family, especially my parents whose emphasis on education allowed her to pursue this career. My husband, Charlie, picked up immeasurable "slack" when my mind and efforts were devoted to the production of this volume. Without his emotional support and, as importantly, practical assistance, I would not have been able to even consider editing this book. I know he tired of my focus on this work, but he also demonstrated incredible patience and fortitude. I am deeply grateful and hope that I never have to put him through it again—although I say that as I embark on another major writing project.

My daughter, Gwyneth, may never understand how much she helped bring this book into perspective. Until now, my examination of human reproduction was theoretical. The physical demands of childbearing and rearing, combined with a professional career, drove home the delicate balance and continual negotiations women make among their various roles. Like most new parents, we never imagined how much time a child required. But, like most parents, we also never imagined how much joy a child brings. That, too, is part of the balance. In many respects, she is my best research project.

From A. L. Zihlman: I would like to acknowledge the colleagues who contributed to the genesis of this volume. The conference on which the volume is based was held at the University of California, Santa Cruz. Chancellor Robert Stevens designated the conference a cornerstone event of the campus' twenty-fifth anniversary celebration. Visits to Caroline Pond at Milton Keynes in the U.K. and to Silvana Borgognini Tarli in Pisa, Italy stimulated me to try to bring together outstanding scholars like these to exchange information and ideas about female biology and evolution. During the two years of preparation for the conference, Amy Nilson taught me logistics; Debra Bolter, Carla Daniel, Beverly McLeod, and Robin McFarland were significant in the planning and execution. Melanie Mayer, Carolyn Martin Shaw, Catherine Cooper, Sheila Hough, and Donna Haraway contributed to the conference discussions. Debbie Neal and Wanda Santos acted as hostesses. Joanne Tanner and Chuck Ernest videotaped the public lectures. Carla Daniel contributed to early stages of the editing process.

In funding the project and forming the inception of the idea, we worked closely with Marilyn Cantlay (UCSC), who assisted in seeking financial support and in grant writing. A small grant from Feminist Studies Research (UCSC) enabled the project to take its first step; the Graduate Division and Dean Geoffrey Pullum provided major funding, supplemented by the Social Science Division and Dean William Friedland (UCSC). The University of Arizona Social and Behavioral Research Institute provided funds for M. E. Morbeck's participation. Private donors covered some additional expenses. A grant from the Wenner-Gren Foundation first for Anthropological Research and the interest and support of Sydel Silverman ensured that the conference happened. Lita Osmundsen, now retired from the Wenner-Gren Foundation, first showed us the value of small conferences.

What Is Life History?

WHAT DOES it mean to be human? How did we get to be the way we are? How do we begin to define "humanness"? We can approach these questions from comparative, functional, and evolutionary perspectives, which allow us to describe and explain variation in biology, behavior, and ecology in ourselves, in our close relatives, and in our ancestors through time and across geographical and ecological space.

In this volume, a number of themes are linked. The most important one is the evolutionary heritage that each organism carries—a heritage that shapes the anatomical, physiological, and behavioral responses possible to environmental stimuli. Throughout the book, we focus, in some way, on our species, humans, the features we share with other primates, and our legacies as mammals.

Life-history theory provides a means of addressing the integration of many layers of complexity of organisms and their worlds. Although most researchers recognize the many facets of what it takes to eat, survive, and reproduce, they highlight only a few traits in their modeling of life history. We broadly define "life history" in this volume to integrate both reproductive and survival features (Morbeck chap. 1), with the goal of examining the complexity itself. Natural selection is a critical component of the optimality models often used in life-history theory, and we recognize its importance as an evolutionary mechanism. We also stress the vagaries of individual life stories and their consequences for survival and reproduction.

We emphasize the life story of the functionally integrated individual, as expressed within its species-defined boundaries, from conception to death, and its genetic and other contributions to its population. This approach requires empirical, multigenerational, natural history studies. Biology and behaviors of known individuals are placed in social and ecological contexts as they negotiate unique pathways through the life stages, mate and, if female, rear offspring.

WHOLE BODIES/WHOLE LIVES

Our emphasis is on the role of the individual in evolution. Each human, like other organisms, brings a genetic legacy that can be passed on to the next generation. Embodied within each person is the record of our evolutionary history. We see it in what makes up our genes and how they work. The genetic and molecular mechanisms for reproduction and guiding development of species- and sex-defined attributes stretch back billions of years to the origin of life itself. Sexual reproduction allows for the recombination of genes to form unique combinations in individuals while retaining the overall pattern of the species (Jolly chap. 19).

We can see our evolutionary past in our bodies, how they function, and how we behave. Our body plan is that of a vertebrate (e.g., head, trunk, four limbs), whereas our basic way of life and how we grow to maturity and reproduce follows the pattern of other mammals. Our large complex brains, body movements, and life-cycle timing are shared with the other primates. Independent use of fore- and hindlimbs, good hand-to-eye coordination based on grasping digits (fingers), and overlapping fields of vision (in-depth vision) are tied to the primate pattern of big brains. Our reproductive pattern that emphasizes one infant and a long prereproductive period of biosocial growth and development, and long periods of gestation, infancy, childhood, adolescence, and adulthood are shared with the

Old World monkeys and apes. Our recent ancestry with these catarrhine primates is evidenced by similarities in both the form and function of our bodies, for example, full stereoscopic vision, thirty-two teeth as adults, and fine-tuned manipulative hands. From our shared history with apes, we have broad, upright trunks, and flexible forelimbs. We are unique as humans—bipedal with large brains, long childhoods, and the ability to organize and explain our worlds via language and culture (Morbeck chap. 9; Zihlman chap. 13; McLeod chap. 20).

We place the organism at the center of the evolutionary process by focusing on individuals as each maneuvers through the life stages. The importance of the individual organism has long been recognized in evolutionary theory in relation to natural selection. As Mayr stated,

> [T]he individual and not the gene must be considered the target of selection. There are many ways of documenting the primary importance of the individual. First of all, it is the individual as a whole that either does or does not have reproductive success. Second, the selective value of a particular gene may vary greatly depending on the genotypic background on which it is placed. Third, since different individuals of the same population differ at many loci, it would be exceedingly difficult to calculate the contributions of each of these loci to the fitness of a given individual. And fourth, accepting the individual as a whole makes it unnecessary to make the confusing distinction . . . between internal selection (dealing with processes going on during ontogeny) and external selection (dealing with the interactions of the adult with the environment). (1988:101–102.)

The individual also serves as the pivotal point for moving into investigations at other levels. As Bennett stated,

> Biologists are usually content to specialize their studies at one functional level of organization (e.g., molecular biology, organ morphology, population biology). They seek the bases for the phenomena they study at one level removed from their expertise and are generally satisfied with explanations at those levels. Organisms, in contrast, are integrated units that encompass all levels of biological organization. They do not make distinctions between such properties as morphology, physiology, and behavior, nor do they develop those properties in isolation from all the other traits they possess. Individual traits do not evolve independently, rather they evolve only in the context of a complete functioning organism. (1989: 191.)

We start with the whole person, you or me, for example (fig. 1.1 chap. 1). From here, for instance, we can move "up" and "out" into the empirical studies that deal with specific individuals as integral parts of populations. This is exemplified in this volume by the work of Fedigan, Morelli, Pavelka, Panter-Brick, Ono, Draper, Reiter, and Vitzthum. We can also move down, into the anatomy and physiology of the individual as seen in the chapters by Morbeck (chap. 9), Galloway (chap. 10), Pond (chap. 11), Borgognini Tarli and Repetto (chap. 14), Zihlman (chaps. 8, 13), McFarland (chap. 12), Panter-Brick (chap. 17) and Vitzthum (chap. 18).

The many dimensions and layers of complexity of the individual and its world weave together features that have evolved to promote survival and biosocial health at each life stage with those associated directly with mating and rearing of offspring in adulthood (Zihlman et al. 1990; Morbeck 1991b). The emphasis on whole bodies and whole lives is more than a theoretical concept. It is grounded in studies detailing how the survival and reproductive "decisions" made at different biobehavioral levels of organization vary throughout life. From the point of view of the organism, such complete integration is essential.

Natural-History Studies

We view *Homo sapiens* (modern humans) as individuals operating as part of the natural world. As such, humans, too, are subject to the same "decision-making" situations that affect all life. Decisions are often illustrated in flow diagrams as an event in the life of the "typical" organism, for example, the trade-off between growth and reproduction. We see decisions as both more complicated and more basic, such as "How do I make it through today?"—what and when to eat, when and with whom to mate, how to avoid danger, etc. It is these activities, made by many individuals, that influence the course of evolution.

We include the study of the natural history of humans within the evolutionary approach. We are interested in (1) what individuals do in particular environments in order to stay alive and healthy, to grow to biosocial maturity, to mate, and to rear offspring; (2) what survival and time-based life-history species- and sex-specific features enable or constrain organisms as they grow to maturity and reproduce; (3) what factors contribute to survival and reproduction, and in particular, from the perspective of the individual's life, what are the short- and long-term consequences of particular biobehavioral and environmental interactions for adult reproductive outcome and individual health and psychosocial well-being throughout the life stages?

Why Females?

When and how often during the life cycle can an individual female or male reproduce? How quickly can an infant achieve physical and, in addition, social independence and reproductive capability? Finally, how much time and energy does each sex devote to activities that promote successful mating and how much to protection and raising of healthy offspring to reproductive capacity?

Females and males share species characters but have different reproductive roles. Sexual reproduction is not new; it has been around for about a billion years. As Jolly (chap. 19) points out, it is the basis for information exchange from generation to generation. Sexual reproduction took an interesting turn more than 100 million years ago with the origin of mammals and a new way of raising young. This new twist, the mammals, is defined, in part, by reproductive asymmetry. The differences between the reproductive roles and "responsibilities" of females and males, especially in placental mammals like ourselves, necessitates slightly different criteria by which to measure their contribution to the vertical passage of genes.

Females have been the primary focus of life-history studies. Our choice to retain this focus in this volume reflects a multilayered set of reasons. Conventional life-history studies grounded in population biology (demography) have concentrated on the timing and extent of energy investment in reproductive events that characterize the females of a species. There are many reasons for this focus. Females are the "limiting" sex in terms of the number of offspring that can be produced. This is seen especially in primates with a single offspring per birth and extended periods of growth and development. Maternity, unlike paternity, is very rarely in question because of the nature of the mammal mother-infant relationship. Females and their offspring are observable and can be documented in field studies. Long-term, multigenerational studies of humans, nonhuman primates, and other mammals, as shown by the following chapters, record both the nature and timing of female reproductive efforts.

In addition to the influence by females on the number and identity of offspring, females are important, at the cellular and genetic level, as the source of the majority, if not all, of the nonnuclear components transferred to the next generation. Part of this component is the DNA housed in the mitochondria, the organelles that generate the cell's energy-storage mechanism. An interesting by-product of this feature is that studies of variation in mitochondrial DNA within and among modern human groups also allow analysis of maternal lineages and times of divergence of various groups in the evolutionary past.

The mother also contributes to her offspring half of the nuclear DNA, which contains the species template for the growth and development. Along with the paternal contribution, these set the parameters by which the organism is formed and functions. Nutrients produced by the mother and carried in the ovum fuel the fertilized egg through its initial divisions until implantation in the uterus.

Placental mammal offspring are totally dependent on the mother for a good start in life. The mother-infant bond, so evident after birth, builds on the biological connections of the prenatal period. Internal gestation allows the mother to transfer energy directly to the offspring through the placenta. The mother is the environment, and the infant relies on her body for its survival and for its continued growth. This does not mean that the infant is without power, as fetal physiology permits capture of maternal nutrients—at times even at the expense of maternal health.

During her pregnancy, the mother must continue to promote her own survival and well-being along with that of her offspring. The requirements of the pregnant female are frequently viewed as the sum of the mother and fetus (i.e., "eating for two"). However, the physiological capabilities of the pregnant female are enhanced to capture nutrients and are combined with behavioral changes in feeding and activity levels that can substantially change the size of the nutrient supply available (Zihlman chap. 13; McFarland chap. 12). Often even an otherwise minimal diet still can sustain the pregnant female with little effect on the success of her reproductive efforts.

After birth, the biological connections are continued and, particularly in mammals such as the primates, social and ecological bonds between mother and offspring are significant. The mother feeds, protects, and socializes her one or more offspring. The primate mother in most species also transports her offspring throughout its infancy.

Among the primates, Old World monkey, ape, and human mothers continue to interact with older offspring even after giving birth to another infant. At the same time, they are still promoting both their own survival and well-being and that of the youngest offspring. In many primates, such as Japanese macaques, mother-daughter interactions apparently are crucial for maintaining group cohesion through generational time (Pavelka chap. 7). These alliances have important benefits for both mother and her offspring. In humans, the mother often also provisions and cares for the older siblings of her youngest offspring although they, in turn, may help the mother in child care.

Temporal/reproductive characters, expressed in individual life stories and aggregated in breeding groups, pattern population dynamics at the population and species level. Females are important since, as suggested above, the investment in time and energy dictates the population replacement rate. This can only be determined by when, how rapidly, and how many offspring a female can bear and rear. Yet, the female does not determine this alone, as it is also governed by the species-specific pattern of growth and development as modulated by the individual offspring itself. Social organization also influences the mating opportunities for females. Availability of males, social context, and individual life story all play a role in how each individual is involved in further reproduction.

Mechanisms that contribute to evolutionary change act on the individual, and their effects are measured in populations and higher taxonomic categories over time. Chance events as well as natural selection are known to influence survival and reproduction of the individual. Gene flow, another source of genetic change, obviously occurs between individuals and affects the extent of genetic variation. Mutation, the source of all variation at the molecular level, occurs in individuals but frequently may arise during the elaborate processes of cell division and replication that accompany sexual reproduction. These changes are measured in populations in terms of gene frequency or as expressed in morphology.

The mothers and infants, within the context of population dynamics, bring us to the level at which we can measure evolutionary change. Monkeys, apes, and humans are social as well as biological beings, and reproduction may be said to be "costly" when measured in terms of energy investment and time. This is especially evident when, as shown by chapters in this volume, (1) reproductive success is defined in its broadest sense as the number of offspring produced during a lifetime that survive to reproductive maturity and produce their own offspring, and (2) when biological and social efforts invested in mating activities and protecting, rearing, and socializing offspring are both recognized as major contributing factors to reproductive outcome.

What Is Life History?

What is the life-history approach, and why is it important? Life-history studies grew out of the natural history tradition. These studies focused on whole organisms and whole lives by examining and documenting individuals through their life stages, survival, and reproduction. This knowledge was derived from empirical observation and some field experimentation (Fedigan chap. 2).

As the biological and social sciences developed, the amount of information that could be collected and analyzed grew dramatically. Individual scientists simultaneously came to specialize in ever narrowing fields—not only in their choice of organisms examined but also in the level at which they were observed. The "whole organism–whole life" perspective faded with the rise of theoretical population biology and quantitative population genetics. The development of mathematical modeling as part of population biology and ecology focused on understanding population dynamics in which the time and energy investments could be calculated based on populational averages or theoretical constraints.

Population Biology and Life-History Theory

Most current definitions of life history are part of the tradition of population-biological studies. Life-history theory can be seen to fall into two different emphases (e.g., Lessells 1991): (1) population biology at the genetic level and (2) population biology at the phenotypic level, generally with the use of optimality models. The former allows for mathematical modeling to the gene or allelic level, examining the effects of natural selection on gene frequencies. The latter concentrates on the timing of the life cycle, generating an average of the features expressed by the individuals. In essence each perspective interprets life-history theory in the manner that may be termed "life-event theory": life is marked by occurrences that are measurable and for which selection works toward the optimal pattern.

These life-history models center on the life cycle or timing of growth and maturity. They emphasize age-specific values of reproduction and death, that is, the reproductive events that pattern the life course (Partridge and Harvey 1988). Stearns (1992:9), for example, stated that "life history theory deals directly with natural selection, fitness, adaptation, and constraint," all of which involve whole bodies/whole lives. Yet, his list of principal life-history traits only includes specific measurable "events," and, in addition, "amounts" such as "size at birth; growth pattern; age at maturity; size at maturity; number, size, and sex ratio of offspring; age- and size-specific reproductive investments; age- and size-specific mortality schedules; length of life" (1992:10). We see problems with linking his definitions and the criteria for measured categories. For example, these events do not occur at a single point in time for the individual in the species or population, nor do they occur at the same time for all individuals. The use of fixed points hinders recognition of the wide-ranging variability of these "events" and the gradual nature of transition between life stages. Furthermore, in this reductionist version of life history, the survival features such as locomotion, feeding, predator avoidance, and social behavior so critical throughout life are assumed or, too often, never considered.

One of the central features in conventional life-history theory as part of populational studies is the concept of "trade-offs" in which the life-history attributes are balanced against each other. These trade-offs link traits that constrain or limit their simultaneous action and evolution. Using this concept, an individual's current reproduction is suggested to imperil current survival and also have an impact on future reproduction. "Because energy used for one purpose cannot be used for another purpose, living organisms face a series of trade-offs through time. The two most fundamental trade-offs, which are at the center of all life-history theory, are those between current and future reproduction and between the number and fitness of offspring." (Hill 1993:80.) Other proposed trade-offs are made, for example, between the number, size and sex of the offspring (Bell and Koufopanov 1986; Boyce 1988; Lessells 1991; R. H. Smith 1991; Roff 1992; Stearns 1992; Charnov 1993; Zhang and Wang 1994).

Trade-offs are not generally discussed in terms of the day-to-day detail of individual's lives, yet life stories are filled with daily decisions about survival and reproduction. Each individual continually confronts and copes with situations that are unique to her/his life. Physiological trade-offs link functions that can be constrained because of reliance on the same energy resource by the individual. For instance, should a female devote energy to mating or to procuring additional food? But then what good is mating if it is unsuccessful because of poor nutrition and body health?

How is our version of life history similar to the population-biological approach? Like population biologists, we focus on females. Females, in conjunction with species-defined growth and development, determine the growth potential of the population by when, how fast, and how many offspring can be produced.

The timing of the life cycle, therefore, is important in all studies of life history. We share this emphasis with conventional life-history theorists. We also recognize the need to look at the population in order to measure aspects of genetic and phenotypic change through generational time. Mathematical models provide us with a framework for organizing information at the populational level or higher. Since we are ultimately interested in how evolution works, we focus on the individual and its links to its population. We broaden our interests beyond counting time to document how each individual's life plays out through the life stages and its survival and reproductive outcome. Our focus on the individual rather than including individuals only in central-tendency statistics sets us apart from the more conventional life-history theorists. Although the importance of the individual is mentioned in virtually all treatises on life history, this view frequently is overshadowed by the emphasis on populational calculations.

Awareness of the individual with its unique life story becomes crucial when we look beyond natural selection to the effect of "luck" or chance events in individual lives. Variability among individuals becomes apparent as researchers have begun to test mathematical models with organisms in the real world. Phenotypic plasticity, the range of variation within a species boundaries, must be viewed as essential, rather than deviations from an "optimal" pattern. As R. H. Smith stated, "Individuals that are "lucky" for some non-genetic reason will tend to perform better in several components of fitness than others that by chance are less fortunate. It is the confounding of effects of luck with those of constraints within individuals that make the simple **E** matrix hard to interpret." (1991:92–93.) Factors that appear critical at one period of the life cycle may be less important later. This can most readily be seen in tracking individuals within populations. As the chapters in this volume show, we can understand the importance of the individual life story and the contributions to population dynamics only with an empirical natural history approach combined with a broad view of life-history variables.

Therefore, we wish to return the primary focus to the whole organisms–whole lives as the appropriate starting point for understanding life history. This viewpoint aligns with a trend toward the reemergence of the importance of the individual in biology and ecology (Mayr 1982, 1988; Bock 1989; Kohn 1989; Wake 1990).

In this context, the emphasis on statistical central tendencies, so prevalent in current life-history models, ignores the complexity of reproductive and survival tactics. By stressing "optimality" and emphasizing life events, the variation in biological and behavioral responses to environmental stimuli is underestimated. Rather than proposing that selection works toward a single optimal strategy, we propose that natural selection may actively favor two or more parallel or complementary effective strategies for survival and reproduction. Each will allow the transmission of genetic material into the next generation but retain flexibility within the species to cope with the environmental fluctuations that may periodically favor one or the other for short periods of time. We see flexibility itself as an evolutionary adaptation.

A New Focus to Life-History Theory

As becomes evident in this volume, we are not fully comfortable with the translation of life-history theory, as it presently stands, into analyses of the long-lived and socially complex mammals such as the primates, including humans. Traditional life-history theory is difficult to apply to organisms with life cycles in which the stages and events are not punctuated by abrupt transitions. Primates are particularly prone to "blurring" of the boundaries between life stages, resulting in great variability in the timing of major life events such as the age at first birth. Much of this variation is due to flexibility in the response to pressures from the physical environment, but the social environment also can impede or accelerate the rate of passage through the life course. Drastic simplification of complex organisms such as primates is required before they can fit a model in which life-cycle timing and energy/time trade-offs can be readily calculated (e.g., Charnov and Berrigan 1993).

How, then, does our approach differ from the traditional approach to life-history theory? First, in the perspective of this volume, which emphasizes individual life stories, the data on each individual are then aggregated to produce a portrait of the population without losing the focus on the individual. In part, our work reflects a more positive approach to life history since our emphasis is on what the animal *does* in order to live and reproduce rather than relying largely on the single features, for example, age at death.

Second, we do not see trade-offs as the best way to understand life-history features. This concept has been traced to the laws of thermodynamics, which provide the principles by which energy is exchanged in *closed* systems (e.g., Hill 1993). The idea that energy cannot be used for two life processes at the same time—a useful world view if one wants to follow mathematical models—does not reflect the situation encountered in real animals living real lives.

As seen in this volume (Pavelka chap. 7; Panter-Brick chap. 17; Vitzthum chap. 18) the life of highly social animals such as primates consists of doing many things simultaneously. Rarely do we see the "choice" as simply one of eating or reproducing. Time and energy, in most cases, do "double duty" or even "triple duty" in that time the mother spends feeding, for instance, also is time spent showing her offspring what to eat, how to find it, and how to claim it.

The use of trade-offs in life-history theory has often been equated with a consistent level of energy and time available to the organism. In real life, the amount of available energy is a constantly shifting quantity. In part, this is due to changes in the availability of resources as a result of seasonal shifts, climatic fluctuations, habitat changes, and so on. The animal, however, also plays a major role in determining the level of energy. This can be done by choosing or getting preferential access to higher-quality foods, or focusing attention on certain food sources. Digestive tracts are not consistent: there is individual variation and, throughout the lifetime, the size and function of the digestive system will vary to accommodate different levels of demand for nutrients. Behaviorally, animals may forgo nonessential occupations such as grooming or resting in favor of meeting the demands of feeding and reproduction. Although the concept of trade-offs may be appropriate in some tightly controlled situations, in the real world the complexity of the choices confounds attempts to force them into a simple dichotomy.

Third, we raise the question of concept of time. Most organisms, including long-lived ones, do not appear to use long-term planning in making decisions. The decisions are based on the opportunities available now, the hazards apparent, the hormonal milieu of the organism, and its energetic resources. Basically, it is asking, "Will I feed or mate?" not "Will I reproduce now or postpone it until a later date when the environment may be better?"

Even for humans who have a concept of past, present, and future, we often live for the present and make decisions about survival and reproduction that are immediate. Hill (1993) pointed out our unwillingness to forgo the pleasures of a high-fat diet or of smoking at the risk of disease later in life. We understand that we have life after the present but infrequently negotiate our pathway with these ultimate goals in mind. For example, few people will admit to being fully aware of how the act of reproduction would alter their lives.

Fourth, to follow the natural history approach we advocate expansion of the current concept that highlights the timing of growth and development markers and reproductive events so that they intertwine with survival features with their biological foundations in functional anatomy, biomechanics, feeding ecology, and so on. Additional features, at an individual level, would encompass well-being—including the influences of disease, trauma, and psychosocial health—and, as aggregated groups, would reflect population demography. This broadly defined life-history approach allows for greater recognition of the multiplicity of factors that constitute the individual's environment. We see natural selection as enhancing the flexibility within species-defined systems. The range of flexibility is seen in hard times; therefore, cross-sectional studies, lacking generational time, often fail to explore this essential aspect of life history.

Fifth, the approach, therefore, should be both multilevel and multigenerational. Within the organism, the shifting roles of biological factors (anatomical and physiological) must be acknowledged. Compilation of data on specific individuals tracked over time will permit an understanding of the mechanisms of populational change. In order to test hypotheses, information on parent-offspring relationships are necessary. Although females will remain the primary focus, the availability of DNA paternity testing in mammals, including nonhuman primates, is providing a window on the reproductive outcome of males.

Finally, returning to our beginning, the need for grounding of life-history studies in natural historical, empirical work is essential. As Southwood concluded: "More holistic studies are needed on communities and organisms along one or other of these axes to test the systems and the predictions that have been made. Combinations of field observations with information from the literature may be a powerful comparative approach" (1988:14.) The use of this approach also permits integration of humans with the natural world.

The population-biological concept of life-history theory has recently emerged in cultural and ecological anthropology (Hill 1993; Worthman 1993; Chisholm 1993; also see Low [1993] on ecological demography). Although the term "life history" has been used by cultural and social anthropologists to refer to recorded biographies, this new usage follows the current but narrow version of life-history theory.

Studies of growth and development have a long history in human biology (Tanner 1978) and has contributed to our understanding of nonhuman primates (Schultz 1956; E. Watts 1985b, 1990; Brizzee and Dunlap 1986). "The problems of human ontogeny and phylogeny will never be solved by the study of man alone, but are largely dependent upon new and more adequate data on the growth and evolution of all the Primates. Since any phylogenetic change has to affect primarily the processes of growth, additional information on the developmental changes in monkeys and apes is one of the first requirements for a thorough appreciation of the peculiarities of human growth, which have separated man and the anthropoids." (Schultz 1933:61.) There has been a recent resurgence of interest in shifting between levels in order to understand the processes of growth and development and apply this to the fossil record (Bromage 1992; B. H. Smith 1992, 1994). Growth and development are now being integrated into biological anthropology with an important evolutionary twist. Data generated from this new research are exciting because combining them with an emphasis on life-history studies on human and nonhuman primates (Hiraiwa-Hasegawa chap. 6; Morelli chap. 15; Pereira and Fairbanks 1993) allows us to begin to synthesize lifeways and life-history features as integrated in individuals throughout their lives. The combination of these studies with life-history theory is a powerful approach to understanding our evolution.

We recognize the phylogenetic as well as the present-day connections among our own species, other primates, and mammals. Via cross-cultural studies, we are able to see and describe the complexities of our lives in the natural, economic, political, and cultural world. We must be evaluating real animals, living real lives, in real time and real environments.

Part I

Perspectives on Life-History Studies

1 Life History, the Individual, and Evolution

Mary Ellen Morbeck

ONE view of genes and individuals:

> A body is the genes' way of preserving the genes unaltered.
>
> R. Dawkins, *The Selfish Gene*

AND a dramatic restatement:

> Bodies are merely the places where genes aggregate for a time. Bodies are temporary receptacles, survival machines manipulated by genes and tossed away on the geological scrap heap once genes have replicated and slaked their insatiable thirst for more copies of themselves in bodies of the next generation.
>
> S. J. Gould, *The Panda's Thumb*

FOR a different view:

> The problems of personal survival and reproduction form a natural hierarchy. In order to contribute genes to the next generation, an animal has to produce offspring that will in their turn breed; to do this, it needs to invest time and energy in rearing its offspring; and in order to do this, it must mate with a member of the opposite sex, which in turn requires that it find and then successfully court such an individual; but in order to be able to do even this, it must ensure that it survives long enough to perform all these activities and this in turn means that it must find sufficient food of the right kinds while at the same time it must avoid falling prey to a predator.
>
> R.I.M. Dunbar, *Primate Social Systems*

LIFE-HISTORY FEATURES, LIFE STAGES, AND EVOLUTION

Evolution is about differences in survival and reproduction: what it takes for organisms to survive thoughout the life stages and to reproduce during adulthood. "What it takes" is more than the genes emphasized by Dawkins—although he also recognizes life's complexities (1976, 1986). And, "what it takes" involves more than Dunbar's (1988) descriptions of behaviors—feeding, escape from predators, courtship and mating, and infant care—by members of a given species in particular environments.

Genes and behaviors, highlighted in these quotations, tell only part of the story. Observed behaviors depend on species-typical characters including their genetic and other biological foundations and how the body plan and way of life develop during the lives of females and males. Growth, development, reproduction, and aging during an individual's life are mediated by interactions within the organism as well as those with external environments. These multifaceted, "whole organism–whole life" features have evolved over many generations and are bounded by phylogenetic history. In my view, all of these components of an individual's life and their expression in populations as species' lifestyle are life-history attributes.

Life-history studies must incorporate the many facets and layers of complexity of biological features as patterned by species, sex, and age. These features combine with their functions, potential behaviors, and the roles they play to promote survival and reproduction of individuals living in particular environments. Survival features characterize lifeways, and reproductive features are time-based and relate to the life cycle (table 1.1). Life histories, as presented here, intertwine lifeways and the life cycle to define lifestyles. This comprehensive view of life-history features, therefore, includes both the pattern of day-to-day life and the

TABLE 1.1. Multifaceted, Multilayered Life-History Features

I. Survival Life-History Features of Different Life Stages
(Functions, ranges of expressed behaviors, and biobehavioral ecological roles)
Organismal integration and self-maintenance
Locomotion
Feeding
Social activity and communication
Predator-pathogen avoidance
II. Time-Based, Reproductive Life-History Features of Different Life Stages
Nature, sequence, and timing of body changes
Patterns of growth, maturation, reproduction (including mating and rearing of offspring), aging, and mortality

NOTE: Time-based, reproductive and survival life-history features are inseparable in organisms as they grow to maturity and reproduce in adulthood. They are distinguished here for heuristic purposes. The terms "primary" (i.e., reproductive/temporal) and "secondary" (i.e., survival) for life-history features were used by Zihlman et al. (1990) as an introduction to their description of the skeletal biology and life stories of Gombe chimpanzees.

TABLE 1.2. Life-History Features as Markers for Life Stages

Life Stages
In utero
< Birth
Infancy
< Weaning
Childhood
< Reproductive Maturity
Adulthood
< Mating
< Rearing of Offspring

associated schedule of growth, reproduction, and mortality (Morbeck 1991b).

Survival life-history characters enable organisms to stay alive and healthy. These biological, physical, and behavioral components of organisms interconnect to coordinate body functions and to allow individuals to interact with other organisms and other aspects of their environments. Modes of locomotion, resting, feeding, social activity, communication, and predator avoidance are examples of survival life-history characters. These social-maintenance activities are expressed in different ways and have different roles during the life stages.

Bodies, bones, brains, and behaviors of females and males unfold in a life course patterned by reproductive/temporal characters. These include the life-history traits traditionally used by population biologists: age and degree of maturation at birth, weaning, dispersal, first reproduction, and death (Cole 1954; Stearns 1976, 1992; Western and Ssemakula 1982; Peters 1983; Calder 1984; Harvey and Clutton-Brock 1985; Harvey et al. 1987; Meszéna and Pásztor 1990; Zihlman et al. 1990; Charnov 1993; also see Hill 1993; Fedigan chap. 2). Such attributes mark the transitions between in utero, infancy, childhood, and adulthood life stages (table 1.2).

Life histories usually are compared and interpreted for aggregates of individuals, in populations, species, and higher taxa (Calder 1984; Partridge and Harvey 1988; Harvey et al. 1989; DeRousseau 1990; Promislow and Harvey 1990; Harvey and Purvis 1991). They also can be investigated at the level of individuals and in terms of the various body systems and those structures and functions that make up whole organisms. Individual life stories are the unique sequences of activities, experiences, and reproductive outcomes that occur during an organism's life. They include where and how long an organism lives, how it behaves, and whether it reproduces successfully.

Many factors, including chance events, affect an individual's life story and its genetic contribution to the next generation and potential role in evolution. As part of reproduction, new variation can be introduced at the level of genetic organization, duplication, and differentiation, and the effects are expressed in individuals. During life, social and other environmental features influence, for example, an organism's physical features, behaviors, and actual ages at first reproduction or death. Knowledge of individual life stories allows us to detect natural selection or drift as measured in populations, that is, which individuals die, which live and reproduce, and whether their offspring survive to reproduce. Thus, life-history features, as I understand them, parallel broadly defined concepts of adaptation (e.g., Darwin 1859; also Bock and von Wahlert 1965; Bock 1980; Mayr 1988).

Why View Life Histories So Broadly?

The conceptual framework for life-history studies that I present here emerges from an interest in human evolution and research on fossil and living Old World monkeys, apes, and humans, that is, catarrhine primates. In the early 1970s I studied limb bones of *Proconsul* and other possible ape and human ancestors that lived in Africa and Europe 10–20 million years ago. Although the fossil teeth resemble those of modern apes, the suite of anatomical features represented in the fossil bones are unlike those of any apes (or monkeys and humans) that we know today.

Sizes, shapes, and other aspects of the anatomy of these fragmentary bones-turned-to-stone allowed me to infer a way of life that included survival features such as locomotor and postural abilities for quadrupedal weight-bearing and movement with grasping hands and feet (e.g., Morbeck 1975). But, fossils provide limited information about individuals, their lives, and their species. Using comparative and functional approaches, I began to look more closely at relationships of anatomy, behavior, and ecology and, later, at patterns of growth, development, and aging in modern primates with the hope of gaining better understanding of the bones and teeth in the fossil record.

Much of my work centered on anatomy and behavior of locomotion and other social-maintenance activities viewed from the perspective of the natural history of primates (e.g., Morbeck 1979). Early in my research I realized that these survival features, in turn, allow for successful mating and rearing of offspring. Survival and reproduction cannot be separated. Although I originally focused on anatomy and behavior of individuals, I also wanted to connect individuals with populations and assess their genetic legacies. This interest led me to the expanding literature on life-history studies in population biology. Current research on the skeletons from Gombe chimpanzees with known life stories adds to a life-history perspective. For example, reproductive profiles are known for a number of females, and we can test hypotheses about the relationships of anatomy and behaviors to reproductive outcome (Zihlman et al. 1990).

Studies of the life histories of catarrhines are complicated. Monkeys, apes, and humans have large bodies, relatively large brains, and good memories. They are long-lived and very social. Prereproductive life stages are long, and females generally produce one infant per pregnancy. Mothers spend a great deal of time and energy in raising infants to weaning age and usually maintain social contact with offspring during adolescence and adulthood. At the same time, mothers must promote their own health and survival (e.g., Fedigan chap. 2; Smuts chap. 5; Pavelka chap. 7; Hiraiwa-Hasegawa chap. 6; Zihlman chap. 13). These intricate, long-term social relationships allowed by long life span and the lengthy period of time prior to achieving reproductive maturity provide many opportunities for unique experiences that can affect an individual's health, survival, and reproductive outcome. Thus, individual lives and group life histories, even in the same species or populations, can vary widely.

My approach to life-history studies as illustrated in this chapter is necessarily broad. It emphasizes: (1) a whole organism–whole life view of biology and behavior of a primate individual as he/she negotiates a pathway through life and reproduces in adulthood; (2) concepts derived from life-history studies of populations that connect the individual with population-level phenomena that can lead to change in overlapping generations; (3) a natural history perspective that places individual monkeys, apes, and humans in social and ecological contexts; (4) the life stages through which each individual passes; and (5) a phylogenetic perspective that emphasizes the past record as well as current evidence of evolution.

Individuals: Whole Organisms–Whole Lives

Survival, Functional Integration, and Life in the "Real World"

What goes into making an organism and how it works depends on the organization and function of the different, but related, body parts. Each component and level of organization has its own properties, operating rules, and temporal dimensions. Each level is greater than the sum of its components. Somatic body systems are more complex than an aggregation of cells. Cell function depends on actions of proteins, which, in turn, are products of DNA. Interconnections are important and the functionally integrated individual at the organismal level relies on continuous adjustments of biology and behavior to accommodate many types of environments in order to sustain life and reproduce.[1]

Humans and other organisms live in physical and biological worlds as well in the social worlds of group life. Organisms can be characterized as open systems that interact with the environment with exchange of energy, matter, and information (Bruton 1989b; see Schneck [1990] for a review of a systems approach to mass, energy, and momentum transport in human biology). We each conform to and take advantage of the principles of physics and energetics as we confront environmental situations at all levels of ecological organization (e.g., O'Neill et al. 1986). These include, for example, the properties of gravity or gaseous exchange, temperature, habitat structure, and distribution of water and food items in time and space.

Living individuals acquire and use energy as part of a network of interactions with their environment to sustain life and to grow, develop, and reproduce. While recognizing the historical constraints of phylogeny and the natural constraints of the physical world, many biologists suggest that selection has favored optimal ways for individuals (or groups of individuals) to maximize energy acquisition and use (Kohn 1989; Parker and Maynard Smith 1990; also see Orzack and Sober 1994). These, in turn, relate to species- and sex-defined life-history characters. Future study of energetics, if it becomes possible to measure directly under natural environmental conditions, would provide a way to synthesize survival and reproductive life-history features and to test ideas about optimality and other expressions of biobehavioral, life-history features (Coehlo 1986; Bronson 1989; Panter-Brick chap. 7; Vitzthum chap. 18).

Growth and Development, Survival, and Reproduction

We still do not understand exactly how the many dimensions and layers of biological structures and functions change while, at the same time, staying stable and integrated in whole organisms throughout the life stages. The hereditary material characterizes a species and functions at the molecular level. It combines with processes of embryonic growth and subsequent morphogenesis under particular environmental conditions to guide the unfolding of life-history features of females and males from a zygote to birth through infancy and childhood to reproductive maturity. (See Wolpert [1991] for a good discussion of pattern formation and rates of change.) Relationships of an organism and its environment form a complicated matrix of features that are intertwined among the different levels of biological and ecological organization (fig. 1.1). Distinctions among the effects of genes (i.e., genotype) or environmental influences on an individual's development, physical features, and behavior (i.e., phenotype) become blurred (also see Ho 1988).

Females and males change during life as part of normal growth (for mammals, see Flower-

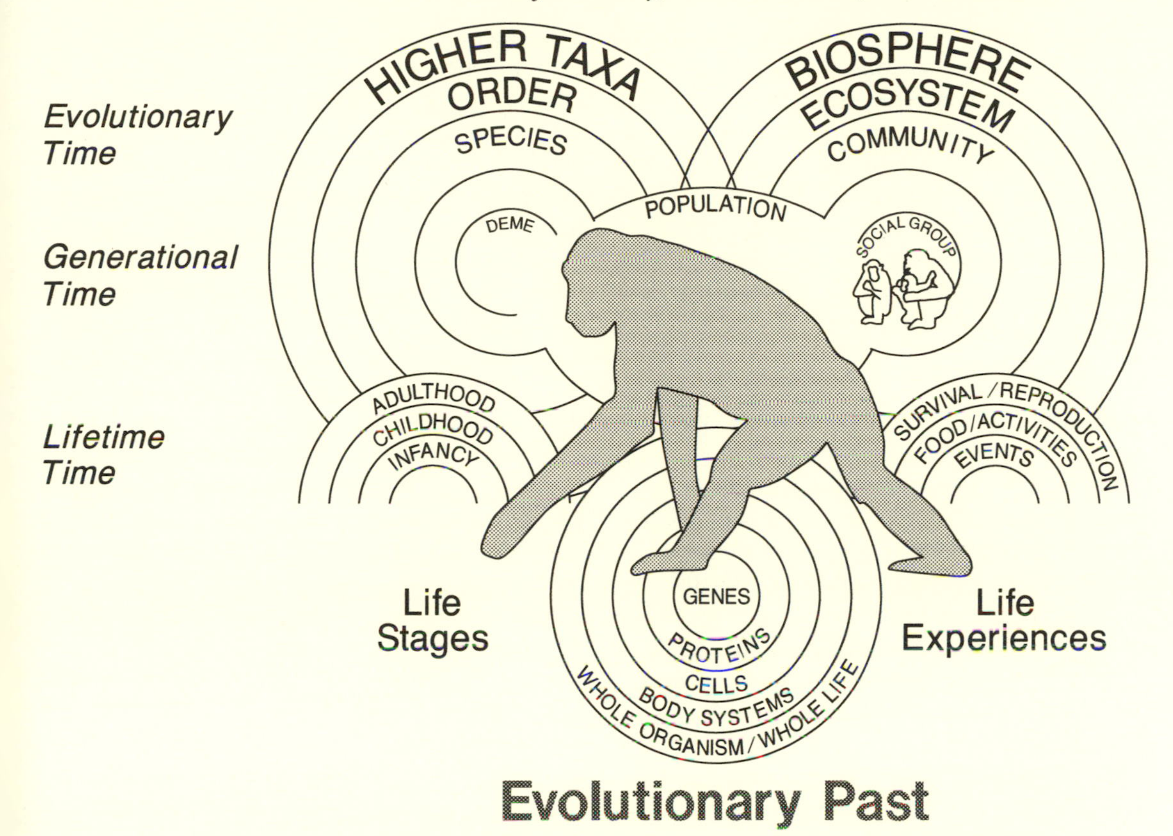

FIGURE 1.1. Life histories: levels and interconnections. The focus is on whole organisms and whole lives as illustrated by a chimpanzee. From the level of the whole organism, we can move "down" or "into" body systems, cells, proteins, genes, and molecules, and "out" from the individual to environmental interactions of social groups and of populations in communities or "up" to populations, species, and higher taxa. Corresponding time frames include: moment-to-moment molecular and cellular activity; lifetime growth, reproduction, and aging; hundreds or thousands of years of accommodation by populations to particular environments; and millions of years of adaptations and evolution documented in the fossil record. The "genealogical" hierarchy, represented on the left, emphasizes individuals throughout the life stages and as part of higher taxa. The intertwined "ecological hierarchy" on the right highlights life experiences and higher-order environmental associations.

dew [1987] and Bronson [1989]). The diversity of different life-history features within individual life stories are patterned by species' ranges of permitted variation. "External" factors affect the actual time of appearance, rate of change, and duration of growth, maturation, reproductive efforts, and aging of different body systems. These include, for example, (a) physical and climatic features of altitude, temperature, sunlight, and so on; (b) quantity and quality of foods; (c) nature, severity, age of onset, and duration of diseases and injuries; and (d) life experiences associated with social behaviors and maintenance activities as individuals grow to maturity and function as adults in ever changing social as well as biological and physical environments (Pavelka chap. 7; Ono chap. 3; Morbeck chap. 9; Borgognini Tarli and Repetto chap. 14; Galloway chap. 10; Vitzthum chap. 18; Morelli chap. 15; Panter-Brick chap. 17; Draper chap. 16).

The potential for variation among individual life stories and their contributions to populations, therefore, is considerable. This range of diversity itself can be viewed as a life-history attribute. Biologists call this variation "norms of reaction" or "phenotypic plasticity" (Ho 1984; Via and Lande 1985; Stearns and Koella 1986; Stearns 1992); biological anthropologists use "adaptability" (Baker 1988), "stress" (Goodman et al. 1988), or "flexible responses" (Vitzthum chap. 18); whereas primatologists prefer "plasticity" and "flexibility" (Clark 1991; Lee 1991; see others in Box 1991).

Individuals as Part of Populations

When individuals are aggregated in populations, reproductive features become "demographic" life-history features. Life stories of individuals pattern the number of births, migrations, and deaths in populations. They contribute to changes in populational phenomena of size and age-sex composition through generational time (Cole 1954; Charlesworth 1980, 1990; Ebenmann and Persson 1988; Dunbar 1988; Łomnicki 1988). Among mammals, in particular, the number of females in a breeding population and their potential rate of reproduction, which are reflections of species characters and phylogeny, determine the maximum number of births. Life-history features that pattern growth and mortality determine the potential frequency of births. Time and energy invested in gestation and lactation by adult females (or, from an infant's perspective, in utero and infancy life stages) define interbirth interval. Ages at first birth and death (except in individuals that experience reproductive senescence) define a female's reproductive span. Dividing the reproductive span by interbirth interval yields an estimate of the possible number of viable offspring that a female can produce during her reproductive years.

Individuals are functional components of populations; their survival and reproductive efforts contribute to demographic variables. Viewed from the level of populations, biosocial phenomena of family or larger groups can affect individual reproduction. Demographic variables at a populational level, in turn, may influence dominance relationships and, thus reproductive efforts (for an empirical example, see Pavelka [chap. 7]; for a simulation model, see Datta and Beauchamp [1991]). Population density, for instance, can affect the allocation of resources necessary for survival or the social opportunities to find a mate or to nurture a growing infant. Whether molded primarily from "below" by the biological characters of individuals and their behavioral-environmental interactions or "up and out" by the influences of their populations (or both phenomena), individual life stories result in different genetic endowments to the next generation.

A review of the literature in ecology and evolutionary biology shows that conventional "life-history parameters" and published data are primarily summaries of general patterns of allocated resources and demographic variables. Life-history models usually assume that energy resources are devoted during an individual's life, first, to staying alive while growing to reproductive maturity and, second, to "trading-off" investment in reproductive effort (e.g., reviewed in Stearns 1992). The "staying alive" components, that is, lifeways or survival life-history attributes, often are assumed or ignored. Most evolutionary theorists, who are interested in life histories, study what I call time-based, reproductive (life-cycle) characters at the level of populations or higher taxa (e.g., average gestation length, age of first reproduction, and so on). They emphasize sex- and age-defined characters of species and higher taxa that estimate when, how fast, and how often females and males can reproduce (Cole 1954; MacArthur and Wilson 1967; Pianka 1970; Stearns 1976, 1977, 1992; Charlesworth 1980; Stearns and Koella 1986; Charnov 1991; Roff 1992; Charnov and Berrigan 1993). Partridge and Harvey (1988:1449), for example, define life histories as "the probabilities of survival and the rates of reproduction at each stage in the life-span." Some biologists who are interested in modeling population dynamics and species evolution recognize that different factors in an individual's life can affect, for instance, age at first reproduction or time of death (e.g., Chepko-Sade and Halpin 1987). However, individual lives often become abstractions as researchers reduce their data to average ages for life-history markers.

Some investigators also have added what I define here as survival life-history attributes to these population and higher-taxa summaries. Body size, metabolic rate, and brain size, for instance, are discussed as part of allometric (i.e., changes in size and their consequences) and statistical explanations of species-typical life-cycle patterning at the level of populations and higher taxa (Western 1979; Western and Ssemakula 1982; Peters 1983; Calder 1984; Harvey and Clutton-Brock 1985; Harvey et al. 1987, 1989b; Partridge and Harvey 1988; Lee et al. 1991; Austad and Fischer 1992; Shea 1992; Smith 1992; Charnov and Berrigan 1993). These studies, therefore, also become removed from real animals and real lives. Yet, it is only among individuals that we actually can see what factors influence survival and reproductive outcome.

Real Animals and Real Lives: The View from Natural History

The most interesting relationships—at least from the perspective of the natural history of organisms like ourselves—are not explained by "traditional" life-history studies. I return to my original question: As biology and behavior change during life, what does it take for an individual to survive throughout the life stages, to mate successfully, and, in female mammals, to continue to survive while rearing offspring?

Natural history emphasizes empirical studies that allow us to ask two major questions: (1) How and why do individual females and males behave as they do throughout their lives in given environments? and (2) What are the consequences of their behaviors for survival and reproductive outcome? Studies of the same individuals throughout their lives (and of their local populations through generational time) begin to answer to the first question. It is also possible to recognize both nonrandom and chance factors operating at different levels of biological, behavioral, and ecological organization that contribute to differential survival and reproduction.

Studies of natural history, in which field researchers record the details of what it takes to stay alive and to reproduce, show evolution in action. They show how life experiences of both females and males influence survival, health, and reproductive outcome. Observations of generations of known individuals, where mothers and infants can be observed directly, allow assessment of a female's lifetime reproductive success. Now, with the recent development of DNA fingerprinting techniques, it also is possible to determine with certainty lifetime reproductive success for males.

Natural history and evolutionary studies represented by chapters in this volume include those by Pavelka, Ono, Reiter, Panter-Brick, Vitzthum, and Draper. They exemplify the importance of knowing about individuals, their layered biosocial functional components, and their whole lives in order to explain the products and processes of evolution. Other studies of well-known individuals in populations of Old World monkeys, apes, and humans also highlight this approach (Altmann et al. 1985; Watson and Watson-Franke 1985; Altmann 1986b; Fedigan et al. 1986; Goodall 1986; Rawlins and Kessler 1986a; Betzig et al. 1988; Nishida 1990; Pereira and Fairbanks 1993) and other species (Boyce 1988; Clutton-Brock 1988; Bruton 1989a).

Life Histories, Lifeways, and Life Stages

Life stages are closely related to both survival and reproductive life-history features (table 1.2). As females and males grow to maturity, the biobehavioral changes from one life stage to the next phase generally are marked by reproductive life-history attributes. These biological features of timing of life stages correspond broadly with development of survival life-history features, for example, behavioral or anatomical features and their suborganismal components (Pereira and Altmann 1985; B. H. Smith 1989, 1991, 1992; Morbeck chap. 9).

Survival at each life stage, as suggested above, depends on a balance of nutritional intake and energy expenditure while fulfilling life-stage roles. Self-maintenance depends on the ability to retain and exchange information about the environment, to travel and feed efficiently, to interact socially, and to avoid life-threatening dangers. Such abilities that are produced in different life stages by the degree of biosocial maturation, combine with an individual's knowledge of how to respond to and how to use a multidimensional environment.

Among monkeys, apes, and humans, survival functions and reproductive outcomes are part of reciprocal social and economic interactions of family relationships and social networks. They can be observed from both female and male perspectives during each life stage (Smuts chap. 5; Pavelka chap. 7; Hiraiwa-Hasegawa chap. 6; Vitzthum chap. 18; Morelli chap. 15; Panter-Brick chap. 17; Draper chap. 16). For example, a mother provides care for her infant while, at the same time, her infant increasingly solicits her attention by communicating its needs (and increases control of its world). In addition, infants may bring to females a change in social status relative to other group members (Pavelka chap. 7; Draper chap. 16).

Birth and Infancy

Birth initiates the first ex-utero life stage with major changes in biobehavioral capabilities and, at the same time, confrontation with new, challenging environments. This distinguishes infancy from the prenatal growth period in which young depend on mothers entirely as a life-support system. Patterns of timing of growth and development of different body systems during infancy establish the foundation for navigating infancy, childhood, and adulthood life stages.

Growth and development in the early life of a catarrhine primate emphasize brain and vision, suckling and digestion, breathing and circulation mechanisms, and skeletal formation. As for locomotor abilities allowed by the nerves, muscles, and skeleton, only the clinging capacities of the hands and feet (except in humans) are well-developed (Grand 1983). Primate mothers still serve as a life-support and transportation system. In addition to supplying milk with species-appropriate nutritional requirements and antibodies that help to establish an immune system, she provides a protected environment in which the infant can experiment and learn about biosocial and physical worlds.

Weaning and Childhood

Weaning, the life-history marker that defines the beginning of childhood, is characterized by a mother's cessation of milk production with corresponding behavioral rejection of her infant from breast-feeding. This occurs at a time in the infant's life when it is possible to shift from dependency on mother's milk to other foods. Brain maturation, emergence of teeth, development of the digestive and musculoskeletal systems, and developing social, feeding, and locomotor skills allow experimentation with new foraging behaviors. Indeed, infants may initiate and often contribute significantly to achieving full locomotor and feeding independence from the mother.

The feeding shift in catarrhines is gradual and includes a period of dietary overlap (Pereira and Altmann 1985) although mathematical models treat "age" at weaning as a point in time (e.g., at about one-third adult weight, Charnov and Berrigan 1993). It marks the transition to more independent living of childhood and active participation in group life. Brains are larger and have more neural connections. Teeth develop and emerge in combination with muscle and jaw development to allow efficient chewing. Growth and development of the trunk and limbs, combined with locomotor practice and social play refine neural coordination and allow independent locomotion. Juvenile humans, apes, and monkeys achieve feeding and locomotor independence in traveling, searching for and preparing food, chewing, and digesting solid foods, and, in addition, manipulation in grooming. Youngsters, however, still rely on adults socially as they increase their knowledge of the environment and establish their social and communication skills in their own networks.

Adolescence and Adulthood

Female and male social and reproductive roles emerge more clearly at adolescence, the transition to adulthood (Pavelka chap. 7; Hiraiwa-Hasegawa chap. 6). Attainment of physiological reproductive ability marks the onset of adulthood. Both females and males must continue to survive and maintain biosocial health in order to mate successfully and, for females, to produce a viable infant. However, they differ in time allocation and energy expended in promoting successful mating (emphasized by males) as compared to biosocial investment in offspring survival (primarily by females). Males may contribute to offspring survival in a variety of direct or indirect ways, whereas females are challenged energetically by pregnancy, lactation, carrying, protecting, caretaking, and socializing infants.

Sex Differences

Adult females and males share species' life-history characters, yet the two sexes function both separately and together as integrated constituents of local breeding groups. They differ in anatomy, physiology, energetics, and biomechanics of locomotor, feeding, and antipredator strategies, as well as activities related to social networks, mating, and rearing of offspring. Sex differences during the life cycle involve expressions of internal programming of growth and reproduction. These include characters associated with metabolism, nutritional requirements, rates of growth and maturation, and aging of different body systems and resulting body weight/size, tissue composition, and distribution (Zihlman chaps. 8, 13; Morbeck chap. 9; McFarland chap. 12; Pond chap. 11; Galloway chap. 10; Vitzthum chap. 18).

Life Histories and Evolution

My discussion thus far has introduced ways to think about life patterns in living catarrhines, including human primate individuals and their groups. My emphasis has been on functional biology, described by Mayr (1982) as the study of proximate causes for how animals "work." This whole organism–whole lives approach to understanding modern animals now must be merged with evolutionary biology, the study of ultimate causes and explanations of phylogenetic history (Mayr 1982). Functional and evolutionary biology, as viewed broadly in terms of evolutionary products and processes, are different perspectives of natural history studies (and those of ecology [Bates 1961]). Kohn (1989) made a similar functional versus evolutionary distinction between what he called "general" and "comparative" natural history compared to "evolutionary" natural history. Discussions of terminology emphasize an important point about evolution: the evolutionary processes we track today in modern organisms are part of populations and species' "fine-tuning" of the historical legacies of organismal designs produced in the past.

Through many generations, evolution is about "chance events plus the persistent biases of natural selection" (Williams 1985:21; also see Williams 1992). This view acknowledges the critical roles of processes that introduce and maintain variation at the genetic level (e.g., mutation) and their various effects in phenotypes and gene flow among populations (Dobzhansky 1937; Simpson 1944, 1949), but emphasizes "shuffling" of variation at the organismal level, that is, differential survival (mortality) and reproduction.

At the levels of species and higher taxa, comparative and functional studies of individuals and of aggregated groups of organisms show that modern catarrhines are closely related and share many life-history features (Morbeck chap. 9). Differences in the timing of appearances of various characters and their rates and durations of growth and development explain much of anatomical and physiological variation among species (Shea 1988, 1990, 1992; E. Watts 1990; Morbeck chap. 9; also see Gould 1977; McKinney 1988; McKinney and McNamara 1991; and for "scale" of evolution, Thomson 1992; also see Allen and Hoekstra 1992). In fact, this idea about the role of changes in timing can be broadened to include explanations of the similarities and differences among all related life forms, for instance, even at the level of the unity of the life process itself based on DNA as the molecule of inheritance (e.g., homeotic genes [De Robertis et al. 1990; Weiss 1990; Melton 1991]).

The apparent importance of temporal changes in life-history features as a way to change species' characters through time can be tested by studying timing mechanisms in modern taxa at the molecular (or genetic) level and their effects on body systems and whole-organism functions (Shea 1990). The fossil record, of course, provides the historical test. Skeletal fragments preserve information about tooth and bone physiology, anatomy, and inferred behavior of past organisms (Morbeck chap. 9).

Life Histories: Biobehavioral and Ecological Hierarchies

We still are discovering the details of how evolutionary processes work within and among different levels of biological, behavioral, and ecological organization, and how these are expressed and preserved in living and fossil organisms. Our current challenge is to synthesize knowledge of functionally integrated, whole organisms and their life stories with historical patterns of organismal design produced throughout phylogeny (i.e., historical constraints or constrained ontogenies [Lauder and Liem 1989; Lauder 1990). Wake and Roth (1989b: 1–2) showed why this has been difficult:

> While diversity and complexity in biology are difficult to comprehend fully even when considered separately, the analysis becomes even more complicated when we try to bring the phenomena together in the framework of one of the few unifying principles in biology: evolution. Acceptance that diversity and complexity of living organisms both result from past evolutionary events leads to a basic paradox. On the one hand, organisms are tightly coupled systems at genetic, developmental, structural, and functional levels, and this constitutes a high degree of integration that is a fundamental prerequisite for self-production and self-maintenance, unique features of life. Such systems are finely equilibrated, but at the same time extremely sensitive to changes and disturbances at critical points of the interactive, self-maintaining network. On the other, evolution has led to fundamental changes in the organization of organisms. Yet they have had to remain integrated and stabilized throughout the entire process of their evolution. Increase in complexity of systems apparently has led to increase in self-stabilization, but this seems to counteract markedly the tendency of evolution to increase diversity. Any modifications in such complex systems must be compatible with the functioning of all stages in the life of an organism, for all changes are basically modifications of ancestral ontogenetic trajectories.

Jolly (chap. 19) makes a similar point. Her comparison of information-transfer systems shows the evolutionary roles of stability and diversity with respect to sexual reproduction and intelligence.

Evolutionary biologists have had a continuing interest in the relationships of phylogenetics, genetics, intrinsic determinants, and external influences on development and resulting phenotypes (Morbeck 1991a). Recent discussions about the mechanisms and products of evolution highlight the uncertainty about exactly how evolution works at organismal, populational, species, and higher-taxonomic "focal levels" within the biobehavioral and ecological hierarchies (or, "strong orderings," Bock [1989]; "criteria for observation," Allen and Hoekstra [1992]). Published schemes are as varied as are their authors' academic experiences and the organisms they study. Most current models are inadequate when applied to the complicated lives of large-bodied, long-lived, group-living mammals, primates, and humans with big brains and good memories.

In contrast to the whole organisms–whole lives approach of natural history, the models offered by paleontologists and others about the nature of potential "introduction" and "sorting" of variation at different focal levels (e.g., including "emergent" levels), like typical life-history studies, are far removed from real organisms and their lives. Arguments about whether genes, species, or other taxa, in addition to whole organisms, can be considered to function as "individuals" in the evolutionary process make the problem of explaining evolutionary processes and products even more complicated.[2]

One description of a hierarchical organization is provided by Eldredge (1985). He defines two sets of nested relationships, a genealogical (reproductive) component focused on "more-making" and an ecological (survival/economic) component focused on energetics. Other researchers also describe these hierarchical relationships. The "genealogical" hierarchy relates to phylogeny and classification, replicators, and reproduction. The "ecological" hierarchy is characterized in terms of control, economics, interactors, and of survival.[3]

In a natural setting, of course, these distinct hierarchies cannot be separated in organisms, populations, or species. Since reproductive and, to a lesser extent, survival characters are included, the two sets of functional layers match, in part, my own view of the life-history approach. In addition, the complementary hierarchies are linked primarily at the organismal level. But, Eldredge and others seem to overlook the complexity of the biological, social, and ecological factors that contribute to individual survival and reproduction. Knowledge of real animals living real lives, especially as related to behavioral flexibility and reproductive outcome, is important during the long life spans that characterize many large mammals and, in particular, catarrhine primates. As shown throughout this volume, such data are critical to understanding the role of the individual in evolution.

Species- and sex-defined sequences and timing of growth, development, reproduction, and aging of individuals are products of evolution and result in observable phenotypes (i.e., features of whole organisms–whole lives) that enhance survival. This, in turn, enables mating and rearing of offspring. All phenotypic products may not always be selected directly. For example, "intermediate structures" that are not "seen" by evolutionary processes are possible (Stearns 1986). But survival and reproductive life-history characters have been "sorted" by natural selection and other evolutionary mechanisms at the level of the organism; that is, individuals that make up local breeding populations. It is the individuals that survive or die, reproduce successfully or do not reproduce; their offspring survive or die, reproduce or do not reproduce.

Life-History Approach: A Review

The broad-based life-history approach used in this volume emphasizes whole organisms and whole lives and their respective roles in evolution. It emerges from studies of known animals, including humans, and incorporates biology, behavior, and ecological relationships of organisms at many levels of functional organization. Although the whole-organism perspective has appeared in anthropological studies of humans (Ingold 1989), our approach differs from that of the social sciences in its strong evolutionary perspective (Fedigan chap. 2).

We now are beginning to find ways to understand the details of what it takes for individuals to survive *and* to reproduce, information that has not always been integrated into previous studies of life history and evolution.

Current research on mammals, primates, and humans that work through some of the connections within and among the operational layers of the biobehavioral and ecological hierarchies are represented by contributions to this volume. Understanding the factors that relate to survival and reproductive outcome of individuals provides a way to integrate the results of evolution with its processes.

Acknowledgments

Scientific meetings facilitate sharing ideas and data and generating new ways of viewing the world. Thanks to all those who participated in "Women Scientists Look at Evolution: Female Biology and Evolution" (1990) and, in addition, to those who contributed to part of its intellectual origins, a conference on "Primate Life History and Evolution" (1987). The Wenner-Gren Foundation for Anthropological Research, the University of California, Santa Cruz, and the University of Arizona provided financial support for these meetings. I am grateful to P. Martin and my colleagues at the Desert Laboratory (University of Arizona) for giving me a stimulating place to think and work; to D. Sample for continued support, and to A. Zihlman, A. Galloway, J. Hoffman, P. Martin, B. McLeod, J. Olsen, and J. Underwood for reading and evaluating versions of this chapter. Ideas about life history and evolution emerged from empirical data collected and analyzed with the support of the Social and Behavioral Sciences Research Institute and the Office of the Vice-President for Research (University of Arizona), the L.S.B. Leakey Foundation, and the Wenner-Gren Foundation. Finally, I thank S. L. Washburn for showing me how to make evolutionary connections.

Notes

1. Related sources include: Simpson 1941; Bates 1950, 1961; MacMahon et al. 1978; Grene 1987; Wake and Larson 1987; Greenberg and Tobach 1988; Ho 1988; Thomson 1988; Bock 1989; Kohn 1989; Wake and Roth 1989a; Liem 1990; M. Wake 1990; and Morbeck 1991a.

2. See Eldredge and Salthe 1984; Vrba and Eldredge 1984; Vrba and Gould 1986; Ghiselin 1987; Grene 1987, 1988; Kohn 1989; Vrba 1989; Gould 1990; Liem 1990; Brooks and McLennan 1991; Ereshefsky 1992.

3. See Bates 1950; MacMahon et al. 1978; Vrba and Eldredge 1984; Eldredge 1985; Salthe 1985; Vrba and Gould 1986; Grene 1987, 1988; Wake and Larson 1987; Greenberg and Tobach 1988; Ho 1988; Thomson 1988; Kohn 1989; Vrba 1989, 1990; Wake and Roth 1989a; Liem 1990; Morbeck 1991a.

2 Changing Views of Female Life Histories

Linda Marie Fedigan

Life history is a term commonly used in the scientific literature on the assumption that everyone understands and agrees upon its meaning; but when one searches for a precise definition, multiple and divergent interpretations surface. This first became apparent to me when I explained to a cultural anthropologist that I was conducting life-history research on female Japanese macaques and discovered that she took this to mean I was writing biographical accounts of monkeys. Later, I realized that even within the biological sciences, the meaning of the term has changed over the course of the twentieth century and that primatologists are now applying and developing the concept in diverse ways. Thus, this paper has three objectives: to describe the use of the life-history concept over time and across disciplines; to review briefly the empirical applications of life-history theory to long-term primate field studies; and to summarize my own research findings and those of others on the life-history patterns of female Japanese monkeys.

Life-History Theory

What is the nature and history of life-history studies? Not long ago I began to ponder the fact that my social scientist colleagues also research and produce "life histories" of human beings, but they have trouble relating my life-history research to their own. Just as social and biological scientists in the nineteenth century used the word *evolution* in two rather different ways so scientists of our own century have come to use the life-history concept.

Much as evolution means "change over time" in a general sense to all scientists, so life-history is commonly understood as the history of changes through which an individual (or organism) passes. But, in the same way that natural selection and sexual selection are theories about mechanisms of evolution specific to biology, so we have structured features of life-history theory in biology that are not present in the social sciences. For example, some standard biological life-history parameters are gestation length, age at weaning and first reproduction, interbirth interval, dispersal pattern, reproductive mode, and life span. Life-history theory in biology attempts to relate these ontogenetic and reproductive traits to population dynamics and evolutionary consequences. To a cultural anthropologist, a psychologist, or a sociologist, a life-history is a biographical account that is elicited from an individual, but which (unlike a biography) focuses on the subject's rather than the author's view of what is important ("life story"; Morbeck chap. 1). A brief history of how this concept has been used and defined in the biological and social sciences should help us to understand the several ways in which it is employed today. (More detailed histories may be found in Stearns [1976, 1977, 1992], Langness and Frank [1981], Watson and Watson-Franke [1985].)

Life-History Theory in the Biological Sciences

In the latter part of the nineteenth and early part of the twentieth century, the term *life history* was used very generally to refer to the changes undergone by an organism in its development from birth to death. A descriptive life-history study (literally "the story of a life") pertained to the individual in the same way that natural history studies pertained to the species. A sense of what was meant is perhaps best captured by titles of studies from this period, such as "Life-histories of northern mammals" (Seton 1909), "A life history study of the California quail" (Sumner 1936), and "The life history of Henslow's sparrow" (Hyde 1937). Around the turn of the century, the biological and social sciences were not as divergent as they are today, and they probably shared a common understanding of the life-history concept.

In 1954, Cole transformed the biological understanding of the concept with the publication of his pivotal paper relating individual life-history phenomena to consequences for populations. He was particularly interested in measuring the effect of iteroparous (repeated) versus semelparous (one-time) reproductive modes on the intrinsic rate of increase of the population. From a historical perspective, his paper is remembered as the pioneering effort to cast life-history phenomena into the selective and mathematical framework of evolutionary biology. In the 1960s and 1970s, several key theoretical papers were published that further developed the mathematical models predicting the relationships between life-history and demographic parameters (e.g., Lewontin 1965; Williams 1966; Gadgil and Bossert 1970; Charlesworth 1973) and also predicting optimal life-history tactics in different ecological contexts (e.g., *r*- and *K*-selection; see MacArthur and Wilson 1967; Stearns 1976, 1977; Horn 1978). The study of life-history tactics is closely related to the study of reproductive success, which as we shall see, has been a primary concern of primatologists.

In the course of developing theoretical models, these papers also formalized a more narrow definition of life history, based largely on standardized and measurable reproductive variables. The evolutionary biologist today usually conceptualizes life-history research in terms of these "traits." For example, an evolutionary biology textbook (Futuyma 1986:272) described such research as follows:

> The "life history" of an organism refers to many features, including dispersal, seed dormancy (in plants), life span, the age at which reproduction begins, fecundity, frequency of reproduction, and parental care. We will consider here the evolution of the chief demographic characteristics that enter into the description of a species' population dynamics, namely the age-specific pattern of reproduction and mortality. We seek to explain why some organisms such as certain species of salmon and bamboo are semelparous, reproducing only once, while others are iteroparous, reproducing repeatedly; why some, such as annual plants reproduce early in life while others, such as trees, delay reproduction; why some have many eggs or seed and others few; and why some are genetically programmed for senescence and death at an early age, and others at a later age.

Between 1950 and 1980, the interest in life-history theory was so intense that theoretical modeling virtually outpaced the empirical research, causing Stearns (1976), for one, to call for more experiments and observations to test the many assumptions and predictions of the various models. Horn (1978) further noted that the theoretical discussions had become so overloaded with "turgid mathematical formalism" that even the specialists found each others' papers unintelligible. I will summarize below the empirical applications of life-history theory to primatological field data. Whereas much of the empirical work before the 1980s was cross-sectional experimental research conducted on short-lived, laboratory colonies of organisms, the past decade has seen the accumulation of sufficient longitudinal field data on long-lived species of primates and other mammals to begin to test some of the predictions of these models across a variety of species, habitats, and life-history circumstances (e.g., Western 1979, 1983; Western and Ssemakula 1982; Harvey et al. 1987, 1991; Clutton-Brock 1988; Partridge and Harvey 1988; Ono chap. 3; Pavelka chap. 7; Reiter chap. 4). For recent overviews of life-history theory in evolutionary biology, see Roff (1992), Stearns (1992), Chisholm (1993), and Hill (1993).

The Life-History Method in the Social Sciences

For the social scientist, the life history is a type of personal document, which takes the form of a retrospective account of a person's life that is elicited and written down by the scientist. It is a specialized form of biography in which the recorder tries to tamper with the informant's view of their life as little as possible. According to Langness and Frank (1981), the biographical case study is a common denominator in the methodology of many disciplines, ranging from psychology and medicine to history and literature. Biographies existed in the fields of history and literature well before the present century, but have only been used as a research method in the social sciences since the early 1900s.

In sociology, Thomas and Znaniecki's *The Polish Peasant in Europe and America* (1918–1920) is often given as the first biographical study. In anthropology, the life-history method emerged out of research on, and popular interest in, the American Indian. Paul Radin's 1926 book, *Crashing Thunder,* is usually singled out as the beginning of rigorous life-history work in anthropology. Radin stated that the aim of his study was "not to obtain autobiographical details about some definite personage, but to have some representative middle-aged individual of moderate ability describe his life in relation to the social group in which he had grown up" (1926:384). From this statement, we can see that there is a fundamental commonality to life history research in the social and biological sciences. Both attempt to relate the life stages and experiences of the individual to higher levels of organization—the culture, the society, or the population. However, unlike the situation in biology, where theory outpaced empirical research for some time, in sociocultural anthropology there has been an emphasis on descriptive data and, until recently, less interest in theory and standardized methodology.

According to Watson and Watson-Franke (1985), there have been three phases in anthropological life-history studies. First, from 1900 to 1930, life histories were regarded mainly as a source of information about culture. As seen in Radin's work, the specific individual life was used to understand the general culture. Then, from 1930 to 1965, the alliance of psychology and anthropology to form the "culture and personality school" resulted in the use of life histories to understand personality development. Also during this phase, Dollard published *Criteria for the Life History* (1935), a treatise that stimulated anthropologists to standardize their sampling techniques and interviewing practices. Growing concern with both method and theory is reflected in the work of Kluckhohn (1945) and Kardiner (1945), for example, as they developed standardized ways of using the individual life to understand both cultural facts and the relationship between culture and personality. The third period, from 1965 onward, is described by Watson and Watson-Franke as the interpretive phase. During this time, some anthropologists, such as Langness (1965) became much more self-conscious about their own interpretative roles as investigator/writers of life-history accounts, and others such as Mandelbaum (1973) set up a standardized analytical framework that could be used across studies. Some later researchers applied Mandelbaum's framework, but others argued that life histories are inherently "emic" and thus a standardized "etic" scheme is inappropriate.

I would suggest that the concern of most sociocultural anthropologists in maintaining the life history as a "subjective" document that presents the informant's rather than the scientist's view of the culture is the major source of the differences between the biological and sociological understanding of the life-history concept. Both seek generalized principles by which they can link the individual to the group through studies of major life events. But anthropologists have by and large let their subjects define these major events, whereas biologists have employed a predetermined and standardized set of life-history traits.

Life-History Theory in Primatology and Physical Anthropology

I have spent some time describing these different histories and understandings of life-history studies because here, as in many other areas, primatologists and physical anthropologists often straddle the social and biological disciplines, and they borrow concepts and approaches from both sides. For the most part, primate life-history studies take place within the framework of evolutionary biology, focusing on quantitative analyses of standard ontogenetic and reproductive traits, using the largest possible sample sizes. However, some primatologists have used a case-study approach, analogous to that of the sociocultural anthropologists to produce a type of "biography" of their subjects (e.g., Goodall 1990; Zihlman et al. 1990; Pavelka chap. 7). I would argue that some recent theoretical articles in physical anthropology (DeRousseau 1990; Morbeck chap. 1; Zihlman chap. 13) are building on the biological model to expand the concept beyond the constraints of traditional life-history traits. I will use three recent descriptions of life-history theory to briefly exemplify the diversity of views.

Dunbar (1987, 1988) is a primatologist who presented a classical evolutionary biologist's point of view on life histories. He noted that demography is divided into two related aspects: life-history phenomena and demographic structure (age/sex composition) of the group/population. Life-history variables (he mentioned primarily birth and death rates, but also migration patterns) are usually available in the form of life tables, and these variables help us to link the behavior of individual animals to population-level phenomena. His review of the life-history data in primatology was primarily focused on longitudinal demographic studies of well-known groups of monkeys, that is, studies of population dynamics. Since he sees life history as a branch of demography, it is perhaps not surprising that the two approaches are not particularly distinguished in his reviews of the topics (see also Altmann and Altmann 1979; Morbeck chap. 1).

In contrast, DeRousseau (1990) devoted a considerable amount of discussion to delineating the distinctions between life-history and demographic studies, a distinction that she traced back to Cole's (1954) paper. Dunbar would agree with her that life history is about the organism, whereas demography explores the population consequences of those organismal traits. But DeRousseau also argued that life-history thinking is distinctive from population thinking in several other ways: life-history studies are case-oriented, ontogenetic, longitudinal (diachronic rather than synchronic, dynamic rather than static, processual and historical rather than structural) and such studies assume that the organism is active in negotiating with its environment, rather than a passive manifestation of genetic processes in a population. According to DeRousseau, the two approaches begin at different ends of the feedback loop: population thinking examines the central tendencies of populations on the assumption that these tendencies express the direction of genetic change over time. Life-history thinking first examines the events and processes that make up an individual's life history, then moves to the population level and evolutionary questions.

DeRousseau also argued that life history is more than a standard set of "traits" and "tactics"; rather, it is an investigative context, or an approach in which whatever is studied is considered in an ontogenetic framework. It is clear from the papers found in her edited volume on *Primate Life History and Evolution,* that she intends to greatly extend the concept to include, for example, studies of growth and developmental plasticity, skeletal biology, embryology, nutrition, health and adaptability. Her view that life histories are distinguished by the case-study approach, and by the consideration of any and all changes through which an individual passes, could be seen to begin to close the gap between the biological and social perspectives. However, given the history and large body of theoretical and empirical research in biology based on a delimited set of life-history variables, it seems unlikely that evolutionary biologists will readily accept such an expansion of the concept. I predict they would argue such breadth of definition renders the concept potentially unworkable. And since, as DeRousseau noted, quantitative methods for dealing with case histories are not well developed, most primatologists conducting life-history studies are still focused, for better or worse, on normative patterns and central tendencies.

Morbeck (chaps. 1, 9) and Zihlman (chaps. 8, 13) also develop a theoretical framework that seeks to expand the life-history concept. They note that, although evolutionary biologists recognize both the survival and the reproductive aspects of life-history adaptations, biologists have focused largely on the reproductive components and neglected the analysis of survival features. A distinction is made for heuristic purposes between time-based, reproductive characteristics of growth and development and those features that promote survival or individual maintenance of life and well-being. Traditional life-history traits (age at first birth, interbirth intervals, and so forth) relate to the timing of life events, whereas survival features relate more to the energetics of physiological and behavioral self-maintenance. (Note that this may be somewhat analogous to the distinction that Altmann and Altmann [1979] made between life-history "means" and "ends.") Through their papers, Morbeck and Zihlman seek to redress the previous imbalance by integrating the survival and maintenance aspects of life histories with the traditional reproductive traits. In so doing, they expand the concept to include the development of cognitive and social abilities in the individual, and the expressed behaviors of the musculoskeletal and dental systems (e.g., locomotion, feeding, and predator avoidance). Zihlman points out that life histories and reproductive success involve surviving to adulthood, surviving while mating, surviving during pregnancy, and surviving while rearing offspring.

Both Zihlman and Morbeck also focus on what they refer to as "whole animals–whole lives" as part of their attempt to integrate the consideration of biological, behavioral, and ecological processes in life-history analyses. As a practical demonstration of this theoretical approach, Zihlman *et al.* (1990) related the events recorded in the skeleton and teeth of known individual chimpanzees from Gombe National Park, Tanzania, to the observational data on these same individuals and the ecological and historical data on the Gombe population in order to produce a sample of detailed individual case studies. Such holistic life histories should help us to understand both variation in individual success and the role of the individual in population dynamics and will ultimately contribute to our understanding of evolutionary processes. These recent overviews of primate life-history research also suggest that physical anthropologists are finding it necessary to develop and adapt the traditional concept as they seek to apply it to the long-lived, developmentally flexible, and socially complex primates.

Empirical Applications: Primate Field Studies

DeRousseau (1990) noted that there are at least three ways in which life-history thinking has been applied to primate and human evolution: hard-tissue studies of ontogeny, cross-species comparisons to elicit broad evolutionary trends, and single-species studies of constraints and plasticity at different points in the life cycle. Only the last type of application to primate data will be reviewed here.

It is probably not a coincidence that life-history research is beginning to flourish at the present stage in the history of primatology. Although some early scientific observations of primates in nature were made in the 1920s and 1930s, field research really only began in the mid-1950s. The longevity of many primates ranges from 25 to 50 years, and thus long-term studies of known individuals started to truly reap their rewards only in the past decade, as the length of the research began to approximate the life spans of the subjects.

There is a considerable overlap between life-history and demographic research in primatology, and many analyses of population dynamics are now available from studies lasting 10 years or more. The latter will not be included in this brief overview, which will be limited to a sample of field studies that either describe themselves as life-history research or directly analyze the traditional ontogenetic and reproductive life-history traits. Although this distinction is, in the case of the primate literature, somewhat arbitrary, the body of demographic research is too large and divergent from the topic to be included here. Readers interested in demographic research on primates are referred to Dunbar's (1987, 1988) reviews of the topic. Those interested in how cross-species comparisons of primate traits can be used to elucidate evolutionary trends in body/brain size and mortality patterns are referred to the work of Charnov and Berrigan (1993); C. Ross (1991, 1992); and Harvey and colleagues (e.g., Harvey and Clutton-Brock 1985; Harvey et al. 1987; Harvey 1990; Promislow and Harvey 1990).

Three of the few primate papers that actually have the words "life history" in the title are Chism et al. (1984), "Life history patterns of female patas monkeys," Sprague (1992), "Life history and intertroop mobility among Japanese macaques," and Sugiyama (1976) "The life history of male Japanese macaques." These three make a good comparative point of departure for this overview of the literature. The Chism et al. study presented data on the standard life-history traits used in evolution biology (e.g., age at first reproduction, infant survival, interbirth interval), and the authors' discussion ties the timing of these reproductive variables both to the unpredictable, seasonal environment of the species and to the polygynous, male-dispersed social organization of patas monkeys. In contrast, the article by Sugiyama was focused on the timing and pattern of male dispersal in relation to the social system of Japanese macaques. Sugiyama presented data from several well-known populations to build a story of the consecutive events in a male's life: the usual age, season, and cause of dispersal from the natal group; the ranging patterns and other behaviors of males during their solitary phase; the age, season, and process of joining a new group; the occasional return to the natal group in old age, and finally, longevity and mortality patterns. Similarly, Sprague tied intergroup transfer to the major events in the developmental and social life history of male macaques, events such as social maturation and changes in dominance rank.

Sugiyama's and Sprague's papers are unusual in several respects. They provide the reader with both a qualitative and quantitative account of a typical monkey's major life events from birth to death, they link intergroup mobility to life-history events rather than to enhanced reproductive success, and they are two of the very few primate articles devoted to male life histories. It is interesting to note that these two accounts are focused on *dispersal* patterns, whereas life-history studies of primate females concentrate on *reproductive* events. Male intergroup transfer is the focus no doubt for the expedient reason that males rather than females disperse in this and most other Old World cercopithecine species. It may also be the focus because, for the most part, we have yet to document the necessary reproductive parameters for males (e.g., genetic paternity). However, this sex difference bears some resemblance to an assertion by Watson and Watson-Franke (1985) concerning implicit assumptions about sex differences in human life-history studies. They pointed out that although, in theory, sociocultural anthropologists are opposed to using standard parameters for life-history accounts, in practice, life histories of women are expected to focus on reproductive events (when and whom she marries, how many children, and whether they survive). In contrast, life histories of men usually center on nondomestic life events and role changes (rites of passage, assumption of leadership roles). One is led to wonder if the implicit assumptions of biologists concerning what constitutes a significant life event for females versus males play any similar role in the structure of primate life histories.

As stated earlier, the vast majority of primate research has presented descriptive and normative data on traditional life-history traits for females of various species. Examples of such single-species studies on free-ranging animals are those on savannah baboons (e.g., Altmann et al. 1977, 1988; Strum and Western 1982; Popp 1983), hamadryas baboons (Sigg et al. 1982), geladas (Dunbar 1980), rhesus macaques (Drickamer 1974; Sade et al. 1976; Sade 1980, 1990; Rawlins and Kessler 1986b), vervets (Cheney et al. 1988), patas (Chism et al. 1984), Japanese macaques (Masui et al. 1975; Takahata 1980), langurs (Winkler et al. 1984), orangutans (Galdikas 1981), gorillas (Harcourt et al. 1981; Watts 1991a), and chimpanzees (Goodall 1983, 1986; Nishida et al. 1990). Although much of the published life-history data is focused on Old World, terrestrial anthropoids, some information also has become available in the past decade on prosimians (e.g., Richard et al. 1991; Sussman 1991a), arboreal Old World species (e.g., Gautier-Hion et al. 1988), and Neotropical monkeys (e.g., Glander 1980; Robinson 1988a,b; Streier et al. 1993).

In addition to presenting descriptive data, most of these papers, like chapters in this volume, attempt to relate life-history information to various social, ecological, and evolutionary processes. For example, testing the models of dominance and reproductive success has been a major concern in primatology, and thus data on reproductive parameters such as fertility and survivorship are often analyzed in view of the individual's likely genetic contribution to future generations (e.g., Glander 1980; Sade 1980; Clarke and Glander 1984; Fedigan et al. 1986; Altmann et al. 1988; Cheney et al. 1988). The primate life-history data have also been analyzed in relation to the type of social system and relations within the group (e.g., Sugiyama 1976; Sigg et al. 1982; Chism et al. 1984; Smuts chap. 5); group size and composition (e.g., Dunbar 1980; Glander 1980); population growth (Altmann and Altmann 1979; Dunbar 1980, 1987); nutrition and food abundance (Altmann et al. 1977, 1988; Cheney et al. 1988; Zihlman et al. 1990); seasonality and rainfall (Dunbar 1980; Chism et al. 1984); and injury and disease (Cheney et al. 1988; Zihlman et al. 1990).

Although it is difficult to characterize such a diverse body of literature, one common theme is the attempt to follow known, free-ranging individuals through successive life events, over periods of time that begin to approximate the normal life span of the species. Given the longevity of large mammals and the logistical difficulties of primate field work, it is clear that life-history research is demanding. However, longitudinal studies of known individuals, such as the studies presented in this volume are, in many cases, essential to the testing of theoretical models in evolutionary biology.

Longitudinal Studies of Japanese Macaques

In the early 1950s, Japanese scientists began to study of number of groups of *Macaca fuscata,* a species of Old World monkey that is indigenous to and widely distributed across Japan, from southerly islands such as Yakushima to the northern tip of the main island, Honshu. From the earliest days of their research, Japanese primatologists used particular methods, such as the identification of all individuals in the group, provisioning of the animals in order to habituate them to observation, and collaborative collection of data over many years. These techniques were employed by Western primatologists only in more recent decades. Although some Japanese ideas, such as "protoculture" and the society of the species ("specia"), never diffused far beyond their country of origin (Asquith 1991), one very influential conceptual approach that they pioneered was the longitudinal, or what Itani (1983) has called the "diachronic" (as opposed to synchronic), approach to behavior. Obviously, long-term studies were not uniquely invented by the Japanese. However, their concern with individuals and how they fit into and influence the ongoing social system, as well as the Japanese habit of passing information on from one researcher to the next, led them quite naturally to a focus on the life history of individuals in relation to the social and demographic history of groups.

One of the monkey groups that was contacted in the early 1950s resided in a mountainous region on the outskirts of Kyoto. The scientists named the group "Arashiyama," after the mountain over which the monkeys ranged. When first contacted in 1953, the group comprised approximately 34 members, but by 1966, after 13 years of study, it had grown to include 160 individuals, at which time a group fission took place. The resulting two daughter groups were named Arashiyama A and B, and soon after the fission, because of the continuing expansion of the monkey population and the human population near Arashiyama, Japanese scientists began to look for a new home for the Arashiyama A group. In 1972, after considerable international negotiation, the entire A group, which by that time numbered 150 individuals, was captured and transported to a large ranch in south Texas. This was the first time that an intact group of monkeys was studied extensively in its native habitat and then translocated in its entirety to a vastly different environment, leaving behind a sister group for comparative purposes. I began to study the Arashiyama A group in Texas (now renamed "Arashiyama West") immediately after the transfer in 1972. I am grateful to have been able to work with a team of people from a variety of institutions in North America and Japan, all of whom have been willing to share their life-history data on these animals so that we can begin to track the entire lifetime of individual monkeys that may survive up to three decades. Thus, when I use the term "we" in this summary, I am referring to these colleagues with whom I have collaborated over the years (see Fedigan and Asquith [1991] for details).

Japanese macaques, like many species of Old World monkeys in the subfamily Cercopithecinae, live in multimale, multifemale (or "polygamous") groups. Females remain in their natal groups throughout their lives, and their closest ties are to their matrilineal kin, whereas most males transfer groups, first at puberty, and again every 5 to 10 years throughout their lives. Thus, a female's life history is considerably different from that of a male.

Life Histories of Arashiyama Female Japanese Macaques

For the past 22 years, my research with the Arashiyama West Japanese monkeys has related to the reproductive patterns of the females, and over the past few years, much of my work has come to focus on a cohort of females born into the Arashiyama group between 1954 and 1967. All of the 79 females in this cohort are now dead, and for all of them, we can trace life-history parameters with considerable accuracy and detail. Much of the following summary is based on this cohort, although the generalizations made in the following paragraph are true of all the Japanese macaque females for which we have published data so far (see Masui et al. 1973, 1975; Enomoto 1974; Koyama et al. 1975, 1980; Mori 1979; Sugiyama and Ohsawa 1982; Iwamoto 1988; Fukuda 1988).

A Japanese monkey usually gives birth to her first infant between 5 and 6 years of age, although she often begins mating a year before her first conception. All Japanese macaques mate during a relatively restricted period in the fall of the year and give birth some 6 months later, in the spring. Most females produce one infant every 2 years; however, there is some evidence that females in nonprovisioned groups give birth only once every 3 years, and that females in heavily provisioned groups may experience yearly births (Sugiyama and Ohsawa

1982; Loy 1988). The average age at death in our cohort is 13.7 years (the oldest age at death is 32). The average number of infants born to a female over her lifetime is 6.4 (range is 0–16), of which 3 offspring usually survive to age at first reproduction. Only one female in our cohort survived well into reproductive age and failed to reproduce before she died at 22 years, suggesting that female sterility in this group is very rare. Forty percent of the cohort lived to be 20 years or older, and two females lived beyond 30, although most of the older females died by the age of 24. Even though reproduction slows down in females over age 20, infant survivorship remains high. In addition, there is no evidence for a universal or standardized cessation of reproduction in otherwise healthy females, corresponding to menopause in humans (see Gouzoules et al. 1982; Fedigan 1991; Pavelka and Fedigan 1991; Fedigan and Pavelka 1994). Most females give birth within 2 years of their death, even when they live to be quite old, and two females produced infants when they were 25 years of age.

Thus far, my colleagues and I have completed two major analyses of the life-history data available on this cohort. The first study (Fedigan et al. 1986) addressed the question of lifetime reproductive success and its component variables, using the portion of the cohort that had completed their reproductive lives by 1984. The second study (Fedigan 1991) addressed differences in reproductive patterns between long-lived and shorter-lived females, using the entire cohort. What follows is a summary of the results of these analyses. Further explanation of methods and results can be found in the references cited.

We conceptualized lifetime reproductive success (*LRS*) for a female Japanese monkey as the product of three components: reproductive life span (*l*, which is the age at death minus the age at first birth), fertility (*f*, the total number of infants divided by reproductive life span) and survivorship (*s*, the proportion of offspring born that survive to adulthood at age 5). We used a program for our analysis that was designed by Dave Brown of Cambridge University and described in detail in Clutton-Brock (1988). We found that the major source of variation among the females who reproduced was differences in life span (*l*), which accounted for 67.3%, or approximately two-thirds of the variance, followed by offspring survivorship, which contributed 30% or approximately one-third of the variance, and that the weakest contributor to variance in *LRS* was fertility, at 20%. These values total to more than 100 because of negative simultaneous variation in the variables, particularly between survivorship and fertility. Such a negative relationship suggests that there may be a trade-off between the production of many infants at frequent intervals (high *f*), and the survival of these infants (low *s*). We tested this relationship further to see if there might be costs of reproduction to either infant and/or maternal survival. Females with lower fertility experienced significantly higher infant survivorship, and tended to live longer, although the latter trend was not significant.

Thus, the lifetime reproductive success of these females depends most strongly on the length of their lives, and what is more, high fertility (i.e., bringing forth babies in rapid succession) results in costs for the survival of the infants. This finding of the overwhelming importance of longevity may seem self-evident to some, but I was initially surprised by the result. Many theories of the evolution of life span (reviewed in King [1982]) have hypothesized that high levels of reproductive output early in life will be selected for, even at the expense of deleterious effects later in the life of the mother, and that the length of life is less important than the ability to produce many offspring quickly (e.g., Hamilton 1966; Williams 1966; Lamb 1977; Rose 1983).

However, Japanese macaques are long-lived, slowly reproducing animals, a life-history pattern that has not been very well studied for obvious reasons. It is not often that biologists have 40 years to dedicate to the study of one cohort (cf. Clutton-Brock's [1988] edited book on life-history studies, a collection that, for the first time, includes several studies of long-lived mammals). Macaques are also seasonal breeders, which means that a female produces a maximum of one offspring at a time (excepting twins) during the birth season, and a maximum of one infant per year, whether or not the previous infant survives. There is little variance in age at first birth (all but two females in this cohort first gave birth within a 3-year range, between 4 and 7 years of age) in comparison to a 27-year range in age at death (6–32) for reproductive females. Thus, female macaques have a limited capacity to accelerate rates of reproduction, early in life or at any time. The key to reproductive success in this species appears to be steady production until death, with adequate interbirth intervals to enhance offspring survival, and the accumulation of many reproductive years.

Again, this may seem obvious after a lifetime analysis has been performed. However, in primatology, fertility and not longevity has been the major measure and often the sole measure of reproductive success. But this may not be free to vary without significant changes/costs to other life-history parameters. Many conclusions about the selective value of particular patterns of behavior have been drawn from short-term (cross-sectional) studies of fertility in monkeys, but little has been said about differential longevity or the variables that might contribute to the length of a female's life. Therefore, our second analysis of the cohort data was an attempt to focus on longevity itself (rather than *LRS*) and its relationship to life-history parameters.

We did this by dividing our cohort into two groups and comparing females that lived long enough to reproduce but died *before* the age of 20 ($n = 31$, "short-lived") to females that reproduced and lived to the age of 20 and beyond ($n = 31$, "long-lived"). Significant differences were found between the two groups in age at first birth, and in the *total number* of infants as well as the *proportion* of infants surviving to age 1. That is, the long-lived females started to reproduce later in life, but they produced more than twice as many surviving infants and exhibited a much higher rate of infant survivorship. They also tended to space births further apart, although not significantly so. As with our earlier analysis, there is an implication of a trade-off between fertility and survivorship. Mothers and their infants survive better when the pace of reproduction is slower.

We also attempted to see if there is any obvious heritability of longevity by comparing the ages at death of 22 mothers with those of their daughters. The mean age at death for daughters of short-lived mothers was 18.5 years, whereas the mean for daughters of long-lived mothers was 19 years. We attempted various methods of expressing the daughters' ages at death (means, medians, only daughters that lived to reproduce, entering each mother-daughter pair as a data point, and so forth), and found no correlations with the mothers' age at death. One problem is that any heritable component to longevity would be masked by environmental factors (accidents, sickness, unique occurrences), and another is that the sample size is very small (human demographers use literally thousands of individuals to answer this type of question). Finally, we compared matrilines for longevity, and at this level, we did find a relationship between average adult weight, life span, and productivity. Although the sample sizes for adult weights are limited, some matrilines are clearly heavier, longer-lived, and more productive than others.

Dominance, Provisioning, and Life Histories of Female Japanese Macaques

An important explanatory mechanism in primatology is the hypothesis that dominance rank influences variation in female life-history parameters, such as age at first birth, fecundity, fertility, and lifetime reproductive success. Thus, we have made three separate attempts to document a significant relationship between dominance and aspects of the life histories of our Japanese macaque cohort, or subsets thereof (Gouzoules et al. 1982; Fedigan et al. 1986; Fedigan 1991). We have been singularly unsuccessful in finding such a relationship, although we tried many different analyses and ways of organizing the data in an attempt to find one. In our Arashiyama West cohort, dominant females do not begin reproducing earlier, or produce more infants in their lifetimes or experience greater infant survival, or produce more sons or daughters, or live longer than do subordinate females. Many of these findings have been corroborated by independent studies of dominance and reproductive success in the Arashiyama East group in Japan by Takahata (1980), Wolfe (1984), and Huffman (1987).

There are at least two mechanisms by which dominance could have an impact on the reproductive success of females. First, dominant females can harass subordinate females to the extent that the latter's menstrual cycles and pregnancies and/or lactation are disrupted. I would suggest that this circumstance is more likely to arise in captivity, where subordinate animals are unable to minimize confrontations with dominant ones, although I do not rule it out as a possibility in free-ranging primates. The second mechanism by which subordinate females can experience decrements in their reproductive success is if they do not have adequate access to resources, a situation that will occur when food, water, or space (e.g., safe sleeping spots) are limited.

Therefore, my present "best guess" as to why dominance is not related to reproduction in the Arashiyama groups, is that these females are provisioned, and thus resource levels are sufficiently high that subordinate females are not experiencing reproductive costs, even though there may still be some dominance gradient in terms of access to food. At the same time, they are not in cages, but either free-ranging or in very large enclosures (50–100 acres) with vegetation, such that subordinate females can minimize confrontations and have access to natural vegetation. In addition, female dominance hierarchies in these monkeys do undergo major reorganizations, even if only once every decade or so (Koyama 1970; Gouzoules 1980; Gouzoules et al. 1982). Thus, over the course of her life history, a female may experience fluctuations in rank, and under such changing conditions, short-term competitive success may not translate into enhanced lifetime reproductive success.

We do not know at what level of relative resource abundance subordinate females start to suffer feeding and/or reproductive costs, and until recently, no one has attempted to examine the effects on life-history variables of providing extra food. Three reviews of the effects of food supplementation are available (Asquith [1989] and Loy [1988] on primates; Boutin [1990] on terrestrial vertebrates). From these syntheses, we know that, among other effects, provisioning increases fertility, decreases infant mortality, and sometimes lowers age at first birth. These effects result in an expanding population, but we do not know for how long (or for how many generations) these effects continue over the demographic history of a group. We also have no idea if provisioning increases life expectancy in any stage of life other than infant survival. I would conclude from what we know thus far that provisioning is comparable to the high-resource circumstances that are found for some free-ranging groups in the wild.

Most importantly, for the issue I have just been discussing, we really do not know yet if the impact of provisioning occurs uniformly across the females and across the dominance ranks of a group, or if it affects individuals differentially. Thus far, we have conflicting analyses of the effects of provisioning on Japanese macaques. Two conclude that provisioning exaggerates the effects of dominance because dominant individuals can monopolize the artificial food sources (Sugiyama and Ohsawa 1982; Iwamoto 1988). Another suggests that provisioning attenuates the effects of dominance because there is plenty for everyone to eat (Mori 1979). Clearly, the impact of provisioning on life-history parameters in female Japanese macaques is a complex issue that will not be easily resolved. Effects depend on factors such as how much and how widely food is distributed and how much access there is to alternative, nutritious food resources. The continued publication of life history data from variably provisioned groups and the anticipated publication of such data on nonprovisioned groups of Japanese macaques from Yakushima (Fa and Lindburg in press) should give us a better comparative base for documenting such effects.

Conclusion

In my own life history as a primatologist, I have moved from short-term, cross-sectional research to lifetime studies of female primates. I can identify three reasons for this shift. First, it is in large part because I had the good fortune to join the cooperative and international team of researchers tracking the Arashiyama Japanese macaques, and because they were willing to share the longitudinal database that has made life-history studies of these monkeys possible. A decade after I began working with these monkeys, I became increasingly aware that many research findings on macaques are constrained because they are based on provisioned and sometimes translocated monkeys, and therefore might not provide generalizable answers about the life histories of primates in nature. To compensate for this limitation in my own research, I established a field site in Costa Rica in 1983 to study the life histories of nonprovisioned female monkeys living in their native habitats. Eleven years later, we have just begun to accumulate enough longitudinal data to compare female life histories across different species (Fedigan and Rose 1995). This work has renewed my appreciation of the research value of the provisioned macaques with their long histories of close study, large sample sizes, and well-documented genealogies, now dating back 40 years.

Second, I was drawn to life-history studies of female monkeys by the animals themselves. Certainly the matrifocal-matrilineal social systems of many monkey societies and the fact that the reproductive events of female lives are so much more observable than those of males lend themselves to a long-term focus on female life histories. Finally, although the difficulties of longitudinal observations of large mammals are many and varied, we have finally reached the stage in the history of our discipline that we can begin to address some of the central questions in evolutionary biology, based on data from the many years of tracking known individuals. In this volume, and in others in the upcoming years, we should be able to use further our hard-earned knowledge of "real animals" with "whole lives" to test and better understand the general principles of life-history evolution.

Acknowledgments

An ongoing grant from the Natural Sciences and Engineering Research Council of Canada (#A7723) supports my research. I thank the editors, Mary Ellen Morbeck, Alison Galloway, and Adrienne Zihlman, for inviting me to write this paper, and the colleagues who have contributed to the development of the ideas presented therein, especially Pamela Asquith, the late Larry Fedigan, Lou Griffin, Harold and Sarah Gouzoules, Mary McDonald Pavelka, and Shirley Strum. I also thank the many members of the Arashiyama project who have contributed to the life-history database. R. Craig Allen contributed his usual excellent editorial assistance.

Part II

Natural History and Life-History Studies: The Mammals

What It Means to Be a Mammal

HUMANS are primates, and primates are mammals. In order to understand ourselves, we must first understand what it means to be a mammal. Evolutionary history includes a layering of biobehavioral features that characterize a species *bauplan* or body plan. Ours is rooted in the mammalian pattern. The way we live our lives, our growth, and our reproductive strategies are those of mammals. How do we know that we are mammals? First, we share readily identifiable life-history features with the other living mammals. Second, we can see the evidence for the evolution of our own unique combination of life-history traits in the fossil record.

MAMMALIAN FEATURES

Mammals live a life that is expensive in terms of energy consumption and expenditure. Therefore, the pattern of interaction among the features and behaviors is particularly critical. This pattern changes throughout the mammalian life span as the fetus grows and is born, and the infant copes with an external environment mediated by its mother, gains locomotor and nutritional independence during childhood, achieves maturity, and, as an adult, mates, reproduces, and rears offspring. These features are manifested in anatomy, physiology, and behavior (Eisenberg 1981; Flowerdew 1987; Boyce 1988; Rowe 1988; Bronson 1989; Clutton-Brock 1991; Novacek 1993). Natural selection can work on the individual at any of these life stages. Changes at any point has repercussions at other levels in the individual.

Anatomically, mammals are distinguished by an internal skeleton that allows for linear growth, which is completed by the fusion of the epiphyses once adult body size is achieved (Morbeck chap. 9). This permits the transformation of the size and shape of the organism between life stages while retaining functional ability. The skeleton supports a locomotor system that is highly flexible and permits adjustments to the different locomotor demands that arise during different periods of life (Morbeck chap. 9). In addition, these bones function as a reservoir for calcium and other minerals needed by the body for growth maintenance and, in females, reproduction (Galloway chap. 10).

Physiologically mammals, along with birds and, perhaps, some of the "hot-blooded" dinosaurs, are capable of internal temperature control. This thermoregulation occurs within a narrow range and is related to biochemical reactions that control both cooling and warming changes and behaviors. Mammals can inhabit many different environments, but thermoregulation also requires a significant expenditure of energy. In order to accommodate this system, mammals not only act to generate heat but must also conserve that heat, for example, with the development of hair and fur. Related to the high activity rates that allow for food acquisition to fuel the thermoregulation, is a complex central nervous system, usually with a large brain relative to body size. This in itself also requires considerable energy expenditure, particularly in oxygen supplies and heat maintenance. As a reward, it allows the mammal to find food, avoid being food, compete with others for that food, and develop and maintain social relationships.

Mammals eat large quantities of food and have a relatively rapid system of digestion. In addition, there is separation of the chewing and breathing apparatus that, in most mammals, allows for simultaneous eating and breathing. To process food, there are two distinctive sets of dentition. In contrast to the reptilian dentition, each mammalian tooth is distinct in shape and size, depending on its function and placement within the mouth. In addition, the switch from a primary to adult dentition allows for survival of the infant/juvenile during an extended developmental period.

The evolution of a system of fat reserves within the mammals allows for conservation of ingested energy (Pond chap. 11; McFarland chap. 12). This ability serves as a complement to thermoregulation, greatly extending the range of environmental niches into which the mammals could move by providing a means of meeting energy demands when nutritional sources fluctuate.

The mammalian reproductive system allows for a gestation period during which the mother retains her mobility despite the increasing weight of the growing fetus and variable or changing environments. In addition, the internal environment provides a buffer for the developing fetus in which the effects of nutritional deficiences, climatic extremes, and diseases are modulated by the mother's physiological responses.

In placental mammals, birth requires passage of the offspring from the uterus, through the birth canal, to the external environment—a transition assisted by the mother. For most mammals, this begins a period of intense interaction between the mother and her offspring. Infants solicit caregiving from their mothers, while mammal mothers promote their own survival and care for their young at the same time. That care includes not only feeding but also protecting, and, among primates, transporting, socializing, and transferring information about the environment. Mothers and infants often inhabit the same environment. Communication between mother and infant is enhanced by the development of both vocal and auditory structures.

Mammary glands provide nutritional support for the infant. In essence they function as a replacement for the placenta during the postnatal period. Milk composition is species-specific, meeting the nutritional needs of each mammal in terms of fat, sugar, and protein content as well as hormonal, immunological, and trace-elemental requirements. In addition to the variation in milk seen between species, milk composition varies remarkably within species and even within the individual. As the offspring moves from a preterm or neonate, the nature of the milk adjusts to the changing needs and absorptive abilities. Even within a single suckling episode, fat content will change dramatically. Both infant and maternal behavior can modify the milk content by altering the intervals between suckling episodes. Finally, maternal diet influences the milk composition, but the amount of variability depends on the developmental stage of the infant.

Not only does the maternal physiology respond to the needs of her offspring, but the infant's own digestive tract and absorptive abilities vary as part of its own growth and development. Initially, enzymes allow for the breakdown of the primary sugar in milk but, after weaning, this ability is lost in almost all mammals. Absorption of larger proteins, including immunoglobins, is higher during the immediate postnatal period.

Measurement of anatomical or physiological aspects portray only a portion of the mother-infant relationship. Beyond the factors that influence mammalian lactation, behaviors also come into play (Vitzthum chap. 18; Reiter chap. 4; Ono chap. 3). The infant is responsible in most cases for initiating suckling after birth. The sucking movement is unlike adult feeding, and while instinctive, can be easily disrupted. Through nipple stimulation, the infant facilitates the milk-ejection reflex ("letdown") of milk. In addition, the experience of the mother in nursing and in accommodating the needs of her infant also plays an important part in the success or failure of even the simple process of nursing.

The reproductive characteristics of mammals, therefore, exemplify the complexity of these organisms. The mammalian revolution in the production and care of offspring had two important long-term effects. First, females invested more time and energy in their offspring than did males. Females and males were able to diverge, to some extent, in physiology, anatomy, and behavior during the life stages. Females faced new demands on their bodies and lives. Second, a new social environment became possible as females stayed with their offspring until they achieved feeding and locomotor independence, leading to increased sociality of individuals at all life stages and a more coherent group life (Pond 1977, 1984; Zihlman chap. 13).

Behavioral aspects on the part of both offspring and mother are critical. Although some are instinctive, almost all can be perfected only by experience and learning. The unique personalities of the infants also affect the nature of the mother-infant bond (Smuts chap. 5). Information exchange in a social context, starting with the mother-infant interaction, becomes important in mammals. This ties into the potential for increasing complexity of the brain and the potential for increasing behavioral flexibility (Zihlman chap. 13). Exploratory play and negotiation of relationships among other individuals is part of the mammalian pattern (McLean 1985).

Life Stages

Mammals experience gradual transitions between life stages. As life stages increase in length and the transitions become blurred, the connection between the genetic substrate and how the animal appears and acts becomes more obscure. This is due to the increased flexibility in how the life story can be played in the individual.

Although these life stages are distinctive, they are also continuous, and the individual must maintain a functioning body during that time—one that feeds, interacts, avoids predation, and so on. These changes are not merely an increase in body size but a mosaic of anatomical, physiological, and behavioral changes that must be coordinated to retain function as well as meet the future requirements for growth, survival, and reproduction. As Coppinger and Smith stated: "The mere fact of an animal's increasing size during ontogeny changes its adaptive relationship with its environment. The niche changes with ontogenetic time and the organism has to change niches. The nerves, muscles, and bones cannot simply extrapolate linearly from one stage to another, but rather at every stage of ontogeny each behavior in an animal's changing behavioral repertoire must be reintegrated with all the others to produce a functional organism that can survive the present as well as the next stage." (1990:340.)

Mammals have complex systems of prenatal development and postnatal care. In placental mammals as discussed above, for example, the placenta transfers nutrients to the fetus allowing prenatal growth. Postnatal growth and development is promoted by production of milk. Feeding infants via the placenta and via lactation allows offspring to concentrate on growth within a protected environment without expending their own energy in feeding. An understanding of the species-specific characteristics therefore requires that both the maternal and offspring perspectives be examined.

Mammals are also characterized by juvenile periods in which the offspring has gained nutritional and locomotor independence from the mother (Periera 1993; Rubenstein 1993). However, the offspring may still use the mother's skills to acquire the social and survival skills needed. From the viewpoint of the offspring, this is a time when it can grow and mature, guided by the genetically based species template. It learns to move around its environment, interact with others, acquire and consume food, and avoid predators. From the parent's viewpoint, it is released from the energy-expensive investment in the young well before the offspring reaches adult size and is not faced with competition from a full-grown offspring.

Following the juvenile period, a subadult period frequently allows for inclusion of mammals in mating systems without the responsibilities of parenthood. This is often when dispersal occurs, as there is basic mastery of survival characteristics even without maturation of the reproductive ones (e.g., Pusey and Packer 1987).

Adult mammals include both reproducing and nonreproducing animals, most, but not all, of which have achieved adult body size. In many cases, there is an overlap of the terminal phases of growth and the beginning of reproduction. In females, this means that the energetic demands of pregnancy and lactation compound the demand for the mother's continued growth. For a time, there may be such heavy increases in the demands on her resources that they preclude successful early reproduction, despite the potential benefits predicted by conventional evolutionary models.

Although other animals exhibit many of these same features, mammals, and especially primates, are noteworthy for the complexity with which they undertake life. Not only are the developmental stages each distinctive in the anatomical, physiological, behavioral, and social interactions with which the organism deals, but there is considerable flexibility in each individual's approach to these interactions. Although behaviors may be characterized at the populational level, the exact actions of an individual animal are less predictable.

The Fossil Record

An examination of the features in the living mammals is not the only place in which we can trace the development of life-history features. They, in fact, can be "read" in the hard data of the fossil record (Morbeck chap. 9, Zihlman chap. 13). Life stages are documented by two sets of teeth and their sizes and shapes (e.g., primary deciduous teeth and secondary, permanent teeth) and by particular patterns of growth and maturation of the skeleton. The presence of a single dentary (one bone in half of the lower jaw) and three ear bones—the stapes, incus, and malleus (the latter two derived from the reptilian mandible)—show, from the perspective of form and function, the importance of vocalizing and hearing sounds.

Natural History and Life-History Studies

In the context of the studies presented in this volume, we can visualize the situation of two "large-bodied" mammals. One is the elephant seal female that comes ashore to engage in all her reproductive responsibilities—giving birth, lactating, weaning her young, ovulating, copulating, and becoming pregnant, and returning to the sea—all within less than a month. The other is the human researcher who watches and records the behavior of the elephant seals. She must negotiate professional and personal roles with responsibilities of bearing and, more importantly in terms of time commitments, rearing children. For her, like women around the world, these responsibilities last for many years.

Empirical natural history studies place known individuals in social and ecological contexts as they negotiate their unique pathway through the life stages. Their aim is to survive while growing to maturity, mate, and rear offspring during adulthood. The goal of these studies is to assess factors and chance events that contribute to determining who dies, who lives, who leaves, who stays, who reproduces, and who does not, and how many of the offspring survive to reproduce. A focus on the individual is best summarized by Marston Bates:

> Natural history is not equivalent to biology. Biology is the study of life. Natural history is the study of animals and plants—of organisms. Biology thus includes natural history, and much else besides.
>
> The world of organisms, of animals and plants, is built up of individuals. I like to think, then, of natural history as the study of life at the level of the individual—of what plants and animals do, how they react to each other and their environment, how they are organized into larger groupings like populations and communities. Other biological sciences take up the study at other levels of organization: dissecting the individual into organs and tissues and seeing how these work together, as in physiology; reaching down still fur-

> ther to the level of cells, as in cytology; and reaching the final biological level with the study of living molecules and their interactions, as in biochemistry. (1950:7.)

Multigenerational-time depth-of-field studies allow researchers to evaluate the roles that behaviors and other experiences of individuals play in survival, in mating, and in rearing offspring. Individuals are studied in two ways: (1) synchronically, by looking at a number of individuals at one point in time, and (2) diachronically, by examining the changes in one or more individuals over a period of time.

The "mammalian model" is characterized by the ability to adapt to environmental extremes. One of these extremes is clearly exemplified by the marine mammals discussed in this section of the book. The features that particularly mark both the sea lions and the elephant seals are (1) the sharply defined life stages, particularly in the elephant seals; (2) the complete separation of breeding and feeding areas; and (3) the short time period of development of the mother-infant bond. Socialization of the offspring is left, to almost a complete extent, to the offspring itself. As Ono (chap. 3:41) states, "The ontogeny of pup behavior and socialization, therefore, is largely independent of the mother who functions primarily as a nutritional base." Within this system, however, the mode of reproduction is not uniform. Sea lion mothers alternate nursing sessions with long feeding trips during which offspring are left to develop locomotor and social skills with other offspring. Elephant seal mothers stay with the offspring but only for a very short time, pumping them full of fat-rich milk that will provide them with fat reserves to fuel them following the mother's departure. Independently, the young elephant seal must learn to swim, hunt, and evade predators. For both species, therefore, experimentation by the offspring itself allows it to establish survival behaviors. Future long-term studies on specific individuals among both elephant seals and sea lions will provide more comprehensive information about individual lives. This will allow us to understand better the relationship between infant experiences and the development of survival skills and reproductive outcome.

The two chapters in this section provide a detailed view of these two mammal species at the extreme edge of the mammalian model, both tied to a marine food-resource base. Both test models of expected reproductive outcome according to various theories of evolutionary process with real animals living real lives. Both find that the models are inadequate to explain fully the observations.

These chapters provide a good comparison with the refinements seen in the "primate model." Both groups consist of large-bodied mammals and confront the same kinds of demands but do so in radically different ways. This section provides a basis for seeing the importance of the primate overlay of being social.

3 Sea Lions, Life History, and Reproduction

Kathryn Ono

PINNIPEDS are mammals that are adapted to live and forage in the ocean. In Latin, "pinni" means wing or feather and "ped" means foot, hence pinnipeds, or "feather foot," a name they earn from their highly modified, paddle-like limbs. Pinnipedia is a subgroup of the mammalian order Carnivora, the carnivores or meat-eaters. All pinnipeds are carnivores, devouring a diverse array of marine life from large fish and squid and even other pinnipeds all the way down to mussels and krill. There is presently a controversy as to whether the two major subgroups of the pinnipeds had the same or different ancestors. They are thought to have evolved from the progenitors of either otters or bears. At any rate, there is complete agreement that their ancestors were land mammals that secondarily adapted to life in the ocean.

Besides walruses, which I will not discuss, the two major types of pinnipeds are phocids, or true seals, and otariids, the eared seals, which include the fur seals and sea lions. The two groups differ in many ways. Phocids have no external ear pinnae (lobes), whereas otariids have very small, pointed ones. Phocids do have perfectly good inner ears and can hear quite well. Phocids have short, blunt, pawlike forelimbs, whereas otariids have elongated, paddle-like foreflippers that they use for locomotion underwater. Phocids use their webbed hindflippers for swimming and move on land by humping along on their bellies. Otariids, in contrast, have narrower hindflippers that rotate under their bodies, so they can waddle on all fours.

One of the most compelling life-history traits shaping the biology and behavior of all pinnipeds is that, although they are adapted for swimming, feeding, and living in the ocean, they must return to land for parturition (giving birth). Newborns of most pinnipeds are not adapted to spend long periods of time in the water, primarily because of the high thermal conductivity of water. Young have to be large enough and well insulated enough to remain in the water for more than a few minutes and still maintain proper body temperature. This constraint makes it difficult to provide for more than one young at a time, since the offspring must be large at birth and gain weight and insulation before becoming an independent forager.

All pinnipeds have at most only one pup per year. In addition, neonates of most species are unable to follow their mothers to feeding areas. Therefore, females with young of that year are tied to the land in order to nurse, but must return to the ocean for food. Pinnipeds tend to give birth on offshore islands (or on sea ice) that do not have many terrestrial predators and have good access to oceanic feeding areas. For many species, pupping takes place on a few traditional sites on these islands, which can be fairly crowded. Nonnursing females are generally scattered over the ocean in search of food, which makes it difficult for males to find them. Therefore, breeding has evolved to take place when females are accessible and concentrated, during the pupping and lactation period. This means that males are also tied to the land for part of the year in order to breed (Bartholomew 1970).

Phocid and otariid females have solved the problem of foraging in the ocean and nursing young on land in two very different ways. Most phocids, such as the northern elephant seal discussed in the next chapter (Reiter chap. 4), stay with the pup during the entire (short) lactation period. They produce milk from the large body stores of energy accumulated during months of feeding before parturition (although there is evidence that some species of phocids feed a little during lactation). Lactation lasts from 4 days to 2 months (Oftedal et al. 1987a), and weaning generally occurs abruptly when the female leaves the pup for a prolonged feeding trip.

In contrast, female otariids, such as the California sea lion I discuss here, commute between the oceanic food supply and the landlocked pup. After an initial period with her pup, the female spends 1 to 9 days at sea foraging, alternating with 1 to 2 days on land nursing her pup (Oftedal et al. 1987a). The length of lactation in most species of otariids is extremely variable. Except for the two species that breed nearest to the poles, most otariid pups are thought to suckle for about a year; nursing yearlings and juveniles have been observed in most of the otariid species.

These differences in lactation strategies have profound effects on the way phocid and otariid pups start out in life. Most phocid pups stay next to their mothers during the entire period of lactation. Pups do not interact much with other animals at this time (except, perhaps, to be threatened by other females), and are protected by their mothers. Suddenly, one day, the mothers leave, and the pups are entirely on their own, without much social experience, in many cases without ever having been in the water (let alone having attempted to swim), and without experience in finding food for themselves. After weaning, phocid pups fast for up to 3 months in some species, learning swimming, feeding, and social skills during this time.

Otariid pups, on the other hand, start out very differently. After birth, a pup stays beside its mother, under her protection, for about a week in most species. During this time, the mother leaves to feed, often when the pup is asleep beside her. Otariid pups are not even close to being physiologically ready for independence at this point; they are small and have little fat to protect them against the ocean waters. Since, for the most part, adult males ignore pups, and adult females are aggressive toward them, they tend to clump together with relatively dense breeding aggregations. Pups interact with one another, and start learning to swim during the long periods when mom is away. When the mother returns, an otariid pup behaves similarly to a phocid pup, staying beside her and alternating nursing, sleeping, and playing beside her. As pups get older, they become better swimmers and start wandering farther from their birth area when mom is away. At this time, they apparently begin to start foraging on their own. Most pups are weaned at about a year of age. By this time they are experienced socially, are competent swimmers, and have had some experience at foraging on their own. However different, both phocid and otariid strategies seem to work, since mortality over about the first 2 years is similar in both families.

Pinnipeds are unusually easy to study during the breeding season because of several attributes: (1) they breed on traditional areas, and the same individuals return to the same area year after year; and (2) many species breed in tight aggregations so that a large number of breeding animals can be observed simultaneously. Since pups are confined to the breeding areas, it is an excellent system in which to study the dynamics of mother-offspring interactions.

Background

My first look at a pinniped breeding aggregation was at the Steller sea lions (*Eumetopias jubatus*) on Año Nuevo Island near Santa Cruz, California. The "aroma" from this very popular breeding island emanates several hundred meters over the water. As the boat neared the landing beach, I wondered how I was supposed to continue breathing while staying there. Fortunately, the Stellers have the sense to inhabit the seaward side of the island which is not so strongly scented. The scene at the breeding area reminded me instantly of my high school math teacher's favorite saying, "can of worms." Animals seemed to be moving and squirming and making a lot of noise all at random. How could I ever make any sense of this? Amazingly, after a few days, everything started falling into place. Each male patrolled and defended a fairly precise piece of land, and each female gave birth, then went to sea, and returned on a regular basis. The "falling into place" had a lot to do with being able to recognize individual animals. Burney Le Boeuf, who has studied northern elephant seals for decades, compares this to watching a football game: The game is more interesting to watch, and certainly makes more sense, when the players are individually identified with large numbers on their backs. It is difficult to get pinnipeds to wear football jerseys, but we still put a lot of time, energy, and imagination into seeing that individuals are well marked.

Having marked animals opens up a whole new world of information. We can determine which adult males return year after year and which are the most successful in reproducing. Marked females tell us whether females have a new pup every year or skip years between pups. Within a season, marked females tell us how often and how long they go to sea to feed, how long they stay with their pups after returning, and which qualities of females and mothering techniques yield the biggest, healthiest pups. Following marked pups throughout the season gives us an idea of the sequence of physical and social development in pups. Lastly, studying marked animals just makes watching them more entertaining. We give each known sea lion a name (usually we try to match the name with whatever identifying marks the animal has) and can quickly recognize idiosyncrasies in the personalities of individuals. Some females, for instance, are very mean to unrelated pups. They go out of their way to chase or bite any pup that dares to come near them or their pups. Other females are much more tolerant and tend to sniff pups, but are not aggressive to them unless they try to suckle.

Male pinnipeds are generally easy to observe because they are usually larger or more ornamented and virtually mark themselves with wounds and scars from fights. It is simple to tell individual males apart. Females are often unscarred and more numerous on the breeding areas so that identifying individuals can be a problem. In species such as elephant seals (*Mirounga angustirostris*), researchers are able to apply hair-bleach marks to animals backs without much disturbance to either the individual or the breeding colony. Other species (mostly otariids) run toward the water when approached by humans. The sight of one panicked animal alerts others, which also run to the water, clearing the entire area of animals.

Since repeated disturbances such as this cause the animals to vacate the area, remote methods of marking are necessary. Females are marked remotely with paint pellets propelled by slingshots or with eggshells filled with hair dye. The paint pellets are aimed at the rocks (this technique does not work on a sandy beach) in front of the animal, and paint splashes mark it (paint pellets just bounce off the side of a soft, cushiony sea lion). The eggshells do not seem to bother the animals either, perhaps because they are accustomed to being climbed over by other animals and harassed by gulls. I would like to note here that a lot of effort was expended in finding suitable ways of marking females; it was a somewhat uncharted technology when I first started studying these animals. We mark pups by causing one big disturbance after most of the pups have been born. The pups are rounded up into a group and given individual bleach letter or number markings (just like the football players).

We study sea lion behavior from an observation blind, a small building in which the researcher sits and watches the animals through windows. The researcher can see the animals, but the animals cannot see (or smell) the researcher. The blind also protects the researcher from the elements and keeps the notepapers from flying away in the breeze. We try to be in the blind from around sunrise to sunset every day during most of the breeding season. Since sea lions breed during the summer, there are 14 to 15 hours between sunrise and sunset. You get to know the animals and the inside of the blind quite well.

There are many different methods of recording animal behavior. For most of the behavioral data I discuss in this paper, we looked for all of our marked mother-pup pairs every hour and wrote down what each was doing when we first spotted it.

The California Sea Lion

The bulk of the data for this paper comes from a study of the California sea lion on San Nicholas Island, Channel Islands, California. I was fortunate enough to be involved in a study initiated by researchers from the National Zoological Park in Washington, D.C. The study took place from 1982 to 1986, and involved Daryl Boness, Olav Oftedal, and Sara Iverson, numerous research assistants, and colleagues willing to work long hours under less than civilized conditions.

California sea lions (*Zalophus californianus*) are the typical zoo or circus "seals" that vocalize by barking. They breed from the Channel Islands to the tip of Baja California and in the Gulf of California. Males and females begin arriving at the traditional breeding areas (generally located on offshore islands) about mid-May. Males defend territories both on land and in shallow water. Each female gives birth to a single pup a few days after arriving on the island. She remains with her pup for about a week before their first foraging trip to sea. A mother returns to nurse her pup after about 3 days, remains on land for about 2 days, and then leaves for another foraging bout. The alternation of foraging and nursing is referred to as a feeding cycle, and females continue this pattern for the remainder of lactation (Ono et al. 1987; Ono 1991).

From an individual pup's perspective, life starts out similarly for males and females. After a comfortable week next to big, warm, protective, and nourishing mom, a pup finds itself suddenly alone. The only other animals on the crowded, noisy breeding area that will allow the pup to sleep next to them, or interact without biting them, are other pups in the same predicament. Throughout the first 3 to 4 months of life, pups alternate between nursing and resting with their mothers and playing and resting with other pups. At 4 to 5 months of age, when they have become proficient swimmers, pups begin venturing out farther and farther from the area of their birth. They generally do this in small groups, swimming up and down the coast, stopping at various beaches and rocks, but usually returning to the place where they last nursed to meet their mothers when they return from their feeding trips. At about this time, mothers also start moving their pups to new areas, from which both mother and pup base their activities. How a female remembers on which beach or rock she last left her pup is unknown.

By about 7 to 8 months of age, pups begin to experiment with catching and eating their own prey. Their major source of nutrition is still probably milk at this point, but nutritional independence is not far off. Most California sea lion pups are weaned by the time the next year's pup is born. We do not know exactly when or how it happens, but pups are still seen with their mothers up until that time. If the pup is lucky (from the pup's perspective), and its mother does not have a new pup, there is some chance that she will continue to nurse it as a yearling. Francis and Heath (1991) found that 33% of surviving yearlings were observed suckling. Most weaned juveniles (juveniles are animals between 1 and approximately 3 years of age) disperse from their natal island. Here is where the differences between males and females begin. Male juveniles tend to migrate north of the breeding islands, and many stay north during the entire year. Females tend to stay around the breeding islands or migrate south (Francis and Heath 1991).

Long-term studies such as those on the northern elephant seal have not been accomplished for sea lions, so we do not have a good database of known-age animals to determine variation in important life-history characteristics. I use sizes of individuals to guess the following ages. Little is known about what young sea lions do until they reach the subadult stage. By the age of about 4 to 5 years, males spend much of the breeding season on "bachelor beaches," areas where young males gather and spar. They also make forays into the breeding areas, only to be chased out by the resident adult males. By, perhaps, 10 to 15 years of age, they are large enough to begin to compete for territories on the breeding areas. During the nonbreeding season, they migrate north, spending most of their time foraging and resting on offshore islands.

Young females first become impregnated at about 3 to 5 years. Most of the females, even if not pregnant, probably congregate on the breeding areas during the breeding season. In our study area, there was always a large number of females that did not have pups, but rested there regularly, and came and went on feeding trips. Breeding areas are most likely the safest terrestrial places for reproductive-age females during the breeding season since the subadult and other nonbreeding males tend to harass them elsewhere. After their first estrus, females tend to spend most of their lives both pregnant and lactating. For mothers, estrus occurs about 27 days after the birth of a pup. Since pups probably are nursed close to the time that the next pup is born, females spend most of their lives near breeding islands, commuting to and from feeding areas located more than 100 km offshore (Feldkamp et al. 1991).

As a group, otariids are among the most sexually size-dimorphic mammals. At the extreme, adult males of the northern fur seal (*Callorhinus ursinus*) are four to five times heavier than females (Gentry 1981). Adult male California sea lions can be up to four times larger than females by weight (Lluch-Belda 1969; Ono 1991). Besides the obvious advantage in competition against other males, large size also is important in enabling a male to fast on his territory over the breeding season (for up to 2 months).

In general, larger male size among mammalian species is correlated with polygyny, that is, a breeding system in which one male mates with many females. The driving force behind the evolution of larger male size is thought to be two aspects of natural selection: intrasexual and intersexual selection (Darwin 1871). Intrasexual selection primarily includes male-male competition in which males compete by fighting or displaying for access to females. Intersexual selection is largely thought to occur via female choice, females preferring to mate with certain males based on where their territories are or how they look, act, smell, and so on.

Given that it is advantageous for males to be larger, what are the mechanisms involved that allow males to grow larger than females? The major theory being explored today was proposed by Trivers (1972) and Trivers and Willard (1973). They predicted that in polygynous species in which the variation in reproductive success (the number of offspring produced) is larger in one sex, and in which an increase in parental investment (roughly, the amount of resources that a parent puts into the young) enhances the ability of member of that sex to compete for mates, parents should invest their resources differentially in offspring of that sex. In otariids, the sex with a greater variation in reproductive success is male. Some males may mate with hundreds of females, but most males probably never breed. The better, bigger male offspring a female produces, the greater the chances that he will be a territorial male and the greater the chances that his mother will leave more grandoffspring.

The picture is somewhat different for female offspring. Females are not restricted in their ability to breed. At least for sea lions, it seems to me that any female that goes into estrus on a breeding area will be mated whether she likes it or not. Even small females can successfully rear an offspring to independence. Females that survive to reproductive age will probably have at least a few pups. At most a female may have, perhaps, 15 offspring. The difference in the total number of offspring between females, then, is much less than the difference between males.

The Trivers and Willard hypothesis also predicts that females in superior condition should bias the sex ratio of their offspring toward males. Females in poor condition should bias the sex of their offspring toward females. Mothers in good condition should have more resources to give to their offspring so that the males they produce should be of higher quality and are likely to outcompete other males. Mothers in poor condition will only have low-quality male offspring that will probably never be able to breed. It is more advantageous, therefore, for these mothers to invest their resources in female offspring that will reproduce regardless of size.

Given that it is advantageous for males to be large and for mothers to invest more in them, the increased investment should start as early as possible (i.e., in utero). We found that male California sea lion pups are significantly heavier than female pups at birth and grow faster than females in the first 2 months (Ono and Boness 1996). The above discussion of parental investment theory shows why we may have expected these differences to be attributable to the larger investment of mothers in male offspring. It is evident, however, that some female pups are just as large as males (i.e., the size distributions of male and female pups overlap) and larger size may confer a survival advantage to pups regardless of sex. Since females have only one pup at a time, they cannot distribute resources differentially to one part of a litter over another. The female's one pup makes up a substantial proportion of her total number of offspring, probably from 6 to 10% if she begins reproducing at age 4, lives to age 20, and has a pup almost every year. It therefore seems reasonable that females may invest as much as they can in any pup, regardless of sex.

There are two major ways in which males could obtain more energy to put into growth. First, their mothers could provide them with more resources. Since females apparently continue to grow in both length and mass throughout their lives (D. P. Costa pers. comm.), female size is somewhat representative of a female's age. If older, larger, more experienced mothers produce more or better milk or have larger pups at birth and bear predominantly male pups, this may explain the difference in resources available to males. Alternatively, females of any size may be producing more milk for males. One way to measure this is by analyzing how often or how long females allow their pups to suckle. More directly, the amount of milk a pup obtains can be measured using isotopic tracers.

The last possibility is that females produce richer milk for male pups. This is not as unlikely a suggestion as it may seem. We know that some species of pinnipeds produce milk with a higher per-volume lipid (fat) content than others. Trillmich (1990) has also shown that variation in milk fat content can occur within a species. For example, female Galapagos fur seals with longer feeding trips have richer milk than those with shorter feeding trips. Perhaps some hormonal or behavioral trigger in male pups can induce their mothers to produce richer milk.

The second major mechanism lies with the pup's behavior, which is somewhat independent from the amount of milk it is getting from its mother. Since there are apparently more reasons for male pups to grow faster (i.e., if early fast growth is coupled with larger size at adulthood), males could try harder to divert energy ingested into growth rather than using it in other ways. Males may alter their behavior in order to conserve energy, that is, spend less time in behaviors (such as play) that require a great deal of energy. Males also may try to obtain extra milk from other females. Since females knowingly suckle only their own offspring, pups must try to obtain milk from unrelated females without the female knowing it (sneak suckling; also see Reiter chap. 4). Basal or resting metabolism—mechanisms needed for basic maintenance of life functions—uses a large amount of energy if you consider that the meter, so to speak, is constantly running. Male pups could save a considerable amount of energy if they could somehow decrease their basal metabolism. Lastly, males could, perhaps, obtain more nutrients from milk by digesting it more efficiently.

The Role of Maternal Resource Investment

Do larger or older (more experienced) females have more male pups? Females in these categories probably have more "resources" to contribute to their pups and therefore should be able to raise larger, healthier pups, which, if male, could better compete for mates at adulthood. Since we have not been monitoring this population of sea lions for the last 20 years, we do not know the ages of individual females. Because size is correlated with age in some species of otariids, we assumed that larger females are older. We assessed relative female size via a size hierarchy constructed similar to a behavioral dominance hierarchy. We compared the sizes of females that are next to each other on the study area, and then, from the relationships we saw, we inferred some of the ones that we did not see. That is, if female A is bigger than female B, and female B is bigger than female C, we deduced that female A is also bigger than female C. Our data show no relationship between a female's size and the sex of her pup. Macy (1982), studying northern fur seals (*Callorhinus ursinus*), found no correlation between weights of mothers and pup sex. Costa et al. (1988) also found no relationship between maternal mass and pup sex in Antarctic fur seals (*Arctocephalus gazella*). In northern elephant seals, Le Boeuf and Reiter (1988) found that older females are more successful in rearing their young to weaning age.

Do females invest differentially in male versus female pups either consciously or unconsciously? How does a female know when her pup is male? There are many reasons to believe that females do not consciously distinguish the sex of their pups. First, mothers do not behave differently toward their male or female offspring. One would expect that mothers of male pups might forage longer or nurse pups longer. The length of their feeding trips (about two days), the length of time present on the study area with pups (about 1½ days), as well as the amount of time spent nursing male and female pups (as a percentage of the time they are present on study area, about 30%) is not statistically different. In fact, we have not yet found any behavioral differences between mothers of male and female pups. Although there is not much information on the foraging habits of females, I doubt if their behavior at sea varies in relation to pup sex, although it is a possibility. In essence, the data we have so far do not indicate that females nurture pups differentially according to sex.

The last parameter in terms of maternal investment is direct energy and nutritional investment. That is, do females actually transfer more nutrients in the form of milk to male pups? The answer is yes, sort of. The deuterium oxide dilution technique (Oftedal and Iverson 1987) tells us that male pups receive more milk from their mothers, on average, than do female pups. But, if milk transfer is measured on a per-pup weight basis, there is no longer any difference. In other words, females transfer more milk to larger pups, not necessarily male pups. Large female pups get the same amount of milk as do males of the same size. This makes sense in that larger bodies require more energy to maintain. A feedback mechanism of increased production on demand has long been known for humans. It seems, then, that females are delivering more milk to male pups, not necessarily because they are males, but because they are larger and need more milk.

The question of milk quality still remains. Do females produce better, richer milk for male pups? We do know that the composition of milk can vary between species or even within a species, depending on the length of the foraging trip. The major change in milk composition is that the lipid, or fat, component is more concentrated in females with longer foraging trips. Since we know this component can vary, it is possible that mothers of male pups are producing higher-energy milk for their offspring. However, as far as I know, there are not enough milk samples from a large enough sample of females of any otariid species to test for a difference. A complicating factor is that milk composition changes with the stage of lactation in otariids, so that to obtain an adequate sample, milk must be obtained from females all at the same point postpartum. Iverson (1988) found that milk extracted from pups stomachs via intubation is easier to get (pups can be captured more easily and do not have to be immobilized to collect the sample). However, digestion quickly alters milk composition so that intubation is a less reliable source of information.

One last component of the maternal investment picture is the length of dependence of young. For sea lions, dependence means nutritional dependence. There is little evidence for consequential mother-offspring interactions after weaning. For most otariids, the length of lactation is somewhat fuzzy. Pups probably start some feeding on their own between 6 and 8 months, but they are not completely nutritionally independent from their mothers until 11 or 12 months (about the time a female is ready to give birth to the following year's young). This system is actually quite useful in that, if spontaneous abortion occurs during pregnancy, or if a female does not become impregnated, she has the option to continue to suckle her offspring from the previous year. A significant percentage of females do suckle yearlings. A small number of females suckle 2- and 3-year olds. If females nursed males longer, or tended to nurse them as yearlings in lieu of reproducing the next year, then there would be an opportunity for a substantial increase in investment in males.

Because of the increased mobility of pups older than 6 months, we have little information about the precise time of weaning for most otariids. Francis and Heath (1991) found no difference between the proportions of male and female California sea lion pups remaining on their island of birth that are suckled as yearlings. If anything, there is a bias toward females. The only otariids that have a more or less well defined weaning time are the high-latitude fur seals. These species (Antarctic fur seal, northern fur seal) wean abruptly at about 4 months. Thus far, I have found no published accounts of age at weaning versus sex. Macy (1982) did, however, describe the process of weaning in northern fur seals. She found that 77% of the weaning observed occurred when pups did not return to the location of the last mother-pup reunion, even though the mothers returned and searched for their pups (i.e., the pup is responsible for weaning, not the mother). The numbers of male and female pups observed weaning in this fashion are approximately equal.

Role of Pup Behavior and Physiology

Pups spend most of the time during the lactation period without their mother. Adult males ignore them, and unrelated adult females are aggressive toward pups, often biting or tossing them when they trespass too closely. Therefore, pups avoid adults other than their mothers. They rest or play alone for the first few weeks and in groups when older. Pups learn to swim and socialize in small groups without the help of their mother (unpubl. data). The ontogeny of pup behavior and socialization, therefore, is largely independent of the mother, who functions primarily as a nutritional base.

Do pups behave differently according to sex? The most obvious difference that one might expect in relation to diverting energy to growth would be behavioral energy conservation. Young pinnipeds appear to have higher basal metabolic rates than do land mammals of similar weights (Thompsen et al. 1987). Since they may be burning about twice the amount of energy as most land mammals, additional activity that would increase metabolic rate even more may be selected against if energy for growth is at a premium. However, we found no differences in the behavior of male and female pups.

Because, according to our observations, pups determine the initiation and conclusion of suckling events, males may extract more milk from their mothers (or induce them to produce more) by suckling more than female pups. We found only one significant difference for the frequency of suckling in scan samples. Males suckled more often than females in the 1984 season. However, since males grew faster than females in all 3 years of data collection, this factor may have contributed to faster growth in only 1 year. Macy (1982) found no differences in either suckling-bout duration or suckling frequency between male and female northern fur seal pups.

In addition, male California sea lion pups do not attempt to steal milk from unrelated females (sneak suckle) at a higher frequency. Attempts at sneak suckling are actually quite rare, except for the El Niño year when about half of all pups tried to sneak suckle (see next section). There are no differences between sexes in any of the years in the proportion of pups that attempted to sneak suckle. Macy (1982) found no differences between sexes in the number of northern fur seal pups that attempted to sneak suckle, although the bias in the distribution is toward female pups sneaking more. She did find, however, that female pups attempted to sneak suckle sooner after the departure of their mother than did males.

There is little information on sex differences in metabolic rates of juvenile animals. Most metabolic rate studies have been performed by physiologists in laboratory settings. Especially for large mammals, this means that the sample sizes are so small that slight differences between sexes would not be significant. Moreover, by definition, basal metabolic rate is obtained from adults. Since metabolic rate in rather exotic species (such as pinnipeds) is most often used for comparison with other species, most of the measurements have been taken from older juveniles and adults. In most adult mammals, metabolic rate is higher in males, partly because of the effects of testosterone. Therefore, there is not a large body of literature supporting the idea that male pups may have a lower metabolic rate than females. Nonetheless, it is apparent that lowering metabolic rate even slightly saves large amounts of energy over time.

In order to get the information we wanted, we needed a large sample size as well as wild rather than captive pups. Since metabolic rate may be affected by environmental conditions and diet, we wanted animals that could offer both natural habitat and social conditions as well as natural diet. Basal (or resting) metabolism is generally measured in the "thermoneutral zone." The thermoneutral zone is the zone of temperatures over which metabolic rate does not change with temperature. Above this zone, metabolism rises with increasing temperature as the body turns on its cooling mechanisms. Below the thermoneutral zone, metabolism increases with decreasing temperature as the body tries to heat up. In order to compare metabolic rates of male and female pups, we first had to determine the thermoneutral zone for pups.

We were able to obtain measurements over a wide range of temperatures for about 40 pups. This was about the minimum number needed to determine statistically slight differences in metabolism. We found that, indeed, male pups have a significantly lower metabolic rate, but only if one first accounts for the amount of time that they were held before measurement. That is, when pups are first taken from the breeding area, they are caught, placed into a pet carrier, transported in a truck, and then handled in order to get them into the metabolic chamber. It stands to reason that pups are a bit agitated after this, but over the course of their captivity, they seem to realize that they are not going to be eaten or otherwise harmed and most become quite tame. If this period is not taken into account, the direction of the trend is correct, but the relationship is not quite significant. If the amount of time animals have been in captivity is taken into account, then the resting rate of male pups is lower than that of females. Therefore, it appears that slower metabolism, along with a larger milk intake, may account for the increased rate of growth seen in male pups.

We do not know whether larger male pups turn into larger adult males and whether those males get good territories and reproduce. The time between weaning and reproduction is long, especially for males, and there are certainly other contributing factors that we have not measured.

Mothers and Young under Stress

Does this picture of greater male growth and milk intake change when the food supply suddenly diminishes, thus placing the health and survival of the mother in jeopardy along with that of her pup? During the course of our study, a major oceanic perturbation, the 1982–83 El Niño occurred. The El Niño is a meteorological phenomenon that occurs about every 3 to 7 years in the tropical Pacific. El Niño varies in intensity, and the 1982–83 event was the strongest in recorded history. During a strong El Niño, warm water from the equatorial Pacific Ocean is swept north up the eastern Pacific coast. The warmer sea-surface temperatures are correlated with decreased upwelling (i.e., the upward movement of colder, deeper, more biologically rich water layers to the surface), increased salinity, and a deeper thermocline (i.e., the region where warm surface water and colder deep water meet and temperature changes rapidly). These changes in oceanic conditions affect the distribution and abundance of sea lion prey. Sea lion prey species that normally dwell in the waters off California, either migrated north where waters were still cold, migrated deeper, dispersed, or died. As a consequence, the normal feeding areas for lactating females were depauperate of prey (Arntz et al. 1991).

Relative body condition is difficult to assess in any species. It is particularly difficult to assess in a species where it is impractical to handle a large number of adults. We can assume, however, that because of the food shortage, females were in poorer average condition during the El Niño year compared to "normal" years. It is an advantage to females not to have a pup at all during lean years because the drain of pregnancy and lactation may decrease a female's ability to survive and the probability of pup survival is low.

If the food shortage began before the breeding season, we would expect fewer females to enter estrus and become impregnated. If the food shortage occurred after the breeding season, as it did in 1982–83, we might expect to observe an increased rate of spontaneous abortions. Throughout the Channel Islands, fewer females gave birth during the El Niño (65% decrease in pups overall; DeLong et al. 1991). In addition, significantly fewer male pups were born on our study area when data from 4 years were pooled in a regression (i.e., the sex ratio switched from 1:1 in a good year to 1.4:1; Ono 1991). In fact, the decrease in the total number of pups born can largely be accounted for by the decrease in the number of male pups. Because females were pregnant (and the sex of the pup was already determined) before the onset of the El Niño, and because males start out substantially larger (~10 kg) than female pups (~8.5 kg), this decrease in males may have simply been caused by spontaneous abortions of larger, more-energy-demanding fetuses. Females in very poor condition, or under poor foraging conditions, may not have the resources to support a male pup.

Females that did have pups during the El Niño year shifted their feeding cycles to compensate for decreased prey. Mothers left about 2 days earlier on the first postpartum feeding trip during the El Niño year, and spent more time at sea feeding, both at a large energetic cost to pups (Ono et al. 1987). Once at sea, they expended more energy on foraging trips (Costa et al. 1991) and consumed different types of prey (DeLong et al. 1991) since many of the normal prey species were not available. This longer, harder foraging did not compensate for reduced prey availability. Pups received less milk from their mothers during the El Niño year and the year following El Niño (Oftedal et al. 1987b; Iverson et al. 1991) and grew at a slower rate (Boness et al. 1991).

Pups also suffered a higher mortality rate during the El Niño year (Ono et al. 1987). On San Nicolas Island, a larger percentage than usual of male yearlings were observed suckling during the El Niño year, whereas the percentage of female yearlings suckling that year remained the same (Francis and Heath 1991). However, since more female yearlings were normally observed suckling, this brought the percentages to about equal. Gentry and Kooyman (1986) hypothesized that the capability of female fur seals to nurse their pup beyond the first year may have evolved in response to fluctuating food supplies such as those prompted by El Niño events. The only two species of otariid that have fixed lactation lengths are fur seals, which live at each pole where the food supply does not undergo large periodic fluctuations in cold waters.

We did not observe females abandoning their pups during the period of decreased food supply. Females could have opted to remain at sea and forage when prey densities did not return to normal. Instead, females always returned to the breeding area, even after very extended feeding trips. We found that a pup usually died after its mother was away for more than 10 days (Heath et al. 1991). Females did not appear unusually thin or emaciated during the El Niño. It is probable that they were catching enough prey to maintain their bodies, but not enough extra prey to produce normal amounts of milk. In the Galapagos, the epicenter of the El Niño event, female feeding trips that exceeded the pups' ability to fast were blamed for the 100% pup mortality observed for Galapagos fur seals (Trillmich and Dellinger 1991). Here, in addition to pups, juveniles as well as adult males and females suffered high mortality.

We may also expect pups to compensate behaviorally for a less-than-adequate milk supply. We found that pups spent less time suckling (there was less milk to suckle) but about the same amount of time resting. They did, however, spend less time in energetically demanding pursuits such as playing. Pups also increased the amount of sneak suckling attempts, although 90% of these were not successful. Although we may expect males to attempt to compensate more than female pups for decreased nutrient intake, male and female pups showed parallel changes in behavior as well as parallel decreases in milk intake and growth (Ono et al. 1987).

Conclusions

The extreme sexual size dimorphism seen in otariid pinnipeds appears to have evolved because larger male size is advantageous in fighting and fasting. In the California sea lion, this difference begins in utero with males being larger at birth than females. Males also get more milk because larger pups obtain more milk from their mothers and male pups are, on average, larger than female pups. However, regardless of size, male pups still grow faster than female pups. This contributes to an amplification of the size difference between male and female pups during the early lactation period. Of the various possibilities for differential energy acquisition and utilization we studied, only resting metabolic rate differed between male and female pups. As pups mature and begin foraging on their own, sex differences in the amounts of milk intake and self-feeding may also contribute to enhanced growth in males.

The female reproductive system also evolved to be somewhat flexible, which allows a female to adjust to environmental conditions through undetermined lactation lengths, and possibly, adjustments to pregnancy (Vitzthum chap. 18). A decrease in food supply affected both pre- and postpartum aspects of maternal investment. The number of pups produced decreased dramatically with the sex ratio skewed toward female pups. Mothers left earlier for longer, more energetically draining feeding trips, and pups received less milk from their mothers. Consequently, there was higher pup mortality.

The El Niño year (1982–83) provides a natural experiment that tests the boundaries of species flexibility and the possible ways in which females can adjust to lowered food supplies. This physiological plasticity apparently has evolved in response to reoccurring but unpredictable fluctuations in nutritional resources and is part of sea lions' life-cycle adaptation.

4 Life History and Reproductive Success of Female Northern Elephant Seals

Joanne Reiter

A VISITOR to an island or mainland elephant seal (*Mirounga angustirostris*) rookery might conclude that elephant seal females lead an idyllic existence, spending most of their lives frolicking along the Pacific coast, eating squid, and participating in short, annual bouts of parental care. A closer examination of elephant seal life histories reveals that those adult females on the beach beat some tough odds to even live to breeding age.

As is true for most female mammals, elephant seals spend virtually all their adult lives bare-flippered and pregnant or lactating. The life histories of female elephant seals, however, represent a departure from that of most large mammals, in large part because they are mammals that have gone back to the sea. Elephant seal females are extremists!

I became a convert to the life of an animal voyeur through a series of events. In a freshman psychology class, after reading *Patterns of Sexual Behavior* (Ford and Beach 1951), which compared all aspects of sexual behavior in mammals from humans to shrews, I decided that whatever this field was, it was the one for me. Two years later, I had the chance to take Burney Le Boeuf's animal behavior course at the University of California, Santa Cruz and pursue this area further.

Le Boeuf had just returned from an exciting sabbatical at Harvard. There he had a chance to talk and teach with Irven DeVore, E. O. Wilson, and Ernst Mayr and participate in seminars with some bright graduate students such as Robert Trivers, Sarah Blaffer Hrdy, Peter Rodman, and Barbara Smuts (see Smuts chap. 5). Le Boeuf returned to Santa Cruz and his elephant seal research imbued with an exciting new perspective focused on the new field of sociobiology. The study of animal behavior was no longer only within the purview of comparative psychology but could now be viewed through the glasses of an evolutionary biologist. I began working on the elephant seal project the very next quarter.

In this chapter I first discuss the natural history of elephant seals and describe their particular variation on the mammalian theme. From this comparative perspective, the lifeways and life cycle of elephant seals are dramatically different from the life history features of primates and especially Old World monkeys, apes, and *Homo sapiens* (Morbeck chap. 1; Fedigan chap. 2; Smuts chap. 5; Hiraiwa-Hasegawa chap. 6; Pavelka chap. 7; Zihlman chaps. 8, 13; Pond chap. 11; Galloway chap. 10; McFarland chap. 12; Borgognini Tarli and Repetto chap. 14; Morelli chap. 15; Panter-Brick chap. 17; Draper chap. 16; Vitzthum chap. 18). I then review the work on sex differences in pups during postweaning development. Finally, I describe important features of the life history of female elephant seals and factors affecting longevity and lifetime reproductive success in females of this species.

NATURAL HISTORY

Because pinnipeds–which include walruses, true seals, and eared seals—are highly modified in their anatomy and physiology, they can live and feed in aquatic environments (Ono chap. 3). They can do things that are impossible for most mammals. For example, elephant seals use pinniped features such as the ability to fast from food and water for weeks or months while breeding. They accomplish this remarkable feat by catabolizing large deposits of stored fat, yielding energy and water. An elephant seal female weighs about 500 kg (1100 lb) and her body contains approximately 36% fat shortly after giving birth (Costa et al. 1986). These mothers have the ability to produce large weight gains in their youngsters within a short, 1-month lactation period by producing extraordinarily fatty milk. They also possess a remarkable ability to dive deeply for extended periods of time that is extreme even among the pinnipeds. They dive deeper (from 200 to 1500 meters) and longer (20 to 70 minutes) than any mammal recorded, including whales, thus, opening a niche not available to other marine mammals (Le Boeuf et al. 1988; Le Boeuf 1994).

Elephant seal social behavior is at the mammalian and pinnipedian extreme in many respects. They are among the most highly polygynous mammals. Males fight to form dominance hierarchies that allow only a few males mating access to large aggregations of females. As a result, only a small proportion of males born do all of the breeding. Only about 20% of males reach breeding age and fewer than half inseminate females. In our studies of lifetime reproductive success of individual cohorts, we found that of the inseminated females in one cohort, 75% were inseminated by 4.4% of the males (Le Boeuf and Reiter 1988).

Interactions of elephant seals are confined primarily to a short intensive period of breeding and lactation during yearly aggregations on traditional rookeries. Although the aggregations may number in the hundreds or thousands, females have no long-term social relationships with their pups or with adults of either sex. Social interactions among females are ephemeral, aggressive, agonistic, and typically for the purpose of adjusting spacing or physical position in the rookery.

Female-male interactions are almost always mating attempts, sometimes violent, and occasionally result in injury or death of females (Le Boeuf and Mesnick 1991). The short mother-pup relationship lasts only through the 1-month lactation period. Nonmaternal female-pup interactions usually result in injury to the pup (Le Boeuf and Briggs 1977). Male interactions with pups are nonexistent unless one counts the unfortunate pups that are killed or injured after being inadvertently trampled by bulls that weigh several tons (Le Boeuf and Briggs 1977).

All pinnipeds that breed annually on islands or along the mainland beaches are polygynous to some degree; that is, only a small proportion of adult males mate with several to several hundred females. The current view is that these pinnipeds have a high evolutionary potential for polygyny because females gather predictably in one place at one time. The females are, therefore, a resource that can be anticipated and easily defended or sequestered by one or a few males to the exclusion of all others (Emlin and Oring 1977; Le Boeuf 1991).

In polygynous animals, sexual selection, an aspect of natural selection that emphasizes traits enhancing the probability of successful mating, leads to differences in anatomy and behavior. Males are selected for features that enhance their ability to gain access to the most females while excluding other males. Females are selected for traits that enhance their probability of producing more offspring of better quality (Ono chap. 3).

Natural and sexual selection for the fittest elephant seal males (i.e., those leaving the most genes in the next generation) thus exclude all but the small proportion of each cohort that survive to adulthood. These males fight well, are large enough to win a place in the dominance hierarchy, and are fat enough to fast for several months during the breeding season.

The fittest females live a long time, begin breeding at an optimal age for their condition and environment, and breed in optimal rookery locations protected by high-ranking males. Their central location in the rookery enhances the chances of pup survival and increases the likelihood of breeding with the fittest males.

With such different selective regimes, the resulting life histories of the two sexes in pinnipeds often differ as much as those in completely different species. This is certainly true of northern elephant seals.

Elephant seals are among the most sexually dimorphic mammals. Adult males can attain weights greater than 2500 kg (7900 lb) and are three to seven times larger than females (Deutsch et al. 1990; Le Boeuf and Mesnick 1991). Males reach puberty at approximately 4 years of age. At this age, however, they look similar in size and appearance to large females and are excluded from breeding by older males. Though sexually mature at age 4, another 4 years are required to develop the distinctive secondary sexual characteristics of an 8-year-old adult male. These include an extensive chest shield composed of rugous, corrugated, callous tissue, a long, pendulous proboscis, and the enormous adult size necessary to maintain dominance status while fasting throughout the 2- to 3-month breeding season. Most females begin breeding much earlier, and produce their first offspring at age 3 or 4 years, several years before they have reached full adult size (Reiter et al. 1981; Reiter 1984).

Members of both sexes fast during the winter breeding season and lose more than one-third of their body mass (Costa et al. 1986; Deutsch et al. 1990). They expend this energy, however, in quite different ways. Breeding females, which average 300 to 700 kg, give birth to pups weighing, on average, 45 kg, and nurse them for 23 to 29 days with milk of 55% fat (Riedman and Ortiz 1979; Reiter et al. 1981). During the last 3 to 5 days of the lactation period, the female comes into estrus, copulates with several males, and then returns to sea after an average stay of 36 days (Le Boeuf et al. 1972; Reiter et al. 1981). Most of a female's lost energy is transferred efficiently to her pup in the form of fat. A pup gains 2 kg for each kilogram spent on milk production by the female and typically weighs 100 to 200 kg (220 to 440 lb) at weaning. There is a large range in the weights of mothers because females begin breeding several years before reaching adult size. Younger, smaller females produce smaller pups, whereas larger females produce larger pups (Reiter et al. 1981; Costa et al. 1986; Deutsch et al. 1994). Weaned pups remain on land, living off catabolized fat for an 8- to 12-week postweaning fast (Ortiz et al. 1978; Reiter et al. 1978). They aggregate in weaner pods near harems and gradually teach themselves to swim.

Survival during the first year at sea is somewhat variable and may be affected by annual variations in oceanic conditions (Reiter et al. 1978; Le Boeuf and Reiter 1991). Juveniles of both sexes follow similar schedules of feeding at sea and molting in the spring until age 3 or 4 years, when females become breeding adults and males join the ranks of the sexually mature but sexually inactive subadults.

Males arrive at traditional rookery sites in late November or early December. They fight for position in a dominance hierarchy and spend the entire breeding season, often more than 90 days, fasting and fighting for access to females that aggregate in harems numbering from a few to several thousand (Le Boeuf 1974). Males of high rank spend even more energy than low-ranking males defending females in their harems from other males (Deutsch et al. 1990).

Males do not discriminate among the females with which they mate; they will attempt to mate with any female, and some attempt to mate with newly weaned pups, yearlings, dead seals, and even fiberglass models of elephant seal females (Deutsch et al. 1990; Le Boeuf and Mesnick 1991).

Females do not appear to "choose" individual males but behave in such a way that only the highest-ranking males have an opportunity to mate with them. Before and during early estrus, a female vigorously and vociferously rejects all mating attempts, thereby calling to the attention of all other males the fact that someone is trying to mate with her. Only a male of high rank has the opportunity to continue mounting her until successful. Lower-ranking males are challenged by males of higher rank and forced to move away. By uniformly rejecting all suitors, in the manner described above, a female incites competition among local males. The result is that she is inseminated by the winner, the male of highest social rank in the vicinity, one of the fittest males present (Cox and Le Boeuf 1977).

Weaned Pups

I began studying elephant seals on Año Nuevo in 1973 as an undergraduate with a new fascination for the field of animal behavior. At that time, Le Boeuf was completing a study on the competition among males and the correlation between social rank in the male dominance hierarchy and reproductive success (Le Boeuf 1974). He and his students had also done descriptive work on female reproductive behavior and pup mortality (Le Boeuf et al. 1972; Le Boeuf and Briggs 1977). Nell Lee Stinson, another undergraduate, and I were interested in the 2- to 3-month developmental period between weaning and a pup's departure from the rookery for its first long trip at sea.

During that first winter, a discussion about the expected sex differences in behavior of sexually dimorphic species piqued my interest in looking at sex differences in the behavior of elephant seal pups. In fact, sex differences in development and behavior emerged as the most interesting aspects of the postweaning period (Reiter et al. 1978).

Male pups tend to weigh more than female pups both at birth and at weaning. In a later study, we also found that they suckle 1 day longer than female pups (Reiter et al. 1978; Reiter et al. 1981; Le Boeuf et al. 1989). Even though male pups appear to be more costly energetically, we have subsequently discovered that females do not have significantly more male pups later in life when they are larger, and apparently the extra cost is not great enough to affect a mother's survivorship (Le Boeuf et al. 1989).

Another sex difference is the tendency of male pups to sneak back into the harem after they are weaned to steal milk from other nursing females. This is a risky behavior since females bite and injure alien pups. However, males that gain additional sustenance by stealing milk while the rest of the cohort is fasting may increase their probability of being one of the few males that breed. In our study, males were six times more likely to become milk thieves than females. In addition, male milk thieves were much more persistent than females (Reiter et al. 1978).

Developmental differences may facilitate male milk thievery. Female weaners begin to molt their natal pelage and erupt teeth approximately 4 weeks after birth, just before or after weaning. Males molt and erupt teeth 2 weeks later (Reiter et al. 1978). They, therefore, look and feel to nursing females like pups instead of weaners, a certain advantage to milk thieves.

We found sex differences in the play behavior of weaned pups; males play-fought with all the components of adult male agonistic behavior save the pulsed threat vocalization. Female weaners interacted with each other in a manner similar to adult females. Males reared up and grabbed, bit, and slammed against each others necks engaging in long-lasting play bouts. Females do not engage in this type of play bout with males. Instead, female weaners interacted with other weaners with the quick bites about the face and growling threat vocalizations typical of adult female aggressive behavior.

Sex differences in behavior, development, and anatomy, even in very young animals, are not unexpected in a species as polygynous and sexually dimorphic as northern elephant seals. Competition among males is so intense in this species that only a few of the fittest males in a cohort will achieve any reproductive success. Variation in reproductive success among males is extremely high, four times greater than that of females. Our studies show that important correlates of reproductive success are survival to adulthood, longevity, social rank, and seasonal tenure on the rookery (Le Boeuf and Reiter 1988).

To achieve high social rank and maintain it long enough to inseminate many females, a male must attain a large size and fight well. A male that maintains a high social rank for several years can inseminate several hundred females in a lifetime. Differences in size at the end of the period of parental investment can be expected to be maintained in adulthood (Trivers and Willard 1973). Stealing milk effectively increases this period of investment by potentially enhancing the size of the milk thief as an adult. The sex differences observed during the postweaning period are likely the result of this intense sexual selection in males to get an edge over others in their cohort in any way possible.

For females, there is much less variation in reproductive success. About 40% of females reach breeding age, and the majority of these produce at least one weaned pup. Therefore, three to four times as many females reproduce as do males (Reiter 1984; Le Boeuf and Reiter 1988; Reiter and Le Boeuf 1991). For females, the greatest correlates of reproductive success are longevity and the rate of weaning success, which increases with longevity (Le Boeuf and Reiter 1988). Females can produce only one pup each year until they die. Few females will ever produce more than 10 pups in a lifetime.

Being large does not carry as great an advantage for females as it does for males. As I discuss later, even if being large increases a female's likelihood of breeding earlier in life, this may not really be advantageous because early breeding lowers longevity (Reiter 1984; Reiter and Le Boeuf 1991). It is not worth the risk of injury for a typical female weaner to become a milk thief to increase her size. Therefore, milk thievery in female weaners would be selected against. In contrast, for a male weaner facing a one in ten chance of ever mating, selective pressure for increased body size or fighting ability may make even a risky endeavor worthwhile.

Adult Females

Evolutionary theory presumes that individuals compete to leave more of their genes in the next generation. When I was studying weaned pups, researchers were observing the way adult male elephant seals compete with each other in order to inseminate the greatest number of females. I, however, became interested in the way female elephant seals compete with each other to produce more viable offspring over a lifetime. Through studies on female competition and reproductive success (Reiter et al. 1981), on lifetime reproductive success (Le Boeuf and Reiter 1988), and on life-history consequences of age at primiparity (Reiter and Le Boeuf 1991), I have developed a better understanding of the life history of female elephant seals.

Upon reaching breeding age, elephant seal females face intense intrasexual competition. For many social mammals, successful competition for food resources is linked with reproductive output. This is true, for example, of many ungulates (Caughley 1970; Clutton-Brock et al. 1982). In contrast, elephant seal females do not breed and feed at the same time or in the same place. They, apparently, feed solitarily in the ocean and fast while breeding or molting on land. It is during the breeding season that females compete with each other, not for food, but for the space and optimal position on breeding rookeries. Good position enhances their probability of nursing a pup to an ample weaning size and their potential for breeding with high-ranking males (Reiter et al. 1981).

In studying females, we found that the most important variables affecting the success of a breeding attempt are a female's age and experience and the density of the rookery (Le Boeuf and Reiter 1988; Reiter and Le Boeuf 1991). Females more than 5 years old, are larger, more aggressive, and more experienced. Though females do not follow a strict dominance hierarchy, older females use their greater size and aggressivity to bite, push, move, and intimidate younger females. These pairwise dominance relationships are maintained throughout a breeding season. Older females tend to arrive earlier in the season and garner good central positions in the harem. They are able to nurse their pups longer, and their pups weigh more at weaning. The largest females produce pups twice the mass of the smallest females. In both high- and low-density rookeries, older females (i.e., greater than 5 years) are much more likely to wean their pups successfully (Reiter et al. 1981).

A pup is weaned successfully when its mother remains with it throughout the nursing period. She transfers her fat stores via rich milk until the pup is fat enough to withstand the postweaning fast. The pups of females that are permanently separated or temporarily but frequently separated from their mothers, or who are not nursed exclusively, tend to die before weaning or do not attain sufficient size to survive after their mothers leave (Le Boeuf and Briggs 1977). Weaning success of young adult females is poor to nil on high-density rookeries where competition for space and position on the rookery are intense.

On Año Nuevo Island where density is high, the average overall pup mortality is 35% and reached 70% in one year with very stormy weather (Le Boeuf and Reiter 1991). Most of this mortality occurs among the pups of the younger females. In dense breeding aggregations, young females are relegated to inferior peripheral locations. This results in fatal mother-pup separations as well as lower probability of breeding with high-ranking males that are usually situated in the center of the harem. The harem periphery is rife with milk-thieving orphan pups and weaners; and marauding subadult males try to sneak copulations out of the view of dominant males. High tides and high surf also cause great disturbance and can result in drownings, separations, and high pup mortality.

The maternal experience of older females also confers an advantage by enhancing weaning success. For example, experienced 4-year-old females had twice the rate of weaning success compared to primiparous 4-year olds (Reiter et al. 1981). Mistakes made by inexperienced females are more severely penalized in the intensity and confusion of high-density rookeries, and younger females are much more successful at low-density rookeries. They tend to colonize newer, less crowded rookeries. Although elephant seal females tend to be philopatric (i.e., breed where they were born) and site tenacious, approximately 25% change breeding sites. A significant proportion of these individuals move after experiencing a reproductive failure, often from a high-density rookery to a low-density rookery (Reiter et al. 1981).

The importance of age and longevity become even more evident when examining optimal life histories for reproductive success in elephant seals. Elephant seal females at the Año Nuevo rookery begin pupping at ages 3 or 4, before they have finished growing. Although natality on the rookery is high (97%) they do not have a high degree of reproductive success until age 6. Only 40% of females live to age 3 and with 20–25% annual mortality, few live past age 14 (Le Boeuf and Reiter 1988; Reiter and Le Boeuf 1991).

Life-history theory predicts that the earlier an animal begins breeding, the sooner its genetic interest begins compounding. Therefore, animals should begin breeding as early as possible. In elephant seal females, however, the putative advantages of breeding early (pupping at age 3) predicted by life-history theory are negated by the tendency of early breeders to die at a faster rate, thereby decreasing the opportunity to breed when they are older, more successful mothers producing larger pups (Reiter 1984; Reiter and Le Boeuf 1991; also see Fedigan chap. 2). In high-density rookeries, the cost of breeding early far exceeded the benefit since very few 3-year-old females ever wean pups in high-density rookeries. Therefore, the optimal life history is to begin pupping at age 4, produce a pup every year until death, and breed at low-density rookeries early in life. Model populations of females constructed from life-table data on survival and weaning success of females at the Año Nuevo rookery show that females that wait until age 4 to produce pups would grow at a faster rate than those that begin pupping at age 3. Furthermore, in very high-density rookeries, females would benefit from delaying reproduction until age 5 (Reiter 1984; Reiter and Le Boeuf 1991).

Summary and Conclusions

The life history of elephant seal females provides an interesting perspective when viewing the life histories of primates and other large mammals. Their social interactions are brief, aggressive, and have only indirect effects on success in weaning offspring and mate quality since they affect their mobility within the aggregation. Social interactions, on the other hand, do not influence females' ability to obtain sustenance for themselves or their young. After a 1-month lactation period, there are no filial bonds. Males do not protect, feed, or even recognize offspring. A female influences which males mate with her mainly by her choice of location and by inciting highly ranked males to compete for her by protesting loudly when she is mounted.

The habit of females to form large aggregations during the breeding season promotes the dominance hierarchy system among males and limits access to females to all but the few adult males that can fast the longest and fight the best. Since males that are three to seven times larger than females can cause injury or death, females and pups benefit by limiting their exposure to them.

The most important components of reproductive success for female elephant seals are longevity and breeding location. Elephant seal females begin pupping while still growing. However, since natality is high, many of the pups of younger females do not survive to weaning age. The putative advantages of breeding early (pupping at age 3) predicted by life-history theory are negated by the tendency of early breeders to die at a faster rate, thereby decreasing the opportunity to breed when they are older, more successful mothers. Therefore, the optimal life history is to begin pupping at age 4, live long, and produce a pup every year until death.

Elephant seals are extreme in their degree of polygyny. They manifest an extreme of the intensive, solely maternal type of parental investment. For these mammals, the most important and primary component of lifetime reproductive success for both sexes is longevity.

Part III

Natural History and Life-History Studies: The Primates

What It Means to Be a Primate

WITHIN the mammalian order, the primates include the living and extinct lemurs, lorises, tarsiers, New and Old World monkeys, apes, and humans. We share an evolutionary history and, therefore, a suite of features that distinguish us among the mammals. In Part III, we emphasize the catarrhine primates (Old World monkeys, apes, and humans).

PRIMATE FEATURES

Most primates live in trees, and almost all reside in tropical or subtropical forests. It is most likely that primates evolved in such a forest setting in tandem with the diversification of flowering plants. Early primates probably moved along the branches of trees in search of edible flowers, fruits, and leaves or along the forest floor in search of insects. From such an origin, the primates evolved two primary behavioral patterns that can be seen in anatomy: (1) grasping with hands and feet, and (2) hand-to-mouth feeding. These biobehavioral complexes overlie the mammalian base and emphasize the (1) differentiation of the limbs, (2) vision, (3) brain development, and (4) reproductive and social strategies (for previous discussion of features, see LeGros Clark 1959; Napier and Napier 1967; Cartmill 1974, 1992; Martin 1990; Sussman 1991b).

Differentiation of the forelimbs and hindlimbs relates to locomotion, feeding, predator avoidance, social grooming, and communicative behaviors. Locomotion and posture emphasize prehensile hands and feet capable of gripping branches and moving on uneven surfaces, often at heights from which falls could be fatal. Early in individual development, grasping ability allows the infant to travel with its mother. Hand-to-mouth feeding emphasizes the use of the forelimbs to gather, prepare, and transfer food to the mouth. Grooming and intensive infant care, also characteristic of primates, depend on the ability to coordinate hand movements for fine manipulation.

Anatomically, there are specializations of the mammalian model to accommodate the above behavioral patterns in terms of limb use. The shoulder is braced by a clavicle (collar bone), the shoulder joint is mobile, and the two bones of the forearm rotate. The hands and feet retain five digits capable of grasping objects, nails permit the sensitive pads of the digits to contact the object being grasped.

Primates have stereoscopic vision in which the fields of vision overlap, providing depth perception. Monkeys, apes, and humans also have good color vision. Even in infants, both the eyes and the neural control and interpretation of visual information are developed, allowing the young to observe their world from the time of birth. This enhancement of the organs of sight and the sense of touch is paralleled by a reduction on the reliance on the sense of smell. Like the changes in the limbs, it is likely that these changes arose from adaptations to movement along uneven branches between trees and from adaptations to grasping food, processing food outside the mouth with the hands, and then transferring it to the mouth. Face-to-face vision also allows intensification of primate social communication.

Crucial to our evolution as an order is the development early in the life story of the individual of a large, complex brain in comparison to the size of our bodies. The large brain relates to eye-hand-motor control in movement and feeding, to vision, as suggested above, and to complex social group behavior, including communication, predator avoidance, recognition and identification of others, mating behaviors, and infant care. Primate intelligence allows flexibility in individual and group interactions with the social, biological, and physical environment. The brain allows many different behaviors in many different environments. Behavioral patterns are not dictated by genes or the environment but contain at least some element of individual choice. Well-developed social intelligence, in particular, characterizes catarrhine primates. We recognize individuals, remember interactions, and manipulate behaviors of others, sometimes even by intentional deception (Jolly 1966, 1988; Byrne and Whiten 1988; Bernstein 1991; King 1991; de Waal 1992).

This enhancement of brain development comes at some expense, however, in terms of the energy investment in early growth. The brain expands rapidly during prenatal or early postnatal life and necessitates a consistent level of nutritional support during this time. In addition, for the infant to begin to develop the mental links by which to organize its behavior, learning must also begin early. In primates, this has evolved by depending upon a close and intense mother-infant bond (Nicolson 1987; Zihlman chap. 13). The mother provides not only food in the form of milk, but also shows the infant how to feed itself, to move through its environment, and, equally important, how to maintain and use the social network. Behaviorally, the mother and the infant adapt to each other, accommodating to each others needs and developing a communication system to inform each other of their desires.

The energetic demands of such an investment may have been instrumental in the evolution of primarily singleton births within the primates. Although multiple-fetus pregnancies occur, in many species, this may result in the loss of all offspring through premature delivery, or loss of all but one through inability or refusal of the mother to maintain more young. A phylogenetic "commitment" to such a birth pattern limits the reproductive options in primates. For example, we are unable to alter litter size substantially.

All primates live in social groups that provide shared information on survival features of feeding, predator avoidance, and reproductive strategies. These groups are composed of both sexes, different age ranges, and social statuses. They are characterized by shifting social subgroups and alliances. The dynamic nature of these relationships, created by the tension between cooperation and competition, is a constant factor in the lives of primates (Smuts et al. 1987; Dunbar 1988). Every expressed behavior has both short- and long-term consequences (Smuts chap. 5; Pavelka chap. 7).

All individuals undergo a period of socialization into this group, possibly because, in most cases, only one infant is born at one time to each female. Parental investment, particularly on the part of the females, is high. In the arboreal context, a mother can leave her infant in a nest or parked on a branch as do some lemurs and lorises, or carry it with her. Most catarrhine primates can grip with their hands and feet, even in the neonate, an infant usually clings to its mother's hair and goes everywhere with her. Physiologically, this close contact assists in maintaining the body temperature of the more vulnerable offspring. In addition, the mother-infant bond strengthens from the continual reinforcement of touch, smell, and sight. The infants learn about the social, biological, and physical environments as they travel with their mothers. They learn about locomotion, predator avoidance, feeding, and social and reproductive behavior by watching their mothers, how their mothers interact with others in the group, and, later, by interacting with their peers and other individuals in the group.

The importance of brain size and complexity early in the life cycle among the primates allows for information exchange in the social context (King 1991). Learning by primate youngsters can be seen as a species-defined survival life-history characteristic that unfolds during growth and development. Storage and retrieval of information allows the offspring to interact with the biophysical and social environments. Similarly, knowledge of the biophysical environment makes the task of foraging more predictable as a means of subsistence. Primates retain detailed information on the distribution and appropriateness of foods in both time and space.

To cope with the heavy burden of care for

the young, catarrhine primates depend on other individuals within the social group (Nicolson 1987). Child care is often shared by related and unrelated kin. From the viewpoint of the mother, she is released for a time to fulfill her own survival and social needs such as feeding, grooming, and visiting. In addition, she is able to conserve some of her normal energy expenditure without necessitating a "trade-off" in neglecting the care of her young. From the viewpoint of the infant, it is still afforded protection although not directly food. The youngster is still within the sphere of information transfer, and it may develop additional social networks. In the long-term, this may be critical for increasing probability of survival should the mother die or abandon the infant prematurely. From the viewpoint of a female caregiver, usually a young adult or subadult, she is able to gain practice and experience in infant handling, which may enhance her ability to successfully raise her own offspring. Male caregivers also benefit from this arrangement, using the infant to help integrate them into the troop. Males have also been seen to use infants as "shields" to terminate aggressive episodes directed toward them.

Foraging, which depends on the ability to manipulate materials with the fingers and to coordinate these motions through the use of vision, permits primates to use otherwise inaccessible food sources. For example, baboons will extract tubers and snails, whereas chimpanzees are known for their extensive "fishing" for termites and also for their use of hammers in cracking nut kernels. Humans, of course, expand this ability to manipulate food by making the land work for us as well (Part V).

Life Stages

With the primates we see considerable blurring of the boundaries of the life-history features that mark the life stages. Compared to other mammals of the same body size, primates take a long time to grow and develop. The gestation period itself is long. This is followed by a lengthy period of infancy and another long period at the juvenile/subadult stages. Anatomically, however, this is not seen as even or parallel development throughout the body. For this reason, we are uncomfortable with the reliance on body weight to gauge growth and development (Hiraiwa-Hasegawa chap. 6). Instead, some features, such as the eyes, are close to fully formed at birth. Others, such as the brain, expand rapidly, whereas linear growth of the trunk and limbs may be much slower. The primate may begin reproducing before completing its full growth potential. This often appears to be a high-risk venture, with many of the firstborn offspring failing to survive.

Portions of the immune system also develop quickly, then subside as other components of the body's protective mechanisms mature. This early expansion maintains a fully functional immune response even after the protection provided by factors in the mother's milk are no longer available. Protection for the vulnerable infant and juvenile is essential because it will inevitably be exposed to diseases from within the social group or the external environment very early in life.

Weaning does not occur abruptly but involves a gradual transition to other sources of food, mediated by observation of the mother's feeding behavior (Peirera and Altmann 1985; Morbeck chap. 1). Primates must also learn to acquire food in a social setting—negotiating access to food sources, maintaining control of food despite competition, and retrieving the maximum possible when the social setting prevents continued feeding. The actual termination of nursing is negotiated by both the mother and her offspring and can be accelerated or delayed in response to environmental factors and the rate of growth of the offspring. Therefore, there is a discrepancy between biological and social weaning, between the time when the offspring no longer needs its mother's milk and when it actually stops nursing.

The juvenile and adolescent stages are also periods of gradual transition, each with slightly different purposes. In humans, the boundaries of the juvenile stage are further blurred by the continued provisioning of juveniles into adolescence. Western societies see further disjuncture with earlier ages of maturation but increasingly older ages of financial and/or subsistence independence.

Play, which is important to the juvenile, allows for exploration of social relationships with peers and other aspects of their environment. It is probable, however, that play is not merely a testing ground for developing alliances but also enhances locomotor and manipulative coordination, brain integration, and communicative skills. In addition, it appears to be quite enjoyable, consuming considerable amounts of time and energy. Primates, in particular, seem to take considerable pleasure in the association with others of their kind.

Within Old World monkeys, apes, and humans, female reproductive function is also blurred. Copulation is no longer tied to clearly defined estrus periods at which ovulation occurs (Martin 1992). Instead, a "desynchronization" has occurred so that mating may occur both before and after ovulation and even while the female is already pregnant.

Natural History and Life-History Studies

In this section of the book, the primate overlay that rests heavily on the sociality of this order is examined. The central features of kinship and friendship, based on social intelligence, appear repeatedly in discussions of primate social structure.

Smuts builds on a long research career on the development of extensive social relationships among baboons in her chapter (Smuts 1985). Her focus is on the social and emotional basis in the primate overlay. Smuts introduces the quantum leap in investment in the young by primate females. In contrast to the relative lack of long-term involvement between the mother and offspring seen in the sea lions and elephant seals (Ono chap. 3; Reiter chap. 4), primate mothers maintain a long, intense, and multidimensional relationship with their young. Smuts focuses on why such intense mother-infant bonds should evolve and on the physiological effects of this relationship.

An overview of the socialization process during growth and development is provided by Hiraiwa-Hasegawa. Her research on macaques and chimpanzees has stimulated her to examine infants, juveniles, and adolescents as each sex moves through the life stages. In this chapter, she shows that social development and learning rather than body size are important in primates.

In her study of Japanese macaques, Pavelka focuses on why females are important for group cohesion. Her longitudinal work on the Arashiyama West macaques highlights the role of females as the initiators and keepers of tradition within the troop. Their relationships with their daughters determine status, and the alliances formed by these kin groups govern troop temperament.

Zihlman expands the discussion of primates to include our closest living relatives, the apes. Rather than concentrate only on behavior or biology, she shows the interplay between biology and behavior in terms of how the sexes use different features to meet both survival and reproductive needs.

Conventional life-history models predict the advantages of passing genes to the next generation as soon as possible. These models, however, discount the complexity of the lives of all mammals and particularly of primates. One common theme that arises from this group of chapters is that there are multiple reasons why delays in reproduction occur and may be beneficial to an individual. Even among the marine mammals (Part II) where maternal assistance beyond food is minimal, delays in reproduction dramatically enhance the ability of a mother to rear an offspring successfully. Even the seemingly simple task of producing and feeding an infant takes skills that are acquired only with practice. Added to this, in the primates, is the integration of the infant into a social unit where relationships of friendship, alliance, and dominance must be respected and cultivated. Experience during the individual life story then appears to be a factor that must be added into any equation for predicting reproductive outcome.

Among primates, the importance of kinship also becomes immediately apparent (Gouzoules and Gouzoules 1987; Bernstein 1991). Both Smuts and Pavelka focus on the utility of kinship groups in enhancing reproduction and survival. As offspring mature, the intense mother-infant bond is not readily broken. Although often one and occasionally both sexes will leave the natal group at maturity, mothers frequently retain long-lasting social links to the remaining offspring. Mothers provide the core of alliances within the kinship group, providing comfort as well as physical support.

In addition to the kin groups, friendships also emerge as a powerful factor in determining the range of primate choices (Smuts 1985; Goodall 1986; Strum 1987). As we have explored the complexity of primate behavior through field studies, we have found that many of the predictions based on a hierarchical model are undermined by primate dependence on other social bonds.

Pavelka (chap. 7:84) noted, "The social animal, living out its life course, is often conceptualized as the end product of interactions between biology and environment." In this group of papers, we aim to show, as does Pavelka, the shallowness of this approach—that primates are negotiating and renegotiating a host of factors, even within their species-specific domains, which may drastically alter the outcome of their life course in terms of both survival, and ultimately, reproduction.

5 Social Relationships and Life Histories of Primates

Barbara B. Smuts

DESPITE dramatic differences in lifestyle that distinguish us from people living in the hills of Nepal, the Kalahari desert, or the mountains of Peru, our lives are similar in one, fundamental aspect: for all humans, the state of a few, intimate relationships with other people strongly influences our well being. If our closest friends, our lover or spouse, our parents, and our children are doing well and feeling good about us, we tend to feel well. In contrast, if any of these people is suffering or if our relationship with one of them is in deep trouble, we tend to feel bad, no matter how well other aspects of our lives are going.

Intimate relationships are the source of our most intense emotions: the greatest joy, the fiercest anger, the most gut-wrenching anxiety, and the deepest sadness. Intimate relationships dominate our thoughts as well as our feelings. We perpetually ruminate and fantasize about our own relationships. We also are fascinated by the relationships of others, as indicated by the popularity of gossip, soap operas, and the prominence of interpersonal themes in literature. Finally, in addition to their force in our emotional and mental lives, most of us also devote considerable time and energy to actively pursuing, maintaining, and repairing intimate relationships—time and energy that could be devoted to other important pursuits.

Preoccupation with intimate relationships characterizes not only wealthy Westerners who can afford the time and energy they demand. It is true for people everywhere, including those engaged in a daily struggle for survival. Marjorie Shostak's book about Nisa, a !Kung woman living in a foraging society in the Kalahari desert, illustrates the importance of such relationships for people whose lives are radically different from ours (Shostak 1981). Nisa's story, told in her own words, is largely a story about a handful of others, her interactions with them, and her feelings about them.

First, Nisa reminisced about her husband: "We lived and lived, the two of us together, and after a while I started to really like him and then, to love him. . . . We lived on and I loved him and he loved me. . . . Whenever he went away I'd miss him, and want him. When he'd come back my heart would be happy." (1981: 166.)

Later, Nisa spoke about the death of her son: "This time I cried for many more months. My son had been the only one left. Month after month I cried, until the tears themselves almost killed me. I cried until I was sick, and I was near death myself." (1981:314.)

These quotations, with which we can so readily identify, illustrate the striking similarities across cultures in the emotions associated with intimate relationships. These emotions appear to be among the most stable, fundamental features of human nature. Nowhere on earth do we find people who fail to delight in the company of their closest friends or who suffer no grief at the loss of a loved one.

I have discussed these facts of life with which everyone is familiar because their very familiarity tends to make us take them for granted. Of course we care deeply about our family and friends. If someone asks why, we'd be likely to answer, "Well, that's just how people are. It's as basic as eating and drinking." And yet, it is possible to imagine an intelligent, adaptable creature that does not form deep bonds with others; consider Mr. Spock from *Star Trek*, or the elusive male orangutan, a close primate relative who spends most of his long life totally alone (Rodman and Mitani 1987). I argue here that students of human behavior should not take human capacities for intimate relationships for granted. Instead, we might ask, as we do of other fundamental aspects of human behavior, Why have these capacities evolved? How did they contribute to the survival and reproductive success of our ancestors?

To answer these questions, we need to broaden our focus to consider close relationships in other species, and particularly in our nonhuman primate relatives. I consider the forms that intimate relationships take in selected primates and how they contribute to survival and reproduction. These can be considered "life-history characters" (Morbeck chap. 1). I begin with the most basic of all mammalian and primate relationships, the bond between mother and child. Then I will consider close relationships among adults. After discussing the evolution of intimate relationships, I will briefly consider some recent research on their development during individual lifetimes that highlights the dynamic, reciprocal interaction between emotions and social life. In a sense then, we are addressing the evolution of emotional capacities as much as we are addressing the evolution of social relationships.

Mother-Infant Relationships in Nonhuman Primates

The mother-infant relationship is intense and prolonged in monkeys and apes (Nicolson 1987). For example, in the baboons that I observed in Kenya for many years (Smuts 1985), a mother and her infant remain in nearly constant physical contact from the moment of the infant's birth until it is at least 3 months of age. Contact is necessary, of course, when the infant is nursing or being carried, whether on the mother's belly as in the first few weeks of life, or on the mother's back from the age of 4 to 5 months until a year or so. But mothers and infants also remain in contact when it is not absolutely necessary, for example, while resting or feeding. And, of course, if the infant is ill, tired, or frightened, it will seek contact with the mother apparently for the sheer emotional reassurance her touch provides.

The importance of contact comfort, first demonstrated by Harry Harlow's experiments with cloth surrogate mothers more than 30 years ago (Harlow 1958), was dramatically brought home to me through the suffering of an infant baboon we can call Pluto. Pluto was the first offspring of a young female who, for unknown reasons, allowed him to touch her only when he was nursing, being carried, and at night. Early in his life, Pluto valiantly and persistently attempted to maintain body contact with his mother. After a few weeks of consistent rejection, he broadened his efforts to be close to and touch two subadult males, one a cousin and the other his mother's closest male friend. These males tolerated the infant's attempts to huddle against them and sometimes even carried him. But, they failed to show the solicitude typical of a real mother, and many times a day, Pluto would find himself sitting alone in the middle of the troop with no one to touch. His acute anxiety at such times was excruciating to witness. Yet, clearly Pluto was in no immediate danger as a result of his lack of physical contact, since he was warm, well fed, and surrounded by other baboons. Why, then, did it cause him such extreme distress?

The answer, in ultimate, or evolutionary, terms, seems obvious: although contact with the mother is not critical to an infant's survival every moment, there are times, such as when a predator draws near, when contact can make the difference between life and death. An infant who lacks a source of contact comfort lacks "life insurance." Thus, natural selection has created an infant with an intrinsic, intense desire for touch that, contrary to the predictions of behaviorism, is very difficult to extinguish through rejection (Bowlby 1982).

Recent laboratory research provides additional insights into the importance of body contact for young primates and other mammals. It seems that the mother's touch is critical, not only for the infant's emotional well-being, but also for regulation of its basic bodily functions and even for growth and survival. Infant macaques whose mothers are removed for brief periods show rapid and dramatic disruption of heart rate, sleep patterns, and body temperature (Reite et al. 1978; Reite and Snyder 1982). This has led to the suggestion that intimate contact with the mother is necessary to establish the infant's physiological rhythms, that without such contact, the infant's system cannot function adequately (McKenna 1990). Some experiments with rats take this notion one step further. Removal of the mother rat inhibits the infant's ability to synthesize growth proteins, so that a rat pup without the mother present will simply fail to grow, no matter how adequate its nutrition (Kuhn et al. 1978). Similar mechanisms may underlie the mysterious "failure to thrive" syndrome in humans, where infants deprived of normal physical and social contact wither away and sometimes die.

As infant monkeys and apes mature, the early need for contact comfort develops into a broader need for social and emotional support. For example, the importance of the emotional bond is illustrated by the responses of wild chimpanzee infants to the loss of the mother. Consider Flint and his mother, Flo, the famous chimpanzee matriarch studied in Tanzania by Jane Goodall and her colleagues (Goodall 1986). Flo was an unusually attentive mother, and Flint an unusually closely attached infant. When Flo, at the advanced age of more than 40 years, gave birth to her last infant, Flame, Flint was still very dependent on his mother for emotional comfort even though he was no longer nursing. He insisted on touching his baby sister at every opportunity and eventually managed to convince his mother to carry him as well as to carry Flame. After Flame died at the age of 6 months, Flint and Flo reverted to their earlier, intense bond, including nursing and even carrying in the baby position. This persisted until Flo's illness and subsequent death when Flint was 8½ old, close to the age when chimpanzee males normally begin to travel independently. Flint stayed near his mother's body and refused his sister Fifi's attempts to lure him away to join her foraging expeditions. He stopped eating and became increasingly lethargic. Within 3 weeks of his mother's death Flint, too, died. An autopsy revealed gastro-enteritis as the immediate cause of death, and studies of his skeleton show he had experienced growth problems (Zihlman et al. 1990). Goodall suggests that, "the psychological . . . disturbances associated with loss made him more vulnerable to disease" (Goodall 1986:103).

This hypothesis is not as far-fetched as it may at first seem. Although Flint's case was extreme, six other Gombe infants have shown signs of severe depression and malaise after the deaths of their mothers (Goodall 1986). Four of the six never recovered, even though two of them were completely weaned and no longer dependent on their mothers for food. Two orphans that did recover showed dramatic retardation of physical and social development.

Even if we grant that these infants experienced emotional suffering when they lost their mothers, why should such a loss lead to disease and sometimes death? Again, recent laboratory work provides the answer. Infant squirrel monkeys and macaques separated from their mothers showed a dramatic and persistent increase in the stress hormone cortisol similar to or even greater than elevations recorded in response to major physical trauma (Coe et al. 1985, 1988). This rise in cortisol facilitates calling and searching by the infant and probably enhances the likelihood of reunion following temporary, accidental separations in the wild. However, when separation from the mother persists, the continued secretion of cortisol depresses the immune system, increasing susceptibility to disease. If the infant is adopted by another female and all of its needs for contact comfort are met, the physiological disruption to the infant's development is ameliorated but does not disappear. As Reite has emphasized (Reite et al. 1978), these results demonstrate that the infant experiences the loss on a psychological level as well as on a physical level. It has not just lost *a* mother; it has lost *the* mother. A large body of epidemiological data from humans indicates that loss of a loved one depresses the immune system and increases risks of disease not just for infants but for adults as well (Laudenslager and Reite 1984).

Fewer data are available on how female primates respond to the loss of an infant. Most of the laboratory studies have focused on the infant's responses to separation. But numerous anecdotes from the wild and captivity indicate that mothers are distressed, sometimes deeply, by infant loss. For example, in my baboon troop, one infant was killed by an adult male. For weeks afterward, every time the troop passed through this part of their range, the mother emitted the plaintiff call baboon mothers give when they are searching for a lost infant (Smuts 1985). The most poignant and direct evidence of maternal grief comes from two examples drawn from studies of captive primates. Female apes that had been taught American sign language each lost an adoptive infant (one a member of her own species; the other a kitten). For weeks, the females behaved in a depressed manner and repeatedly asked the researchers where their babies were.

The mother-infant relationship is primary both phylogenetically and ontogenetically. Thus, lessons drawn from the study of this earliest relationship may hold relevance for our understanding of other intimate bonds. I would like to emphasize three lessons here.

First, to understand fully the mother-infant relationship, we must acknowledge its emotional nature and be willing to incorporate emotions into our analysis. Mother-infant research indicates that it is possible to study emotions objectively—as Darwin (1872) forcefully argued—for example, by defining depression in terms of behaviors identifiable across species, such as decreased activity levels, lowered appetite, a decline in muscle tone, adoption of hunched-over postures, and so on. Physiological monitoring greatly enhances the study of emotion, because we can compare external changes with internal evidence of altered states. Furthermore, people working with apes that have learned symbolic language can actually ask their subjects what they feel, opening up heretofore undreamed of possibilities for understanding emotions in other animals.

Second, research on mothers and infants clearly demonstrates the adaptive significance not only of the relationship as a whole, but of very specific aspects of the bond. For example, as Bowlby (1973) pointed out many years ago, infant monkeys, apes, and humans all respond in similar ways to brief separations from the mother: first they call and show agitated activity, then they lower activity levels and appear very depressed, and finally, after reunion, they increase their proximity to the mother. The first, agitated phase functions to reunite the mother-infant pair. The second, depressed phase conserves the infant's energy and decreases the chances that it will become lost or preyed upon while the mother continues to search. Finally, the increased clinging after reunion is likely to decrease the probability of subsequent separations. The discovery of mechanisms so finely tuned to the circumstances a mother and her infant encounter in nature reminds us that many aspects of intimate relationships can be understood only when we consider the natural history context in which they evolved.

Third, while research on mother-infant relationships illustrates the survival value of close bonds, it also demonstrates that such bonds entail costs as well as benefits. As we have seen, the rise in cortisol that facilitates reunion when the mother and infant lose one another temporarily can lead to a depressed immune system if the stress of separation is prolonged. Chimpanzees have evolved one of the most intimate mother-infant relationships in the animal world. Yet, the same responses that keep this bond intact while the mother is alive may lead to the infant's death if she dies. This suggests that natural selection has not found a way to design an organism that can both form an intense bond and easily relinquish that bond when the partner is no longer available.

These three lessons should be kept in mind as we consider some other examples of intimate relationships.

Female-Male and Male-Infant Relationships

Until recently, most primatologists assumed that in species lacking long-term, monogamous pair bonds, male-female relationships were ephemeral and strictly sexual. We now know that this is incorrect. Research on savannah baboons, for example, demonstrates the existence of bonds between adult females and males that persist throughout the female's pregnancy and lactation when she is unwilling to mate (Seyfarth 1978; Altmann 1980; Smuts 1985; Strum 1987). These relationships are discernible to any experienced observer and can be demonstrated through quantitative measures.

My data, for example (Smuts 1985), show that female-male friends spent much more time near one another than did other female-male pairs. A female's interactions with her male friends differs dramatically from her interactions with other males. Females frequently approach their friends but nearly always avoid other males. They groom their friends and allow these males, in turn, to groom them, whereas they only rarely groomed with other males. Females often indicate their ease in the presence of a male friend by ignoring him, but when a nonfriend male draws near, females express their discomfort through nervous glances and appeasement gestures. Only with friends are lactating females willing to engage in (nonsexual) intimate physical contact, and they do so often. Finally, females not only allow but actively encourage interactions between their male friends and their infants, while they zealously protect the infants from the attentions of other males. As a result, close bonds develop between the female's male friends and her infant, even if, as is often the case, he was unlikely to be the infant's father.

As the infant grows older, the bond with the male friend becomes increasingly important. By about 9 months of age, infants actively seek out their male friends, especially when the mother is *not* nearby, indicating that the male may function as an alternative caregiver. The males seem genuinely attached to their diminutive companions. During feeding, the male and infant express their pleasure in each other's company by sharing spirited grunting duets. If the infant whimpers in distress, the male friend is likely to cease feeding, look at the infant, and grunt softly, as if in sympathy, until the whimpers cease. When the male rests or feeds, the infants of his female friends huddle behind him, one after the other, forming a "train," or, if feeling energetic, they may use his body as a trampoline (Smuts 1985).

With male-female and male-infant relationships in mind, we can briefly return to the three lessons mentioned earlier in connection with mother-infant relationships. First, I want to stress again the importance of emotions in these relationships. Baboon friends appear to be strongly attached to one another, as discussed in detail elsewhere (Smuts 1985).

The second lesson is that intimate relationships provide selective advantages to the individuals that form them (Smuts 1983, 1985). Male baboons protect their female and infant friends from harassment and attacks by other baboons, and they share favored feeding spots with them. Females, in turn, are more willing to mate with their friends in the future, increasing those males' chances of fathering subsequent offspring. Even the tiny infants provide important benefits to their male friends, allowing themselves to be carried into the fray when the friend is involved in a tense interaction with another male. The presence of the infant inhibits the rival's aggression because if the infant appears to be in danger, the mother and her friends and relatives are likely to mob the offending party, chasing him clear out of the troop. Males also use their female friends as buffers in this way, and males also sometimes attack the female friend of a rival in order to challenge him. These last two examples show that the third lesson also applies; that is, intimate relationships entail costs as well as benefits.

FEMALE-FEMALE AND MALE-MALE RELATIONSHIPS

Female primates also form intimate bonds with one another. In many Old World monkeys, females remain in their natal groups and form lifelong bonds with their female kin (Gouzoules and Gouzoules 1987). Closeness between females is expressed in many of the same ways as closeness between males and females: sitting together, frequent grooming, physical contact, and, in general, frequent relaxed interactions. The reproductive advantages of these bonds among close female kin are clear-cut. Female relatives consistently support one another in competition with conspecifics (Walters and Seyfarth 1987). This within-group support from her relatives determines a female's position within the female dominance hierarchy, which, in turn, has been shown to influence her reproductive rate (Harcourt 1987).

In my baboon troop, for example, the interbirth interval for high-ranking females was, on average, 6 months shorter than the interbirth interval for low-ranking females (Smuts and Nicolson 1989). A series of events in this troop illustrated the essential connection between close bonds with kin and female reproductive success. With the exception of one mother-daughter pair, the four females who ranked at the top of the dominance hierarchy did not have strong relationships with one another. These four were challenged by a large, tightly knit group of related females ranking just beneath them. The challengers won, and the high-ranking females fell to the bottom of the dominance hierarchy. Their reproductive rates dropped accordingly (Smuts and Nicolson 1989; see also Pavelka chap. 7).

Among apes, in contrast to most monkeys, females usually do not remain in their natal groups and so fail to form extended female kin networks (Wrangham 1989). If mothers and daughters end up in the same group, however, they maintain an intimate bond for life (Goodall 1986; Stewart and Harcourt 1987). In captivity, female common chimpanzees, gorillas, and pygmy chimpanzees also show a capacity to develop lifelong friendships with unrelated females (de Waal 1982; A. Parish pers. comm.).

Male primates also form bonds with other males, but I am not sure that it is appropriate to label these relationships "intimate." For example, in my baboon troop over a 6-year period, only one pair of adult males, Alex and Bz, formed a stable, long-term coalition. They used their alliance to protect one another from other males and to acquire sexually receptive females from younger and stronger rivals. Yet, despite the importance of their partnership, I never saw them touch one another except during highly ritualized greeting ceremonies. These greetings, during which both males typically remain very tense, are as close as male baboons ever come to "intimacy" with one another (Smuts and Watanabe 1990).

Within the primate order, in addition to humans, chimpanzees form the closest male-male bonds (Nishida and Hiraiwa-Hasegawa 1987). In the wild, males remain in their natal groups and cooperate with their male relatives against males from other groups, much as many Old World monkey females cooperate against females from other groups (Cheney 1987). And again, like many female monkeys, males form coalitions within the group that they use to improve dominance rank. However, in a detailed study of male-male relationships in a large group of captive chimpanzees, de Waal (1984) showed that these coalitions were unstable. This is because males opportunistically changed partners when a new ally proved more effective in helping a male improve his status. Both males and females formed stable friendships with members of the same and the opposite sex, but only females consistently supported their friends during agonistic interactions. The males were much less predictable, because political expediency rather than interpersonal loyalty dictated their patterns of support. In his discussion of these results, de Waal suggested intriguing parallels with our own species. Certainly in our culture, male-female and female-female bonds generally appear to be more intimate than male-male bonds. This is apparently true in some other cultures as well. I raise this possibility not because I am convinced of its validity but because it is an issue worthy of further investigation.

Development of Emotional Capacities and Intimate Relationships: Human Attachment Research

Recent research on the development of intimate relationships among human children illustrates one of the points I have emphasized: to understand either social relationships or emotions, we need to consider each in the context of the other. They are intimately linked, and research on child development illuminates the nature of this linkage.

The research I will summarize was stimulated by John Bowlby's evolutionary theory of mother-infant attachment (Bowlby 1982). This research involves numerous investigators and diverse methods, including detailed home observations of mothers and infants in the first year of life, a variety of laboratory procedures for documenting aspects of the mother-infant relationship, and longitudinal studies linking these data to the child's social behavior several years later. The results indicate the existence of striking individual differences in the mother-infant relationship by age 1 year that strongly predict the child's future socioemotional functioning (Ainsworth et al. 1978; Arend et al. 1979; Sroufe 1984; Main and Kaplan 1985; Grossman and Grossman 1990).

These differences, of course, fall on a continuum, but for heuristic purposes they have been classified into several different types of relationships. Two main types of mother-infant relationships will be considered here, "secure" and "insecure-avoidant" relationships. "Secure" relationships typically characterize more than half of the infants sampled. Infants with a secure relationship interact warmly and frequently with the mother, but they are also curious and exploratory. During play, they frequently return to the mother to "touch base" and then resume their activities. Securely attached infants become very distressed if the mother leaves them, even very briefly, in a strange place or with a strange person. When the mother returns, they greet her warmly, are quickly comforted, and then return to play and exploration (Ainsworth et al. 1978).

Infants who have an "insecure-avoidant" relationship with the mother act very differently. They tend to play by themselves and do not interact with the mother as often as the securely attached infants do. They do not repeatedly "touch base" with the mother, although they may monitor her presence visually. If the mother leaves the infant in a strange place or with a strange person, the infant often fails to show overt distress. Most striking of all, when the mother returns after a brief time away, instead of greeting her, the infant will turn away and actively avoid her (Ainsworth et al. 1978).

Home observations during the first year of life revealed dramatic differences in the behavior of the mothers of secure versus avoidant infants (Ainsworth et al. 1978). The mothers of infants later identified as secure were very responsive to their babies' needs. They comforted them when they cried and seemed very comfortable touching and holding them. The mothers of infants later identified as avoidant, in contrast, were consistently rejecting, frequently failing to respond to their babies' signals. Furthermore, they found close body contact aversive, touched their infants less often, and held them stiffly and awkwardly. Subsequent studies by others have added to these characterizations. Mothers of secure infants seem adept at stimulating fluid, reciprocal interaction with their infants. Mothers of avoidant infants are controlling and insensitive to infant responses, so that the infant lacks opportunities to develop the skills of social give-and-take (Main and Kaplan 1985).

Longitudinal studies reveal striking parallels between attachment classifications at 1 year and social, emotional, and cognitive functioning 3 to 5 years later (Sroufe 1984; Main and Kaplan 1985; Grossman and Grossman 1990). Here, I highlight a few of these differences. In preschool, 4-year-olds rated secure as infants are much more successful at establishing and negotiating social relationships than are children who have been rated avoidant. The secure children are outgoing and friendly. They are liked by other children and their teachers. They are independent and self-reliant, but they also seek help from others when the situation requires it. They do not pick on other children, but they are not submissive either, standing up for themselves when others try to bully them. They are empathetic and helpful when other children are hurt or sad. Most striking to the researchers are the ways they handled emotionally difficult or frustrating situations. They show affection openly, express a wide range of different feelings, and are able to modulate their emotional responses.

As preschoolers, the children who have been rated avoidant at age 1 have few friends and are not well liked by either teachers or other students. They tend to play by themselves, and when they do interact with other children, they often alienate them by not following the unspoken rules of give-and-take. Some avoidant children are actively hostile; they are the class bullies. They show no empathy for other children in trouble. They appear to have difficulty controlling their emotions, and their range of emotional expression is limited.

In-depth interviews with the mothers of secure versus avoidant children reveal consistent patterns. In brief, mothers of secure infants trust others and create trusting relationships with their children. In contrast, mothers of avoidant children do not trust others and are, in turn, not trusted by their own children (Main and Kaplan 1985).

These studies and many others support Bowlby's (1973, 1982) notion that children develop "working models" of intimate relationships based on early experiences with caregivers (Morelli chap. 15). These models include a set of assumptions, not necessarily conscious, about how other people will treat you and strategies for coping with others, given those assumptions. The models also include emotional components: ways, often unconscious, of experiencing or repressing or showing or hiding feelings associated with close relationships. For example, the avoidant child assumes that others cannot be depended on for support and inhibits or represses desires for closeness. Attachment researchers stress that these working models are not simply passive reflections of the child's experiences; rather, they are active constructs that influence future experiences (Main and Kaplan 1985). Children bring these working models to their interactions with new people, which helps to account for the consistency of social and emotional patterns from ages 1 to 6 and beyond.

Evolutionary Perspectives

I conclude by considering working models from an evolutionary perspective by returning to observations of nonhuman primate societies. Evolutionary theory predicts that individuals will behave in ways that, on average, increase their own reproductive success. From this perspective, all individuals are selfish competitors. Yet, paradoxically, formation of cooperative relationships is sometimes the most effective way to increase individual reproductive success. However, the genetic interests of individuals are not identical, and conflicts of interest perpetually endanger the survival of these relationships, particularly when they involve unrelated individuals. This familiar tension between individual self-interest and the well-being of some larger social unit reflects the selection pressures that underlie all social life, both in humans and other animals. Thus, animals could not evolve stable, long-term, mutually dependent, reciprocal, intimate relationships with nonkin without simultaneously evolving mechanisms to ensure that, on average, each member of the cooperating unit received benefits greater than she or he would receive if acting alone or in cooperating with others instead.

These mechanisms, I argue, determine the nature of intimate relationships. They involve three essential elements: reliability, empathy, and reciprocity. Reliability simply means that one can predict what someone else is going to do based on past behavior. Clearly this is essential to the development of any mutually beneficial, cooperative relationship. Empathy means that an individual can somehow determine—not necessarily consciously—what his or her partners need from the relationship, so that she or he can offer them benefits that make the relationship profitable to them. Reciprocity means that the relationship involves an exchange of benefits, so that both partners acquire a net advantage. These are the principles that appear to underlie intimate relationships among humans as well as among nonhuman primates.

Let us now return to attachment research. Particular aspects of early interactions with caregivers seem to function as essential elements in the construction of the child's working models of social reality. These include the reliability of caregiver responses, the caregiver's sensitivity in understanding and meeting the infant's needs, and the development of social rituals that involve reciprocal give-and-take. I suggest that the social elements most salient to the human infant—reliability, empathy, and reciprocity—are those that are most critical to the adaptive functioning of intimate, cooperative relationships throughout life.

Natural selection apparently has favored special sensitivity to these elements and the ability to employ them in the construction of working models. In other words, perhaps childhood development of working models is guided by some very specific, structured mechanisms. Such mechanisms could be adaptive if they allowed individuals to focus more quickly and with greater ease on those aspects of their interpersonal world most critical to establishing effective cooperative relationships—relationships that would, in turn, contribute to survival and reproduction.

Our ancestors spent their whole lives in small, face-to-face groups where the basic dynamics of social relationships probably changed very little, if at all, from one generation to the next. Yet, social relationships probably varied a great deal across groups, or at least across different geographical areas, because of variation in demographic, ecological, and historical circumstances. Thus, the most adaptive developmental plan might be one that began with a flexible capacity to generate rules of social interaction based on early exposure to a few, critical people whose patterns of social interaction, in turn, reflected the social milieu in which they themselves grew up.

I have moved from concrete examples of nonhuman primate social relationships to hypotheses about social development, which, as yet, remain largely untested (but see Draper and Harpending 1982; Chisholm 1993). Whether or not the details of my speculations are correct, it seems clear that the universal preoccupation with intimate bonds reflects the adaptive significance of close relationships for our ancestors. Clearly, there is much to be learned by studying intimacy from a comparative, evolutionary perspective.

6 Development of Sex Differences in Nonhuman Primates

Mariko Hiraiwa-Hasegawa

IN THIS review, I summarize some aspects of the development of biological and behavioral sex differences among nonhuman primates. About 5 years into my research, I became interested in sex differences. I had avoided this subject because, in those days, too many popular books on sex differences in animals described stereotypical images of a female as a good mother and a male as a good fighter. Lacking functional explanations, they seemed to help justify discrimination against women in human societies.

I began my career in primatology in the mid-1970s by studying mother-infant relationships in free-ranging Japanese macaques (*Macaca fuscata*) in Chiba Prefecture, a suburb of Tokyo. These monkeys live in a troop containing several groups of females and their offspring—related to each other through their matriline—as well as several adult males who come into the troop from outside. The troop I followed had about 100 individuals. They lived in a temperate, deciduous forest, and it was very difficult to find and follow the monkeys because of the rugged terrain. But, whenever I managed to observe them, it was fascinating: they seemed to know every tree with their staple foods, every place to stop and feed, how to survive a severe winter by huddling with each other, and how to endure the hot summer by bathing in a stream.

At this time, scientific papers and popular books on monkeys and apes often explained social organizations only from the male point of view. Although the importance of female lineages and lineage-dependent rank system had been well known since the early days of Japanese primatology (Kawai 1958; Kawamura 1958), both researchers and popular writers exaggerated the role of the alpha male in the troop and much attention was paid to the male hierarchy. However, my observations suggested to me that the alpha male was not leading the troop. I also became aware of how male and female Japanese macaques were different. They moved around, played, and associated with other members of the troop in different ways. One sex was not superior to the other, but females and males apparently were interested in different things and led different lives.

I finished my master's thesis in 1977 and then went to the forests of the Mahale Mountains, Tanzania, to study other infant behavior in common chimpanzees (*Pan troglodytes*). My research projects provided interesting comparisons of two primates that are closely related to ourselves, *Homo sapiens*. Macaques and common chimpanzees (and humans) share catarrhine primate biobehavioral features. They differ in significant aspects of both their lifeways and the timing of life cycles (Morbeck chap. 9). For example, survival life history features such as modes of locomotion represent two variations of the catarrhine primate plan. Chimpanzees are much larger in body size than most primates. They have long forelimbs, broad chests, short trunks, and no tail. They climb by reaching for branches in the trees and quadrupedally knuckle-walk when traveling on the ground (Zihlman chap. 8). Japanese macaques, like other Old World monkeys, have larger hindlimbs, which drive their locomotion, and quadrupedally walk with their palms flat on the ground. They have narrow chests, long, flexible trunks, and a tail. Both species eat a variety of plant foods, but chimps make and use tools to capture social insects. Chimps also hunt mammals, including other primates, and eat the meat.

Common chimps have a very different social structure from that of Japanese macaques. They do not have a tightly packed group like macaques, but, instead, form small groups or parties in which individuals leave and join as they like, a so-called fission-fusion social structure. Females disperse from their natal groups and, unlike Japanese macaques, adult females do not have close, affiliative relationships. Males, in contrast, form a tight bond based on their kin relationships. Finally, compared to Japanese macaques, common chimpazees are quite intelligent. I had many opportunities to be surprised by their cleverness (also see Jolly chap. 19). Indeed, chimps and humans have a longer period of shared ancestry than do chimps and macaques.

When I began my observations of chimpanzees, many of the studies of mother-infant relationships among primates were done from the viewpoint of psychology. As my research on mother-infant relationships continued, however, I struggled to set my research in the framework of evolutionary biology (Reiter chap. 4). I then realized that how primate mothers raise their offspring is a fundamental question of reproduction and that females and males differ in their activities related to mating and rearing of offspring. In primates, females usually produce a single infant at one birth event, and this offspring has a long period of dependency on its mother. Therefore, each offspring represents a lot of parental investment by a mother; life clearly is quite different for the male primate. It sounds obvious now, but during my early training in Japan, I was not taught a theoretical framework for integrating biology and the evolution of behavior.

It was more difficult to understand infant development in an evolutionary context. I then recognized the importance of life-history strategies. Species- and sex-defined features of growth and development are evolutionary adaptations. Individuals vary as they grow to maturity and reproduce within these "constraints." Their growth and changes in behavior also reflect an individual's interactions with all aspects of their physical and social environments. All of these factors contribute to an individual's survival and reproductive outcome (see Morbeck chap. 1).

Evolution of reproductive behavior has since become the central theme of my research. I came to see my studies of mother-infant relationships as a part of the broader study of survival, mating, and rearing of offspring by females and males, that is, parental investment patterns and life-history strategies.

General Assumptions on the Interpretation of Sex Differences

Females and males sometimes differ considerably in size, anatomy, physiology, and behavior. A male gorilla weighs almost twice as much as a female gorilla, for example, and male mandrills have vivid facial colorings that females lack. Other nonhuman primates show sex differences in body size and body composition, coloration, vocalization, and so on (Zihlman chap. 8).

From an evolutionary perspective, sex differences usually are attributed to the outcome of different kinds of selection pressure operating on the individuals of two sexes (Darwin 1871). Typically, it has been assumed that males compete for the opportunity to mate in order to increase their reproductive success. Female mammals and especially primates with long periods of gestation and lactation cannot increase their reproductive success by increasing the number of matings (Trivers 1972). Instead, females are choosy about mate quality and compete for food in order to ensure the survival of offspring. Therefore, male anatomy and certain behaviors are expected to be selected primarily for successful competition for mates and those of females are expected to be selected primarily for successful rearing of offspring. However, this is a very basic, rough generalization. In fact, there are a variety of phylogenetic (i.e., species life-history features) as well as ecological constraints that affect these selection pressures in quality and in quantity. Therefore, the manifestation of sex differences will vary from species to species. Indeed, in some mammalian species, females are larger than males, and the sexes are quite similar in body size in many monogamous species such as gibbons (Ralls 1976; Clutton-Brock et al. 1977; Zihlman chap. 8).

Sex differences become apparent over the course of an individual's development. Within a species, this timing is regulated by the development of sex hormones, but how and when the two sexes differ may vary from one trait to another. Information about individual animals throughout the life stages, especially during infancy and childhood, is vital for understanding the selection pressures operating to mold adult female and male life-history patterns related to reproduction.

Size is probably the most important single factor affecting animals' life-history patterns and ecological adaptations (e.g., Peters 1983; Calder 1984). In many primates, where sex differences in size are apparent, males are usually larger than females. If males and females are to reach different body sizes as adults, their respective growth patterns should differentiate at some point during the early life stages (also see Ono chap. 3, Reiter chap. 4).

Maternal Investment before Birth: Gestation and Birth Weight

If there are no special constraints on gestation period or on fetal growth, it may be advantageous for a larger sex to be born larger. However, available evidence, although limited, suggests that the sex difference in birth weight is not significant both in dimorphic and nondimorphic primates. Although there is a tendency for the larger sex to be born slightly bigger than the smaller sex, birth weights of males are not significantly greater than those of females in captive common marmosets (*Callithrix jacchus*), Taiwan macaques (*Macaca cyclopis*), long-tailed macaques (*Macaca fascicularis*) (Willner and Martin [1985] for these three species), rhesus macaques (*Macaca mulatta*) (Small and Smith 1984), bonnet macaques (*Macaca radiata*) (Willner and Martin 1985), and chimpanzees (Gavan 1953). The same results are reported from free-ranging vervet monkeys (*Cercopithecus aethiops*) (Lee 1987) and chimpanzees (Pusey 1990).

Infancy and Growth Rate before Weaning

Fewer data are available for the growth rates than for the birth weights of male and female nonhuman primates. However, the growth trajectories for males and females are similar until adolescence in captive common marmosets, long-tailed macaques (Willner and Martin 1985), talapoin monkeys (*Miopithecus talapoin*), grey-cheeked mangabeys (*Cercobus albigena*), de Brazza's monkeys (*Cercopithecus neglectus*) (Gautier-Hion and Gautier 1985), chimpanzees (Gavan 1953), and gorillas (Dixson 1981). The same tendency is reported for free-ranging Japanese macaques (Hiraiwa 1981), vervet monkeys (Lee 1987), and chimpanzees (Pusey 1990). Therefore, the size difference is apparently not very conspicuous from the time of birth through the period of lactation among primates.

Childhood and Growth Rate after Weaning

As there is no marked difference between males and females in birth weight and early growth, the larger sex will have to grow at faster rate after weaning or take a longer time to attain its adult size. Evidence suggests that both processes occur. The growth curves for the above-mentioned species show that males grow at faster rate after weaning until puberty. In addition, full adult size is reached later in males than in females in species in which males are larger. In many dimorphic primates, age at sexual maturity is reported to be 1 to 3 years later in males than in females (Harvey et al. 1987).

Adulthood: Energetic Requirements and Feeding Behavior

Different energetic needs should be revealed with different body sizes. For example, feeding behaviors such as feeding time, bite rates, and food choice may differ. If males grow faster during the juvenile and adolescent stages, the energetic needs of males of these stages may be greater than those of females of equivalent ages. It would be interesting to compare the feeding behavior of still-growing adolescent males with that of adolescent females that have attained almost full size but have not reached the age of reproduction.

In many species, adult females feed longer and prefer higher-calorie food items than do adult males. This is probably due to their higher caloric need for maintaining pregnancy and lactation. Adult female siamangs (*Symphalangus syndactylus*), for instance, feed faster and longer than adult males (Chivers 1977). Similarly, adult female mangabeys spend the highest percentage of time feeding among all age-sex classes. In addition, members of age-sex classes differ in the extent to which they use different food types (Waser 1977). Adult female Japanese macaques feed longer than do adult males, whereas adult males feed faster than do adult females (Iwamoto 1987). Female squirrel monkeys (*Saimiri oerstedii*) feed on arthropods more frequently than do males in all seasons, and females feed on fruits and flowers more often than do males in the early wet season (Boinski 1988).

In chimpanzees, adult females spend more time feeding on termites in Gombe National Park, Tanzania (McGrew 1979) and on wood-boring ants in Mahale (Uehara 1986). In both cases, these sex differences are evident from the juvenile stage (Pusey 1983; Hiraiwa-Hasegawa 1989). Fishing for ants and termites is a time-consuming activity with low nutritional return per unit of time. In addition, the prey acquisition rate cannot be improved by the consumer's effort because the maximum acquisition rate is set in part, by the behavior of the ants. It is conceivable that growing and adult males cannot afford to spend time feeding on ants or termites. Instead they feed on foods that have higher energetic return per unit of time.

Development of Some Behavioral Sex Differences

Sex differences in behavior and their development in nonhuman primates have been a focus of attention mainly because of the interest in the biological meaning of human sex differences (Zihlman chap. 13). However, as Smuts (1987) correctly pointed out, attempts to make a simple generalization based on behavioral sex differences across species neglects the inter- and intraspecific variation in male and female behavior and the functional explanation of the differences. The following section reviews some examples of behavioral sex differences of functional interest observed among immature primates.

Formation of Affiliative Relationships

It is well documented that social relationships among individuals differ according to the societies in which they live. In species in which males disperse from their natal groups, females form closely knit social relationships with kin as well as with nonkin, whereas males usually are out of this network. Among these species, the ability to form a coalition with the members of the group against a competitor over a variety of resources is more crucial to females than to males (Watanabe 1979; Walters and Seyfarth 1987).

The tendency for the sex that stays in the natal group to form a close bond among themselves is reported to be apparent from the time of the juvenile stage. For the species in which females stay in their natal groups, young females groom others, particularly adult females, more frequently than do immature males (Cheney [1978] for baboons; Silk et al. [1981] for bonnet monkeys; Missakian [1974] for rhesus monkeys; Fairbanks and McGuire [1985] for vervets; Mori [1975] for Japanese macaques; Rowell and Chism [1986] for patas monkeys; Wolfheim [1977] for talapoin monkeys), and immature females are more active in the formation of alliances than are immature males (Walters 1987).

On the other hand, for the species in which females disperse from their natal groups, relationships among adult females are less affiliative, whereas adult males support each other and exchange grooming more than do adult females (Goodall [1986] for chimpanzees; Kummer [1968] for hamadryas baboons; Fedigan and Baxter [1984] for spider monkeys). Reflecting these tendencies among adults, in these species, grooming interactions of immature females are limited to immediate family (Pusey [1983, 1990] for chimpanzees; Kummer [1968] for hamadryas baboons), whereas immature male chimpanzees seek contact with adult males (Pusey 1983, 1990; Hayaki 1988), and considerable affiliation between adult and immature males is seen in hamadryas baboons (Kummer 1968). These patterns demonstrate that immature males and females behave differently and learn social skills differently. This depends on their future relationships with the group members, probably in a way that will accrue long-term benefits for their continued survival and reproduction as adults.

Play-Fighting

In many primates, the duration of play sessions is longer for young males than for young females, and young males engage in play-fighting (rough-and-tumble play) more frequently than do young females (Symons 1978; Fagen 1981; Hayaki 1985; Meder 1990). This behavior often is interpreted as adaptive because play-fighting is thought to provide an opportunity to learn fighting skills, which are more important for males as adults than for females. In many monkey species, male dominance rank seems to depend on fighting ability, whereas female dominance rank depends on lineage or age. At least in rhesus monkeys, play-fighting serves to promote the acquisition of adult fighting skills (Symons 1978). However, exactly how fighting skills are important for male reproductive success and not for female reproductive success is not known. In talapoin monkeys, adult females are dominant over adult males, and adult females are more aggressive than adult males. In this species, too, Wolfheim (1977) found that immature males tend to engage more frequently in play-fighting than do immature females. She attributed this tendency to the effect of androgen. However, this may not be the case for all primates. Sex differences in play-fighting are not found in immature patas monkeys (Rowell and Chism 1986) and in immature common marmosets (Box 1975).

More research is needed in order to avoid circular reasoning in the argument of the functional significance of play-fighting. First, we have to gain precise knowledge on the relative importance of fighting ability to the reproductive success for males and females of the species in question. It would be interesting if we could investigate the frequency of play-fighting among juveniles of the species in which fighting ability is more important among females than among males.

Infant-Handling

It is well demonstrated that young female primates engage in infant-handling more frequently than do young males (Cheney [1978] for baboons; Berman [1982] for rhesus monkeys; Lancaster [1972] for vervets; Hiraiwa [1981] for Japanese monkeys). Usually this tendency is already apparent in infancy and continues until females give birth to their own offspring. Infant-handling is interpreted as advantageous in females for successful rearing of their own infants (Lancaster 1972; Hrdy 1976).

Among cooperatively breeding callitrichiidae, immature males and females assist equally in carrying infants (Goldizen 1987). If infant-handling has some functions for future infant-caring behavior, the immature males of these species may exhibit a greater interest in infants than do the immature males of species in which male care is absent. It would be interesting to know the amount of infant-handling in young males and females of the species for which biparental care is the norm.

Maternal Care and Mother-Offspring Relationships

In primates, there seem to be no conspicuous and consistent differences in the various aspects of behavior of mothers directed toward female or male offspring before weaning. Researchers could not detect significant differences between the two sexes in the amount of time spent suckling (Gomendio [1989] for baboons; Rowell and Chism [1986] for patas monkeys; Hiraiwa-Hasegawa [1990] for chimpanzees), nor in the amount of time an infant was carried by its mother (Rhine et al. [1984] for baboons; Hiraiwa-Hasegawa [1990] for chimpanzees). Finally, there is no significant difference in the age at weaning between male and female infants or in the birth intervals after a mother successfully reared sons versus daughters in baboons (Altmann 1980), vervets (Cheney et al. 1988), rhesus monkeys (Small and Smith 1984), and in chimpanzees (Pusey 1983; Hiraiwa-Hasegawa 1990). Considering the above-mentioned similarity in growth in males and females before weaning, these results are not surprising.

Primates apparently differ from other mammals. Studies on developmental patterns of dimorphic mammals have demonstrated heavier birth weight and faster growth rate for males. This indicates heavier maternal investment before weaning in males than in females (Clutton-Brock et al. [1982] for red deer; Lee and Moss [1986] for elephant; Reiter et al. [1978] for northern elephant seal; Trillmich [1986] for Galapagos fur seal; Wolff [1988] for American bison). This life-history character, as viewed from the perspective of the mother, emphasizes greater investment in sons, ultimately the larger sex. From her infant's viewpoint, species- and sex-defined rates of growth in body size also are greater in males during the prereproduction life stages.

An exceptional case is the spider monkey. Symington (1987) reported that male offspring of high-ranking female spider monkeys in Peru receive more milk and are transported by the mother for a longer period of time than are female offspring. Although the sample size is small, she argued that heavier investment toward sons might be adaptive in spider monkeys because (1) males remain in their natal group, and high-ranking females are able to facilitate their sons' integration into male hierarchy; (2) body size seems to be an important determinant of male rank, and size may be correlated with maternal investment; and (3) high rank confers unusually high reproductive success by facilitating monopolization of all the estrus females.

However, in many species of primates studied so far, body size in males does not seem to be a good predictor of mating success. Rank is not always correlated with body size, and the relationship between male rank and reproductive success is still unclear (Bercovitch 1989). However, age (Hasegawa and Hiraiwa-Hasegawa 1990), coalition with other males (de Waal 1982), friendship with a particular female (Smuts 1985), or duration of residency in one group (Altmann et al. 1988), all considerably affect male mating success.

In the species in which males disperse from their natal group, sexually mature daughters maintain close contact with their mothers. However, this does not seem to hold for mothers and sons in species in which males remain in their natal groups. In all the species studied so far, daughters spend more time in proximity to their mothers. In addition, mother-daughter grooming becomes reciprocal as daughters mature, but mother-son grooming is always one-sided from the mother, regardless of the dispersal patterns of the offspring (Fairbanks and McGuire [1985] for vervets; Kurland [1977] for Japanese monkeys; Missakian [1974] for rhesus monkeys; Rowell and Chism [1986] for patas monkeys; Cheney [1978] for baboons; Pusey [1983] for chimpanzees). In all cases, mothers groomed sons and daughters for equal amounts of time. The tendency for immature males to be away from their mothers may be related to inbreeding avoidance. Future research should center on the association patterns of sons and daughters with their male and female parents in monogamous or family-living species such as marmosets, titi monkeys, night monkeys, and gibbons.

Conclusions

In nonhuman primates, there seems to be no conspicuous difference between the sexes in growth from birth until weaning. The growth trajectories for females and males are similar, even in dimorphic species. As a result, greater maternal investment of time and energy in sons generally is not evident before weaning. In primates, male body size is not a single important factor affecting male reproductive success. Many other factors derived from their complicated social relations can have considerable effects. In many species, the mortality rate is higher before weaning, and the energetic costs of growing quickly during this vulnerable period therefore may outweigh the benefits of getting bigger as fast as possible. Maternal cost may also, set a limit on growth in primates in which mothers transport their offspring until weaning.

Juveniles and adolescents are independent from their mothers for nutrition and locomotion but still free from the burden of reproduction. During this period, females and males start to differentiate their behavior in accordance with their different future reproductive strategies. For animals like primates—especially, humans—that have large brains and long lifespans, the acquisition of skills to cope with both the physical and social environments is essential for their survival and reproduction. The juvenile and adolescent periods in primates are of crucial importance in learning these skills for the future (Morelli chap. 15). This important phase between weaning and maturation may be a specific character of primate life-history patterns.

Thus far, we know that immature males and females differ in specific behavioral patterns like play-fighting, infant-handling, and formation of close bonds with group members, as described above. However, the behavioral sex differences that exist among immatures may not always have a direct effect on their future reproductive success. They may be adaptive in that particular stage of their life history.

Much more attention has been paid to infancy and mother-infant relationships than to juvenile and adolescent periods in nonhuman primate studies (see reviews by Pereira and Altmann [1985] and Pereira and Fairbanks [1993]). In addition, our knowledge of nonhuman primates is heavily skewed toward species like baboons and macaques that have multimale, multifemale social groups and male dispersal. More information is needed on species with different social structures; for example, lemurs in which the female is the larger and dominant sex; langurs with one-male groups; and monogamous, biparental species. Future research on the immatures of various species of primates will contribute greatly to the understanding of the development of sex differences and the selection pressures operating on this important primate life stage.

7 The Social Life of Female Japanese Monkeys

Mary S. McDonald Pavelka

SOUTH TEXAS is hot, dry, and covered in thorny brush. It is home to cactus, rattlesnake, and javelina, to cowboys, cotton, and cattle. Certainly an odd place to find snow monkeys, and, yet, here they are. Transplanted from Japan in 1972, the Arashiyama West Troop of Japanese macaque monkeys (*Macaca fuscata*) has tripled in size and now thrives on a large ranch outside the small agricultural town of Dilley, halfway between San Antonio and Laredo.

The monkeys are essentially free ranging, as some years ago they ceased to respect the fence that is intended to enclose them. They are provisioned once a day with grain and monkey chow scattered along a half-mile stretch of road. This food-enhanced environment has contributed to the large size of the troop, which now numbers more than 500 individuals. The animals are trapped and given a facial tattoo early in life, but beyond this they exist with a minimum of human interference. In spite of the south Texas heat, observation conditions are excellent.

Since May 1981, I have had the great privilege of being able to visit the colony, and I have spent many hours sitting quietly in the hot sun, captivated by the complex social lives of these animals. A typical scene would have an extended family, two or three generations, sitting together in the shade of a thorny mesquite tree. A mother grooms her adult daughter, both females with nursing infants at their breast. A couple of feet away, another pair of adult daughters sit sleeping, their fur touching. A 3-year-old female, a juvenile, sits nearby watching a small group of 1- and 2-year-olds play. The play suddenly turns rough, and the kids are screaming at one another and looking to their respective mothers for help. The mothers look up, ready to rush in, but suddenly all is calm again and the mothers go back to their grooming and resting. It brings to mind the goings-on of a group of human mothers having coffee as their children play nearby. But there are many differences, including the fact that monkeys do not converse. In addition, there are no husbands and fathers to return to at the end of the day.

Japanese macaques live in multimale, multifemale societies characterized by females that stay in their natal groups and males that move to other groups. At puberty most males will disperse from the group into which they were born and spend the rest of their lives in the company of animals other than their kin. Females, on the other hand, remain with their close female relatives throughout their lives. Social groups, therefore, generally comprise related females and unrelated, immigrant males. Female kin groups form the basis of the social group, and interactions with adult males are comparatively limited. The social life history of males and females is thus very different.

Because we cannot ask the animals to explain to us the nature of their social lives, it has taken years of intensive observations on individually identified animals to know the female kinship structure of Japanese monkey society. When Japanese scientists first started to recognize individual monkeys in the early 1950s, they gave them individual names. All of the descendants of these original females were given names beginning with the name of the original matriarch. All of the Bettas are descendants of the original Betta. All of the Matsus are descendants of the original Matsu. There are 12 such matrilines currently represented in Texas in the Arashiyama West troop. As a visitor to the Arashiyama West monkeys, I am indebted to Lou Griffin, who as the manager is there day after day, observing the monkeys and then generously sharing some of her knowledge with me. It is she who remarked to me, as we sat silently fascinated by the intricate interactions of these monkeys, that to the extent that it is possible, we have been able to experience life as it is lived by members of another species.

Japanese Macaque Social Life

Kinship

Being able to identify individuals and, then, knowing how these individuals are related to one another is what makes the social interactions make sense to the human observer. In 1981, during my first visit to the colony, I felt lost and overwhelmed by the sea of identical monkey faces and the seemingly random and chaotic buzz of interaction. I spent 2 or 3 days trying, first, to learn to read the facial tattoos on the animals, then to be able to tell the difference between a male and a female, and to distinguish a juvenile from an adult. Still I was lost and convinced that it would never make any sense to me. Lou Griffin and the instructors, Linda and Larry Fedigan, chatted back and forth about the monkeys. ("Did you see little Deko this morning? She seems to be trying to get close to Summa" "Really? She used to be so intimidated by him. Maybe she is trying to cash in on her mother's friendship with him," and so on). I was filled with frustration. I even wondered if they might be making it up! On about the fourth day I was assigned a small group of animals to watch. I was to read their tattoos and then look up the number on my census sheet to get the animal's family name, which has all the information needed to tell who is related to whom.

The group contained four adult females and a number of kids. They were sitting peacefully together grooming one another, with the kids playing all around (and sometimes on top of) them. The first tattoo that I read was #125. The tattoos are a series of dots on the face that correspond to certain numbers. Every animal has a tattoo number and a family or genealogical name. I looked it up. According to my sheet, the family name for animal #125 was Rheus6775. Okay. So far this means nothing. I could have read the tattoo incorrectly, and there was no one around to help me. The next tattoo that I read, after some difficulty, was #58. I looked it up. The family name for animal #58 was Rheus67. Aha! In the genealogical naming system of the Japanese scientists, each animal is given their mothers name with the year of their own birth added on to it. Rheus67 would be the mother of Rheus6775! Rheus67 was born in 1967 to the female named Rheus. In 1975, Rheus67 had a daughter who was thus given the name Rheus6775. Here was a mother and daughter pair, #58 was the mother, #125 was the daughter, and I knew the exact age of each. I was filled with excitement. The fog was lifting—suddenly things were not so random and chaotic anymore. Anxiously I strained to read the tattoos of the two remaining adult females. I identified these two animals as Rheus6772, who would be another daughter of #58, sister of #125, and Rheus6371, who, after a few moments of concentration I figured out would be the cousin of #125 (see fig. 7.1). When Larry came along I proudly asked "Is this #125 with her cousin #22 and her sister #20 and her mother #58?" Obviously pleased, he confirmed my identification. "Yes" he said, "this is Toklas, #125, and her family. We call her Toklas because in her first estrus season she courted females only. Her sister, #20, is 9 years old and yet has never had a baby. We wonder if she ever will. Her cousin, #22, seems to be ill. She has lost weight and is coughing a lot." It was amazing how the overwhelming sense of frustration gave way at that moment to a longing to get to know more animals and to know the stories about their lives. I now have come to know hundreds of

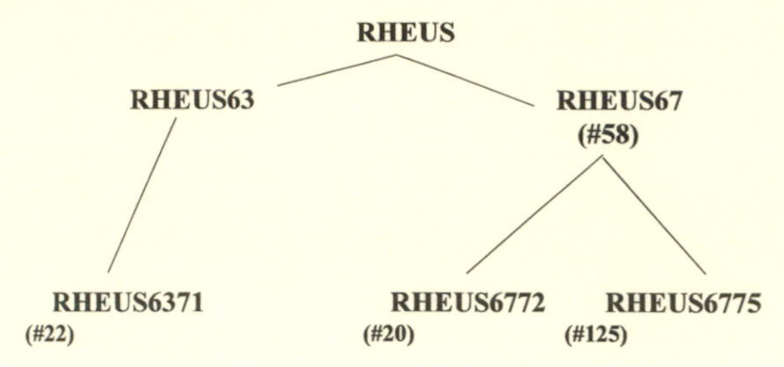

FIGURE 7.1. Some members of the Rheus lineage.

animals. Toklas is not a particularly interesting monkey to me anymore, but for some reason, she remains my "favorite" to this day. Her sister, #20, finally started having babies at age 11, which is very late, and she was, thus, nicknamed "Bloomer."

At one time it was thought that sexual attraction between unrelated adult males and females was the bond that held primate groups together (Zuckerman 1932). Now we know that the most important bond in any primate group is the mother-offspring bond (Altmann 1980; Gouzoules and Gouzoules 1987; Nicolson 1987; Smuts chap. 5), and more specifically, the mother-daughter bond. This is particularly true in Japanese macaque society where the mother-daughter relationship persists throughout life.

Kinship is the fundamental organizing principle of Japanese macaque societies. Once you can identify individuals and if you know how the animals are related to one another, much of the social interaction becomes, if not predictable, at least somewhat regular and understandable. Knowing who is related to whom—and how—reduces the chaos and confusion to a murmur. Kinship influences almost all aspects of social life. Even the effects of age and dominance rank are much better understood within the framework of kinship. Family members spend most of their time together. They eat and sleep together; they groom one another; they fight among themselves; they make up; and they support each other in conflicts with nonfamily members.

Who are the important family members? Actually when we talk about relatives, we are only talking about maternal relatives. We do not yet know the paternal relatedness of the animals, and they give no indication that they do either. The important relationships are those that derive from the mother-daughter bond.

The principal social partner that any female will have is her mother, early in her life, and her daughter, later in her life. Sisters are friendly, but they are secondary to the relationship that any one female has with her mother. For example, in a family containing many daughters, each appears to be bonded first to her mother, and then to each other. The highest-ranked female in the troop, tattoo #49, had four adult daughters, and these five animals were together almost all of the time, along with each of their immature offspring. It was clear that the matriarch of the group was the preferred partner for each of the daughters, and that they groomed one another only as the second choice, when mom was busy with one of the others.

The significance of maternal kinship as the fundamental organizing principle of macaque society could only be appreciated once large numbers of animals were individually identified and then studied for many years (Fedigan and Asquith 1991). This became possible when Japanese scientists in the 1950s began to provide free-ranging troops with extra food. The time that the animals spent in the feeding grounds was the time that was needed to learn to tell the animals apart. A new baby could be followed for years, and the basic female kin structure of the society eventually became known. It is to be expected that a food-enhanced environment alters some behaviors, such as the amount of time spent searching for food, but this seems a small price to pay, because provisioning made possible the kind of longitudinal research on known individuals that revealed the pattern of male dispersal and female philopatry (Asquith 1989).

Dominance

Before this, knowledge of primate societies was based on short-term observations on groups where individuals were not necessarily recognized. Researchers focused mainly on the large, easy to observe males, and primate societies were thought to be organized around rigid male dominance hierarchies (e.g., Hall and DeVore 1965). With the recognition of the female-bonded nature of many primate societies, dominance fell by the wayside as an important organizing principle. I think this has fallen too far in its recognized importance. Making sense of the complex social interactions of the Japanese monkeys in Texas begins with being able to identify individuals. Then comes knowing the family relations of those individuals. For me, the next most important feature to know is something about their dominance rank. Social dominance is complex, highly contextual, and very difficult to reduce to a measurable commodity. Yet, from a monkey's viewpoint it is nonetheless a very real and important element of Japanese monkey social organization. Not surprisingly, it is very much tied to kinship.

What does it mean to be high or low ranking? To try to explain what dominance means to a monkey, I will begin by describing what it is like at the top, with the highest-ranked female in the troop.

From 1974 until her death in 1990, Betta5966, tattoo #49, was the alpha female of the Arashiyama West troop. She deferred to only one animal in the troop, the alpha male. What did her high status mean to her? Sometime around 1980 she was given the nickname Hatchet. Her nickname is suggestive; she was very tough and very sharp, although not large and not particularly mean or aggressive. By monkey and human standards, Hatchet also was an excellent mother.

It goes without saying that what Hatchet wanted, Hatchet got. She had priority of access to all desired resources, including food, water, shade, and social partners. Any animals that she approached were happy to have the opportunity to socialize with her. However, this was mostly limited to her daughters, a few other high-ranking Betta females, and the top-ranked males. Other animals stayed out of her way, and kept their kids out of her way, not wanting to risk a violation of appropriate behavior. Lower-ranking animals, including her own very high-ranking daughters, often fear grimaced (i.e., a submissive facial gesture) to show their acceptance of her position. From the behavior of their mother, and because they are treated as essentially the same as their mother, babies learned early on to whom they were dominant and to whom they were subordinate. Even those who might rarely, or even never, have occasion to interact with Hatchet would know who she was.

Hatchet's status was without question. She had a presence that inspired respect in monkeys and humans alike. Once when we were trapping yearlings to tattoo, Lou jokingly suggested that I try to trap Hatchet. A novice student helper was enthusiastic about the prospect, but Lou was most definitely only joking, and we both knew it. For some difficult to describe reason, it seemed to Lou and me unthinkable that we could mess with Hatchet in anyway. She was the Queen! It was akin to the feeling one might have when a friend/colleague suggests in a whisper that you should play a practical joke on the university President. Delightfully outrageous idea, but unthinkable to carry out. But Hatchet was only an 18-pound monkey! Lou and I were not so much worried about our physical safety; it had more to do with the thought of invading the space of, or infringing the freedom of, someone so important. It had to do with our internalized respect for authority. The fact that she was "only" a monkey was lost on us as it is on anyone who knows these animals well enough to know, at a gut level, that they are complex, intelligent individuals who could never be dismissed as "only" monkeys. Especially not Hatchet.

Our physical safety was very much jeopardized when we trapped any of her offspring or their offspring. At the top of the troop, she could count on the support of every other animal if she wanted it and this includes the alpha male and all of his allies. At times like these, the sea of faces became a sea of canine teeth, and we did not enter into these situations lightly.

Lou used to say that, to her, Hatchet was the

ideal woman. She was a super success careerwise and a super success at home with a huge, thriving family on which she doted. Of course, for monkeys, there is no public versus private domain to master, but Lou's point is well taken. Although the norm for these females is to have a baby every other year, Hatchet had a baby every year, and they almost always survived. Hatchet was a kind and caring mother, always ready to defend her children from any dangers. Actually this was seldom required. Hatchet's position—and, thus, that of her children—was so secure that she rarely needed to show aggression. If she did perceive any infringement on the part of other monkeys toward herself or her family, a single facial threat was enough to send the offenders screaming away, trying to flash appeasement gestures as they ran. Likely it was the fear that if she was not quickly appeased, a transgressor would soon find him or herself facing all of the Bettas and all of their canine teeth.

This social security allowed Hatchet to be a very permissive mother. She didn't need to worry that her infant would get itself in trouble by committing a social faux pas. For lower-ranking mothers, this is a very real concern and they are comparatively restrictive of their infants' movements. When one of Hatchet's infants went for a toddle, it was all of the others who were wary. On more than one occasion, I watched as an infant of hers would approach a high-ranking central male. The males would find themselves in an awkward position. They did not want to take a chance on interacting with the kid, lest there be any misinterpretation, yet they did not want to upset the infant by shooing it away or by abruptly leaving it. It could be quite humorous to watch. A tiny dark infant would innocently try to touch a huge adult male, who would look horrified and try to ease himself away, as though the infant were a deadly spider. An infant of a lower-ranking female would get a very different reception, possibly even getting batted away, thus drawing the mother and her kin into quite a conflict.

As is the case with all Japanese macaque mothers, Hatchet always supported the youngest in an in-family dispute. Thus, her daughters ranked just below her, in reverse order of their ages. This is why dominance and kinship are so closely related. Family groups share a relative dominance rank and whole matrilines can rank above or below other matrilines. The Bettas are the top-ranked family. Dominance is both acquired and maintained via the support of the family group. This is true for both male and female infants, but since males generally leave the matrifocal unit, the long-term effects are greatest for females. A female will almost never rank above her mother, but will be inserted into the hierarchy just below her. As more daughters are born, they are inserted just below the mother, and the older sisters are bumped down a spot. All members of the family will support one another in any conflict with a nonfamily member, and thus each new member reinforces the position of the matriline as a whole.

There is, of course, only one alpha female. All of the other animals are dominant to some and subordinate to others. And, while the rule of ranking just below your mother, but over all of your older sisters, generally holds, each actual dominance interaction will be affected by a host of contextual variables. First and foremost, is your family nearby? How much higher- or lower-ranking is your opponent? Is her family nearby? What adult males are nearby? What is the nature of the conflict? How motivated are the players to actually win the dispute? All of these factors affect a dominance interaction and thus dominance, a complex form of social power, cannot be measured by any one interaction. Even very low-ranking animals will win some disputes with higher-ranking others, so contextual phenomena are not easy to reduce and measure. In fact, a typical fight in a Japanese macaque society involves lots of animals and lots of noise, but little actual contact. Often it is not clear which side "won."

Dominance often is mistakenly conceptualized as something that an individual inherits biologically from its mother or father. In order for something to be biologically inherited, it must, at least, be something that an individual can possess. Since dominance is a characteristic of a relationship, it cannot reside in any one individual. Certainly males, who may change social groups repeatedly throughout their lives, likely will experience a variety of different status positions. The biological influences on dominance are tenuous and complex. For females, it has much to do with who your mother is, but this clearly is more of a social effect than a biological one. Hatchet experienced a very high level of reproductive success; however, dominance and reproductive success are not statistically correlated in this troop (Fedigan et al. 1986). Many low-ranking females also enjoy great reproductive success. Researchers of the Arashiyama West monkeys have pondered the question of what dominance is for, if not for reproductive success, but no easy answer is forthcoming. Certainly at the level of social rewards, it seems that to be higher-ranking is better, although in the wild high-ranking animals of another species have been seen to incur rank-related costs as well, with deaths caused by predation concentrated among high-ranking individuals (Cheney et al. 1981). At the group level, dominance seems to make good sense in that it provides some predictability, some social rules for the animals to follow. Certainly it may be damaging to the fitness of all individuals if constant fighting is needed to settle the ongoing competition for desired resources.

Personality

Is the dominance hierarchy of females absolutely fixed and unchanging, given that it is so closely tied to kinship? Is an animal born into a low-ranking family always low ranking? Actually no. The family dominance that an animal gets is like a hand of cards it is dealt. Personality seems to play a major role in determining how these cards are played, and whether the hand is accepted at all. The biological basis of personality remains an open question.

One striking success story of personality and opportunism I come by secondhand. In 1972, when the monkeys first arrived from Japan, one of the Japanese scientists commented to one of the Western scientists to keep an eye on a young female from a very low-ranking family, as he suspected that she had "plans." The female was only 6 years old, just barely an adult, and the American researcher dismissed as unlikely the idea that this little female could amount to anything. In 1974 this little female staged a coup that ultimately lifted her entire low-ranking lineage to the very top (see Gouzoules [1980] for a full description). The family was the Bettas, and the female was Betta5966, later to be nicknamed Hatchet. This story has always struck me as an extreme example of the actions of personality, motivation, and opportunism, but I suppose it also says volumes about the insights of some of the Japanese scientists.

Age and the Life Course

Thus far I have discussed the social life of female Japanese monkeys by describing three important variables: kinship, dominance, and personality. The other important variable to take into account in making sense of the social life, and social life history of these females, is age. As female Japanese macaques move through the life course, they face an ever changing set of challenges, restrictions, and opportunities.

Both females and males enjoy a very strong bond with their mother in infancy. From the mother, the infant will learn most of what it needs to know in order to survive. This is especially true for females, who will live out their lives in their natal group and in the same habitat to which they were born. This continuity in the physical and social environment is a major distinction between the lives of females and males.

Major sex differences begin to show themselves when the animals are about 3 years old (also see Hiraiwa-Hasegawa chap. 6). Females continue to be closely bonded to their mothers and to spend most of their time in close proximity to their maternal family. They learn to interact appropriately with the members of their mothers' social network. Young males spend more and more time away from their mother, in favor of the company of other young males. Many will move to the periphery of the troop and join the band of males. From here they may move off to try to join another troop. Of course this is a bit of a challenge in south Texas, but that is another story.

Mortality for young males is high at this time, as they leave the safety and security of their natal group. Although females remain in the troop, they too may face a major challenge to survival at this stage of their lives. Interestingly, this challenge comes with their introduction to the world of reproduction. First estrus brings hormonal changes that affect behavior, but some learning is required if these behaviors are to be used in a socially acceptable manner. Mortality is high for first-estrus females as they do not have the experience necessary to interact safely with adult males.

Another favorite monkey of mine was Imo. Named after the famous wheat-and-potato-washing Imo of Koshima Island (Nishida 1987), she was a precocious little monkey, always trying to get into the pockets of the manager and the researchers, even though this kind of habituation is discouraged. When I was "introduced" to Imo, it was the fall of 1982. I was told that she might come into estrus soon as she was now 4½ years old. Japanese macaques are seasonal breeders that mate in the fall. During the 1982 mating season, I was collecting data on the courtship behavior of the females for my Master's thesis, and Imo was one of my youngest subjects. I was interested in how age affects a female's courtship behavior, so I selected females of all ages for the study. By watching Imo and her age mates during their first mating season, I realized that this is a dangerous time in the life of a young female. Subjected to the sudden influence of internal hormonal fluctuations, these young females often behave quite erratically. They are nervous and hyperactive, advertising their new status as potentially mating partners, yet often unprepared for the male attention that it brings. They seem at once motivated to approach unrelated adult males, but still fearful of them. The fear is not unwarranted. Until this point in their lives the young females would have had little occasion to interact with these males and would have avoided them. Certainly they would have had no occasion to engage in the sustained cooperative interaction required of a successful consort.

Male courtship behavior involves "courtship chasing," which, while usually harmless, would be understandably terrifying to an inexperienced young female. Moreover, these young first-estrus females often behave in an abnormal manner, since they have not yet learned the appropriate use of courtship behavior. The female repertoire of courtship behavior, for example, includes an "estrus hack." This vocalization is somewhat like a bark and is used by more mature females to prompt a male to begin mounting. First-estrus females can be seen running around frantically, estrus hacking and screaming for extended periods of time, as though they suddenly found themselves living in a different body, one with which they were awkward and unfamiliar. I found this incessant vocalizing and running around annoying, and the males responded in a manner that suggests that it also was annoying them. In fact, monkeys will generally respond to abnormal behavior from other monkeys with aggression, and this is what these young females faced. They were frequently chased, and often these chases ended in contact aggression. I learned from Lou Griffin that one or two young females die each mating season as a result of male aggression. Apparently the age-related endocrinological changes of first estrus contribute to behaviors that put these animals in situations that can threaten their very survival.

Most young females do survive first estrus, and many get pregnant in their first or second estrus season. The transition to female adulthood generally is fairly rapid and abrupt. At her fourth birthday in the spring, she is still clearly a juvenile. At 4½ years, she will quite suddenly experience estrus and a whole host of new behaviors and social interactions. At 5 years old, she may become a mother. Both biologically and socially she is now an adult.

Males, on the other hand, may begin mimicking adult sexual behavior when they are 3, begin engaging in sexual interactions at 5 or 6, begin producing live sperm at 6 or 7, and still not reach skeletal maturity and full adult status until 10 or more years of age. Unlike females, males experience a very gradual transition to adulthood, and even the experts disagree over when a male should be classified as an adult.

A female Japanese monkey attains adulthood quite abruptly, but after this enjoys a life of relative stability and continuity. Most will live out their lives with more or less the same dominance status and, of course, with the same family. Unpredictable occurrences, however, may end this continuity at any time. For example, a female with a small family, perhaps with only one adult daughter, may find herself quite alone if anything happens to that daughter. High reproductive success for the females has very definite proximate benefits. The key to social success, regardless of dominance rank, seems to be the production of many daughters. Daughters are real social security in Japanese macaque society. In fact, failure to produce any daughters may result in lowered dominance rank for a female after her mother passes on.

Female-female relationships clearly are the fundamental bonds of Japanese macaque societies; however, relationships with adult males are an important part of the adult social life of many females. Each of the troop's central adult males (there may be 10 such males in a troop of 200 animals; a ratio of one adult male to four adult females) has a number of females that can routinely be found in his proximity. Special male-female friendships have been described for baboons (Smuts 1985; Strum 1987), and they exist in this macaque troop as well. Hatchet was very good friends with the second-ranked male. Rocky, the alpha male, is very close friends with a medium- to high-ranking female called Adrienne. Her association with the alpha male has increased the dominance ranking of her whole family, as she will always support them in a battle, and Rocky will usually support her. This increase may depend on Adrienne's family staying close to Rocky. There is no evidence yet that their higher rank is fixed, independent of Rocky's support in a conflict.

These special male-female friendships do not normally cross over into mating relationships in the fall, although there are exceptions. During mating season, each male will mate with any number of different females, consorting with them for a few hours to a few weeks. Hatchet and Rocky are not particularly friendly during the year, although they do have each others support if they need it, but they routinely mate during the mating season. They each also mate with many others of varying rank.

Estrus is a period of physiological change that leads the female to become interested in mating activity. Estrus is sometimes described as the time in which a female is sexually receptive; however, this terminology can be misleading. Receptivity refers to female acceptance of male sexual advances. This describes only a small part of the behavioral repertoire of estrus females. A female in estrus differs behaviorally from a nonestrous female in many ways, one of which is that she may respond positively to the advances of males. Estrous females also behave proceptively—actively soliciting males to engage in sexual behavior. Female Japanese monkeys are very active partners in the mating game, but the sexual bonds, while obviously critical to an individual's reproductive success, are relatively unimportant to the overall social network of the female. After all, she will spend only a tiny fraction of her total lifetime involved in this kind of relationship. We have no evidence that the monkeys recognize paternity and, thus, there are no extended interactions with the "father" of the offspring.

The maximum recorded life span for a female Japanese monkey is about 32 years. The longest living animal in the Arashiyama West troop in Texas was 30. Most females will die by the time they are 23. Old age for Japanese macaque females is very different from that of human females. In fact, they do not seem to experience any real social changes that make the later years of their adulthood different from the earlier ones. There are at least three reasons for this, the most important being the absence of menopause. Japanese macaques generally continue to give birth until the year preceding their death. Thus, there is no postreproductive life span as exists for women. Grandmothers can be old or relatively young, but in either case, they are still mothers, first and foremost, with their own infants. Furthermore, because all animals must be able to feed themselves in order to survive, there is no possibility that an old female could become dependent and need to be cared for by others. Dependent animals, other than nursing infants, are not an element of macaque social life at any age. Finally, there is no evidence that these animals have an awareness of mortality in the sense that they anticipate the end of their lives as they get older. Old females, therefore, are very much like middle-aged females, and they seem to have little in common with aged women, for better or for worse. Obviously old female monkeys are undergoing biological changes, but they apparently do not, either as individuals or as a group, give meaning to these changes. The gradual biological changes of old age are apparently not sufficient to set in motion changes in social behavior, social relationships, or social organization in this society. A female's dominance rank, because it is much more tied to kinship than to any physical prowess, does not decline in her later years.

A model of the human life course, biological or social, will show distinctions that demarcate the aged as a distinct group, however blurred the actual boundary. Nonhuman primates, on the other hand, will show a life course with no clear demarcation past the attainment of adulthood, which is essentially continuous until death. The mother-offspring bond has long been heralded as the basic unit of primate society, and its endurance to the end of the life span is probably fundamental to the essential continuity in the life course of female monkeys.

Beyond the Monkey

In this chapter I have described the social life of female Japanese macaques with particular attention to what I consider to be the important sociodemographic variables: kinship, dominance, personality, and age. I focus on the social dynamic of individual interactions and individual lives. But, the animals can be studied from a variety of levels (Morbeck chap. 1). Some researchers investigate monkey bones. Others investigate disease and epidemiology. Others focus their attention on demography and reproduction. Some appear to have little interest in the animals themselves and prefer to ponder the theoretical questions of how monkeys ***should*** behave.

The most fruitful research is likely to integrate the various levels of analysis. Kinship, dominance, personality, and age are critical to understanding the social life of female Japanese monkeys. But an even better understanding will come with the integration of information from other levels. An interdisciplinary approach to a group of animals, with integration of knowledge gained, is an exciting prospect.

The social animal, living out its life course, is often conceptualized as the end product of the interactions between biology and environment. An animal is endowed with certain attributes from birth, and this forms the core of her being. Influences from the social and physical environment are laid on top of this core, and together these produce the whole animal throughout its life span. The society is often regarded as an irrelevant abstraction "above." In fact, this is a simplistic and misleading way to approach the study of social dynamics and the social life history of individuals. An animal's social life story is molded by multidirectional interactions among its anatomy, physiology, health, personality, family, age, sex, dominance, reproductive performance, individual history, and the dynamic nature of the larger society of which the individual is a part. All of these different factors pose opportunities and restrictions, and the individual animal makes choices, seizes opportunities, and works around limitations. Each monkey is an intelligent, but highly individual social strategist that uses the resources available—or not.

Many interdisciplinary questions, looking at the interactions of the various levels, remain to be explored. Starting at the "bottom" and working "up," we may ask about the genetic basis of intelligence, personality, and social skills, as these are critical to understanding the lives of individual animals. Or about changing hormonal influences across the life span. It appears that hormonal influences pose a serious survival challenge in early adulthood. Social learning is necessary to meet and survive this in order to go on to pursue reproductive success with its proximate and evolutionary benefits.

What about the skeletal system and the limitations and capabilities it provides (Morbeck chap. 9)? Using bipedal humans for comparison, the quadrupedal pelvis of these monkeys enables them to give birth easily to clinging infants, and thus the production of infants does not appear to slow down a female in her other life pursuits (Zihlman chap. 13). What about bones and old age? Do animals without menopause suffer from osteoporosis? At Arashiyama West, I have observed no apparent limitations in locomotor performance with old age; however, in habitats where food is high in the trees, the story might be different. Mariko Hiraiwa-Hasegawa described to me old Japanese macaque females she had observed who were unable to climb tall trees and thus lost access to some food sources (also see Zihlman et al. 1990).

At the level of the individual, each is affected by the lives of each other, by reproduction, and by the social choices made. All of the Bettas found themselves facing very different social lives as a result of the actions of Hatchet in 1974. Was their lifetime reproductive success affected by the new priority of access to resources that they would have enjoyed? At Arashiyama West, there are enough food resources for everyone, but could ecological factors such as the characteristics of the food supply interact with changes in the social structure and affect the reproductive success of individuals differentially?

The interaction between biological ("reproductive") life-history variables—such as age at first birth and infant survivorship—with social circumstances may have important consequences (Morbeck chap. 1). For example, how do the size, dominance, age, and personality of a female's social network affect her reproductive performance? Can you have more babies if you have certain kinds of social support?

The integration of theory with what is known about real animals and their lives is essential. Frans de Waal has pointed out how theoretical musings have led to images of animals that are completely unlike what many observers know about primate societies (de Waal 1987). The interaction between an animal's proximate motivations and the evolutionary effects of its behavior need further attention.

Reproductive success is the widely accepted bottom line in evolutionary explanations of animal, including primate, behavior, but what are the immediate benefits? In Japanese macaque society, lots of daughters is the way to go, but what about the evolutionary importance of many sons?

Understanding the intricate complexities of Japanese macaque social dynamics begins with intensive long-term observations on individually identified animals. Thanks to the efforts of Japanese scientists and individuals like Lou Griffin in Texas, we have come a long way in our knowledge of the social lives of individual animals. But there are many levels, both "below" and "above" the individual, that interact with each other, and these interactions must be pursued in a more multidisciplinary approach to the study of female lives.

8 Natural History of Apes: Life-History Features in Females and Males

Adrienne L. Zihlman

LIFE HISTORY, as defined broadly in this volume, offers a comparative, functional, and evolutionary framework that reflects the complexities of individual female and male lives. I first recognized the mosaic nature of sex differences in the 1960s while studying fossil bones of our probable ancestors (the human family, hominids who lived 2 to 5 million years ago). Sex differences involve many biological levels: from the genetic base to the anatomy and physiology of the brain, bones, and teeth, body size, shape and composition, external features, and expressed behaviors. What does it "mean" to be a female? What does it "mean" to be a male? Sex differences are life-history features that reflect species-defined variation in the pattern and timing of growth, development, reproduction, and aging.

The study of sex differences has been a major focus in my research. How can we understand sex differences in humans, fossil hominids, and nonhuman primates? When I first asked this question in the 1960s, I was doing research on fossil hominids. I was attempting to sort and explain species variation from variation between males and females. I thought this would be an easy task, but as my research progressed, it became apparent that defining and explaining sex differences in fossil populations and, in addition, determining whether these differences increased or decreased during the course of human evolution, are problematic. Furthermore, explaining sex differences in living species is also difficult; most explanations generally are overly simplistic and rely too heavily on one or two traits. I now recognize that a life-history framework more accurately clarifies female-male similarities and differences and how females as well as males survive and reproduce within the wider context of the species' adaptation.

I will briefly review the steps I took in coming to my current framework. The hominid fossil record is fragmentary, and it is difficult at best to distinguish variation that results from species differences from variation attributed to sex differences. The distinctions used by other researchers largely rest on canine size and estimates of body weight (e.g., Wolpoff 1975). With few complete bones of individuals and no way to draw a population sample (Morbeck chap. 9), it is not possible to establish body-weight ranges for presumed females and males within, or even between, species (Zihlman 1976, 1982, 1985). Even with sophisticated statistical techniques, most researchers accept that smaller teeth and bones are female, and larger, more robust ones, male. Other skeletal features are assumed to follow a similar size pattern.

Research on monkeys, apes, and humans also led me to conclude that sex differences are expressed as a mosaic and are species-specific. For example, body weights differ between female and male of both pygmy chimpanzees (*Pan paniscus*) and common chimpanzees (*P. troglodytes*). Yet, *P. paniscus* individuals show no sex differences in cranial capacity, limb bone length, or joint size, whereas *P. troglodytes* do (Cramer and Zihlman 1978; Zihlman and Cramer 1978). Because of the variation in differences within each species, I concluded that, even in closely related species, species-defined sex differences consist of a mosaic of features.

Body weight alone (or, for fossils, estimates of weight) cannot predict the *degree* of sex difference in other features *within* or *between* species.

Finally, my research led me to conclude that traditional explanations for sex differences in body size have little bearing on other differences between females and males (Zihlman 1981). Larger body size was first proposed by Darwin (1871) to give males an advantage in competing with other males for access to females. He proposed the mechanism of sexual selection to account for increases in male body size associated with the presumed variation among males in successful mating. Thus, sexual dimorphism in body size, often based on a single measurement (e.g., Alexander et al. 1979), usually is interpreted in terms of male-male competition, sexual selection, and mating patterns. In other words, in many species where males are twice the body size and weight of females (i.e., extremely sexually dimorphic for this feature), the males establish and maintain territories through fighting and breed with many females (e.g., see Reiter chap. 4). In species with little size difference, males may breed with only one female.

The functional meaning of body weight differs between species, among age classes, and between females and males. It is not a uniform feature (Grand 1977a,b, 1983, 1990). Linking body weight only to male-male competition (for which there are few direct measures [Fedigan 1992]) focuses exclusively on the mating behavior of males. This ignores female body weight as related to her behavior and, especially, reproduction. Consequently, sexual selection cannot account fully for variation in body-size differences within and between species (Ralls 1976, 1977). Finally, sexual dimorphism is not one invariant entity on which selection operates uniformly (Fedigan 1992).

My recent research continues to explore the mosaic nature of sex differences in human and nonhuman primates. Dozens of anatomical dissections of apes and intensive study of the skeletons of known chimpanzees combined with new observational data on free-ranging groups allow me to begin to answer my original questions, What does it mean to be female, or to be male? (e.g., Morbeck and Zihlman 1988; Zihlman et al. 1990).

Humans are primates. Primates are mammals. Large-brained, long-lived animals like humans and other primates respond throughout their lifetimes by altering behavior. There is no simple or direct relationship between biological sex differences and differences in survival and lifetime reproductive outcomes. Long-term observations of free-ranging primates document reproductive outcomes of females (e.g., Fedigan et al. 1986; Altmann et al. 1988; Fedigan 1991), and time-allocation methods illuminate how females and males spend their time and energy over their lifetimes.[1] Whether females are ovulating, pregnant, or lactating influences the frequency of activities. Social group variables such as rank and social network, as well as environmental variables such as food availability, predators, and group size also affect behavior and, consequently, survival and reproductive outcome of females and males, each with their own life story.

To explore several dimensions of sex differences, this chapter incorporates case studies of gibbons and great apes, our closest living relatives, into a life-history perspective. Each "profile" looks at the species adaptation and female and male variation within it. A life-history framework means considering individual as well as species histories (Morbeck chap. 1) and helps to clarify and quantify female-male variations on a species theme. Because humans also are hominoids and have shared a long evolutionary history with the great apes, this framework also provides a way to broaden understanding of what it means to be women and men.

Females and Males: Profiles of the Apes

Old World monkeys, apes, and humans share catarrhine primate features (Morbeck chap. 9) and are characterized by long, socially complicated lives (Smuts chap. 5; Fedigan chap. 2; Hiraiwa-Hasegawa chap. 6; Pavelka chap. 7). Apes and humans, together referred to as hominoids, are variations on the catarrhine theme. In this chapter, I highlight gibbons (lesser apes) and orangutans, gorillas, and chimpanzees (great apes). Humans, in many ways, are another kind of ape, a large-brained, talking, two-legged ape. Lifeways and life cycles and some of the biological foundations of these life-history features of humans are emphasized in the following chapters (Parts IV, V).

In the following ape profiles, I describe for each species (1) a mosaic of anatomical and behavioral features; (2) the range of survival life-history features, that is, the daily (and seasonal) activities for each sex, such as locomotion and foraging, other social-maintenance behaviors, and mating and rearing offspring during their lifetimes. These case studies illustrate similarities and differences among the genera and between species, and female and male variation within species. The shared features of hominoid life history underlie the adaptations of apes as expressed in body weight, group composition, and ecology.

Shared Phylogeny

The evolutionary history of the apes and humans goes back some 20–25 million years when the hominoids diverged from the Old World monkeys (cercopithecoids) (Sarich and Cronin 1976). Hominoids have similarities in dentition, limb and body proportions, timing of growth and development, and most notably, trunk and upper limb specialization. Hominoids share a vertical orientation of the trunk; enhanced shoulder mobility through positioning of the large, well-developed clavicle on a broad chest; a stable elbow joint but with forearm rotation; wrist flexibility; and "finely tuned" grasping hands (Washburn 1951, 1968; Schultz 1968, 1969a). Human beings, compared to other hominoids, have relatively larger brains and longer life stages, as well as an emphasis on hindlimbs related to bipedal locomotion (see Morbeck chap. 9; Zihlman chap. 13).

Life History Features

Time-based reproductive life-history characters for five species of hominoids (table 8.1) are more similar than one might predict given the wide variation in body size of lesser versus great apes. These similarities reflect the common evolutionary history of the hominoids. Gibbons are among the smallest catarrhine primates, and gorillas are the largest; they also show considerable variability in diet and habitat and in social organization. However, female body weights that range from 5 kg in gibbons to 85 kg in gorillas are not paralleled in gestation lengths (210 vs. 256 days). Among free-ranging populations, the small-bodied gibbons take 6 to 8 years to reach sexual maturity (Geissmann 1991), and age at first reproduction is about 9 years, close to the average for mountain gorillas (Watts 1990a). In contrast, a 7-kg monkey (*Macaca nemestrina*) has a gestation of 170 days and age at first reproduction at 4 years. The similarities in gestation lengths, ages at weaning, sexual maturity, and age at first reproduction reflect the shared evolutionary history of complex social and survival behavior and of anatomical features (such as a large brain) of the hominoids, despite extremes in body weight.

Each example reviews the main features of anatomy and behavior of free-ranging populations and integrates anatomical, functional, and behavioral features. This approach does not preferentially treat one anatomical feature, such as body weight, or one behavioral pattern, such as male-male competition or female caretaking of infants. Female-male differences are part of a total pattern that contributes to the reproductive success of each species and are more than the sum of single or unrelated characteristics. The mosaic of anatomical and behavioral differences illustrates, not only the shared way of life of females and males, but also

TABLE 8.1. Ape Life History

Species	*Female Weight (kg)*	*Female/ Male Weight (%)*	*Birth Weight (kg)*	*Female's Age at First Reproduction (years)*	*Gestation Length (days)*	*Age at Weaning (years)*	*Birth Interval (years)*	*Life Span (years)*
Lar Gibbon *Hylobates lar*	6–7	92	0.4	9	210	3	3+	30+
Orangutan *Pongo pygmaeus*	40–50	45	1.5 (1.3–1.6)	12 (9–13)	260	6	8	40+
Gorilla *Gorilla gorilla*	80–90	51–55	2.0 (1.6–2.3)	10 (9–13)	256	4–5	4+	40+
Common chimpanzee *Pan troglodytes*	30–40	75–85	1.8 (1.4–2.4)	15 (11–23)	228	4–5	5+	40+
Pygmy chimpanzee *Pan paniscus*	30–35	75–80	1.3	13–15	230+	4–5	5+	40+

SOURCES: Data were compiled from information on free-ranging populations whenever possible: Schultz 1956; Willoughby 1978; Harcourt et al. 1980; Goodall 1986; Harvey et al. 1987; Kuroda 1989; Morbeck and Zihlman 1989; Galdikas and Wood 1990; Nishida et al. 1990; Geissman 1991; Kano 1992; Tutin 1994; Palombit 1995.

NOTE: The values given are the means. Numbers in parentheses are the reported ranges.

their divergence from each other, to a greater or lesser degree, into distinctive and separate lives.

Gibbons

The gibbons (*Hylobates*), including siamangs (sometimes designated as a separate genus *Symphalangus syndactylus*), comprise nine species (Marshall and Sujardito 1986) inhabiting the rain forests of Malaysia, islands of Malaysia, Indonesia, and the Indo-China mainland. The hylobatids diverged from other hominoids about 15–18 million years ago (Cronin et al. 1984). The species within the genus *Hylobates* share similarities in locomotion, feeding, and social life (Preuschoft et al. 1984). These highly arboreal apes, renowned for their spectacular "brachiating" abilities, rely on fruit as a major dietary item and feed effectively on terminal branches (Grand 1972). Their social groups, unusual among catarrhine primates, consist of one adult female, one adult male, and several offspring. Females and males are similar in body size. Each group physically and vocally defends the boundaries of its small home range.

Anatomy

Most species of lesser apes weigh about 5 kg, although *H. hoolock* weigh slightly more (6–8 kg), and siamangs are twice as heavy (10–12 kg) (Schultz 1973; Willoughby 1978; Leighton 1987). Among wild shot lar gibbons, males tend to be slightly heavier than females (5.7 vs. 5.3 kg) so that females are 92% of male body weight. In trunk length, females are slightly longer, although the difference may not be significant (271 mm–269 mm) (Schultz 1944). In some species of gibbon, pelage differs in color between females and males.

Cranial capacity averages are similar within each sex (101 cc vs. 104 cc), although there is a considerable range (Schultz 1944). There is little sex difference in jaw morphology (Lucas 1981) or canine size and shape, though male canines apparently are slightly longer than females in lar gibbons (Schultz 1944; Frisch 1973). In all hylobatids, both females and males have *pronounced* canine teeth. Among siamangs, both sexes have laryngeal sacs and vocally announce their presence and defend their territory.

In the limbs and trunk, a mosaic of differences is apparent. In lengths and proportions of limb bones, female gibbons are 99% of males, whereas female siamangs are 96% of males (Schultz 1973). However, in body proportions, male siamangs, and to a lesser extent male gibbons, have somewhat larger chest girth and shoulder breadth relative to females, but there is little difference in relative hip breadth (Schultz 1956, 1973).

The bones and teeth, although similar in size and shape, may nonetheless "record" relative proportion of trauma experienced by each sex. For example, in adult dentition, 41% of males compared to 20% of females had lost, severely damaged, or broken canines (Frisch 1973). Similarly, in a collection of 233 gibbon skeletons (*H. lar*), 37% of adult males, but only 28% of females showed healed fractures of the long bones (Schultz 1944). These differences in trauma apparently mirror differences in behavior of females and males during life.

Behavior

Young individuals, females without young, and males can be difficult to distinguish from anatomical features alone unless the observer manages a close look at the genitals. Sex differences in feeding and travel are not marked, although some do exist. In all species, females carry the young exclusively for at least the first year of life. In some species, adult females travel first in group progressions, which may give them first access to fruit. Among lar gibbons, a female in late pregnancy and in the early postpartum period becomes dominant to the male during feeding (Ellefson 1974). Among the larger-bodied and more folivorous siamangs, females feed faster than adult males and for about 30 minutes longer each day (Chivers 1977). Males participate very little in direct infant care although male siamangs may carry the young in its second year (Chivers 1974), apparently as a result of the infant's initiative (Alberts 1987).

Both females and males are aggressive and intolerant of other adults of the same sex encountered in neighboring groups. As the young mature, they become peripheral in the family unit, then leave the natal group and establish their own territory. Both sexes play a role in defending their territory and in maintaining the integrity of the pair (Raemakers and Raemakers 1985). In intergroup interactions, however, males engage more often and for longer in calling and chasing than do females. For example, in encounters at territorial boundaries, lar gibbon males threaten and chase neighbors (Ellefson 1968). In conflicts between siamang groups, females hide while males chase each other (Chivers 1974). Males of *Hylobates agilis* spend 13% of their activity in territorial behavior and are the protagonists in territorial disputes 76% of the time, compared to the females' 10% (Gittins 1980). Female (*H. agilis*) gibbons apparently take a less active role than do males in territorial behavior when carrying a small infant. Moloch gibbon (*Hylobates moloch*) females engage in indirect aggression through vocalizations toward neighboring groups, but the males engage in direct aggression toward intruders that have entered the territory (Kappeler 1984).

Vocalizations

Field data document the importance of vocalizations in the overall adaptation of all gibbon species, both in their species-specificity and in sex differences (Marshall and Marshall 1976; Mitani 1992). As a social activity within and between groups, calling takes up about 4% of the activity period (Leighton 1987). Among gibbons and siamangs, as in other primate species, individual calls are recognizable to neighboring groups and thus provide information on age, sex, and number of individuals in the group. These calls also function to attract mates, to reinforce the group's female-male bond, to identify the caller's sex to neighboring groups, and to locate, define, and maintain territorial boundaries.

In most species, females and males vocalize together in duets and, except for hoolock gibbons, there are sex differences or sexual "divocalism" (Haimoff 1984). The "great call" produced by the females of all species is the most easily identified part of gibbon songs and appears to be neurologically programmed and strongly determined by inheritance (Brockelman and Schilling 1984). The frequency of calling differs among species, and within species differences may exist between females and males. For example, among Kloss's gibbons, males sing once every 2.5 days, whereas females sing once every 5 days (Tenaza 1976).

Orangutans

Today orangutans (*Pongo pygmaeus*) are found in rain forests that vary from hilly or mountainous areas to swampy lowlands of Borneo and Sumatra (Galdikas 1988) but remains from Pleistocene and Holocene times have been discovered in southern China, Vietnam, and Java (Groves 1986). Orangutans diverged from the African group some 9–11 million years ago (Cronin et al. 1984). They inhabit the forest canopy, rely on fruit as a main dietary item, and exploit these resources through their climbing, hanging, and reaching skills (Chevalier-Skolnikoff et al. 1982). Their prehensile hands and feet and extreme fore- and hindlimb joint mobility allow orangutans to be both large bodied and highly arboreal. In fact, they are the largest arboreal mammal, although orangutans have pronounced sex differences in body weight and proportions. Socially, they are unusual in their low level of association (except between mother and offspring) and lack of formal group organization (Galdikas 1979).

Anatomy

The features that distinguish females and males clearly demonstrate the mosaic nature of morphological variation (Morbeck and Zihlman 1988). In body weight, females average about 37 kg (range 31–45 kg), and males 80–85 kg, although they may weigh more than 90 kg (Rodman 1984). Females are about 45% of male body weight, or stated another way, males weigh 223% of females. I use a ratio of male-female values to express differences. (Males can be interpreted as "built-on" the species plan [McCown 1982].) Orangutans differ in features other than adult body weight and canine tooth size, but until recently these features have not been emphasized in assessing sex differences (Morbeck and Zihlman 1988).

Substantial sex differences occur in skull dimensions (Schultz 1962), and brain and tooth size differ, but to a lesser degree than does their body weight. Brain weight in females, for example, is 84% of males (119% male/female, calculated from Willoughby 1978); consequently, females have a larger brain to body index compared to males (94.2 vs. 88.5). Canine length and breadth in males are 135% of females', whereas molar teeth are less than 10% larger, demonstrating a mosaic within the dentition itself (Schultz 1973; Oxnard et al. 1985; Morbeck and Zihlman 1988). In males, chewing muscles are twice as large relative to body weight as those in females (Morbeck and Zihlman 1988).

There are differences between females and males in several linear and volume dimensions of the trunk and limbs. In relative chest girth (chest circumference relative to trunk length), for example, males are 118% of females, but both sexes have similar hip breadth (Schultz 1956). Long bone lengths of males average 116% of females, and bone weights 210% of females. Females, on the other hand, have larger acetabular and femoral head joint surface areas relative to body weight than do males (Morbeck and Zihlman 1988).

Based on dissections of two animals, proportions of muscle, bone, and fat also differ between females and males. The female has less muscle tissue relative to body weight and more body fat than the male. Relative limb weights also differ. Forelimbs are similar (female: 16%; male 17% of total body weight) whereas hindlimbs differ (female 18%; male 12%) (Morbeck and Zihlman 1988).

Growth and Development

Sex differences are apparent in patterns and timing of growth and maturation. Among immature orangutans, significant sex differences appear in skull length, breadth, and height (Schultz 1941; Winkler 1987). Females achieve adult body weight about 10 years, males at 15 years (Fooden and Izor 1983). Females complete skeletal growth at 8 to 9 years but males continue until age 15. Similarly, the distinctive adult features of males—inflatable throat sac, prominent fatty cheek pads, and a fatty neck region (Schultz 1941; Rodman and Mitani 1987; Morbeck and Zihlman 1988)—are not completely developed until about 15 years of age. These features distinguish adult males visually and vocally from other age/sex classes (MacKinnon 1974; Galdikas 1979; Schurmann and van Hooff 1986). By about 10 years of age, female and male orangutans also have diverged behaviorally in daily and seasonal activities.

Behavior

Field observations suggest that sex differences in locomotion, feeding, and social behavior exist, although research at different study sites is not consistent in the extent or in the direction of the differences (e.g., Rodman 1977, 1988; Galdikas 1988). Some differences might be predicted based on body weight, whereas others are not.

To some extent, body weight correlates with mode of locomotion, use of the trees, and feeding postures. Adult females use "quadrumanous scrambling" as their most frequent mode of locomotion and hang more and sit less than do males (Sujardito 1982; Sujardito and van Hooff 1986; Cant 1987a, 1987b). Females and adolescents generally sleep and travel in the higher levels of the forest than do the males. Males engage more than twice as often in tree swaying to travel across branches to new trees and less travel by quadrumanous scrambling. In some areas, males, but not females, travel or feed on the forest floor, where leopards and pigs present a threat.

When females are carrying young, they travel more frequently in the lower canopy than do the adult males, apparently in response to the additional weight of the young that requires caution during locomotion. Females with infants have, no doubt, honed their arboreal skills to compensate for a changed center of mass. Although females with offspring travel in lower levels, they return to higher zones to rest and sleep, whereas males do not. The additional weight of offspring could account for females' relatively larger hip-joint surfaces.

During feeding, females hang under branches more frequently and use smaller branches, whereas males sit and stand on larger branches (Cant 1987b). Females pluck the food from branches, whereas males more often pull in branches to detach food using their long and strong arms and hands.

A 4-year study from Tanjung Puting, Central Borneo, documents diet, range, and activity with comparisons between females and males (Galdikas 1988). Here orangutans consume an estimated 400 food types, almost all found in the canopy. Fruit (mostly ripe) accounts for 61% of observed foraging time; other items include flowers, leaves, vines, and insects, and bark and sap of various trees. Diet varies with time of year. For example, in Galdikas's study, orangutans eat 51 species of bark. During 6 months of year they eat none, whereas in 1 month they feed on it 47% of the time.

Sex differences exist in proportions of resources in the diet and in ranging patterns. Females consume more food types per day than males (mean for females 9.6 food types; for males, 7.1) and spend more time eating bark and young leaves, whereas males spent more time eating termites mostly from the ground. In terms of activity pattern, adult females spend from 57%–68% each day foraging; males forage for 41%–65%.

Adult females with their dependent offspring occupy smaller (5–6 km^2) and more stable home ranges than those of adult males. At Tanjung Puting, exact figures for adult male ranges are not available because no adult male remained in the study area throughout the 4 years. Their movements indicate that male ranges encompass several adult females' home ranges (Galdikas 1979). Daily distance traveled is smaller for prime adult females than for prime adult males (females, 710 m; males 850 m). However, there are no differences in mean daily travel time because males travel faster than females.

Orangutan social life is shaped by a combination of a predominantly arboreal, frugivorous, and opportunistic diet, which entails large body size and monitoring and using seasonal foods from trees. Sociality is expressed as sustained proximity during traveling or resting and by mutual tolerance during feeding (Galdikas and Vasey 1992). Focal adult females spent 13.5 % of their time in social groups. They associate with all age and sex classes, with other adult females more often than with adolescent females and, depending on reproductive condition, with males (Galdikas 1984, 1985a). Day ranges increased considerably when adult females traveled together or contacted other units. The use of permanent resources such as bark enhanced the occurrence of social groupings among adult females; when sharing preferred foods, adult females ate fruits faster than when alone. During consortships, adult females had larger day ranges than usual, and males had shorter ones, suggesting each sex compromises to remain in consort (Galdikas 1988). Focal adolescent females were in groups 40.8% of the time over and above the time spent with their mothers.

Immature males are gregarious and participate in groups about 41% of time when observed as focal animals. Their sociality centers around females; they associate with them 86% of the time, versus 3% exclusively with other males (Galdikas 1985b).

Adult males are the most solitary of any age or sex class. They spend less than 2% of their time in contact with other individuals and only associate with females; nonconsorting adult males are, almost without exception, solitary (Galdikas 1984, 1985ab). Their loud calls, which carry over long distances, distinguish them from other age and sex classes (Galdikas 1983). Playback experiments demonstrate that these long calls do not function in attracting mates but seem to regulate spacing between males through avoidance (Mitani 1985).

Adult males avoid each other, and males seem to be avoided by others (Galdikas 1979, 1981). Healed wounds on males indicate overt fighting occurs between them, and aggression can be severe in the presence of an adult female. Galdikas (1985b) suggested that direct aggression between males may be more prevalent than indicated by the violent incidents observed.

Male anatomy may enhance their presence and visibility: their inflatable throat sac emphasizes resonating vocalizations; the facial configuration of prominent fatty cheek pads serves as a visual signal denoting age and status. These features may serve to attract females, to repel other males, or to enable animals to choose whether to associate. Their large body size and canine size enable them to inflict serious wounds on other individuals.

Gorillas

Gorillas (*Gorilla gorilla*) inhabit the mature lowland tropical forests and swamps of West and Central Africa and the montane forests of Rwanda, Uganda, and eastern Zaire (Schaller 1963; Casimir 1975; Tutin and Fernandez 1985; Fay et al. 1989). Evolutionarily, after orangutans split off, gorillas shared a common ancestor with chimpanzees and humans before branching off some 6–8 million years ago (Miyamoto et al. 1988). Gorillas comprise three distinct populations or subspecies: *Gorilla gorilla beringei* (mountain), *G. g. graueri* (eastern lowland), and *G. g. gorilla* (western lowland) (Groves 1986). These large-bodied apes are generalized opportunists and feed on a diversity of foliage, fruits, and insects (Schaller 1963; Goodall 1977; Tutin and Fernandez 1988). Gorillas live and forage in cohesive groups with several adult males, females, and young. Using their hominoid ability for climbing and reaching, they feed and sleep in the trees but also travel, feed, and sleep on the ground. When moving quadrupedally they knuckle-walk and bear weight on the mid-dorsum of flexed fingers.

Information differs according to subspecies, which is an added difficulty for assessing sex differences and life-history characters. On the one hand, information on anatomy and physiology (such as birth weights) and social behavior of captive animals derives almost exclusively from lowland gorillas (e.g., Willoughby 1978; Dixson 1981; Maple and Hoff 1982). On the other hand, knowledge about individual lives, social behavior, and organization among free-ranging populations comes almost exclusively from long-term studies on the mountain gorillas (*G. g. beringei*) in the Virunga Volcanos (e.g., Schaller 1963; Fossey 1983; Harcourt 1979; Stewart 1987; D. Watts 1990a; Tutin et al. 1991). Ongoing studies at Lopé, Gabon have yielded information on foraging profiles of free-ranging lowland gorillas (*G. g. gorilla*) and the sympatric chimpanzees (*Pan troglodytes*) (see, Tutin et al. 1991).

In contrast to the foraging profile of mountain gorillas, lowland gorillas frequently eat almost 100 species of fruit and regularly search for and consume several species of insects. Even silverback males climb trees up to 30 meters high in order to feed on fruit (Tutin and Fernandez 1985, 1992; Williamson et al. 1990). That gorillas eat so many species of insects and fruits came as somewhat of a surprise, because gorillas—based on the mountain gorilla data—were traditionally characterized as "folivorous." These findings highlight the influence of the environment on diet, since gorillas seem to prefer fruit if it is available.

Anatomy

Adult females are half the body weight of adult males (Willoughby 1978; Groves 1986). In mountain gorillas, males average 190 kg; whereas females weigh an average of 100 kg and are 53% of male body weight (190% male/female). Lowland gorilla males are 156 kg and females at 85 kg are 54% (calculated from Willoughby 1978).

A number of other features differ between the sexes, though to a lesser degree than does body weight, and serve once again to illustrate a mosaic pattern. For example, among western lowland gorillas, cranial capacity in females is 85% that of males (117% male/female), and brain/body weight index is 92 in females and 90 in males (Willoughby 1978). In dentition, degree of sex difference varies among tooth types. Male canine teeth are 60% larger than those of females, whereas molar teeth are only 6% larger than those of females (McCown 1982). Adult males are significantly larger than females in cranial dimensions and have more prominent sagittal and nuchal crests. On these crests, attach large temporal and nuchal muscles that move the large jaws and canine teeth.

The length of each of the long bones (humerus, radius, femur, tibia) in lowland females is 85% that of males (117% male/female). The greatest sex difference lies in the upper body, in chest, arm, and forearm girths (80% female/male). The least difference is in wrist and ankle girth (88% female/male), and 82% ratio in hips, thigh, and knee girths (calculated from dimensions on live lowland animals reported in Willoughby [1978]).

Fully mature males, called "silverbacks" because of the silver pelage on the dorsal part of their trunk, also possess a prominent "axillary organ" that is more well developed than in females. Large and numerous apocrine sweat glands are concentrated in the armpit and are responsible for the pungent odor of silverbacks (Ellis and Montagna 1962; Dixson 1981) that has been noted by field researchers (e.g., Fossey 1983).

Most female and male differences develop postnatally (Watts and Pusey 1993). At birth, body weight and girth are similar; at about 5 years of age, differences appear in body weight (80% female/male). Arm and leg length, which are less different in adults, diverge somewhat later, at about 7 years of age (based on measurements from captive lowland gorillas reported in Willoughby [1978]). The size differences in the skull remain insignificant until the last stages of growth, when the canine teeth and third molars are fully developed (Schmid and Stratil 1986). Males acquire their silver pelage after age 12, about the time they reach adult size and weight.

Behavior

There is some information from field studies regarding sex differences in locomotion, foraging, and feeding among mountain gorillas at Karisoke, Rwanda. For free-ranging lowland gorillas in Gabon, there are fewer observations.

Sex differences in locomotion do exist, but they are not marked. For example, except for juveniles, females with or without infants climbed more frequently than other age and sex classes, and silverback males climbed the least (Schaller 1963). Females may travel more slowly in the final stages of pregnancy, and the group may slow down to accommodate them (Fossey 1983). Only adult females are reported to travel tripedally (Tuttle and Watts 1985); during an infant's first month of life, the female tightly supports it with one arm while using the other for locomotion (Schaller 1963; Fossey 1979). Silverbacks have the highest frequency of bipedal behavior because they beat their chests in displays (Tuttle and Watts 1985).

Observations on captive lowland gorillas tend to support the field data that females alter locomotor and social behavior during reproduction. During the first three months of pregnancy, these activities declined, remained low throughout pregnancy, and were more reliable indicators of pregnancy than physical appearance (Meder 1986).

In feeding and foraging, sex differences among Karisoke mountain gorillas are slight and less pronounced than those found in orangutans with a similar degree of difference in body weight. Absolute distance traveled per day tends to be short for mountain gorillas (D. Watts 1991b). Silverback males feed longer and spend less time moving than do females (Harcourt and Stewart 1984; Watts 1988). Between silverback males and adult females the difference in feeding time varies among females at different reproductive stages. Nonpregnant and nonlactating females fed for less time on average (53.2%) than pregnant or lactating females (55.8%) (D. Watts 1988). Although this difference among females is small, the direction of the difference is consistent across all vegetation zones. According to Watts (1988), this supports the interpretation that nutritional demands of reproduction may account for the small disparity in feeding behavior between females and males. Among lowland gorillas at Lopé, there is no indication of significant differences in locomotion or foraging profile, except that silverbacks eat insects at a slightly lower, but not significant, rate than other age-sex classes (Tutin and Fernandez 1992).

Gorillas live in relatively cohesive social groups of variable size, from about 6 to over 20 individuals; each group consists of at least 1 and up to 4 silverback males, 1 or 2 blackback males, several adult females, juveniles, and infants (Schaller 1963; Fossey 1972; Casimir 1975; Stewart 1981; Yamagiwa 1987; Tutin et al. 1992). Both female and males leave the natal group to form new groups and male silverbacks may be solitary (Stewart and Harcourt 1987; D. Watts 1990b, 1991b). Adult females in a group are usually not related. The cohesion of mountain gorilla groups seems to depend primarily on the relationship between the dominant silverback and the adult females rather than on the relationships among the females. Females and young animals tend to spend more time near the silverback than to any other adult (Harcourt 1979; Stewart and Harcourt 1987).

Among mountain gorillas, adult males have been observed to confront other male gorillas and females and infants in other groups, even injuring and killing them (Fossey 1983; D. Watts 1989). At the same time, adult males seem to contribute a great deal to group stability and cohesion. The use of bipedal display combined with vocalizations makes males effective sentries, watchful of movements of other gorilla groups, possibly predators, and also humans. Leopards do attack gorillas (Schaller 1963; Tutin and Benirschke 1991; M. Remis pers. comm.), and males attempt to defend the group against perceived threats, whether other male gorillas, carnivores, or humans who seek to capture the young (Fossey 1983; Tutin and Fernandez 1988). Adult females and other group members also act to protect the young, though males seem to be first to confront danger.

Although mountain gorillas are generally silent, there are sex differences in types and frequency of vocalizations. The screaming roar, the clear hooting preceding the chest beat, and the staccato copulation call seem to be unique to adult male gorillas (Schaller 1963; Fossey 1972; Harcourt cited by Marler 1976). Female mountain gorillas apparently emit the largest variety of sounds of all age/sex classes, and the pant ho-ho-ho produced during displays may be peculiar to them alone (Schaller 1963). Male gorillas dominate in vocal output (Marler 1976). Whether similar differences hold for western lowland gorillas is not yet known.

Chimpanzees

Chimpanzees (*Pan*) are most relevant to humans because the two lineages shared a common ancestor after gorillas split about 6 million years. The chimpanzee-human divergence took place about 5 million years ago (Sibley and Ahlquist 1984; Miyamoto et al. 1988; Ueda et al. 1988; Hasegawa 1992). Chimpanzees comprise two species, *Pan paniscus* and *Pan troglodytes*; the two separated at 2 and 2.5 million years ago (Cronin et al. 1984).

Chimpanzees inhabit a wide range of habitats in Africa, from Senegal on the West African coast, across the northern part of the equatorial rainforest to the eastern part of Lake Tanganyika, a range that stretches some 6000 km (McGrew et al. 1981; Collins and McGrew 1988; Kano 1992). Chimpanzees are opportunistic omnivores and feed on numerous species of fruit, herbs, leaves, insects, and vertebrates (including other primates and mammals); they forage in the trees and on the ground, and like gorillas, bear weight on their knuckles when walking quadrupedally. Chimpanzees live in communities of about 50 individuals with remarkable tolerance among males compared to other hominoids (Nishida and Hiraiwa-Hasegawa 1987). Within the community, daily group size consists of "parties" that vary in size and composition. Their social system is described as "fission-fusion" and contrasts with the more stable and relatively cohesive gorilla groups.

Prominent genitals distinguish adult male chimpanzees, unlike gibbons and gorillas. Females have pink sex skin that swells during estrus. Although the two species overlap in overall body weight, surprisingly, they differ in the degree of sex difference in other features. For this reason, each species will be treated separately.

Pan troglodytes

Pan troglodytes comprises three subspecies: *P. t. schweinfurtheii* in the easternmost part of the range in Tanzania, *P. t. verus* in central Africa, and *P. t. troglodytes* in the western range. The habitats are diverse, ranging in altitude up to 3000 meters, and in climate from high humidity and annual rainfall to more arid habitats. Long-term studies of free-ranging populations continue at several localities including (1) Gombe since 1960 by Goodall and colleagues (Goodall 1986); (2) Mahale Mountains in Tanzania since 1965 by Nishida and colleagues (Nishida 1990); and (3) the Tai Forest, Ivory Coast since 1980 by C. Boesch and H. Boesch (1989). These studies document variation between populations in habitat, diet, group size, social behavior, and tool using (Collins and McGrew 1988; Boesch and Boesch 1990; Nishida 1990; McGrew 1992), as well as the variation that occurs within populations between females and males and is emphasized here.

Anatomy

Overall, sex differences vary from moderate to slight. In body weight, females are similar to female orangutans and range from 25 to 45 kg; males range from 40 kg to considerably more than 50 kg. The Gombe population is the lightest, where females average 30 kg, and males, 40 kg (Morbeck and Zihlman 1989). In other populations, individuals are, on average, heavier (e.g., Mahale: females, 42 kg; males, 48 kg; Uehara and Nishida 1987). In body weight then, female weight is usually between 75% and 85% of male body weight (or 133–134% male/female).

In a Zaire skeletal series, male cranial capacity averages 404 cc, and females 375 cc (males 108% of females), and some skull dimensions are larger in males (Cramer 1977). Male canines are larger, but there is little difference in postcanine tooth area (Kinzey 1984; Oxnard et al. 1985).

Details of the postcranial skeleton also differ, but only minimally. For example, in the Gombe population, long bone lengths of females and males are relatively similar, whereas in aggregated populations, males have longer limb lengths, on average and larger joint diameters (Schultz 1969b; Zihlman 1976; Zihlman and Cramer 1978). Females have a wider pelvic outlet than males and a wider pelvis relative to thigh length (Kerley 1966). The pubis is longer, on average in females (Morbeck et al. 1992). Males have a larger average chest circumference and longer clavicles (Schultz 1956; Zihlman and Cramer 1978).

Females and males in captivity exhibit an adolescent growth period with the female spurt occurring earlier (E. Watts 1985a). Behavioral differences are not marked before adulthood, but some sex differences are observed in anteating behavior, daytime bed-making behavior, and greeting (Hiraiwa-Hasegawa 1989; chap. 6).

Behavior

Sex differences exist in daily travel and range. Females have shorter daily ranges than males at Mahale (Hasegawa 1990), in the Tai Forest (Doran 1989), and at Gombe (Wrangham and Smuts 1980). Females in estrus have longer daily ranges than females in late stages of pregnancy, and for a few weeks after giving birth; females with young have smaller ranges than do cycling females (Goodall 1986; Hasegawa 1990). These sex differences in travel suggest that females' energy budgets vary during their reproductive cycle. The greater amount of solitary foraging by females at Gombe has been interpreted as a strategy to reduce competition for food but could also relate to reducing time (and energy) spent in social interaction with other than immediate family members (e.g., Altmann 1980; Dunbar and Dunbar 1988).

Among Tai chimpanzees, average daily range differs between adult females and males, although they do not differ in frequencies of overall locomotor activities (Doran 1989). During feeding, however, females are more quadrupedal, whereas males climb and scramble more and are more bipedal than females (Doran 1989).

Sex differences have not been reported for locomotor mode as for orangutans and, to a small degree, for gorillas. However, the contribution of effective locomotion to raising offspring is illustrated by two adult Gombe females afflicted with poliomyelitis that partially immobilized muscles of the wrist or arm. This affliction inhibited the mothers' ability to move tripedally in order to give added support to newborns, to forage and feed effectively, and to protect offspring adequately. No offspring of these mothers survived beyond a few months or a year, and their mothers' locomotor disabilities must have been a contributing factor (Zihlman et al. 1990; Morbeck et al. 1991).

In composition of diet, food processing, tool using, and predatory behavior, some differences between females and males have been documented from long-term studies. For example, females are more insectivorous at Mahale (Uehara 1986). At Gombe, chimpanzee females engage more frequently and for longer periods in using implements to extract termites hidden in mounds throughout the year (McGrew 1979; McGrew et al. 1981; Goodall 1986). Among Tai Forest chimpanzees, females use stone and wooden hammers to crack open nuts with more efficiency and with greater frequency than do males (Boesch and Boesch 1981, 1984). Females are active in catching prey; at Mahale, females engage in predation more frequently than do females at Gombe and take different prey (e.g., Takahata et al. 1984; Goodall 1986). But in both populations as well as in the Tai Forest (Boesch and Boesch 1989; Uehara et al. 1992), males engage in predatory activity more frequently than do females.

Social relationships between females and males differ to varying degrees among populations. Adolescent or cycling females change communities, whereas females with young tend to be more solitary and associate less with each other than do males (e.g., Wrangham and Smuts 1980; Wrangham et al. 1992; Nishida 1989). However, among Tai Forest chimpanzees, unrelated females form alliances and share food to a greater extent than do females in other populations. Boesch (1991a) attributed the more cohesive and mixed associations of the Tai chimpanzee social system to ecology and to the demonstrated predator pressure from leopards.

Males engage in aggressive displays (Kuroda 1980; Nishida and Hiraiwa-Hasegawa 1987; Kano 1992). Sex differences are apparent in affiliation; males associate closely and travel together to patrol the territorial boundaries of the community.

In vocal communication, there are no sex differences comparable to those of other hominoids; the basic categories of all chimpanzee vocalizations are uttered by both sexes (Marler 1976). However, gestural communication involving greeting behavior does differ between females and males (Goodall 1986; de Waal 1989). At Mahale, sex differences in greeting behavior appear early in development (Hiraiwa-Hasegawa 1989; chap. 6). Males attack more than females and consequently have more ways to deal with aggression. Attacks by males are more often followed by reassurance toward the victim than are attacks by females (Goodall 1986; de Waal 1989).

Pan paniscus

Pygmy chimpanzees (*Pan paniscus*) inhabit the rain forest in a small region of Central Africa in the Zaire River Basin and have been studied at four localities (Uehara 1990; Kano 1992). Two field sites have sustained studies since the 1970s, at Wamba (e.g., Kuroda 1980; Furuichi 1989; Kano 1992) and at Lomako (Susman 1984; Thompson-Handler et al. 1984). Some differences may exist in social organization between the two sites. Nonetheless, because of their fewer numbers and their more limited geographical distribution, *P. paniscus* is less well known in behavior and anatomy than *P. troglodytes.*

The term "pygmy" was initially applied to the small cranial, facial, and dental features of *P. paniscus* before research on its postcranial anatomy and body weight (e.g., Coolidge 1933). Later studies demonstrated that *P. paniscus* and *P. troglodytes* are distinguished by body proportions (e.g., humerus/femur and intermembral indices) but not body weight, which overlaps a great deal in the two species (e.g., Zihlman and Cramer 1978; Morbeck and Zihlman 1989). Species distinction has been demonstrated in DNA and proteins with estimated divergence time of 2–2.5 million years (Cronin and Sarich 1976). Distinguishing species' features in social behavior of *P. paniscus* include the strong relationships that develop between unrelated females, the low level of agonistic and intimidation behavior of males, more elaborate sexual behavior, and distinct vocalizations (e.g., Mori 1984; Furuichi 1989; Idani 1991; White and Wrangham 1988; de Waal 1989)

Anatomy

Based on small samples, body weight ranges from 27 to 38.5 kg for females and 38 to 61 kg for males; females average 33 kg and males 45 kg (reviewed in Morbeck and Zihlman 1989). Females are 73% of male body weight, close to that of Gombe chimps. Canine teeth are relatively small in both sexes, though canine lengths and breadths show some discernible sex difference. No significant difference occurs in postcanine dentition (Johanson 1974; Kinzey 1984). In spite of a fairly significant sex difference in body weight, there is no difference in cranial capacity, cranio-facial dimensions, limb bone lengths, and joint sizes (Cramer 1977; Cramer and Zihlman 1978).

Behavior

At present there is little indication of notable sex differences in travel, foraging, and ranging at Lomako or Wamba. During travel, males climb, scramble, and leap more frequently than females, and females use more quadrupedalism than males (Doran 1989). Tool use has not been documented from any population, and there are only a few reports of predation, mostly on flying squirrels (Inhobe 1992). Because social groups are most frequently mixed age-sex classes (Kano 1980, 1982; Kuroda 1980), sex differences in foraging or ranging would therefore not be expected.

In social behavior as in anatomical features, few sex differences stand out. Rates of aggression are low for both sexes, and males do not engage in the types of aggressive displays characteristic of *Pan troglodytes*. The small canine teeth of both sexes and especially those of males, and the associated musculature, are consistent with the behavioral findings. Food sharing is widespread, and both sexes share (Kuroda 1984). A distinctive feature of this species is that females tolerate feeding close together. They actively bond and interact in grooming and food sharing and engage in genital-genital rubbing, an expression of greeting and friendly behavior (Thompson-Handler et al. 1984; Furuichi 1989; Idani 1991). High levels of affiliation also exist between females and males but not among males.

The Apes in Review

Hylobatids, comprising gibbons and siamangs, are small-bodied, fruit-eating apes living in groups of one male, one female, and their young. Few if any significant sex differences exist within each species in anatomical features such as weight, proportions, and skeletal, and dental dimensions. However, field observations indicate slight sex differences in activities including (1) calling, (2) relative engagement in territorial behavior, (3) duration of feeding time at different reproductive stages, and (4) engagement in aggression as reflected in behavioral accounts and supported by findings on damaged canine teeth and broken bones. The slight differences in trunk proportions noted by Schultz may relate to the female's role in carrying additional weight during pregnancy and while carrying a nursing infant.

These minor differences in anatomy and behavior emerge by adulthood. They are important in the lives of the animals and suggest that females adjust their activities to accommodate pregnancy, lactation, and carrying offspring in addition to self-maintenance. At the same time, females also contribute to defense of the territory and its resources. They do so while expending less energy than the males. Male care of infants is rarely direct. However, a continuous male presence in the group indirectly contributes to the survival of the young through territorial defense and promotion of group integrity.

In the large-bodied, fruit-eating, opportunistic, highly arboreal orangutans, the sexes diverge considerably in a number of anatomical features, such as body weight and proportions, skull dimensions, canine tooth size, facial cheek pads, and to a lesser degree, molar tooth size, limb length, and joint sizes. Because their diet depends a great deal on fruit that is dispersed throughout the forest canopy, adults must forage independently. Females and their young associate with each other, though other social interactions are infrequent. Adult females interact more frequently than do adult males. Females and males become less social as they pass adolescence. Consequently, adult females and males live most of their lives independent from each other.

Compared to other species of apes, orangutan females and males differ to the greatest extent from each other anatomically and in their individual life histories. Anatomical differences relate in both general and specific ways to everyday survival behaviors (e.g., locomotion and feeding), to social interaction including mating, and to females' caretaking of offspring.

The large-bodied gorillas consume foliage, fruit, and insects, feed and sleep in trees or on the ground, and travel exclusively on the ground. Gorillas live in relatively cohesive groups, usually with several adult males, adult females, and young. Sex differences in body weight are marked; differences in cranial capacity, limb bone lengths and girths are less pronounced. There are not marked differences in foraging profile and locomotion. The main sex differences in displays and vocalizations (and canine teeth) seem to relate to the aggressive protective function of the males toward other group members, a pattern distinctive in anatomy and behavior from orangutans.

Pan troglodytes is a moderate-sized omnivore that forages on a wide range of plant and animal species, using tools to exploit them, in communities of changing party size. Sex differences in anatomy are moderate in some features (body weight, canine size) and slight in others (cranial capacity, postcanine dentition, limb length, joint size). Overall, the mosaic of anatomical features is distinctive in *P. troglodytes* compared to other living hominoids and to *P. paniscus.*

The society is hierarchical, where adult males are generally dominant over adult females, and are aggressive to females and young and to each other, but the aggression is mitigated through communication. Females may travel to neighboring communities to mate and return or to stay; they tend to be less social than males and, with their offspring, range less widely. Females are more insectivorous, and males engage more often in predatory behavior. Females and males in all populations use implements of organic materials or stone to exploit a range of insect species and kinds of nuts. Sex differences are observed in frequency and skill level. These behaviors no doubt reflect possible dietary differences and emerge during adolescence. These differences seem to have little to do with anatomical features per se and instead seem to be influenced by resource availability, predator pressure, learning and practice, social patterns, and individual bioenergetics.

Pan paniscus, like *P. troglodytes*, forages on a wide range of plant and animal species and lives in communities of changing party size. Tool use among free-ranging animals is rare, and predatory behavior infrequent. The society is nonhierarchical, and males do not dominate females. The most obvious sex differences observed are in social behavior, in the close bonding of nonrelated females, and in the close bonding of adult males with their mothers and not with other males. Ongoing research on this species offers future possibilities for delineating in more detail possible sex differences in anatomy and behavior.

Lessons from the Apes

Taken together, the apes illustrate several points. Sex differences within species comprise a mosaic of features in anatomy, growth and development, and behavior. The study of whole animals and whole lives in a social and environmental context more clearly highlights the ways the two sexes diverge or converge in order to survive and reproduce, than if a single trait (body size or canine size) or a single behavior (mating or fighting) is used. There is no simple correlation between anatomy and behavioral expression, within or between species. In comparing sex differences (e.g., body weight difference) across species, the ecological setting and adaptation of the species are important variables.

Of the apes, female and male gibbons diverge least in their life pathways. Although they differ little in body size and proportions, the sexes differ in frequency of damaged or broken canines. Except for distinct vocalizations, only direct study of the animals' lives reveals that behavioral variation is not pronounced.

Both species of chimpanzees exhibit moderate sex differences in body weight; however, the two species express different patterns of female and male features of canine teeth, cranial capacity, and limb bone lengths.

In *Pan paniscus*, sex differences are most pronounced in social interactions. Females apparently take priority during feeding, and even unrelated females bond with each other. Bonds between females and males are strong. Males maintain strong bonds with their mothers into adulthood, but bonds between males, even brothers, are very weak. The small canine teeth of both females and males may reflect the overall sociable relationships among various age/sex classes.

Female and male *P. troglodytes* diverge more than do female and male *P. paniscus* in canine size, cranial capacity, and limb bone length and joint size. Some of the species differences, such as canine size, may be due to the greater habitat variability of *P. troglodytes,* and more time spent on the ground, with the consequently greater risk of predation. It is also possible that *P. troglodytes* males must guard the territory and resources (food and females) from other community males. This could account for the pronounced aggression as well as the affiliation between males, and somewhat less intense bonds among females whose travel and foraging patterns differ from those of males. Females and males differ in tool-using behaviors that would not have been predicted on the basis of anatomy nor discovered without long-term field studies.

Among orangutans and gorillas, the feature that has received the most attention is the much larger body size and weight of adult males. To explain large male body size, the mechanism of sexual selection, which presumes male-male competition and differential mating success of males, is most frequently invoked (e.g., Rodman and Mitani 1987). The problem here is that sexual selection overemphasizes the significance of male body size and underemphasizes female body size. The significant body-weight difference between the sexes in orangutans and gorillas must be considered from two points of view: (1) constraints on female body weight, and (2) advantages for larger-bodied males.

Orangutan females and males live most of their adult lives separate from each other and are distinct in anatomy, as well as in foraging, locomotion, and social behavior. For mountain gorillas, on the other hand, in spite of the large difference in body weight, females and males do not diverge as much in diet and foraging, travel, and range size as do the chimpanzees and diverge considerably less than do orangutan females and males.

In orangutans, female body size (within the chimpanzee range) is probably constrained by the demands of gestation and lactation in a highly arboreal habitat and in associating with juvenile offspring. Females forage for seasonal foods and travel with infant and juvenile offspring through the forest canopy, which requires locomotor skill, time, and energy. Maintaining small body size might help reduce energy requirements and minimize potential feeding competition between females and juveniles. Furthermore, most of the time, adult females forage independently from other adult females and males.

The larger body size of orangutan males may have fewer constraints and may be advantageous in several ways. In spite of their large body size, adult males retain arboreal skills and do so because of their extreme hominoid features—large, heavy, and well-muscled hands and feet, long upper limbs, and mobile limb joints (Zihlman 1992). Larger body size permits greater fat storage in the cheek and neck regions that may cushion against seasonally reduced food abundance or variable quality. (In turn, these features combined with vocalizations distinguish adult males from other age/sex classes). Their large size also permits relative safety during occasional terrestrial travel and makes possible a larger seasonal home range that provides increased resources and greater opportunities for finding and mating with females. Finally, large size may be advantageous for aggressive encounters with other males.

Thus, the variation between adult female and male orangutans is an expression of *both sexes'* living within the constraints of being a highly arboreal, frugivorous, large-brained, large-bodied hominoid, with a long period of infant dependency.

In contrast, among mountain gorillas, terrestrial travel and foraging combined with a somewhat folivorous diet allow females to become large. From the point of view of females, it may be desirable to have at least one and preferably more adult males available in the group for mating, for promoting group stability, and for protecting resources and infants from other males and from outside threats. Unlike adult male orangutans, adult male gorillas can live together in the presence of adult females in spite of their large body size and potentially lethal aggression. Therefore, the presence of several males may increase reproductive outcome for both females and males, as it does for cebus monkeys, which live in groups larger than 15 (Robinson 1988a).

This comparison between orangutans and gorillas highlights the overriding *species' adaptation,* within which the pattern of sex differences is expressed. Given the larger absolute body weights of female and male gorillas versus those of orangutans, one might expect an even greater degree of difference between the sexes in gorillas than in orangutans. Perhaps there is less difference than expected because of social interactions, divergent environments, and the distribution of resources.

Two factors relate to species' adaptation and may influence sex differences. The first is terrestriality. Female gorillas travel primarily on the ground, which imposes fewer constraints on gorilla body size, in contrast to female orangutans who, with dependent offspring, travel exclusively through the forest canopy. Furthermore, given the terrestriality of gorillas and the potential danger of dwelling on the ground, large size may be advantageous for both sexes.

Second, female and male gorillas of both species have a somewhat more folivorous diet than orangutans or chimpanzees. Gorillas forage on or near the ground on herbaceous and leafy foods that are more available and more evenly distributed than fruit and require less travel between food patches. Even when eating fruit, lowland gorillas (Tutin et al. 1991) do not travel through the forest canopy, but on the ground, ascending and then descending trees. This feeding and foraging pattern allows several gorilla females and their juvenile offspring and mature males to remain together with minimal competition.

Male gorillas also experience few constraints on body size. In fact, large size may be a distinct advantage. Males, the first to meet danger, provide formidable threat vocalizations and displays to protect other group members from aggression by other male gorillas or attacks by predators. Fighting abilities of male gorillas may be enhanced by their large size, but it does not necessarily follow that sexual selection alone can account for large body size of male and female gorillas.

For both orangutans and gorillas, the mechanism of sexual selection and emphasis on male-male competition and differential mating success of males, although perhaps one element, offers only a limited way of looking at body size and other features. Explaining female body size is also a consideration. Within the life-history perspective, it is apparent that body size and weight may be more or less constrained depending on the species' way of life and how each sex accommodates to it. For instance, because of exclusively arboreal travel, the necessity of carrying dependent offspring, and a diet of mainly ripe fruits, female orangutans must be constrained in body size. Female gorillas, in contrast, given their more terrestrial traveling and foraging over short daily ranges and a relatively folivorous diet, are less constrained. Male gorillas appear even less constrained in body size than females—or male orangutans—and for similar reasons. For male orangutans, perhaps their adaptation *is* a large body size that gives them greater flexibility in dietary composition and foraging, larger home-range size, and body proportions that contribute to their locomotor abilities for arboreal travel.

Summary

Hominoid sex differences in anatomy and behavior range from minimal to extreme. Too often sex differences have been dealt with simplistically (only body weight, body size, or canine size) with the male as the standard; too much attention has been focused on mating and too little on the entire life.

Behavioral studies on free-ranging apes significantly document the relationship between anatomical and behavioral features. To underscore this point, note that female and male gorillas are more similar to each other in how they live their lives than might be expected given the large body-size difference. Conversely, among chimpanzees (*Pan troglodytes*), behavioral variation between females and males (e.g., tool use, predatory behavior) perhaps is greater than would be expected given the moderate or minimal differences in anatomical features.

A life-history framework takes into account the species (or genus) adaptation and therefore goes beyond body weight or canine tooth size, and beyond form and function, to consider sex differences in the context of what each sex has to do to survive and reproduce throughout the life course. The integration of survival with reproductive life-history characters gives a more complete picture of adaptation and possible mechanisms of evolutionary change. It offers a wider framework for interpreting sex differences during early and recent hominid evolution (Zihlman 1993; chap. 13; Morbeck chap. 1; Borgognini Tarli and Repetto chap. 14).

In reproductive life-history features, as illustrated in table 8.1, the hominoids share more similarities than is expected given their divergent body weights. Consequently, the function or evolutionary significance of body size between females and males and between species do not lend themselves to simplistic explanations because body weight can account for only part of the variation.

The perspective presented here highlights the behavioral flexibility that is part of the life-history profile of catarrhine primates and its contribution to survival in each species. This flexibility, especially as expressed in females, emerges during an individual's life. The long period of development and late sexual maturity is beneficial to both females and males. At the same time, this long period of individual experience before sexual maturity allows for the variation expressed during adult lives, and finding simple correlations or causal relationships between anatomy and behavior is more difficult.

Acknowledgments

For comments and discussion on the ideas and on the paper, I thank M. E. Morbeck, R. McFarland, B. McLeod, and M. Remis. For financial support, I thank the University of California, Santa Cruz, Division of Social Sciences and the Faculty Research Committee.

Note

1. Time-allocation studies assist in indirectly assessing energy budgets and recognize that energetic output cannot be accurately measured directly (e.g., Coehlo 1974, 1986; Altmann 1980; Bernstein and Williams 1986; also see Panter-Brick chap. 17).

Part IV

Anatomy, Physiology, and Variation: The Catarrhines

What It Means to Be a Catarrhine

NESTED within the mammalian and primate life-history features are those that apply to the Old World primates, known as the catarrhines. This group includes species of monkeys, apes, and humans. We are identified by a distinctive nasal apparatus, but the importance of this classification rests with our shared ancestry. When we look at this group of mammals, which includes ourselves, we see that it allows us to make some interpretations of life-history characteristics based on the skeletal material. Because of the shared ancestry, we can predict how some events will have repercussions on the skeleton. We also can see in contemporary species how the physiological reactions form a similar set of responses. In this section, we are looking at both of these aspects: how we can interpret life-history features from bone and other hard tissues, and how we can look at comparative physiology to explore the evolution of life histories within the catarrhines.

THE "OUTSIDE VERSUS INSIDE" VIEW: A CHIMPANZEE CASE STUDY

Chimpanzees are the closest living relatives of humans, and, like humans, are a large-brained and long-lived species. They have long prereproductive life stages that emphasize brain growth and information exchange. Chimpanzees and humans share common physical features of bones, teeth, and muscles as well as biological and social patterns of growth, development, reproduction, and aging.

Our research on biology and life history provides case studies of individual chimpanzees from the well-studied populations of Gombe National Park, Tanzania. We weave together the "outside" and "inside" views of these chimpanzees including: (1) real events that happened to real animals in the course of their lives and influences on their survival, health, reproductive outcome, and mortality; (2) how these events are reflected in their skeletons; (3) the contribution of bones and teeth to whole functioning animals; and (4) application to interpretation of the fossil record of our early ancestors.

Our evolutionary framework, of course following themes in this volume, emphasizes a life-history approach and a view of reproductive success that includes survival throughout an individual's lifetime, as well as mating and rearing offspring in adulthood. We study skeletons of known individuals combined with detailed biological and behavioral information collected during their lives by Jane Goodall and her colleagues (Goodall 1986). With data on the factors that shape bones during life and that relate to survival and reproduction, we can understand the anatomy and its range of variation. We, then, can more realistically assess possible behavioral correlates of ancestral hominids as well as modern humans.

An individual's skeleton preserves both phylogenetic history and biosocial and ecological interactions during life. As part of her study, Goodall retrieves chimpanzee bodies after death; skeletons are cleaned by natural decomposition. Unlike many other studies that rely only on bone lengths, data are derived from a variety of methods and give information at several levels of biological and behavioral organization. Our research team has analyzed hundreds of raw measurements and calculated variables for each adult, juvenile, and infant (Morbeck and Zihlman 1989; Sumner et al. 1989; Zihlman et al. 1990; Morbeck et al. 1991, 1992, 1994). These noninvasive data include (1) volume/weight, linear, and area dimensions of skulls, faces, teeth, limbs, shoulder and pelvic girdles, and vertebrae that define bone size and shape to determine body proportions, movement, and weight-bearing; (2) CT (computed tomography) scans that show cross-sectional shape and indicate biomechanical loading conditions of "slices" of upper arm and thigh long bones; (3) bone mineral content measured via photon absorptiometry at the same scan sites and in the forearm for comparison with live, captive chimpanzees and with humans; and (4) nonmetric evidence of growth, disease, injury, possible female reproduction, aging, and cause of death determined from whole bones and via radiographs.

Combined with individual life stories, Gombe skeletons reveal the factors that influence the size, shape, internal structure, and mineral content of bones and teeth. They also illustrate components of both survival and reproductive life history of female and male chimpanzees. Survival characters are reflected in the configuration of the joints, distances between joints, and skeletal markers of muscle placement, as well as internal structure and composition of bones. For example, these features indicate movement capabilities associated with walking and climbing for food, grooming, or avoiding predators. Similarly, size, shape, and wear on teeth relate to the properties of food items and chewing mechanisms in feeding. Reproductive characters, especially the timing of life stages, are delineated by general patterns of growth and maturation (i.e., eruption of the third molar, long bone epiphyseal closure).

Unlike museum collections or series of fragmentary fossils, the Gombe data make it possible to integrate information from skeletal analyses with details of each life story. A 68-page set of data sheets, organized according to the life stages, constructs life-story profiles for each animal. It allows us to assemble, cross-reference, and retrieve published data about individuals in the skeletal series. Gombe field-study records emphasize social interactions and survival activities. They also provide biological data on growth, diet, illness, injury, aging, and death; live body weights; and female cycles, pregnancies, and mother-infant behaviors (Goodall 1983, 1986).

We work "down" from skeletons to investigate, for example, species, sex, and population life-history features, as well as individual life stories (e.g., malnutrition, disease, or aging as causes for loss of bone mass). We also work "up" to study whole organisms as they function in a local breeding group. Only with field observations can we understand the importance of what we read in bones. For instance, timing of the disease, injury, or stressful social situation relative to the life cycle influences how such "insults" affect survival and reproduction (Morbeck et al. 1991).

Individual skeletons reveal normal growth, maturation, and aging and, in addition, causes and effects of malnutrition, disease, injury, and age at death. This information contributes to understanding how whole organisms function in a local breeding group. Many biosocial factors influence the demographic variables of age at first reproduction, interbirth intervals, migration, and death. The field observations confirm and elaborate what we read in the bones.

Disease, for instance, paralytic poliomyelitis and disuse osteoporosis as observed in Gilka, may have a greater impact on the subadult's life and, especially eventual reproductive outcome compared to the effects on reproductive efforts in adults Madam Bee or Melissa. Prime adult male activities associated with patrolling community boundaries and establishing consort relationships pose risks for injury. These can be viewed as "costs of reproduction." For example, high-ranking, prime male Hugo severely injured his ankle and foot. He lived a long life after recovery and often was observed with females. In contrast, Charlie, also a prime male and high-ranking, at about age 26 years was injured fatally, thus truncating his potential reproductive span by at least a decade. Adolescent Flint's early death resulted from probable physiological stress caused by emotional trauma at the time of his mother's death combined with poor physical health. As these examples indicate, individuals respond to disease, injury, and stressful situations in different ways at different stages of the life cycle, and each of these negative occurrences affects survival and reproduction (Zihlman et al. 1990; Morbeck et al. 1991).

When we move "up" to the level of the population and species and "out" to environmental interactions, we see that the Gombe adult chimpanzees, when aggregated into groups, show a mosaic of sex and age differences within the population and size differences among other chimpanzee populations. Females weigh less than males and have smaller canines, but cranial capacities and skull lengths are similar (Morbeck 1991b).

Our data also suggest that both older females and males apparently have less bone mineral at the time of death than do adults that died at younger ages (Sumner et al. 1989; Morbeck unpubl. data). We also compare data for the aggregated group at the levels of populations, subspecies, and species by using previously collected data for *Pan* and other hominoids. At these levels, brain and tooth size are like those of other common chimpanzees, but Gombe adults weigh less and have shorter limb bones (Morbeck and Zihlman 1989).

The Gombe population and its skeletal series, therefore, provide a natural experiment for evaluating the influences of both genetics and ecology. We can begin to assess the patterns of variation compared to closely related populations and species in terms of both phylogenetic data (e.g., historical "constraints" of sizes of the brain and teeth, which grow early in life) and, at the same time, ecological explanations for small body size (e.g., limited resources at Gombe National Park that may slow limb bone growth rates).

Chimpanzees, like humans, are hominoids, catarrhines, primates, and mammals. Each natural group shares a theme composed of associated anatomical, physiological, and behavioral features. For example, Flo, the most famous of the Gombe chimpanzees, lived a long life and gave birth to five known infants, three of which survived to maturity. Her reproductive life was similar to that of the other catarrhines. During most of her adult life she was either pregnant or lactating and caring for her offspring. We can read in her bones that she had undergone a reduction in bone mass (i.e., osteoporosis also seen in older women), but her pelvis shows no signs of "scars" of birth. We can read in the field observations that she had no cycles toward the end of her life (Zihlman et al. 1990; Morbeck et al. 1992). Flo confronted the same challenges as all catarrhine females, challenges (discussed in Part IV) that center on reproductive physiology.

Ovarian Function and the Hormonal Link to Anatomy

Reproductive physiology of catarrhine primates, like Flo and ourselves, has evolved as part of mammalian life-history features. Ovarian function among the mammals is carefully coordinated so that ovulation occurs when fertilization is most likely to take place and, as importantly, when the uterus is receptive for implantation of the fertilized egg. This complex of physiological, anatomical, and behavioral processes, developed from a long evolutionary history, has produced a highly flexible set of responses that can fluctuate in response to a host of stresses (Bongaarts and Potter 1983; Bronson 1989).

Technical distinctions are made between fertility and fecundity. Fertility is the actual production of offspring and can be seen by fieldworkers. In contrast, fecundity is a female's ability to conceive, carry, and rear offspring, whether or not that event actually ever occurs. Although fertility is easier to document, actual fecundity is a more reliable indicator of the ability of the body to respond to environmental and internal stresses.

"Typical" Ovarian Function in Catarrhines

Females are born with a large supply of eggs; these are held in primordial follicles in the ovaries. Among catarrhines, these provide the ova that will mature periodically, usually producing only a single fertilizable egg with each menstrual cycle. This process begins at sexual maturity and continues throughout life. In humans, the life span now frequently exceeds that of the ovarian function, and we experience a long postreproductive phase.

Menarche

Menarche is noted by the onset of menstrual flow. Despite the representation of this first menstrual cycles as a point in time, menarche, like most reproductive events, is actually a gradual process. It is the transition from infecundity and subfecundity, toward full fecundity. Hormonal changes, such as the increase in estrogen, occur well before the actual start of menses, whereas actual ovulation may be delayed until well after menstrual flows appear on a regular cycle.

In humans, the appearance of menstrual cycling also seems to be coordinated with the linear growth of the skeletal system. The peak of growth is achieved just before the first menses. Broadening of the pelvis through the lateral expansion of the ischium (lower portion of the anterior pelvis) also occurs at puberty and may be an important factor that determines when menarche occurs (Ellison 1989).

Menses and Estrus

The menstrual cycle in mammals, including catarrhine primates, coordinates ovulation and, to some extent, a female's interest in sexual activity. This does not mean that mating behavior is strictly limited by hormonal cycles. Especially in primates, sexual activity performs a variety of social functions, only one of which is reproduction. Thus, mating is frequently observed at times other than when it may result in a conception.

During the evolutionary history of the catarrhines (including humans), menstrual cycling with ovulation was probably an unusual state for the female. Like Flo, females probably spent the majority of their lives either pregnant or lactating. Therefore, the criteria used today for establishing "normalcy" in menstrual function of women (e.g., regular menstrual periods, ovulation, lack of spotting, etc.) may not have played a major role in the evolution of reproductive ability.

The "typical" cycle lies under hypothalmic control through the production of gonadotropin releasing hormone (GnRH). In the anterior pituitary, this stimulates release of two hormones: follicle-stimulating hormone (FSH) and luteinizing hormone (LH). The former promotes development of the egg from a series of follicles in the ovaries, which are activated with each cycle. These follicles also are hormonal regulators, producing estrogen. Estrogen promotes thickening of the uterine walls to facilitate implantation of a fertilized egg. As the levels of estrogen increase in the blood stream as a result of follicle maturation, production of FSH diminishes. When estrogen reaches sufficiently high levels, the anterior pituitary begins abrupt secretion of the second hormone, the LH surge. This induces ovulation, the release of the egg from the ovarian follicle, which marks the end of the follicular phase of the cycle.

In the later portion of the cycle, known as the luteal phase, the released egg is taken up by the fallopian tubes that will transport it from the ovary to the uterus. The ruptured follicle is transformed into a structure called the corpus luteum that continues to secrete both estrogen and another important hormone, progesterone. If implantation does not occur, the corpus luteum disintegrates and secretions of the two sex hormones drop precipitously, prompting loss of the uterine lining in a menstrual flow.

In some catarrhine primates (e.g., chimpanzees and baboons), ovulation is marked by the appearance of pronounced sexual swellings. These brightly colored swellings of the tissue surrounding the vagina peak at mid-cycle. Although, in many catarrhines, such swellings are an outward sign of ovulation, in some species, swellings occur during pregnancy or are so minor that only close inspection can determine their presence. In humans, obvious sexual swellings are not present. One of the theories proposed for this "concealed estrus" is that it forces the male to monitor his female constantly. A more logical proposal is that large sexual swellings are completely incompatible with bipedal locomotion: the affected tissues are now situated between the hindlimbs rather than to the rear. The fact that they are no longer visible to the "monitoring male" could be only incidental to the locomotor needs of upright, striding biped.

Reproduction

If fertilization is accomplished, the egg will attempt to implant on the uterine wall. Once established, hormonal changes, brought about by the collaborative physiological and anatomical efforts of the embryo and of the mother, prevent the disintegration of the corpus luteum. As the embryo grows, the placenta itself becomes a major hormonal producer, particularly of estrogen and progesterone. Together, the physiologies of the mother and fetus work to supply the nutritional needs of the growing infant and maintain the mother. Species-defined pattern formation and timing guide growth and development of the fetus (Wolpert 1991).

As the gestation progresses, hormonal changes also promote the preparation of the body for delivery and later lactation. Prolactin, thought to be instrumental in milk production, rises as do placental lactogens for the development of maternal breast tissue. Relaxin increases, which softens the joints surrounding the birth canal. Birth itself is prompted by abrupt shifts in the hormonal profiles.

In most catarrhine primates, birth is a relatively easy process. However, in some species, the close fit between the pelvic dimensions and the size and shape of the neonate head and shoulders makes delivery physically more difficult. In humans, birth is complicated with the reshaping of the human pelvis by upright posture and the increase in neonate head size. The passage through the human birth canal, therefore, necessitates not only adequate maternal dimensions but also an elaborate rotation of the infant in the birth canal (Tague and Lovejoy 1986; Trevathan 1987).

After birth, the baby can suckle. The mother initially produces a clear fluid in the breasts called colostrum, which is high in nutrient value, immune factors, and substances to stimulate the newborn's digestive system. Within a short time, however, colostrum is replaced by mature milk, the composition and amount of which will vary by the age and needs of the infant. Milk from primate mothers is low in fat, protein, and calcium but higher in sugars in comparison with that of other mammals (Oftedal 1984). Milk composition parallels the extended period of growth and development of the infant. Slowing of the growth period has the consequence of reducing the required rate of nutrient transfer from the mother.

Much of a catarrhine female's reproductive years are spent in lactation. This was also probably true of humans before the introduction of wet nurses and artificial infant foods. During this period, some mammals, including humans, reduce female fecundity through changes in the hormonal milieu. This phenomenon, known as lactational amenorrhea, is particularly important in mammals in which the young depend on their mothers for an extensive period of time because it extends the interbirth interval. In contrast, some other mammals, such as elephant seals, mate shortly after giving birth to their young and prior to the termination of lactation.

Menopause

For most catarrhine primates, reproduction continues throughout the remainder of life. Toward the later years, interbirth intervals may become longer and reproductive outcome more uncertain (Pavelka chap. 7). Human life spans have increased. Women, unlike most other primates, now routinely experience menopause, which terminates their reproductive capacity long before the end of their normal life spans. This transition can be lengthy. It is seen initially as the perimenopause, a period of increased irregularity in menstrual length, irregular bleeding, and a number of changes that result from fluctuating hormones. Eventually, menses cease altogether, and menopause is completed.

Flexibility in Ovarian Function

Flexibility is a hallmark of mammalian reproductive physiology. The primates stand out as adapting their reproductive behavior and their reproductive capabilities to a very great extent. Throughout the reproductive years, the efficiency of ovarian function will fluctuate in ovum production. Although fetal loss, neonatal death, and inability to find suitable mates may play a role, a major, but often underappreciated, factor in determining total fertility is fluctuation in female fecundity (Ellison et al. 1993b). Genetics probably play a role in determining the level at which "normal" function is possible, but environmental circumstances are monitored by the reproductive physiology of all mammalian females to increase further the variation among females within any population.

Menarche

This perception of environmental factors is seen in the females, first, by the timing of the onset of ovarian function, which can be accelerated or delayed, depending on the situation in which the female matured. For instance, in humans, earlier menarcheal age is seen with better nutrition, urban living, and even changes in family structure such as the presence of unrelated males in the house. Menarche can be delayed similiarly by a poor diet or generally poor health, or such things as heavy physical exercise in maturing adolescence.

Once apparently normal menstrual cycling has been achieved, the period of postpubertal subfecundity may also be lengthened even though the female may be, by all appearances, fertile and capable of producing an offspring. In women, we know that the subfecundity period can be lengthened by the same factors that extend the period to menarche.

Ovarian Function in the Adult Female

We presume that these same influences continue to play a role in the adult female. Ovarian function moves through a continuum in which one extreme—what we usually consider "normal"—is the monthly release of a mature ovum for fertilization. As environmental stresses increase, the normal or eumenorrheic pattern begins to change, usually with alteration of the height of the hormonal peaks or the lengths at which they change. There may be lengthening of the follicular phase and shortening of the luteal phase. This does not necessarily result in any obvious disruption in menstrual flow since the changes in the length of one period are compensated by changes in the duration of the other. The luteal phase becomes inadequate, and the preparation of the uterus becomes less thorough. Fertilized eggs no longer implant in the optimal places in the uterus and may be more frequently lost. The height of the LH surge also may be diminished, becoming insufficient to stimulate ovulation. Despite this, menstrual cycles still occur with relative regularity. Toward the opposite extreme from "normal" function, therefore, are oligomenorrhea and amenorrhea. In women, oligomenorrhea is defined as fewer than nine cycles per year with cycle lengths varying from 7 to 11 weeks, whereas amenorrhea is complete lack of menstrual cycling (Rosetta 1993).

In women, a series of factors—including body composition and shape, physical exercise, diet, energy balance, and disease—are known to interfer with the level of fecundity (McFarland chap. 12; Vitzthum chap. 18). Similar responses are also believed to exist among the nonhuman primates, although determination through hormonal analysis is more difficult, especially in the free-ranging animal.

In addition to the modulations in female fecundity, actual fertility can also be affected by other factors such as mate availability, predation, infanticide, and so on. In women, marital status, spouse separation, coital frequency, and a number of social and cultural determinates may be involved (Panter-Brick chap. 17). For instance, women may be isolated from their husbands for long periods of time because of taboos against sexual contact during lactation. Even in these cases, however, physiological changes may play a role; for example, the drying and atrophy of the vagina during lactation may make intercourse more uncomfortable for the female.

Body Composition and Shape. Overall body form follows a species-defined plan. Among the catarrhines, there is commitment toward certain anatomical and physiological features. Despite this, there is still considerable within-species variation in body size, proportions, tissue accumulation, and integrity. In females, this appears to interact with ovarian function. A body of literature has linked alterations of ovarian function to specific body types, specifically those with low adipose tissue content. In humans, leaner women who are physically active apparently are more prone to menstrual disruptions. This hypothesis, originally proposed by Frisch and Revelle (1971), was based on work primarily on the onset of menarche among Western women. The relationship between fat and ovarian function, however, appears to reflect energetic balances rather than absolute values (Pond chap. 11; McFarland chap. 12).

Physical Activity. Episodes of strenuous exercise have been linked to a significant rise in a number of hormones such as androgens, aldosterone, cortisol, luteinizing hormone, and prolactin, each of which is known to affect ovarian function. Although other hormones such as estrogen and follicle-stimulating hormone (FSH) are not altered, the ratios between hormones will vary. The FSH/LH ratio is significantly lowered. Beta-endorphins, the factor linked to the "runner's high," also increases and, in women, is associated with menstrual dysfunction.

As the regularity of physical activity increases, the disruption of ovarian function becomes more apparent. Women who regularly partake of such activity often are anovulatory despite having regular menses. Reductions may occur in the pulse of the LH surge in the follicular phase (Rosetta 1993) and in levels of progesterone in the luteal phase (Ellison and Lager 1986), both indicators of failure of ovulation.

When a female engages in extreme physical activity, ovarian function terminates completely, and menses does not occur. Among catarrhine primates, this condition is rare but it does occur in humans, particularly among athletes such as dancers, runners, and gymnasts. Such women have lower average levels of estrogens, FSH, and LH than exercising but cycling women.

In much of the world, women are the primary stay of agricultural activities, which often require heavy labor throughout much of the year. In addition, women frequently must process and prepare the food, activities that often involve strenuous work (Rosetta 1993). Much of this work occurs during the peak of their reproductive years (Hurtado et al. 1985), but women consciously and physiologically balance this work with reproduction (Panter-Brick chap. 17; Vitzthum chap. 18).

Diet. Catarrhine primates are eclectic omnivores, consuming a wide variety of foods including meat. Of this group, humans have evolved into superb omnivores. We consume many foods, prepared in a multitude of ways. No longer are we confined to the foods that grow or are obtainable in the immediate environment. We acquire, through technology or trade, foods produced throughout the world. In lieu of the strictly ecological variables with which most primates cope in terms of food, humans have added socioeconomic, cultural, and religious factors. This combination influences women's ability to reproduce by affecting primarily their ability to conceive and the amount of weight their fetuses can gain.

A brief fast may disrupt the LH secretion but rarely causes long-term disruption of ovarian function. Chronic malnutrition, however, may have profound effects on ovulation and menstrual regularity. Women living with chronic malnutrition are also often subject to fluctuations in ovarian function. Usually these women must subsist on diets that are higher in carbohydrates, very low in proteins and fat, and often deficient in specific nutrients. A vegetarian diet is associated with decreased estradiol (Hill et al. 1984). In particular, low-fat and high-fiber diets seem most frequently linked to cyclical irregularity. Despite this, many women are able to conceive, carry, delivery, and nurse children even in extremely marginal nutritional situations (Prentice and Prentice 1988).

Energy Balance. The problem of explaining the variation in ovarian function in the face of physical activity and dietary insufficiencies has led many to consider energy balances (Ellison 1989). In essence, this states that the body monitors the consumption and expenditure of energy. When there are increases in activity without compensating increases in energy intake, ovarian function is disrupted to prevent conception because a pregnancy would further drain the mother's resources. When food is drastically reduced, even when activity is consistent, a similar response would ensue.

Disease. The external environment can also impinge on the ovarian function of the female by way of exposure to pathogens (McFalls and McFalls 1984). With disease's effects on fecundity, we see two major problems. The first is the stress on the body induced by infectious and parasitic diseases. These conditions change the energy balance within the body, and although the female may be cycling normally, she is not capable of achieving ovulation and therefore will not be able to conceive.

In addition, there are many problems with venereal disease, particularly with pelvic inflammation diseases that cause scarring and blockage of the fallopian tubes (Peacock 1990). In these cases, passage of the ovum into the uterus is prevented as is movement of the sperm into the fallopian tubes. Similarly, diseases in the vagina that disrupt the fluid balance may hinder sperm passage or prevent changes in sperm that are necessary for fertilization of an egg.

Fetal and Infant Loss. Once a woman has conceived, a number of factors may reduce fertility. Most of the factors that cause early loss of the fetus are related to genetic problems in the fetus itself but also may be due to incompatibility between the mother and the fetus. In the latter case, many fetuses are lost well before the female knows that she is pregnant or manifests any changes in the body associated with pregnancy. Such losses simply appear as an extended menstrual cycle.

Later-term pregnancy loss is more likely caused by injury to the placenta or inappropriate placement of the placenta resulting in insufficient nutrients reaching the fetus. In some instances, obstruction of the blood vessels supplying the infant may occur causing fetal death. Throughout most of the pregnancy, the fetus is quite resistant to loss resulting from insufficiency of the maternal diet. This seems to be a characteristic of the female primate, and although fetal loss does occur, it is not a widespread response to dietary inadequacies. However, dietary restrictions in the latter part of pregnancy do cause insufficient growth, particularly weight gain, for the infant, which may be born small for gestational age. Low birth weight is a serious threat to infant survival.

In addition, postnatal factors can determine the fertility of the female in producing offspring that survive to weaning. These factors, of course, include social aggression, disease, and malnutrition. Females are remarkably resistant to the termination of lactation before the age of weaning. In most cases, nutritionally stressed females continue to lactate, although at reduced levels. The consequence is lengthening of the period of lactation. Thus, in the catarrhine primates that breed seasonally, the subsequent birth may be delayed for two or more years. This provides extra time to insure that offspring will attain sufficient body size to survive on their own. In nonseasonal breeders, subsequent pregnancies may be delayed until the current offspring has reached sufficient body size to be weaned. This will reduce reproductive success in the long-term but does produce more reliable survival of the offspring produced.

Ovarian Function in the Later Years

In older females, ovarian function changes suggest that the hormonal patterns are faltering as the availability of viable follicles is lessened. In nonhuman catarrhine primates, the interbirth interval is often longer, possibly because of a lowering in fecundity necessitating additional cycles before conception. Greater early fetal loss as a result of chromosomal abnormalities also may occur. In some isolated cases, complete cessation of ovarian function with shrinkage of the ovaries has been documented in nonhuman primates.

Actual menopause, as defined by deterioration of reproductive organs and decline in estrogen and other hormones in combination with termination of menses, may be limited to humans. The age at menopause in humans appears to be remarkably consistent in timing. Unlike the age at menarche, which varies in women, menopause seems to have an almost universal terminal age of 48 to 50 years. Obviously, some women undergo an early menopause, but remarkably few have a much later menopause. Therefore, it is felt that this process functions under fairly strict genetic control. This seems to be a capacity of primates and possibly all mammals. Pavelka and Fedigan (1991) have asserted elsewhere that this probably has to do with termination of viable follicles. The body is no longer capable of continuing to produce ova. In humans, the expanded life span has revealed the possiblity of this post-reproductive period.

Although the age at menopause does not appear flexible, the symptoms that accompany menopause, such as hot flashes and irregular bleeding, are not as pronounced a problem in many non-Western cultures. In part, the reason for the difference may be that non-Western women continue breast-feeding their youngest children and may be lactating throughout the perimenopausal years. Lactation will obliterate many of the changes Western women experience as a result of the abrupt changes in estrogen and progesterone.

This variation in ovarian function seen within the catarrhine primates also affects all of the tissues with which the sex steroids and the related hormones interact. Within the mammalian body, no system stands alone. This is most certainly true of the reproductive system, and changes in ovarian function will have repercussions on other organs within the body.

The Inside View

In this portion of the book, we take the opportunity to examine how both anatomy and physiology are integrated into the behavior and social and reproductive lives of catarrhine primates. It is important to remember that the context of the individual and its whole life is retained. We have selected two aspects of human biology, our skeleton and our fat (adipose tissue), features that we have already mentioned to some extent in the context of both mammals and primates, to explore this relationship.

Morbeck discusses the importance of the skeleton and its limitations for interpretation of lifeways and life cycles as life-history features. She sets the stage for understanding variation in skeletal anatomy, growth, and development, and life experiences as part of individual life stories aggregated in both contemporary and fossil groups.

The remaining papers in this section focus on anatomy, physiology, and the energetic and nutritional requirements of reproduction. In many respects, the female sex hormone, estrogen, can be used as the connection between the skeleton and adipose tissue discussed in these papers. In females, estrogen facilitates the accumulation of both bone mineral and fat in coordination with the reproductive cycle. Estrogen receptors have been located in bone, and estrogen sufficiency is associated with maintenance of bone mass. Loss of estrogen results in abrupt losses in bone. Estrogen receptors have also been located in adipose tissue. Similarly, estrogen is linked to the accumulation of fat depots at menarche and may play a role in the additional storage of fat during pregnancy.

Galloway focuses on the dynamic nature of the skeleton through the life course. Using data generated from studies of contemporary humans, she provides a possible evolutionary explanation for the development of a change in skeletal morphological characteristics of mammals that has serious implications for modern humans. This scenario focuses on the role of calcium mobilization in maintaining an adequate milk supply. Her work draws attention to the importance of individual ability to respond to a diversity of environmental conditions and how that may be measured at the populational level. This mammalian feature of calcium mobilization, played out over the greatly expanded human lifetimes, probably made postmenopausal osteoporosis inevitable. In the course of human evolution, this modern day "cost of reproduction" cannot even be seen as a delayed expense because it was probably rarely experienced.

The use of studies on contemporary humans and comparisons of these results with work on nonhuman primates and other mammals is also critical in the investigation of the role of adipose tissue in humans. Fat forms an important resource at each life stage, beginning in primates with early growth and development of the brain. Energy is stored in adipose tissue, which enables such long-lived animals as the catarrhines to move into more marginal areas, carrying with them an energy reserve. It also plays a role in assuring energy for reproduction.

Pond discusses a comparative approach to adipose tissue, focusing on the cells and depot anatomy across mammals. The placement and expansion of fat depots is intricately linked to overall body shape. Unlike elephant seals whose mobility on land is limited, primates have more constraints on where fat can be located without unduly interfering with the biomechanics of locomotion. The relatively recent shift to bipedalism with humans also raised the specter of inappropriate distribution of fat which necessitated exaggeration of certain depots. The anatomical location of fat depots, however, seem to be a mammalian characteristic restricting the potential variation in redesigning the placement and extent of the depots.

McFarland introduces the hormonal controls by showing how fat influences the ability to conceive, carry a fetus to term, and lactate successfully. Unlike some mammals, humans are not seasonal breeders and therefore must have a steady energy source for their young who are nutritionally dependent on the mother for an extended period of time. Her work on macaque monkey females shows that the ability to accumulate fat may be an important factor in predicting reproductive outcome.

This series of papers, therefore, demonstrates how moving "down" into the anatomy and physiology of individuals allows us to shed light on species and primate features. Without knowing how the body works—the hormonal interactions, energetic demands, morphological possibilities—we can make only limited interpretations.

9 Reading Life History in Teeth, Bones, and Fossils

Mary Ellen Morbeck

ANTHROPOLOGISTS are interested in human evolution, and I first studied fossil apes and humans. Fossil teeth and bones are mineralized, leftover hard parts of once-living individuals. They provide a glimpse at the lifeways and life cycles of our ancestors and closest relatives. Teeth and bones make up the skeleton and function in real animals living real lives in real places. The skeleton contributes both to survival throughout the life stages and to reproduction during adulthood. It records the evolutionary story of humans and all other primates, mammals, and vertebrates.

SURVIVAL- AND TIME-BASED, REPRODUCTIVE LIFE-HISTORY FEATURES

What can teeth and bones reveal about life-history features? What cannot be revealed? This chapter focuses on what we can determine about phylogeny and species' features, age and sex characters, and life experiences from the visible anatomy and physiology of teeth and bones.

The skeleton tracks a species' way of life and its life cycle by recording the life stories of individuals. Species-defined movement and weight-bearing abilities, for example, allow locomotor and feeding behaviors. These, in turn, are determined by size, shape, internal structure, and organic and mineral composition of the trunk and limb bones, and the head, jaws, and teeth. Teeth and bones also document the sequence and timing of the life stages, which are marked by tooth development and emergence and by the pattern and extent of bone growth, mineralization, and fusion of joints.

The skeleton can be studied both in living and in past individuals. Calcium and other minerals make teeth and bones into "hard" tissues that remain intact after an individual's death longer than other parts of the body. In addition to showing the species' features, these "dry" teeth and bones, and those fully mineralized as fossils, summarize an individual's life. The skeleton records age at death and experiences during life. It preserves the effects of nutrition, loading conditions in locomotor, feeding, and other activities, disease and injury, and, in some cases, reveals the influences of pregnancies, birth events, and breast-feeding in reproducing females (table 9.1; Morbeck 1991b).

Teeth and Bones Record Evolutionary History

The skeleton mirrors our phylogeny. Our skeletal plan is that of a vertebrate layered with mammalian, primate, and, within the primates, catarrhine (i.e., Old World monkeys, apes, and humans) and, most recently, unique human features (fig. 9.1). The anatomy, physiology, and behavior shared by living species and preserved in fossilized teeth and bones allow broad phylogenetic interpretations.

Our mobile skeletons with well-formed joints, two sets of teeth of different sizes, shapes, and functions, and our jaw, face, and brain case are skeletal life-history features shared with other mammals. We also share patterns of mammalian growth, development, and reproductive

TABLE 9.1. Life-History Features in the Skeleton: What We can "Read" in Bones and Teeth

9.1.A. Species Characters: Evolved Lifeways throughout the Life Stages

- Sex
  - Female: features related to survival, mating, and adult social roles, especially, to rearing offspring
  - Male: features related to survival, mating, and adult social roles
- Age
  - Survival life-history characters during in utero, infancy, juvenile, adult life stages
  - Biosocial growth, development, reproductive potential, aging, and death as expressions of time-based reproductive life-history characters
  - Bone and tooth tissue organization, maturation, mineralization; and bone epiphyseal fusion, tooth emergence; age, and condition at death
- Skeletons
  - Teeth
    - Upper and lower jaws with teeth, face, relation to braincase
      - Mechanical processing of food items
    - Teeth meet teeth
      - Number, size, shape, internal structure and tissue distribution, and organic/mineral composition
      - Positions and jaw-muscle attachment sites
      - Abilities to cut, crush, and grind
      - Directions, ranges, speed, and power of movement of jaws
  - Behaviors
    - Chewing in feeding and fighting, or other expressions of survival life-history characters
- Bones
  - Whole skeletons
    - Support, movement, protection of organs, and mineral storage, red blood cell and immune-system cell-production sites
    - Body plan, size, proportions, braincase
    - Number, kind, size, shape, internal structure, and tissue distribution, and organic/mineral composition
  - Joints and connecting bones or levers
    - Size, shape, position, and orientation of joints; stability and directions and ranges of movement; weight bearing and transference of weight
    - Shafts with muscle attachment sites; speed and power of movement
  - Behaviors: locomotion, posture, manipulation, and chewing as expressions of survival life-history characters

9.1.B. Individual Life Stories: Unique Sequence of Lifetime Experiences

- Genotype
  - Species-defined, sex-associated, and individual genetic potential
- Environmental interactions
  - Biological and behavioral interactions associated with growth, maintenance repair, and reproduction
    - Energy acquisition and use: nutrition, vitamins, minerals, activities
    - Physiological disruptions
      - Severe or long-term "insults," disease
      - Genetically based "under" or "over" growth in size and mineralization (metabolic bone disease)
      - Malnutrition: decrease in rates of growth and maturation and immune system disfunction leading to more disease
      - Vitamin and mineral enhancements or deficiencies
      - Infections
        - Generalized disruptions to growth, tissue maintenance, and repair
        - Local markers of particular infections
      - Injuries and repair
    - Reproduction by females
      - Possible indicators of giving birth
      - Bone mineral decrease related to breast-feeding
  - Biophysical and biochemical interactions associated with life's activities
    - Load-bearing in locomotion and posture
      - Biomechanical geometric properties
      - Cortical and trabecular orientation; cross-sectional size and shape
      - Material properties: organic and mineral content
    - Fracture injuries and repair
    - Wear on teeth and joints, dental caries and abscesses, bone deterioration and arthritis
  - Sociophysiological interactions associated with complex negotiations in social group
  - Preserved physiology related to environmental interactions

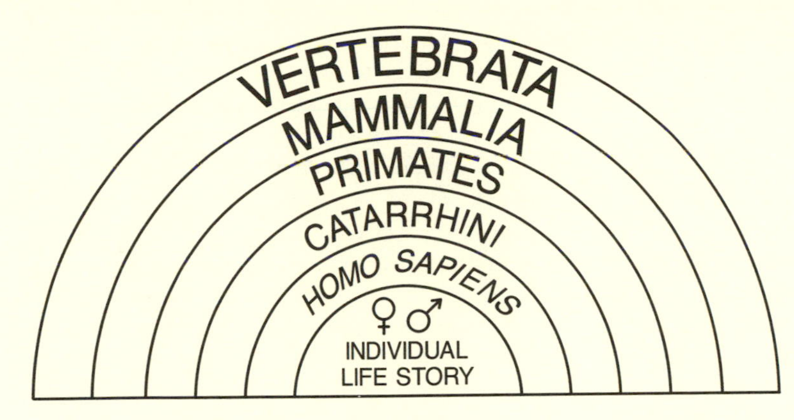

FIGURE 9.1. The skeletal plan combines features of a series of layers linked to the hierarchical classification of humans, *Homo sapiens*.

potential. Features include how the skeleton functions to maintain life; the ways in which our hard parts form and mature; how teeth and bones mark distinctive life stages; and, in females, how the skeleton contributes to the production of healthy offspring during pregnancy and, after birth, to incorporation of calcium in breast milk.

Mammalian functional biology and behavior provide the basis for the primate skeleton. Primates, in turn, have a large brain case (relative to the face or overall skeletal size) that protects a big, complex brain; generally forward-facing orbits that position eyes with overlapping fields of vision; and sizes, shapes, and structures of trunk and limb bones that enable hand-to-mouth feeding and climbing by grasping behaviors of youngsters and adults. This ability allows a (nonhuman) primate infant to grip its mother's hair and ride along with her, thus promoting a close mother-offspring bond.

Old World monkeys, apes, and humans provide another historical overlay. All have large brain cases (with brains that allow considerable memory storage and potential behavioral-environmental interactions); a fully stereoscopic (and colorful) way of seeing the world; sizes and shapes of bones and joints that allow more-refined hand movements (with increased coordination of hand-eye-brain control); and long periods of growth and development prior to skeletal maturation. Humans, in turn, expand these catarrhine features with even larger brain cases, more finely tuned hand movements, and longer life stages and life span. Our *Homo sapiens* skeletons also distinguish us as two-legged walkers with small front teeth and faces.

What Teeth and Bones Do

What it "takes" to survive and stay healthy during growth, development, reproduction, and aging is similar among Old World monkeys, apes, and humans (Morbeck chap. 1). The skeleton is integrated into all aspects of an animal's life from infancy to old age as an individual interacts with physical, biological, and social environments. It houses the brain and protects other organs, allows us to move, eat, and engage in other social-maintenance behaviors such as communication, grooming, fighting, mating, and infant care. Although teeth and bones work together in the skeletal system, they have different properties, functions, growth patterns, and biological roles.

Teeth

Teeth are the front end of the digestive tract. As mechanical food processors, they have a vital role in transferring energy from the external environment to the body for tissue growth, maintenance, repair, and reproduction. Different tooth sizes, shapes, and chewing surfaces cut, grind, or crush food, and, thus, prepare food for digestion.

The dentition develops early in life, mostly in utero and during infancy. After crown growth is completed, teeth emerge through the gums from their sites of development in the jaws. Tooth anatomy has been suggested as being under fairly direct genetic control since there apparently is relatively little variation within species (B. H. Smith 1989, 1991).

"Teeth are born fossils . . . they hardly remodel during life . . . [and] are hardly influenced by environmental factors" (Boyde 1990: 229). The outer coating of enamel is the hardest of the body's hard parts. Once it forms, this highly mineralized tissue (i.e., mostly calcium phosphate) does not change except as a result of wear or other external factors. Both the tooth's inner chamber—that is, the less calcified dentine with its nerves and blood supply—and the roots that anchor a tooth in the jaw, are somewhat more responsive to the body's ongoing physiological functions and environmental influences.

Once emerged, teeth confront the physical forces of chewing. Tooth-food-tooth interactions, use of teeth as tools, and impacts that produce injuries can modify enamel surface anatomy. Because teeth are part of the biochemical environment of the mouth, chemical erosion, and bacterial infections also may erode enamel surfaces. Compromising any part of the food-processing mechanisms of teeth, jaws, and their functions can affect feeding efficiency and lead to nutritional problems, breakdown of the body's self-maintenance system, and even death.

Bones

Bones support and allow movement of body parts. In addition, they protect body organs, serve as mediators of mineral use and its storage, and provide places for red blood cell and immune-system cell production.

Bone is both rigid and resilient. The architectural design and material properties expressed in bone shapes and structures of the head and jaws, limbs, and trunk underlie the mechanics of movement and stability and weight-bearing abilities associated with expressions of various survival behaviors. The sequence and timing of species' growth and maturation (i.e., skeletal modeling) associated with changing life-stage roles are linked genetically. This growth pattern guides (with varying ranges of diversity in different body regions and among individuals) the time of onset, rate, and duration of increase in size and mineralization combined with changes in shape and internal structure.

Bone, unlike tooth enamel, continues to grow and mineralize until full adulthood is achieved. After its organization and formation, bone tissue also can respond to environmental conditions (i.e., skeletal remodeling). Bone continuously adjusts to the amount of energy and minerals obtained from diet, to physiological disruptions, and to physical forces produced as part of an individual's lifelong activities. Bone size, shape, internal structure, and mineral content, therefore, can reflect life experiences in a variety of fluctuating environments (table 9.1B).

Life Stages

Teeth and bones preserve information about life stages. Reproductive characters set the tempo and result in the growth and maturation of skeletal anatomy and behavior that allow expression of social and maintenance behaviors. Rhesus macaque monkeys, common chimpanzees, and humans provide examples of what the skeleton tells us about changes in biosocial roles typical of each life stage (Morbeck chap. 1).[1]

In Utero

Prenatal growth and development of teeth and bones are similar among Old World monkeys, apes, and humans. Like other placental mammals, the fetus grows in the mother's uterus. From the perspective of this new individual and its developing teeth and bones, complete life support and minimal weight-bearing requirements characterize its environment (although some load-bearing associated with muscle actions occurs with fetal movement). Long before a monkey, ape, or human needs to chew food or move around independently, teeth begin to form and bones begin to grow and develop. From the beginning of development, therefore, teeth and bones and their relations to other body systems are integrated fully into whole animal function.

Primary (or deciduous) teeth begin to calcify before birth and have no precursors. Toward the end of the in utero life stage, the first permanent molars also begin to form. Tooth crowns develop by weekly and daily growth cycles that reflect regular changes in cellular activity. Secretions by specialized cells are punctuated by periods of little or no deposition. This growth and maturation process is captured in tooth structure by incremental markings in the dental tissue.[2] Dentine and roots continue to develop, and teeth emerge later during infancy.

Bones also first appear as organized tissue and begin to form and mineralize before birth. Growth is characterized by size increase (number and size of cells), whereas modeling and, later, remodeling involve change in shape by adding or taking away bone. Early skeletal growth and modeling of humans, apes, and monkeys follows that of other mammals (i.e., with the skull bones forming over the membranes that cover the brain and using the cartilaginous precursors for the limb bones).

Skeletal growth and mineralization require that fetal body systems function at an early stage. A mother supplies calcium, phosphorous (which is essential both for calcium deposition and for growth and development of "soft" tissues such as muscles and other organs), and the various forms of vitamin D (for intestinal absorption of calcium) via the placenta. Her developing fetus contributes the regulating hormones (i.e., parathyroid hormone, calcitonin). By the time of birth, the skeleton of the growing individual has become better suited for the biomechanical demands of life outside its mother.

Birth, Infancy, and Weaning

Birth marks the transition to infancy. Infants must independently acquire, store, regulate, and mobilize minerals and nutrients required for growth and maturation. Skeletal changes during infancy include (1) brain growth and development with increase in skull size; (2) a shift in feeding strategies from suckling mother's milk to chewing solid foods with a change from a condition of no emerged teeth to one with a battery of functional primary teeth; corresponding growth in upper and lower jaws that house the teeth; and development of chewing muscles that move them; and (3) transformation of a musculoskeletal system from a condition of minimal load bearing to independent movement and weight bearing correlated with increase in size, development of joints and muscles, change in body proportions, and enhanced neural coordination of chewing, locomotion, posture, and manipulation.

As an infant's primary teeth continue to mature, roots develop with eventual closure at the root tips. The secondary and molar teeth also begin to form. In addition, the two parts of the lower jaw or mandible fuse. Growth, development, eruption, and emergence of primary teeth during infancy in combination with growth and development of the face, jaws, and chewing muscles (muscles of mastication) permit mechanical processing of solid food.

Among most mammals, primary tooth emergence corresponds with age at weaning and serves as a reproductive life-history marker. In primates the story is more complicated. Infants begin to eat solid foods while they still rely on their mothers for milk, transportation, protection, and social life. Metabolic and social weaning are separated. Thus, age at weaning cannot be inferred only from teeth.

In contrast, bone composition, at least in modern humans, provides a more reliable way to infer feeding behavior and biological weaning. Constituents of mother's milk differ from that of the usual weaning foods. For instance, bone composition may record a shift in strontium-calcium ratios from lower ratios while an infant is nursing to higher ratios when individuals eat solid, especially plant, foods.

During infancy, bone and its cartilaginous precursors in the postcranial skeleton continue to increase in size and calcium content. They also change in shape and structural strength as part of the species-typical growth process. As in the teeth and jaws, changes in body proportions and the bones of the trunk and limbs parallel species life ways as infants increasingly become independent and fully weight-bearing animals.

Childhood

Juvenescence is characterized by the emergence of secondary front teeth and molars. Bones of the trunk and limbs continue to grow, and joints become mineralized and well formed as the body increases in size and locomotion becomes more coordinated.

Permanent incisors and canines—that is, successional teeth—develop and emerge; their development may be influenced by primary teeth. Molars, on the other hand, are primary teeth with no precursors. These teeth emerge in combination with continued growth and maturation of the face, jaws, and braincase. The first adult molars, for example, appear at the same time as the face and jaws lengthen and increase in mass. This maintains an efficient chewing mechanism.

During childhood, the bones of the trunk and limbs continue to increase in size, shape, and mineralization. Bones and their joints continue to model as a result of species- and sex-defined intrinsic genetic and physiological factors related to survival features. At the same time, they remodel in response to forces generated during movement and load bearing, especially with practice of locomotor skills and increased travel distances.

Adolescence and Adulthood

Mature monkey, ape, and human adult skeletons have fully emerged teeth, united sutures of the face and the brain case, and fused joint epiphyses in the trunk and limb bones. Although adulthood is achieved at different chronological ages among catarrhine species, skeletal maturity proceeds in similar ways. The beginning of adulthood is marked by initiation of limb bone fusion of shafts with their joints (epiphyses). This indicates the cessation of linear growth. Adult teeth usually emerge before epiphyseal closures of all bones are completed.

Among the limb bones, the distal humerus (elbow joint) is the first epiphysis and the proximal humerus (shoulder joint) is the last epiphysis of the long bones to complete closure and stop growing in length. Thus, fusion of the elbow and shoulder ends of the humerus generally encompasses adolescence. This transitional period between childhood and adulthood includes a "growth spurt" that is measured in living individuals by increase in body weight and size and by genital development and achievement of reproductive maturity. Skeletal maturity follows, and bones continue to record individual life-story experiences throughout adulthood, including aging or senescence.

Sex Differences

Skeletal features of females and males show a mosaic pattern of similarities and differences. Biosocial roles of females and males are expressed as part of their individual as well as species-defined lifeways and life cycles (Zihlman chaps. 8, 13). Differences in the extent of these and other sex differences vary among species and among populations of the same species. The extent of female-male differences can be measured in human skeletal series from historical and archeological sites. For example, Borgognini Tarli and Repetto (chap. 14) show how sex differences relate to cultural, ecological, and socioeconomic factors associated with different resources bases, technologies, activities, and nutrition. Similarly, behavioral and ecological factors pattern variation associated with sex in other primates (Morbeck unpubl. research).

Lifeways and Life Cycles of Monkeys, Apes, and Humans

The general body plan, female and male biosocial roles during the life stages, and the nature and extent of variation among individuals, between sexes, and across populations and species are similar among catarrhines. These features reflect our common evolutionary history (see above; also see Schultz 1956, 1969a; Shigehara 1980; Swindler 1985; Watts 1986a, b, 1990). Details of lifeways and the timing of life cycles differ within the boundaries (i.e., historical constraints [and licenses, Thomas and Reif 1993]) of this catarrhine framework.

Survival, Lifeways, Skeletons, and Biomechanics

Animals live in a physical world. We conform to and take advantage of the principles of physics and energetics (Vogel 1988; Bennett 1989; Schneck 1990; Thomas and Reif 1993; Morbeck chap. 1). Linear, areal, and volumetric relationships are well documented, but some relationships are still debated among primate anatomists and skeletal biologists (Morbeck unpubl. research; Swartz 1989; Godfrey et al. 1991). Mechanical functions determine and, in turn, are patterned by species-defined capabilities and an individual's real-life behaviors (see Carter et al. [1991] for discussion; also see Hurov 1991). These are survival life-history features and reflect both past and current lifeways.

Studies of the biomechanics of the skeletal system, combined with natural history studies of observed behaviors in particular environments, provide a way to read species- and sex-typical characters and their expressions in individuals throughout the life stages. Old World monkeys, for example, are habitual quadrupedal climbers, leapers, and runners. In contrast, apes include arm-hanging/swinging gibbons and siamangs, slow-climbing orangutans, and knuckle-walking and climbing gorillas and chimpanzees; whereas humans are habitual terrestrial bipeds.

The skeletal view of survival features such as locomotion, posture, and manipulation focuses on bone size, shape, structure, and mineral content. These bony features combine with soft tissues (e.g., cartilage, ligaments, joint capsules, and muscles) to determine movement and load-bearing potentials of the skeleton.

Body movement occurs in joints, the areas where two or more bones and corresponding body segments meet. Size, shape, area, and orientation of joints permit particular types, directions, and ranges of movement in both growing subadults (with primarily cartilaginous joints) and in adults (with well-formed and fused bony epiphyses). Muscles that cross joints attach to levers (i.e., bones between joints) and move the skeletal system. Placements of origins and insertions relative to joint centers affect power and speed of movement (Morbeck unpubl. research). Finally, cross-sectional shape, the amount and distribution of mineralization, and the orientation of internal bone bars or plates (trabecular bone) record the directions and suggest the magnitude of forces that result from bone produced by an individual's musculoskeletal system activities and the influences of gravity during body movement and weight bearing (Lanyon and Rubin 1985; Martin and Burr 1989; Ruff 1989; Smith and Gilligan 1989; Ruff and Runestad 1992).

Getting and eating food illustrate the importance of biomechanics as part of survival life-history characters. The size and shape of joints, bony levers, and muscle attachment sites, as suggested above, allow the locomotor and manipulative behaviors of obtaining food. The mechanics of feeding involve tooth size, shape, and placement in jaws as well as attachment sites of the muscles that move them. Food items differ in physical as well as nutritional properties. Sizes, shapes, internal structures, and mineralization of nipping incisors, tearing or slicing canines, and shearing, grinding, and crushing premolars and molars are species characters that have evolved in tandem with size and structure of preferred food items. In big-brained, manipulative catarrhine primates, these form-function features also have evolved in relation to use of the hands during food preparation outside the mouth (e.g., Burton 1972; Morbeck 1979).

Time-Based Life-History Features and the Life Course

Timing is crucial for understanding animals' lives. Times of onset, rates, and durations of change in growth and development of different body regions are associated with anatomy, behavior, and ecology. These features distinguish closely related species (for discussion of this idea in other contexts, see Gould 1977; McKinney 1988; Shea 1989, 1992; McKinney and McNamara 1991; Wolpert 1991).[3] Comparative skeletal studies also show the importance of timing of growth and maturation, that is, reproductive life-history characters, as contributing to variation within species, among populations, between the sexes, and among individuals.

Growth to maturity is not uniform; it involves a mosaic of temporal trajectories in different body systems (Tanner 1978; Grand 1983; Sinclair 1989). Patterns of tooth eruption and emergence, skeletal development, and brain growth among monkeys, apes, and humans illustrate this idea. "Fast" or "slow" growth and development of different skeletal components result from earlier or later initial tissue organization, shorter or longer durations, and faster or slower rates of change, or a variety of combinations of these processes (Watts 1990).

Teeth and Brains

Old World monkeys, apes, and humans have the same number and kinds of teeth. Young catarrhines have 20 primary teeth that emerge during infancy and include 2 incisors, 1 canine, and 2 molars on each side of upper and lower jaws. Adults have 32 teeth; each quadrant of the tooth row has 2 incisors, 1 canine, 2 premolars, and 3 molars.

The processes of tissue organization, enamel and dentine formation of the crown and subsequent root formation, eruption through the jaw bone, and emergence through the gum mark the life stages. As described above, these developmental processes are similar in catarrhines (and, perhaps, in all primates). We also share (with some interesting variations) the general order of tooth development, molar cusp formation, and emergence, although the end products of occlusal surfaces differ (i.e., Old World monkeys with four-cusped, bilophodont chewing surfaces in both upper and lower molars; apes and humans with four-cusped upper molars and five-cusped lower molars).

Periods of tooth formation in rhesus monkeys, chimpanzees, and humans form a sliding time scale when measured by chronological time (Swindler 1985). Tooth emergence is completed at about 7 years (or earlier) in rhesus monkeys, 12–15 years in chimpanzees, and 18 years or later in humans. Emergence times of the first adult molars show this temporal shift among catarrhine species. These molars appear in rhesus monkeys, chimpanzees, and humans at different chronological ages: about 1 to 2 years of age in rhesus monkeys, at a little more than 3 years in chimpanzees, and at about 6 years in humans. When these emergence times are compared to the lengths of the combined preadult life stages or to adult anatomy, however, the relative times of emergence are similar.

A "big picture" look at skeletal markers that match time-based, reproductive life-history features seems to confirm a time shift among species that correlates with other life-history characters. Statistical analyses of cross-sectional samples of many species, for example, show that the average time of emergence of first molars correlates with a brain case of about 95% of adult size (data from Harvey and Clutton-Brock 1985; B. H. Smith 1989, 1991, 1992). These data are drawn from studies that use a variety of methods and samples and include variable expressions of environmental influences on individual growth schedules. This kind of broad-based, statistical correspondence at species or higher taxa levels still needs to be evaluated at the organismal level with longitudinal studies of individual life stories.[4] Yet, the link between the timing of molar tooth emer-

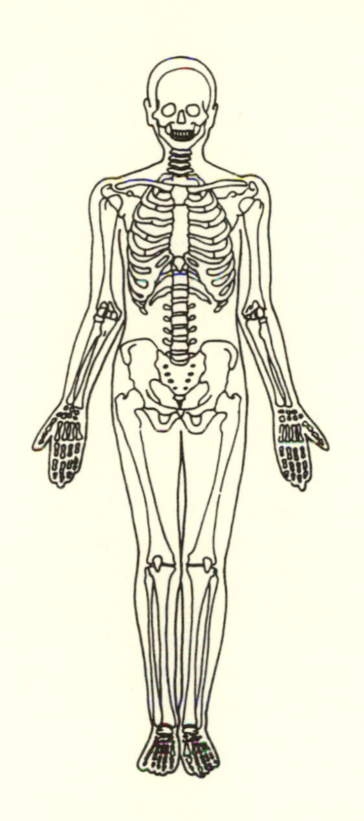

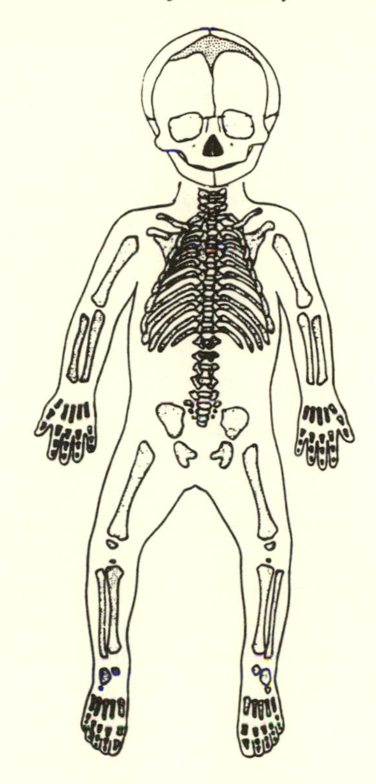

FIGURE 9.2. Skeletons of adult (left) and newborn (right) humans, *Homo sapiens.*

gence and brain case size relative to the achievement of adult status (i.e., connections among time-based reproductive and survival life-history features) is apparent. These features reflect shared ancestry with a conservative pattern of growth and maturation of teeth and brains, but with shifting time frames among these catarrhine species. Other body parts show different patterns of variation in timing of growth and maturation that relate to life cycles and lifeways (Zihlman chap. 8).

Bones and Big-Brained Humans

Humans have the longest prenatal period among the catarrhines (i.e., the oldest chronological age at birth). This is evident when measured in real time, if skeletal maturity of human newborns is compared to adult anatomy, or against a relative scale based on changes between life stages and the length of life span. Yet, as newborns, we have the least ossified skeletons, for example, with little organized tissue in our wrists and hands. In contrast, we have large brains and skulls (fig. 9.2). The growth pattern relates to importance of human cognitive and linguistic abilities during subsequent life stages. These are impressive, and their biological foundations and behavioral expressions are essential to most survival and reproductive activities.

In general, primate brains grow early in life. This is especially noticeable in long-lived monkeys, apes, and humans. The frontal bone of the skull grows as two bones that unite after birth. A "soft spot" is formed by a membrane on the front of the head that connects the separated skull bones before the bones fuse. Compared to other catarrhines, the timing of birth relative to the extent of skeletal development of the skull in humans, but not the developmental sequence, is "specialized." From an evolutionary perspective, a "delay" in fusion of the two parts of the frontal bone facilitates increase in the size of brain and brain case during the in utero life stage relative to adult female pelvic size. It also permits skull bones to move during birth and allows more brain growth after birth (for an appreciation of timing of different aspects of neurological development, see Falk [1992]).

The skeletons of newborn rhesus monkeys and chimpanzees have relatively smaller braincases when compared to humans. In addition, they have relatively larger and more mature hands (and feet). In rhesus monkeys, for instance, the absolute length of time devoted to prenatal growth is several months shorter than in humans. But, at the time of birth, beginnings of many hand and wrist bones are present (Watts 1990). Unlike modern humans, infant monkeys and apes use well-developed forelimb muscles to cling to their mothers (Grand 1983; Zihlman chap. 13; Morbeck and Zihlman unpubl. research). Furthermore, grasping hands and forelimbs are part of independent locomotion in older infants, juveniles, and adults. Thus, like the big brains of humans, monkey and ape forelimb and hand skeletal size, proportions and degree of mineralization at birth relate to lifeways that become important in later life stages. Species' life histories, therefore, combine a mosaic of survival and time-based features associated with the unfolding of particular lifestyles.

Teeth and Bones Tell an Individual's Life Story

Teeth and bones, in addition to expressing species characters, record an individual's life story as it grows to maturity, reproduces, and ages. Individuals constantly adjust body functions to meet the demands of their social as well as physical and biological environments. Life stories of monkeys, apes, and humans are played out during long prereproductive and adult life stages. Thus, numerous opportunities arise for many kinds of organismal-environmental interactions that potentially can affect whole organisms and their skeletons in different ways and at different levels of functional integration.

Teeth and bones document ongoing physical and chemical interactions with the environment and, of particular interest to skeletal biologists, skeletal responses to environmental "insults." Bones, for example, remodel according to physical activities (see above; Kennedy 1989; Merbs 1989) and incorporate the chemistry of particular foods into their structure (as revealed by biochemical and trace-element analysis; see above; Ortner and Aufderheide 1991; Schwarcz and Schoeninger 1991; Saunders and Katzenberg 1992). Thus, weight bearing and diet can be read in their structure and composition.

Tooth wear reflects species-typical geometric structure and composition, how many times a tooth has "worked," and the physical properties of food or other objects (Lucas and Corlett 1991). Microwear studies of frequencies and orientations of scratches (i.e., soft items) and pits (i.e., hard objects) on enamel record the kind of physical properties of food items (but not the actual species or its parts) chewed between the teeth (Walker and Teaford 1989). Phytoliths (mineralized plant cells) attached to teeth, if present, record what plant groups were eaten before death (Piperno 1988). The extent of mechanical forces that can crack enamel and the patterning of missing or broken teeth also suggest the nature of the habitat structure or social environment (e.g., tooth breakage resulting from falls or fights [Lovell 1990a; Zihlman chap. 8]).

Table 9.1 shows that an individual's skeleton reveals its health and experiences throughout the life stages in addition to its species-defined features (including sex and age). Skeletons that show species features and do not exhibit the effects of disturbances to growth or maintenance are assumed to represent healthy individuals. But we cannot be sure of a lifetime of good health. Bone is always changing, and the configurations of previous modeling and remodeling can disappear. Furthermore, certain conditions indicate successful coping with environmental insults either by evidence of repair or by compromising other skeletal components.

From the time of conception to the time of death, life-history attributes of each species allow a range of potential interactions with different aspects of the environment. This plasticity allows organisms to continue to function in many situations (table 9.1B; Morbeck chap. 1; "flexible response system," Vitzthum chap. 18). The extent of this adaptability can be determined from teeth and bones.

When the body's normal physiological functions no longer can adjust easily or quickly to particular environmental conditions, an individual is said to be "stressed." Disturbances are defined in the skeleton especially when they occur early in an individual's life, are severe, or long-term. Disturbances to normal growth or body maintenance result from, for example, physiological reactions to difficult social situations, disease (including malnutrition), and injury (table 9.1B).

The skeleton responds to life's activities, including health problems, with both generalized and specific responses. Acute or chronic health problems produce systemic physiological adjustment, that is, a generalized response. For instance, tooth development or bone growth may slow down or stop during illness as a youngster's immune system confronts infection or during starvation when the digestive system cannot process adequate food to supply energy for normal growth (Simmons 1990). In contrast, particular biological (including genetic) and environmental conditions are marked by distinctive bony reactions. Many genetically caused diseases (see, e.g., Brighton 1988) and certain kinds of nutritional, mineral, or vitamin deficiencies leave "diagnostic signatures" (Goodman et al. 1988; Kleerekoper and Krane 1990).

Stress markers, even when they show tissue repair had occurred, often are said to be pathological. They do, in fact, indicate poor health during one part of an individual's life. However, from the perspective of "whole organisms-whole lives" (Morbeck chap. 1), these features also demonstrate an individual's capacity for survival by overcoming environmental insults. Skeletal response and recovery become part of "the costs and limits of adaptation" (Goodman et al. 1988; an example is porotic hyperostosis as a response to increased pathogen load, see Stuart-Macadam [1992]). If the individual's chances of survival and reproduction (e.g., achievement of adult status) are enhanced, these life-story markers become adaptive responses to general environmental conditions or to particular genetic conditions and life events. The extent and pattern of stress responses among individuals in a population or skeletal series emphasize the importance of a species' range of possible physiological and anatomical adjustments.

How can we link the wide variety of individual skeletal features with biosocial and other environmental interactions, physiological function, behavior, and reproductive outcome? Studies of physiological function are beginning to show connections among hormones and social and other survival and reproductive behaviors (e.g., Selyean stress concept used by Goodman et al. [1988] and Sapolsky [1990]; also see Dukelow and Dukelow 1989; Ziegler and Bercovitch 1989). Future studies that combine behavioral, physiological, and skeletal data from living individuals and analyses of their skeletons after death will allow us to match skeletal indicators with real-life experiences in a direct way. At present, we only can try to make these connections in retrospective studies of individuals that make up the few skeletal series including information about sex, age at death, and lifetime behaviors of known individuals (e.g., Rawlins and Kessler 1986a; Lovell 1990b; Stringer et al. 1990; Zihlman et al. 1990).

What We Can Read about Populations, Demography, and Potential Evolutionary Change in Teeth and Bones

Populations and Demography

We can describe and explain skeletal features in relation to individual life stories as an overlay of population, species, and higher-taxa survival and time-based, reproductive life-history characters. Just as teeth and bones are functional parts of living organisms, collections of skeletons sample individuals that once were integral parts of populations. It seems possible to reconstruct some demographic characters of populations from skeletal series.

Information from living populations, of course, is more comprehensive. Skeletal series allow study of only the individuals who died during the period from which the series was assembled and, in addition, only the ones for which (usually incomplete) skeletons are preserved and recovered. Another problem exists: human skeletons from archeological sites and those of nonhuman primates that died of natural causes often include a disproportionate number of individuals that died earlier than an expected species-defined life span. Still, when individuals are analyzed as part of a skeletal series, they can provide a wealth of information about past groups (see Wood et al. [1992] for further discussion).

We count each individual in a skeletal series and—from its teeth and bones—determine its sex, and evaluate its age, health immediately before death, health history, and reproductive potential at the time of death. Morbidity and mortality patterns for a skeletal series can be described using data from these individual life stories. Reproductive profiles, in contrast, are more difficult to determine from only teeth and bones.

Individuals that are subadults at the time of death are assumed to have left no offspring. Adults are presumed to have achieved reproductive maturity. However, whether an individual had produced offspring cannot always be determined from visual inspection and traditional skeletal measurements.

Adult female skeletons meet the demands of nurturing an infant before and after its birth. These reproductive efforts sometimes are preserved in bones. Skeletal biologists have suggested two ways to reconstruct reproductive profiles. These include first, the presence of "pitting" or bone resorption depressions near the pelvic joints (e.g., pubic bones and sacroiliac joints) that may be associated with high estrogen levels, relaxation of ligaments, and joint mobility related to giving birth (Rawlins 1975; Krogman and Iscan 1986; Tague 1988, 1990); and second, low bone mineral content as evidence for bone calcium loss that can be linked to long periods of breast-feeding (Galloway 1988).

Recent studies have tested ideas about what can be read in the pelvic and other bones by using individuals of known sex and female reproductive profiles. Results show that these skeletal markers are not reliable indicators of birth events and may reflect other life experiences (Galloway in press). For example, although more women than men have resorption areas, they do not always occur in human mothers. Resorption areas occur in males and also vary with age in both sexes (Suchey et al. 1979; Tague 1988, 1990). Among other catarrhines, skeletons of free-ranging chimpanzees show that females that were known to have given birth did not have resorption areas associated with the pubic or sacroiliac joints (Morbeck et al. 1992). In addition, skeletons of free-ranging rhesus monkeys from Cayo Santiago show that older individuals have greater degrees of bone resorption (Tague 1990).

Calcium mobilization during breast-feeding diminishes bone mineral content of mothers (and increases calcium in the bones of infants; see above) and may cause changes in strontium/calcium ratios (Sillen and Kavanagh 1982; Galloway 1988). Both older women and female chimpanzees show apparent decrease in bone mass (Sumner et al. 1989). However, other factors such as poor nutrition, decreased locomotor activity and load bearing, physiological changes related to disease and injury, and, finally, decrease in estrogen during menopause in women affect bone mineral status. These body processes operate at the same time. When apparent decrease in calcium content and bone mass are evident, it may be difficult to confirm infant feeding as its only cause, even in well-known individuals.

Although, we cannot evaluate reproductive status with accuracy—even in females—from external features or bone mineral content, recent advances in DNA analyses allow extraction and analysis of genetic information from bone tissue. It now is possible to identify mothers, fathers, and their offspring in a skeletal series (Hagelberg et al. 1991). Such studies, of course, are useful only if these individuals are preserved and if they can be placed in a context of generational time.

Sampling of past groups will always be a problem for demographic studies. Even with "high-tech" methods, we can describe population phenomena only in general terms. Population size, structure, and composition through generational time and possible evolutionary dynamics still must be reconstructed.

Leftover Hard Parts Preserved as Fossils

Mineralized teeth and bones are fossils. They represent once-living individuals after a transition from the biosphere to the lithosphere (i.e., taphonomic processes). Fossil teeth and bones preserve many aspects of physiology, anatomy, and behavior. Paleontologists traditionally have assumed that remains of mineralized individuals they study are representative or "typical" of the species. More importantly, fossils allow us to interpret individual stories as well as species characters. Life stories, in fact, may be responsible for many attributes expressed in now-fossilized elements.

Fossils, when the nature of observed features are understood, provide opportunities for testing hypotheses about when, where, and under what circumstances humans and other organisms evolved. Fossil monkeys, apes, and humans are widely dispersed both geographically and through geologic time. The chances of hard parts "surviving" to become part of the fossil record are slim. A lot can happen both before and during the period that organic materials in teeth and bones are exchanged for minerals (Martill 1991; McKinney 1991). Yet, once we strip away the effects of fossilization processes, we can read fossils in the same way that we read modern skeletons.

Paleoanthropologists interpret "osteobiographies" (Saul and Saul 1989) from teeth and bones. We can determine what an animal could do and did do during its life. But, as shown above, teeth and bones, even of known individuals, tell only part of the life-history story. Most biological, social, and environmental information about individual lives and their groups is lost at the time of death. We must reconstruct details about feeding and social behaviors or growth, development, reproduction, and aging. We must estimate the kind and extent of variation in local populations or species and how individuals and their populations functioned as parts of ecological communities. Fossils are even less informative. Many represent species with no living analogues. We can work out the historical overlays, species-defined characters, and individual life stories only in general ways. We will never know with certainty whether our interpretations are correct.

Paleoanthropologists also try to understand lineages, that is, ancestral-descendant relationships. The geologic, "deep" time of the fossil record provides a context for interpreting hard parts and inferred behaviors that are shared among groups in terms of possible generational or evolutionary sequences. In fact, "virtually all of our knowledge of the evolution of vertebrate morphology, including that of our own species, is based on bones" (Hall 1988: 175).

These vertebrate morphologies (including those of catarrhine primates) provide only small clues about genetic and phylogenetic "family tree" relationships. They are products of an ongoing weaving together of genes and the environment at many levels of biobehavioral organization throughout an individual's whole life. Connections among genes and whole organisms are still elusive (Morbeck chap. 1). For instance, an estimated 1 to 5% of the human genome determine structural proteins. Furthermore, only a small portion of the proteins that make up the cells of the skeletal system and enable it to function seem to code for anatomy, including teeth and bones (Nei 1987).

Direct evidence of DNA or complete proteins that could confirm phylogenetic lineages still are missing or rare in mineralized fossils. Methods for extraction of noncontaminated, organic material for molecular assessment of its evolutionary meaning both in terms of lineages (Pääbo 1993; P. Ross 1992) and lifeways (Sillen 1992) continue to be probed. We still need a more thorough understanding of the living links within life's hierarchy (fig. 9.1, Morbeck chap. 1). This includes better ways to investigate past organisms, especially those represented in the fossil record, and ways to delineate evolutionary processes in order to explain lineages. The time is ripe for integrating molecules and morphology.[5]

Summary: Teeth and Bones Are Life-History Characters

Teeth and bones tell us about an animal's: (1) way of life or survival life-history features, its abilities and environmental interactions including expressed behaviors; (2) sequence and timing of the life cycle, that is, life stages and time-based, reproductive features of growth and mortality; (3) and, as a result of shared features with other groups, probable phylogenetic relationships both in terms of historical anatomical overlays and, eventually, shared molecular histories. Teeth and bones, therefore, document the genetic substrate and the structural effects of biobehavioral processes that allow organisms to survive throughout the life stages and to reproduce during adulthood. Starting with the lifelong biobehavioral roles of teeth and bones of functionally integrated, whole individuals, we can move "down" to investigate, for example, the role of the skeletal system, at the levels of cells, proteins, and molecules. We can aggregate individuals to "move up" to characterize age-sex classes, populations, species, and so on, or "out" to explain environmental interactions of individuals and their groups in a community and ecosystem (Morbeck chap. 1).

Our current challenge is to explain how different intrinsic and extrinsic factors contribute to the skeleton and how these relate to life history and evolution. We can never know *exactly* what happened during the course of catarrhine evolution, but, if we understand life-history characters and how to read them in teeth and bones, we improve our chances of telling reasonable evolutionary stories.

Acknowledgments

Ideas presented here are based on research supported by the Wenner-Gren Foundation, The L. S. B. Leakey Foundation, and the University of Arizona Social and Behavioral Sciences Research Institute and Office of the Vice President for Research. Thanks to A. Galloway, A. Zihlman, P. Martin, and conference participants for clarifying my views on teeth and bones, to J. Underwood, D. Sample, and J. Hoffman for continued support, and to P. Martin and colleagues at the Desert Laboratory (University of Arizona) for providing a wonderful place to think and write.

Notes

1. Data concerning life stages are from human (*Homo sapiens*) growth studies and those of captive common chimpanzees (*Pan troglodytes*) and rhesus macaque monkeys (*Macaca mulatta*). Nonhuman primate data are presented and interpreted by Schultz (1935, 1937, 1940, 1956, 1960, 1969a, b); van Wagenen and Catchpole (1956); Hurme and van Wagenen (1961); Nissen and Riesen (1964); Gavan and Swindler (1966); Kerley (1966); Krogman (1969); Gavan (1971); Michejda (1980); Shigehara (1980); Watts and Gavan (1982); Swindler (1985); Watts (1985a, b, 1986a, b, 1990); Brizzee and Dunlap (1986); Dean (1989); Aiello and Dean (1990); Anemone et al. (1991); Conroy and Mahoney (1991); Schwartz and Langdon (1991); and Siebert and Swindler (1991). Human data are primarily from Tanner (1962, 1978); Krogman and İşcan (1986); Molleson (1986); Weaver (1986); Albright and Brand (1987); İşcan (1989); İşcan and Kennedy (1989); Aiello and Dean (1990); and Hillman (1990). Information about weaning is from Pond (1977) on mammals; Pereira and Altmann (1985) on primates; and Sillen and Smith (1984) on strontium-calcium ratios in humans. Sources centered on aging include Frost (1987, 1991); Riggs and Melton (1988); İşcan (1989);

Finch (1990); Galloway (chap. 10). See Zihlman (chap. 13) and Borgognini Tarli and Repetto (chap. 14) for more on sex differences.

2. Scanning electron microscopy now allows us to see the cyclical physiological processes of tissue formation and maturation. Counting these species-defined structures as "time bands" gives a good estimate of chronological tooth age, that is, how long an individual has owned its skeleton at the time of death, even when we do not know about its life story (Dean 1989; Aiello and Dean 1990; Boyde 1990). However, these "normal" growth cycles represented by time bands first must be distinguished from biomarkers that document interruptions of growth caused by "abnormal" physiological disturbances produced by various life events, for example, birth, disease, malnutrition, or injury (Lukacs 1989; Goodman and Rose 1990). Once incremental markings are identified and counted with accuracy, other skeletal features, including those used to assess biological or skeletal age relative to adult status, can be interpreted in a "real time" framework (for a different view, see Mann et al. 1990; Simpson et al. 1990, 1991, 1992). If time bands are interpreted correctly, chronological ages provide more-reliable information about time-based, reproductive life-history characters, and about age distribution of deaths in an aggregated group (in some cases, breeding populations), and thus, may be used for reconstructing generation length and explaining long-term change through generational time.

3. What I call time-based reproductive life-history characters at species and higher levels of organization (Morbeck chap. 1) are often equated with average adult body weight (or a skeletal substitute for body size) and, more recently, brain size or cranial capacity.

Body weight has been described statistically in terms of a "size-time axis" among primates (Harvey et al. 1987) and in terms of "body-size strategy" in other organisms (Calder 1984; also see Lindstedt and Swain 1988). These statistical analyses, however, become abstractions of functioning animals, their lives, and their environments. Among catarrhines and, perhaps, other closely related groups with shared ancestry in broadly similar environments, ages at birth, weaning, and reproductive maturity, for instance, are not necessarily tied directly to weight or skeletal size (e.g., gibbons vs. gorillas, Zihlman chap. 8). Furthermore, although body weight or size "summarizes" many aspects of a primate's life, functional biology and behavior still must be "dissected" to show what contributes to the expression of life-history characters. Body proportions, that is, distribution of body weight in segments (e.g., head, trunk, limbs), tissue composition within segments (e.g., skin, muscle, bone, fat), and the joints that connect bones and their segments provide a foundation for survival life-history characters that depend on movement and weight-bearing abilities.

4. The timing of the adolescent growth spurt and achievement of menarche in female macaques, chimpanzees, and humans relative to life-history characters recorded in the skeleton provides another example. The timing of the adolescent growth spurt relative to adult skeletal status and to attainment of reproductive maturity could be a species character as implied by E. Watts (1990; Watts and Gavan 1982) or, more likely, a reflection of species' ranges of variation and proximate environmental conditions (Tanner et al. 1990; also see McFarland chap. 12). An inherent range of potential variation among individuals provides "flexible responses" that link anatomy and physiology with behavior and ecology and highlight the mosaic nature of variation among life-history characters as expressed in individual life stories.

5. There continues to be discussion about differences in the pattern and extent of variation as measured in populations at the molecular level (the genetic substrate) and variation expressed at the levels of body systems (especially teeth and bones preserved as fossils) and whole organisms (e.g., Thorne and Wolpoff 1992; Wilson and Cann 1992; Melnick and Hoelzer 1993). Debates center on phylogenetic relations and possible separation times of lineages. Humans (*Homo sapiens*) and chimpanzees (*Pan troglodytes*), for instance, share almost all of their DNA and, in addition, many physiological, anatomical, and behavioral features. But, individuals are easily recognized as different kinds of animals. Molecular and anatomical features are often said to be "decoupled." However, if a decoupling exists, then it is in a disjuncture in scientists' understanding of genetics, ontogeny, mechanisms of evolutionary change, and phylogeny within and among the various levels of biological organization, not in functionally integrated, living organisms (Morbeck 1991a).

10 The Cost of Reproduction and the Evolution of Postmenopausal Osteoporosis

Alison Galloway

HUMANS are unique in experiencing a long period of senescence or old age. Why do some people age rapidly whereas, for others, old age brings the freedom to do as they please with little loss of bodily function or appearance? What is it that allows some to "grow old gracefully," others to appear to resist all aging, whereas others are "old before their time"?

Skeletal deterioration is among the multitude of changes that take place in the aging body. Some women, for instance, quickly lose height and acquire the classic "dowager's hump." Such a woman may no longer engage in her normal activities for fear of falling—a fear tinged with the pain of previous fractures. It may be difficult for her to breathe because of the hunched shape of the spine, and coughing may result in broken ribs. She may limit meals to small amounts, as the chest cavity is hunched against the abdomen. She may move slowly and cautiously and take care to avoid any sudden motions, slick surfaces, or uneven paths. Complete hip replacements are common among such women. Fractures, when they do occur, take a long time to heal and may necessitate long and expensive hospitalization. Hospitalization itself increases the exposure to infection, particularly pneumonia, and can lead to an earlier than anticipated death.

In contrast, other women are active throughout the aging process, playing tennis, gardening, or taking care of grandchildren. In non-Western societies, such women continue the pattern of daily activities that characterized their adult lives with minimal accommodation for their advancing years. Many women may continue the hoeing, harvesting, and carrying tasks commonly performed by women in nonindustrial agricultural societies. For these women, falls are associated with painful bruises but little fear of breakage. Even when fractures do occur, usually from some major trauma, healing is quick and uncomplicated.

The process of aging is highly variable. Each individual brings to her older years a unique history of combined life experiences (i.e., life story; Morbeck chap. 1). These experiences interact with the genetic framework each woman has inherited—genes that can increase or decrease her susceptibility to disease, aging changes, and deterioration. How can these extremes exist even within one cultural group, that is, people who may share much of the same genetic stock? In part, as suggested above, the aging process reflects how a woman's body has functioned throughout her life. The past and present environment, including social, biological, and physical aspects, in which the woman finds herself affects how she can and will cope with the aging process.

This problem of variation in human aging intrigued me as I completed my dissertation research on whether reproductive history contributed to the status of a woman's bone mineral in her postmenopausal years. My results suggested that, indeed, reproductive history was important, but it was lactation rather than pregnancy that had lasting effects. Specifically, breast-feeding for long periods of time with little time before the subsequent birth appeared to be most detrimental (Galloway 1988).

Like many researchers, at the time of conducting this project, my focus was on the immediate situation. I was not yet fascinated by the evolutionary implications. As I was able to distance myself from the work, however, I began to wonder why a woman's bone should be so subject to loss. What could explain the differences in the incidence of postmenopausal osteoporosis between contemporary populations? How much derived from our mammalian and primate heritage, and what was uniquely human? Was there some reason why the ability to mobilize calcium from the skeleton would be beneficial? As I became more familiar with the specifics of bone physiology, which has been rapidly detailed in the last 10 years, some possible answers began to emerge.

An individual's life experiences are determined, in part, by the evolutionary history of the human species. We are mammals and primates and follow these models in the growth and maintenance of our skeleton. Bipedal locomotion, that human characteristic, also affects the way in which the bones are designed to bear weight. Sex-defined patterns of growth, development, sexual maturity, and reproduction form parameters within which variation can occur. Our individually unique combination of genetic inheritance also affects skeletal growth, development, and maintenance. Bone loss is most common in women and particularly acute among those whose ancestry is of Asian or European origin. The rate at which a woman's long bones developed, her physical activity during childhood and adolescence, the speed and magnitude of her adolescent growth in relation to the onset of puberty, the foods that she had eaten at various periods during her childhood, adolescence, and adulthood, the number and timing of her pregnancies, and the rearing of her children all play a role in setting the stage for the condition of her skeleton in her final years. Superimposed on these factors are changes in nutrition and health care. Decreased physical activity also permits degeneration of skeletal integrity.

Hormones control many of these changes, and bone loss itself has hormonal consequences. The interaction among the female sex steroids, other hormones, and bone takes place throughout a woman's lifetime. Because the efficiency with which this hormonal interaction works may affect the lifetime reproductive success of the individual, it should also have strong evolutionary foundations. During most of the reproductive years until menopause, levels of estrogen fluctuate monthly in conjunction with the menstrual cycle. During most of human evolution, however, such repeated cycling was probably unusual and, instead, cycles of pregnancy and lactation were the norm. Hormonal fluctuations during pregnancy and breast-feeding (lactation), as well as the calcium demands for the nourishment of the fetus and young infant, have a major impact on maternal bone metabolism. Therefore, the discussion of the evolution of calcium metabolism in women must center on women's lifetime reproductive activities.

The critical loss of bone, now seen after menopause, became apparent with the evolutionarily recent extension of the human life span. In some women, this bone loss is compounded by other bone-depleting conditions, leading to an increased risk of osteoporotic fractures. However, I argue that reproduction rather than senescence is the key to understanding the evolution of postmenopausal osteoporosis. Postmenopausal bone loss may be the unfortunate result of a long evolutionary history that favored a female's ability to mobilize calcium for lactation. The same physiological trigger that prompts calcium mobilization, the loss of estrogen and progesterone, occurs in menopause, as it does during lactation. However, the hormonal preparation that has allowed for earlier accumulation of calcium during pregnancy is not present nor is the compensating accumulation that occurs at weaning. As a result, there is only unrelieved loss of bone mass.

Functions of Bone: Beyond Support

The skeleton is of obvious importance in any vertebrate animal. The skeleton forms the framework around which the muscles are positioned and makes movement possible. In this capacity, the skeleton withstands and monitors both compressive forces and the bending or tensile forces that occur when we walk, run, or carry loads. Bone responds by alteration of shape and size when these forces are increased or decreased. It experiences extensive remodeling during growth while maintaining its role of structural support.

Throughout the life stages, the function of bone tissue, however, is not limited to simply providing a structure for the body. Bone tissue plays a deeper and more integrated role in the daily function of the body. It forms a reservoir of calcium and other minerals from which to fulfill the body's needs. Calcium is particularly important for a variety of bodily functions. (1) It acts as a critical intracellular messenger. (2) It is vital for nerve function, mediating the release of neurotransmitters that communicate the impulses sent by one nerve to the next. (3) Its presence maintains the electrical potential that permits even the most basic of cell functions. (4) It is also a key component in muscle contraction. (5) In addition, it plays roles in digestion, blood clotting, and a host of other activities.

Because of its importance, ionized calcium, the form needed by the body, is maintained within strict limits in the blood stream (Norman and Litwack 1987). Too much calcium can lead to calcification of soft tissue, whereas too little sends the body into tetany, a response that can lead to death. Binding calcium to other components, primarily blood proteins, keeps calcium reservoirs readily available. When serum calcium is insufficient, blood-borne reservoirs of calcium are first activated. However, bone is broken down to release calcium in emergency situations and to cope with daily needs in excess of these readily accessible resources.

Calcium is acquired from the diet through intestinal absorption. The ability of the body to extract calcium from the food depends on the production of proteins that line the intestinal wall and capture calcium as it passes through the digestive tract. This protein synthesis is hormonally regulated. Once in the blood stream, the calcium is available for use by the body with excess amounts being stored in bone or excreted into the urine. At the kidneys, calcium initially lost to the urine can be reclaimed. Other losses of calcium may occur through the sweat glands, and there is a mandatory loss of calcium from the body associated with digestion.

To coordinate this overall process, an elaborate hormonal transfer system has been developed to move calcium back and forth between bone and the rest of the body (fig. 10.1). As the availability of calcium from other sources increases or diminishes, hormonal regulators correspondingly store or remove calcium from bone by dissolving portions of the bone matrix. This transfers calcium through the primary bone mineral, calcium hydroxyapatite.

Osteoporosis: When Old Bones Break

Osteoporosis is the severe loss of bone tissue, often accompanied by fractures of the bones (fig. 10.2). The incidence of osteoporosis is one of the major health risks for older people and a financial concern in health care management (Cummings 1987; Stini 1990; Riggs 1991). Although osteoporosis is the clinical condition in which the risk of fracture is high, bone loss appears to be a common feature of the aging process in older mammals. With the loss of bone tissue, the bones are less able to resist the compressive and tensile forces to which they are subjected through normal use. The most common sites of breakage in humans are the vertebrae, proximal femur (hip), distal radius (Colles' fracture of the wrist), and proximal humerus. These areas are particularly prone because the bipedal posture of humans places much of the weight along the spine and hips (Zihlman chap. 13). Breaks in the arms usually result from bracing oneself during a fall.

The social impact of osteoporosis is also profound. People afflicted with weakened bones are at high risk for fractures, often attributable to little or no traumatic event. For many, this

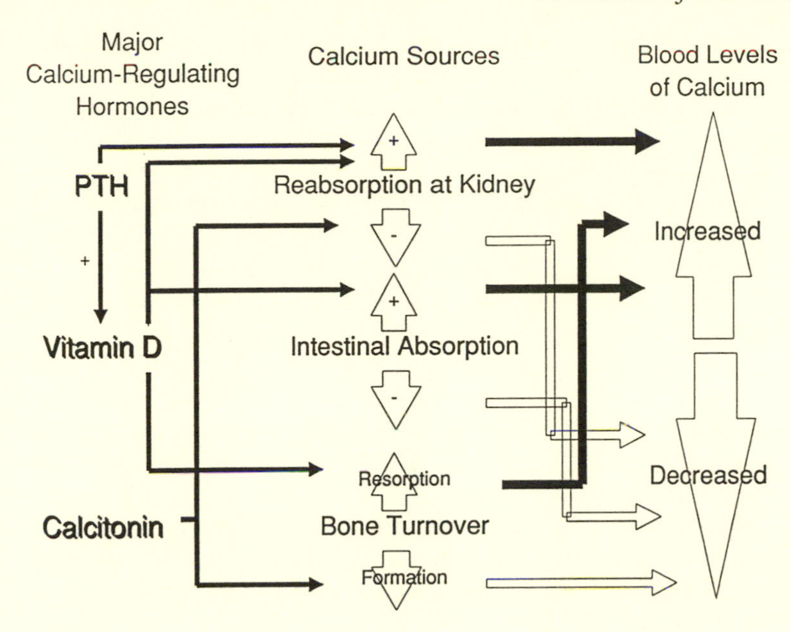

FIGURE 10.1. Major calcium-regulating hormones; the sites of the activity and the responses in the levels of calcium in the blood are shown.

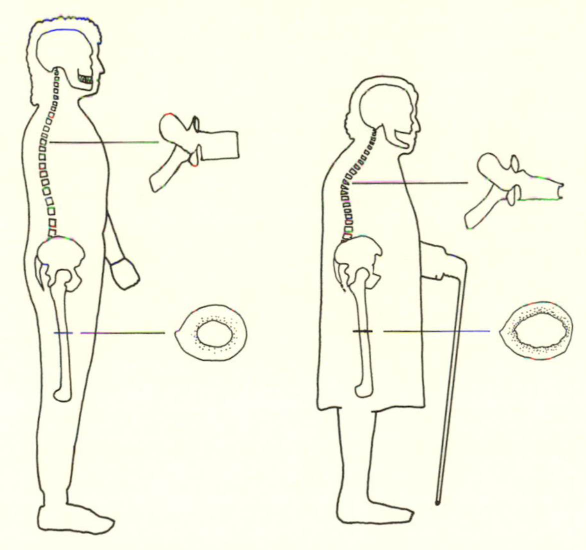

FIGURE 10.2. Differences in height, posture, form of the vertebrae, and the cross-section of the cortical bones between a premenopausal woman and a woman suffering from the combined effects of menopausal and senile osteoporosis.

means living in constant fear of future breaks, limiting physical activity, and, possibly, being confined to a nursing home. The mortality rates associated with some of these fractures are also considerable.

Although the separation cannot be sharply drawn, two major expressions of the processes leading to the clinical condition of osteoporosis are commonly recognized (Riggs and Melton 1983, 1986; Stini 1990; Riggs 1991). These have been termed "senile" and "postmenopausal" bone loss. Senile or age-related bone loss is found in all older mammals and has been identified in numerous primates (DeRousseau 1985; Sumner et al. 1989; Zihlman et al. 1990). It is found in both sexes and may, in part, be due to changes in activity patterns in older individuals as well as to changes in hormonal control over calcium metabolism.

Postmenopausal bone loss is a human characteristic seen only in women. It is a period of rapid resorption that dissolves both the organic and inorganic components of bone, releasing calcium to the blood stream. This resorption occurs particularly from the interior of the bones and within spongy bone, following menopause (fig. 10.2). During their lifetimes, women do not, on average, develop bone tissue to the extent that is found in men, so this additional period of bone loss is particularly detrimental.

Research on the causes of osteoporosis has attacked many different potential sources of dysfunction in the calcium metabolism of older humans. No single pathological change can account for all instances of this disease. Furthermore, the complexity of the hormonal control of calcium metabolism would argue against a simple cause. Two specific avenues of interest are the impaired ability to attain bone mineral during growth and early adulthood and the accelerated rate of loss in the later years. It is likely that both processes contribute to the development of osteoporosis.

The delicacy of the control over this critically important aspect of our physiology would argue for long and strong selective pressure during evolution. Therefore, we must view calcium metabolism and bone tissue maintenance within the context of mammalian evolution. Here I discuss the hormonal changes seen during pregnancy, lactation, and menopause and draw parallels between the adjustments made in bone tissue at these different life stages. Within the framework of this evolutionary approach, a female's ability to gain access to calcium quickly and easily can be seen as a "cost" of successful reproduction, one that has little detrimental effect until it is reactivated, in humans, in the postmenopausal years.

Hormones of Calcium Regulation

Blood calcium levels and extraction of calcium from intestines, urine, and bone are regulated by a complex of hormonal factors. The principal hormonal regulators are parathyroid hormone, various forms of vitamin D, and calcitonin (Garel 1987). In females, estrogen and, probably, progesterone play a major role in maintaining skeletal integrity (Horsman et al. 1977; C. C. Johnston et al. 1985; Christiansen et al. 1987). Fluctuations in the sex steroids appear in concert with alterations in the levels of the major calcium regulatory hormones.

Major Calcium Regulatory Hormones

Parathyroid hormone (PTH) responds to decreases in serum calcium. It stimulates an increase in the cells that resorb bone tissue, releasing calcium (Garel 1987). It limits bone formation and prompts retraction of the cells that form a protective layer over bone, exposing the bone to resorptive activities. Calcium reabsorption at the kidneys also appears to be increased by PTH, and it promotes the synthesis of vitamin D at the kidney.

Vitamin D is synthesized in a three-stage process at the skin, liver, and kidney (Garel 1987). The most potent form, ($1{,}25(OH)_2D$) can also be absorbed from the diet. The primary function of vitamin D is to increase intestinal absorption of calcium by prompting the production of calcium-binding proteins. In bone, vitamin D allows normal calcification by increasing the amount of calcium available in the circulation and, along with PTH, appears to increase the scale of bone resorption.

Calcitonin (CT) is released in response to increases in serum calcium (Garel 1987). Its primary role is to limit these increases, which normally follow digestion, by increasing excretion of calcium into the urine. It also actively inhibits bone resorption (Chambers 1982).

Female Sex Steroids

The female reproductive cycle is largely governed by two steroid groups, the estrogens and the progesterones, which are, in turn, governed by hormones secreted by the pituitary. Both estrogens and progesterones appear to interact with the calcium-regulating hormones and affect bone formation and resorption. Bone turnover normally functions as a carefully orchestrated cycle of resorption and formation, usually with net loss of bone in maturity. The combination of estrogen and progesterone may work to uncouple this cycle in such a way as to preserve or increase bone mass.

Estrogen's effects include an increase in bone and collagen formation and an increase in absorption and retention of calcium (Silberberg and Silberberg 1976). Estrogen is associated with a reduction in PTH (Lindsay 1987) and may inhibit the sensitivity of bone to that hormone (Heaney 1965; Nordin et al. 1976). Estrogen is also linked to an increase in vitamin D metabolism (Riis and Christiansen

1984). Some studies have linked estrogen medications with increased secretion of calcitonin (Morimoto et al. 1980), but others have found that estrogen had no effect (Hurley et al. 1986). Calcium absorption increases in the presence of estrogen (Nordin et al. 1976, 1985; Riis and Christiansen 1984).

Progesterone, the other major female sex steroid, has been credited with increasing bone mass in postmenopausal women (Christiansen et al. 1985). Experiments show that progesterone will increase bone formation through proliferation of osteoblasts (Tremollieres et al. 1992).

Calcium and Hormonal Changes during Reproduction

In mammals, the process of reproduction has been divided into that of gestation and lactation. Because the young depend on the mother for nutritional support both before and after birth, reproduction must be viewed as including *both* aspects. The investment in the first portion (pregnancy) without fullfillment of the succeeding portion will not result in successful reproduction. Only in humans do infants commonly survive even though their mothers have never nursed them at the breast.

During pregnancy, the maternal calcium supplies must support fetal bone development while allowing the continued survival of the mother, all without seriously compromising her future health. In humans, as in most primates, pregnancy itself does not exert overwhelming demands on the maternal skeleton. The average human newborn skeleton contains 25–30 grams of calcium, compared to approximately 1200 grams in an adult skeleton (Christiansen et al. 1976; Gertner et al. 1986; Garel 1987). About 80% of this is accumulated in the last trimester.

Pregnancy alters the blood calcium. Ionized calcium levels are usually maintained without significant fluctuation, but there is a small decrease in the total calcium, partly because of a reduction in the proteins to which they bind (Pitkin et al. 1979; Conforti et al. 1980; Colussi et al. 1987). Some studies show a slight reduction in calcium ions at mid-second trimester, corresponding to the increase in demand for the fetal skeleton (Drake et al. 1979). Because calcium readily crosses the placenta and binds within the fetal blood, the calcium concentration in the fetal blood rises above that of the maternal circulation (Reitz et al. 1977).

Hormonal changes result in much greater calcium absorptive ability of the intestines among mammals. This is true of humans as well (Kumar et al. 1979), even when dietary calcium is low (Shenolikar 1970; Heaney and Skillman 1971). Intestinal absorption rises by the second trimester.

Calcium also can be reclaimed from the urine, but this does not appear to be a major factor in human pregnancy. Urinary losses appear to increase during pregnancy (Heaney and Skillman 1971; King et al. 1992), probably because of an increase in urine production, although calcium losses may fall in the third trimester (Shenolikar 1970).

After birth, during early lactation, calcium levels in the woman's blood are usually below nonpregnant levels, but they rise slowly and within 6 months they are at normal, prepregnancy levels (Greer et al. 1982b). This rise may be due to the return of normal levels of blood protein to which the calcium can bind. Calcium content in milk is kept in a constant ratio with the phosphorus content to enhance the availability of the calcium in the infant's digestion (Akre 1989). In humans, from the time of childbirth through the first 3 months of lactation, the milk calcium levels increase (Greer et al. 1982a; Karra et al. 1988). After that, they gradually decline, partly dependent on maternal calcium intake. Calcium losses into milk of lactating women are about 220–340 mg per day during the first 3 months. Since this must be combined with normal calcium losses, the total loss is about 400 mg/day (Hillman et al. 1981).

Animal models show that maximization of calcium accumulation during lactation is a strong mammalian characteristic (Garel 1987). It is particularly apparent in rapidly growing, large-littered species. In contrast, with humans,

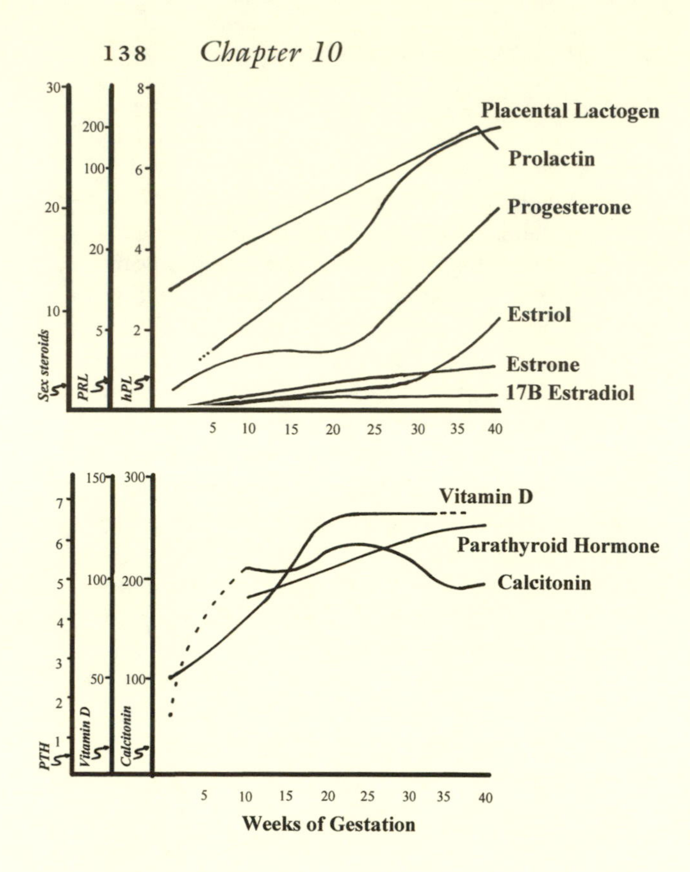

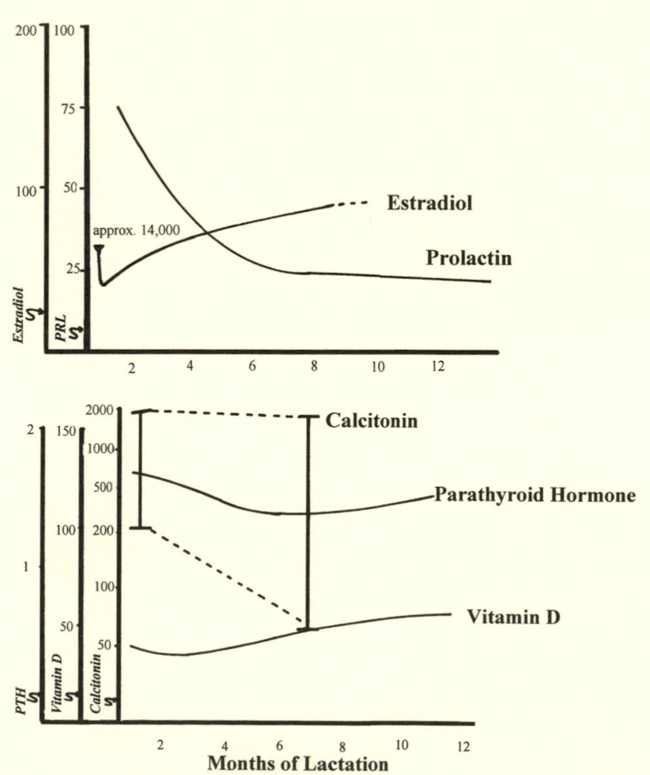

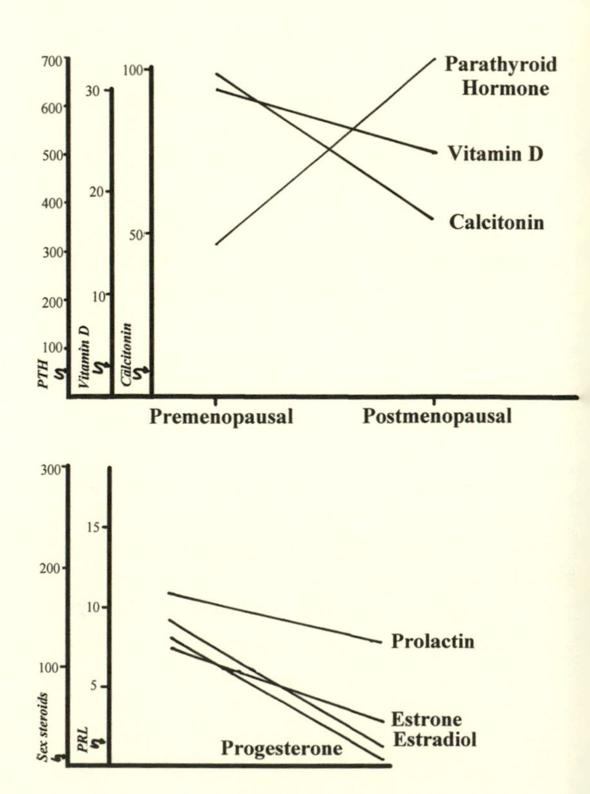

FIGURE 10.3. Comparison of the hormonal changes during pregnancy and lactation with those following the menopause.

intestinal absorption declines after parturition and does not appear to stabilize substantially above nonreproducing levels in early lactation (Kent et al. 1991).

Urinary excretion of calcium is reduced during lactation in mammals (Huq et al. 1988; Kent et al. 1990). This is secondary to a drop in urine production, although it is probable that some other hormonal change is also involved (King et al. 1992).

During the reproductive cycle, significant changes in hormonal production occur (fig. 10.3a). During pregnancy, the placenta, whose primary purpose is transfer of nutrients between maternal and fetal blood, also has an important role as a source of additional hormones. Hormones are also generated in an increasing number of other locations, including the fetal adrenals and liver supplementing the changing production at the maternal ovaries, adrenals, and liver (Norman and Litwack 1987). Both fetal production of hormones and alterations in maternal production affect maternal calcium metabolism. In addition to the sex hormones, other hormonal changes regulate the movement of calcium from the diet, into and out of bone, into fetal circulation, and into breast milk. The major hormonal factors include the three calcium-regulating hormones, placental lactogen, and prolactin.

Estrogen and Progesterone

Estrogen levels are elevated during the pregnancy, especially as the third trimester begins (Norman and Litwack 1987). Just before parturition, estrogen levels drop rapidly. Following the birth, estrogen will eventually increase and, over time, the menstrual cycles will emerge. The onset of this transition appears affected by the species-defined parameters and by the intensity and duration of lactation (Vitzthum chap. 18).

Progesterone also shows fluctations during pregnancy and lactation. Progesterone continues to be secreted until just before parturition, at which time levels fall, slightly preceding those of estrogen. This fall initiates the birth contractions.

Parathyroid Hormone

In mammals, parathyroid hormone (PTH) fluctuates during reproduction. Some women exhibit a reduction in PTH during pregnancy, although it remains within the normal range (Cushard et al. 1972; Whitehead et al. 1981; King et al. 1992). This drop is most pronounced during the twentieth to thirtieth week of pregnancy and may be followed by a sharp rise until PTH is twice normal levels (Cushard et al. 1972; Conforti et al. 1980). Although not completely explained by fluctuations in serum calcium, this rise does coincide with the greater calcium demand by the growing fetus. At birth, the levels begin to fall.

During lactation there appears to be great variability in PTH levels. Indeed some animal studies indicate that this hormone is not even needed to mobilize bone at this time (Hodnett et al. 1992), while others suggest that enhanced dietary calcium absorption during lactation may be secondary to increases in PTH (Garel 1987). Human studies, however, suggest that PTH levels quickly return to normal following birth even if the woman is breastfeeding (Huq et al. 1988; Kent et al. 1990). A variation on the mammalian pattern may arise in humans and, possibly, other slowly maturing primates, where PTH may modulate vitamin D levels in prolonged lactation (Cushard et al. 1972; Specker et al. 1991).

Vitamin D

An increase in the active forms of vitamin D has been reported early in pregnancy in most mammals. Both the placenta (Whitsett et al. 1981) and the fetus (Wieland et al. 1980) play a role in increasing vitamin D synthesis. In humans, this increase is seen in the first trimester, well before there is a demand for calcium by the fetal skeleton, and remains high throughout the pregnancy (Kumar et al. 1979; Whitehead et al. 1981; Breslau and Zerwekh 1986; Gertner et al. 1986). This could bring about an increase in the intestinal tract proteins that capture and help transport calcium into the circulation. It also could enhance calcium reabsorption by the kidneys. A complication is that there is also an increase in vitamin D-binding proteins that could bind most of the active forms maintaining the bioactive portion at roughly prepregnancy levels (King et al. 1992). This continues until the last 4 weeks of pregnancy (Bouillon 1983; Gray 1983; Bikle et al. 1984).

Animal studies have shown dramatic increases in calcium absorption during nursing—up to six times greater than usual (Toverud et al. 1976)—which have frequently been attributed to increased vitamin D activity. Such an increase continuing into the lactation period has been reported in some (Kumar et al. 1979) but not all human studies (Hillman et al. 1981; Greer et al. 1982b), raising doubts as to its role in calcium absorption for breast-feeding. In humans where the lactation period is prolonged over years, vitamin D may play a major role in maintaining access to calcium. Greer and colleagues (1982b) found an increase in vitamin D levels after about 6 months of lactation; the increase continued in the small number of women in their study who nursed for 1 year. Similar findings were made by Specker and colleagues (1991).

Calcitonin

Animal studies suggest that levels of calcitonin, the third major hormonal regulator of bone, are substantially higher during pregnancy. Calcitonin during pregnancy may inhibit the resorption of bone while still permitting PTH and vitamin D to activate absorption of dietary calcium and kidney reabsorption (Lewis et al. 1971).

Studies of pregnant women suggest that, while still within the normal range, calcitonin levels are significantly higher (Drake et al. 1979; Stevenson et al. 1979; Wieland et al. 1980; Kovarik et al. 1980; Whitehead et al. 1981). This increase is variable and may be seen in only about 50% of pregnant women (Pitkin et al. 1979). These increased levels appear early and do not differ substantially through the latter two-thirds of pregnancy. This increase may partly be in response to the enhanced calcium absorption, which, if uncontrolled, would dangerously increase blood levels of calcium. Calcitonin is actively transported to and retained by the fetus (Samaan et al. 1975), suggesting that this may form part of the fetal calcium-capturing mechanism.

During lactation, animal model studies suggest that increases in calcitonin may provide a protective mechanism for maternal bone calcium (Toverud et al. 1976, 1978) in the face of higher prolactin, PTH, and vitamin D levels that might otherwise lead to bone loss. In humans, results have been more debatable. Hillman and colleagues (1981) and Greer and his colleagues (1982b) reported calcitonin levels in lactating women comparable to those in nonlactating women. Toverud and Boass (1979), however, were more frequently able to obtain measurable levels of calcitonin in lactating women than in controls. The confusion over measuring calcitonin in nursing mothers may be due to the possible addition of breast tissue itself as a source of calcitonin (Bucht et al. 1986).

Placental Lactogen

Placental lactogen increases throughout pregnancy but is confined to the maternal circulation. Here it may enhance bone-formation processes. Animal studies have shown that periods of enhanced bone formation coincide well with placental lactogens (Miller et al. 1986). This hormone falls rapidly after birth and is not a factor during lactation.

Prolactin

Lactation is initiated by prolactin, which rises sharply during pregnancy. This helps prepare the breast tissue for the production of milk, but the full effects are limited by the presence of high levels of estrogen and progesterone. An important consequence of increased prolactin during pregnancy may be increased calcium absorption at the intestine. Prolactin helps stimulate synthesis of the active forms of vitamin D (Spanos et al. 1976; Lund and Selnes 1979; Pike et al. 1979; Robinson et al. 1982). Animal studies have shown, however, that prolactin can stimulate intestinal calcium absorption, despite vitamin D deficiencies (Gray 1983; Barlet 1985).

Prolactin continues to be secreted as long as regular suckling occurs (Vitzthum chap. 18). According to results from animal studies (Pahuja and DeLuca 1981; Miller et al. 1982), high levels of prolactin produced during breast-feeding are expected to deplete maternal bone mineral. As weaning approaches, prolactin levels diminish, lessening the body's ability to mobilize calcium from bone. It may be at this time in humans that the PTH and vitamin D complex take over some of the tasks of releasing calcium for breast milk (Cushard et al. 1972).

Effects of Reproduction on Bone Metabolism

In mammals, calcium demands of the fetus and of milk production are met by a combination of enhanced intestinal absorption, renal conservation, and bone resorption. Since the needs and rate of transfer for different species vary, so does the reliance on each of these different avenues of acquisition (Garel 1987).

Pregnancy appears to be associated with small but measurable increases in bone formation. In humans, this is supported by biochemical indicators of bone formation that increase even early in pregnancy (Valenzuela et al. 1987). By the third trimester, however, it is possible that the maternal skeleton is being mobilized to supplement the dietary sources. Again, biochemical indicators support this concept of increased maternal bone turnover in the terminal stages of pregnancy. Direct assessment of bone is more difficult, but some studies suggest preferential loss of bone from trabecular sources such as the vertebrae (Cann 1989).

Bone resorption is more prevalent during mammalian lactation. Losses in nonprimates can be significant, amounting to 15–25% loss of the calcium from thigh bones in rats (Garel 1987). In humans, the scale of transfer of calcium from bone to milk is difficult to measure, but, as with the losses in late pregnancy, the extraction appears to occur preferentially from trabecular bone (Hayslip et al. 1989; Kent et al. 1990; Sowers et al. 1993). Biochemical indicators are consistent with high rates of bone turnover during lacation (Huq et al. 1988; King et al. 1992).

After weaning, the mother experiences a period of bone accretion. In humans, this process appears to take place over a few months to years (Cann 1989; Kent et al. 1990; Sowers et al. 1993). Increased PTH levels after weaning may continue urinary conservation of calcium while the restored estrogen levels would inhibit PTH-stimulated bone resorption. Insufficient periods between pregnancy and lactation cycles for recovery of bone may be responsible for overall lower bone mineral status (Galloway 1988).

The hormonal stimulus behind these changes in bone is open to debate. The calcium-regulating homones do not seem to be the primary mechanism of control. Instead the shift in the hormonal spectrum specifically related to reproduction is the more likely regulator. Of these hormones, prolactin has attracted considerable attention with its known ability to stimulate increased intestinal absorption and, also, mobilization of calcium from bone (Pahuja and Deluca 1981).

Estrogen is critical to the transfer of calcium between mother and offspring in mammals. The absence of this sex steroid may facilitate the bone resorption mediated by the reduced amounts of PTH and vitamin D and by prolactin (King et al. 1992). In the presence of estrogen, these slight changes in the calcium-regulating hormones may be inconsequential, but, without it, they can have significant consequences.

The importance of the ability to mobilize calcium during pregnancy and lactation lies in the flexibility it affords the mother in her reproductive capacities. As Pond (1977) has asserted, lactation allows the infant and female to occupy the same habitat. By drawing on the maternal reserves of fat and calcium among other ingredients, lactation also allows the mother to continue her reproductive efforts despite temporary interruptions in her food resources. This flexibility in calcium mobilization also may respond to the differing needs of the mother herself.

The extent of the variation in this complex can be seen in a brief examination of some factors that produce modulations in calcium acquisition during reproduction. The scale of these changes is more pronounced in mammals that produce many rapidly developing young or young that need high levels of mobility quickly after birth. In humans, difficulties arise, not because of the rate of transfer which, following the primate model, is relatively slow, but because lactation traditionally lasts 2 to 3 years per child. During that time, the mother must cope with many unpredictable demands on her reserves.

Low Calcium in the Diet

If the mother has a low-calcium diet, nursing proves a major drain on her resources. Regardless of maternal calcium intake, the calcium content of the milk is largely maintained at normal levels. When the mammalian mother cannot supply adequate calcium for milk production, the quantity of milk decreases (Rasmussen 1986). In humans, it is likely that the maternal skeleton provides an excellent reservoir for calcium to maintain both milk quality and quantity (Moser et al. 1988). Without this reserve, the growth period of the child would be prolonged as its sources of nutrition were lessened. The period of breast-feeding would be extended until the child was large enough to subsist on solid foods. Longer lactation and slowed infant growth would increase the interbirth interval, lowering the lifetime reproduction of the mother.

Maternal nutrition during lactation may be partially responsible for the relative maintenance of bone mineral status in humans. Women whose calcium intake is high and who maintain adequate synthesis of vitamin D are less likely to lose substantial bone mineral and more likely to recover quickly from any losses following weaning. A lactating woman may alter her calcium absorption by increasing food consumption and selecting calcium-rich foods (Brommage and DeLuca 1985). In addition, structural changes in the digestive tract, including increases in total gastrointestinal size and activity, may also change calcium absorption and are seen in association with lactation (Toverud and Boass 1979). These may not completely compensate for lactational losses, however, and subsequent pregnancies may mean accumulated reductions in bone mass.

Prolonged and Intense Lactation

The duration and intensity of lactation may also be a factor. Among women who nursed over 6 months, bone mass, especially at more trabecular sites, was lowered (Wardlaw and Pike 1986; Sowers et al. 1993). Women with higher calcium intake during lactation did not exhibit effects as severe. Postmenopausal women appear to continue to show the effects of lactational requirements for mobilization of calcium, particularly if their breast-feeding has been prolonged or the pregnancy/lactation cycles were in rapid succession (Galloway 1988).

Calcium and Hormonal Changes with Menopause

The above discussion shows that, even in humans, the hormonal changes seen during pregnancy and nursing are critical for maintaining adequate supplies of calcium to support and nurture an infant and to allow the mother to continue activities necessary for her survival. Bone becomes an important auxiliary source of calcium during reproduction. This results, however, in costs that are paid by the mother in depletion of bone mass. A large part of these costs are, in effect, a "loan," recovered after the child is weaned.

In women, menopause signals the termination of reproductive function. Because of the close relationship necessary between reproductive and survival functions in all mammals, this event cannot be viewed as exclusive to one bodily system. Changes in the levels of sex hormones trigger massive changes in calcium and bone metabolism.

Postmenopausally, total serum and urinary calcium increases (Eisenberg 1969; Sokoll and Dawson-Hughes 1989), and almost all women are unable to replenish the calcium normally lost daily (Heaney et al. 1977; Nordin et al. 1979; Heaney and Recker 1986). Much of this increase is a consequence of bone resorption. The mechanisms that maintain the tight limits on serum-ionized calcium must stimulate an increase in urinary excretion. There is a gradual reduction in the total serum calcium with age, but as in pregnancy, this is, at least partly due to a reduction in the blood proteins to which calcium can be bound. At the same time, there is a decline in calcium absorption related to inadequate production of calcium-binding proteins in the intestines (Armbrecht 1989).

A woman's bone undergoes substantial alteration at the end of the reproductive years (Riggs and Melton 1986; Stini 1990; Riggs 1991). This period is characterized by appreciable loss of cortical bone—the dense bone that makes up the shafts of the long bones—lasting for 8 to 10 years. Although of shorter duration, the intensity of trabecular or spongy bone loss is approximately three times greater. Trabecular bone occurs in large proportions of the interior in the ends of the long bones and in the spine. These same regions are subject to more rapid turnover during pregnancy and lactation. The rapid postmenopausal trabecular loss is associated with heightened risk of vertebral fractures, often leading to "dowager's hump" (Melton and Cummings 1987).

During this period, there are also substantial changes in the woman's hormonal profile (fig. 10.3b). Estrogen and progesterone levels fluctuate during the perimenopausal period and fall following menopause. The loss of estrogen is associated with the precipitous reductions in bone mass (Richelson et al. 1984; Ruegsegger et al. 1984) that have been shown to correlate with circulating estradiol levels (Horsman et al. 1977; C. C. Johnston et al. 1985; Riis et al. 1986; Christiansen et al. 1987). This phenomenon has led to the separate diagnosis of postmenopausal osteoporosis in contrast to, and overlying, senile osteoporosis (Riggs and Melton 1983, 1986), as discussed earlier.

Along with changes in the levels of the female sex steroids, there also are changes in the levels of calcium-regulating hormones with age. When the most active forms of PTH are examined, most elderly show no significant increase. In contrast, people confined indoors and those diagnosed as osteoporotics often had elevated levels (Wiske et al. 1979; Saphier et al. 1987). This was not found in all elderly who suffer from severe bone loss (Milhaud et al. 1978), suggesting that PTH response varies individually and may correspond to the severity of the bone loss.

Older individuals may have lower serum levels of the various forms of vitamin D even though they may be within the normal range (Fujisawa et al. 1976; Parfitt et al. 1982; Riggs and Nelson 1985; Hordon and Peacock 1987). Vitamin D supplementation has been shown to increase calcium absorption in osteoporotics dramatically (Caniggia et al. 1984). Other research has indicated that some osteoporotic females may have serum vitamin D levels much lower than age-matched subjects (Riggs and Nelson 1985), suggesting that, for at least some osteoporotics, an impairment of vitamin D activity may be partly to blame for their bone loss.

Calcitonin appears to decrease progressively with age in both sexes. Comparison of menstruating women and those undergoing early menopause show that the latter group has significantly lower calcitonin levels (Corghi et al. 1984; Perez Cano et al. 1989). Some studies suggest that calcitonin secretion in postmenopausal osteoporotics is not deficient in the active forms (Tiegs et al. 1985), although others report that some women are no longer responsive to calcitonin therapy (Wallach 1987).

Prolactin is a normal component in the circulation of nonlactating women, although not at the levels that occur during lactation. Following menopause, prolactin drops in parallel to estrogenic secretion (Ben-David and L'Hermite 1976).

Parallel Bone Mobilization during Reproduction and Menopause

In general, pregnancy and lactation alter the normal physiological responses that modulate calcium levels in the circulation. The addition of prolactin and placental lactogens and their effects on bone resorption and formation also cause a major change in calcium regulation. These adjustments meet the rise in estrogen and progesterone, the current requirements of the fetus, and the need for a functioning skeletal system for the mother. The result is a maximization of intestinal calcium absorption. The onset of this complex hormonal interaction appears to be triggered, not by the fetal demand for calcium, but by the changes in the hormonal controls over pregnancy. Since the needs of the fetus for calcium are relatively low during most of the pregnancy, much of this increased calcium is stored. Late in pregnancy, as the fetal demands escalate, calcium must be moved across the placenta where it is captured by the fetal circulation.

After the child is born, the rapid transfer of calcium to milk production during early lactation causes hypocalcemia, a lowering of the calcium present in the blood. The normal calcium mobilization responses to restore the calcium levels are enhanced by the presence of prolactin. Prolactin may be more influential during early lactation, whereas PTH and vitamin D become important closer to weaning. These combined changes greatly enhance intestinal calcium absorption of the mother through increased synthesis of activated vitamin D. To compensate for the huge influx of calcium that will then occur following a meal, calcitonin may increase to place an upper limit on the blood calcium. This has the additional effect of helping to protect the maternal skeleton. As weaning nears, suckling is less frequent and prolonged as the infant's diet is supplemented by other foods, resulting in a reduction of the prolactin levels.

During lactation and at menopause, the levels of estrogen and progesterone drop, loosening the restrictions on mobilization of calcium from bone. The body appears to read the relaxation of the restrictive forces of these two hormones as a signal that calcium is needed from storage. Prolactin does not increase at menopause as it does during lactation, so the hormonal avenue of calcium acquisition from the intestine is limited. Instead, the body resorts to "borrowing" calcium from bone as it did during lactation. Unfortunately, this is a debt that cannot be repaid.

The calcium-regulating hormones also may change at menopause and with advancing age. Parathyroid hormone does increase in many women prone to osteoporosis after menopause, paralleling the increases seen in prolonged lactation. This emphasizes the perceived need for calcium, increasing the levels in the blood and the demand for calcium mobilization from bone. Since vitamin D production is not enhanced, the second avenue for intestinal absorption is seriously limited. Calcitonin levels may be maintained at premenopausal levels or may even drop, depriving the woman's skeleton of the added protection provided during lactation.

The most accessible target for calcium acquisition is bone tissue. This raises the blood levels of calcium, which must then be removed to prevent inappropriate calcification. The body responds with increased urinary excretion of calcium. In essence, menopause results from the same stimulus as lactation, the cessation of estrogen and progesterone secretion, without laying the groundwork for providing alternative sources of calcium or having the possibility of future recovery.

Why Humans Suffer from Postmenopausal Osteoporosis

Why, at this time, are women faced with severe bone loss that results in the clinical condition of postmenopausal osteoporosis? A major factor must be the expansion of the life span beyond the reproductive years, which exposes this otherwise hidden cost of reproduction. In most primates, reproduction continues throughout life (Pavelka chap. 7). In humans, postreproductive life is greatly extended. This may be a relatively recent feature in hominid evolution, probably appearing with the advent of anatomically modern humans. Until this last century, however, average life expectancy did exceed the reproductive span.

The selective forces that would shape health in old age have only had a short period in which to exert any pressure. Old age can, therefore, be viewed as the "by-product" of the forces that affect younger individuals (Washburn 1981; Kirkwood 1985). Survival while the body is in decline necessitates freedom from predators and access to suitable nutrition (Cutler 1976, 1981). Although there are some selective advantages for life beyond the reproductive years, such as grandmothering effects, these are minor compared to those in force during earlier life stages.

In contrast to the weak selective forces that have acted on the postreproductive human, those acting on the efficiency of the reproductive event are much stronger. During most of the course of human evolution, and as still occurs in many parts of the world, women spent much of their lives involved in reproduction, especially in prolonged lactation. For the infant, breast milk alone provides the means of support and is still a major food source to the child, often through the second and even the third year of life. With long lactation comes lactational amenorrhea and an increase in interbirth intervals (Vitzthum chap. 18). This phenomenon is particularly apparent if maternal nutrition is deficient (Prentice and Whitehead 1987). At these times, infant growth may also be slowed. Fewer pregnancies and lower survival rates for offspring are likely to result, lowering the lifetime reproductive success for each woman.

Strong selective forces have been at work throughout mammalian evolution to allow for the mobilization of calcium from bone during lactation. When calcium can be obtained only through diet, any dietary deficiencies would be directly reflected in a reduction in milk yield. Therefore, the availability of an alternative source of calcium, primarily from the maternal skeleton, would be a major factor in maintaining the levels of milk needed to sustain the infant. Without such yields, the infant would not only be delayed in development but would also be susceptible to much higher rates of mortality than the well-nourished infant. Extension of the infant and juvenile stages in primates reduces the rate of calcium transfer via milk but increases the possibility that the mother will encounter dietary deficiencies during the time she is nursing an infant. The time and effort invested in each offspring is also greater so that the mother cannot afford to abandon her young at such times. The long period of dependency of human infants increases the need for maintaining adequate calcium sources across a range of seasonal changes and through periods of unanticipated shortages.

The selective forces acting on the ability to mobilize calcium to produce and maintain milk yield would, therefore, have had a greater role in determining female bone physiology than the possibility of forestalling osteoporotic fractures during senescence. The rapid and overwhelming response to increase calcium availability in the face of declining dietary sources during the period of lactation would be critical. The abrupt decline in estrogen/progesterone at parturition signals activation of this enhanced sensitivity. A similar decrease in estrogen and progesterone, although less abrupt, occurs at the termination of the reproductive years at menopause. Again, the body responds by dropping its guard over the calcium reserves, allowing them to be mobilized at an accelerated rate. The negative effects of this response have only been evident for a relatively short period in the evolution of our species, and it is unlikely that these effects would outweigh the benefits of calcium mobilization for milk production, before the development of nonendogenous milk sources.

Summary

The process of calcium mobilization during reproduction may play a central role in understanding the evolution of female bone metabolism. Both lactation and menopause are periods in which estrogen and progesterone levels are rapidly reduced and cyclic secretion is halted. In both cases, calcium is mobilized. In lactation, however, hormonal compensation provides the means of acquiring calcium from the diet and after weaning. In menopause, these options are not enhanced. Furthermore, during lactation there may be hormonal protection of the maternal skeleton limiting the depletion of this source of calcium. This feature is also lacking in menopause.

The ability to mobilize calcium may be critical to the ability to reproduce successfully. Without adequate calcium, the amounts of milk produced may be insufficient to sustain the infant, even though the milk is within the normal range for calcium content. Therefore, it is likely that selective forces would favor a hormonal response that greatly enhances calcium accumulation and mobilization. In lactation, this hormonal complex is designed to enhance the exploitation of dietary calcium, with utilization of the calcium reservoirs in bone implemented when calcium absorption proves insufficient. In menopause, the fine-tuning of the hormonal response is lost because of the lack of hormonal and anatomical changes of a preceding pregnancy, the mechanism of hormonal control, and the absence of a recovery period. The resorption of bone becomes the primary and chronic response. Postmenopausal osteoporosis therefore may be the negative side effect of successful adjustment to maintaining adequate milk yield despite uncertain dietary calcium by resorting to calcium stores in the bone.

As women, our bodies are shaped by an evolutionary history that has emphasized the features that allow us to reproduce successfully. The unfortunate costs of these adaptations may be apparent only when we enter that part of our lives in which selective forces have had little opportunity to work for more beneficial adjustments. If postmenopausal bone loss is to be included as a "cost of reproduction," then the expenses are high. The positive side is, however, that the condition may not be a pathological one requiring a "cure." We may carry within our hormonal systems the ability to readjust our bone metabolism to maintain skeletal integrity after the close of our reproductive years. The key will be to understand the entire process within the context of our evolutionary history.

Acknowledgments

I would like to thank those who helped in the development and production of this paper. Dr. Patricia Hoyer attempted, with limited success at the time, to interest and familiarize me with the function of the endocrine system and its control over bone. To her I owe considerably more than I realized at the time. My initial interests in reproduction and bone arose from the work I did for Dr. William A. Stini, to whom I also owe thanks. Dr. Mary Ellen Morbeck has been a critical editor and wonderful colleague, and I am and will be eternally grateful for her assistance and patience throughout the years. I would also like to thank Charles Miksicek for allowing me to discuss this stuff interminably, and my daughter for bringing it all to life.

11 The Biological Origins of Adipose Tissue in Humans

Caroline M. Pond

THE CAPACITY to store nutrients, minerals, and protein, as well as energy, has been one of the major factors in the evolutionary success of mammals. Its critical role in both survival and reproduction is also part of primate and human evolutionary history. The major storage tissue in vertebrates, fat or adipose tissue, is unique to this group of animals. It normally occurs as one or a few pairs of locations, called depots, in the abdomen or, in a few fish species and in many amphibians and reptiles, the tail. Its main functions in these animals are first, to store energy during winter cold or summer droughts in order to survive when food is scarce, and second, to allow rapid synthesis of large, yolky eggs in the breeding season.

Mammals need large quantities of stored energy because they are endothermic and must expend energy to maintain their body temperatures within a narrow range. They also depend on stored energy, for example, during hibernation and during temporary food shortages. Mammalian adipose tissue also has a different and, in many ways, more demanding role in reproduction, especially lactation. Mammalian eggs always are very small and few in number, rarely more than about 20 per litter.

The energy requirements of pregnancy are not very great, mainly because mammalian fetuses, particularly those of primates, grow quite slowly and most species (except humans) are lean at birth. The main energetic cost of reproduction is lactation or milk production after the birth of an infant when newborns grow quickly and accumulate fat very rapidly. Females of many species meet the demands of lactation by eating more or higher-quality food with adipose tissue serving only as a supplement and reserve. Reproduction is often timed so that the young are born when newly grown grass or a glut of insects can provide abundant food for lactating mothers. A few species, for example, seals, whales, and polar bears, meet much of the requirements of lactation from energy stored in adipose tissue (Reiter chap. 4). For these species, and to a lesser extent for mammals in general, adipose tissue plays a central role in reproduction. Before giving birth, up to 50% of the mother's body mass can be fat, and she loses it rapidly during lactation.

Modern Western humans are obese continuously, not just while breeding. People classified medically as "normal" are much fatter than most wild mammals, and severe obesity is common. In spite of the popularity of dietary and exercise regimes aimed at modifying the abundance and arrangement of adipose tissue, there is little understanding of the "optimum" body composition and shape.

Biochemists and psychologists sometimes ask in despair, "What is the right amount of fat for a normal human?" In any population in which diet and habits recently have undergone major changes, "average" body dimensions may not correspond to "optimal" or "natural" values. The few remaining groups of human hunter-gatherers have been strongly influenced by contact with other cultures and may not be typical of the prehistoric condition. Modern medical opinion almost unanimously agrees that severe obesity is harmful (Krotkiewski et al. 1983; Björntorp 1987). Although most students of human adipose tissue (e.g., Bouchard and Johnston 1988) assume that its condition in modern populations is abnormal, there is no clear exposition of how its structure or arrangement differs from that of other mammals, or when, in evolution, the deviations arose. As Tanner and Whitehouse (1975:142) stated, after more than 20 years of measuring superficial adipose tissue in children, "standards represent what is, not what ought to be." Studies of wild animals with different natural diets and capacity for exercise help to define the "ideal" human body composition. This chapter is about wild animals as well as humans, particularly those that naturally become obese at some time in their lives and remain healthy and active. Studies of such "professional fatties" help to understand, and perhaps avoid, many of the adverse consequences of obesity.

The prevalence of obesity among all humans has led to the notion that human fat is atypical, but it is impossible to be precise about how human adipose tissue differs from that of other mammals because there is so little scientific information about the comparative anatomy of adipose tissue. There are few studies of the natural functions of adipose tissue in humans or other primates. Consequently, there are few objective criteria to establish which features of human adipose tissue are unique and which are shared with other species (see McFarland chap. 12).

This chapter describes the anatomy of human adipose tissue compared to that of wild and domesticated mammals and reviews the theories that have been proposed to explain exceptional features of adipose tissue in humans. A thorough understanding of the distribution and abundance of adipose tissue in nonhuman mammals, including primates, helps to assess whether there really are fundamental anatomical or functional differences of adipose tissue in humans.

Background

I began my studies in biology by studying control mechanisms in insect flight and other aspects of invertebrate neurobiology. I was teaching general and evolutionary biology in the mid-1970s and became interested in energy storage as part of the evolution of reproductive strategies and its ecological consequences (Pond 1977). Comparative anatomy of storage tissues seemed to be relatively unknown. This impression was confirmed when I took a temporary job teaching gross anatomy at a veterinary school. Many of the elderly dogs and horses that we dissected were fat. In fact, there was so much adipose tissue in some individuals that we had trouble finding the "real" anatomical structures (i.e., nerves, blood vessels, and so on). It seemed strange that the textbooks of anatomy devoted only a few sentences to describing what can be one of the most abundant tissues in the body. Anatomists clearly dismissed adipose tissue as "having no anatomy." True, while many specimens were fat, others, such as stray dogs and old milking cows, had almost no visible adipose tissue. This variation among specimens in the amount of fat, compared to less variable bones and blood vessels, made it difficult for anatomists to identify the important aspects of adipose tissue. I wondered, was this attitude really fair or does it amount to a case of tissue chauvinism?

I scoured the literature for information about the anatomy of adipose tissue in vertebrates and synthesized it as best I could (Pond 1978). The task was not easy. Nearly all authors were too coy to mention "adipose tissue" in the title. Most of the information was tucked away in the fine print under discussions of methods. Above all, there was no universally recognized nomenclature to describe adipose depots. Authors invent their own terminology; reading the literature was like discussing urination and defecation with elementary school children. A wide range of almost synonymous terms exist. Many are imprecise and of uncertain etymological derivation and have different meanings for different people.

When I returned to England in 1979, I decided to look at fat in greater detail. I settled in Milton Keynes, a new city that has been under construction for the last 25 years. The land was used previously for sheep ranging and fruit farming with small areas of uncultivated woodland. As new roads were built through farm land, many wild mammals (e.g., rabbits, hedgehogs, badgers, deer, and foxes) became victims of accidents. I started to collect these casualties in order to examine their adipose tissue. Bus drivers and delivery people would inform me of the locations of fresh cadavers. Fortunately for my research, the main kind of cell in adipose tissue, the adipocyte, is robust and retains its basic size and shape for many hours after death.

I measured, with the assistance of Christine Mattacks, the gross mass and adipocyte volume of adipose tissue in as many depots as could be found in these wild animals. We also measured adipose depots in laboratory rodents and a variety of zoo specimens. Our research "breakthrough" came when we dissected a guinea pig, a wallaby, and a fox on the same day. Comparing the anatomical arrangement of the adipose tissue and the site-specific adipocyte volume of these different mammalian species allowed us to see a consistent anatomical pattern. The observations formed the basis of studies of a wide range of different species, including humans (Pond and Mattacks 1985a,b; Pond et al. 1993a,b; Pond 1986, 1987, 1991).

Facts about Fat: Comparative Anatomy of Adipose Tissue

Body shape seems to have been of significance to prehistoric humans, as it is to modern people. Many aspects of body shape arise from the abundance and anatomical arrangement of adipose tissue. Although 35,000 years is a very short time in evolutionary terms, artifacts and drawings of the human body dating from the Pleistocene Ice Age to modern times (fig. 11.1) tell us about the prehistoric human form and, perhaps, even more importantly, how people viewed themselves. Adipose tissue is represented conspicuously in only a minority of the thousands of prehistoric representations of the human form that have been described, so there is no reason to suppose that the obesity on the scale represented in Palaeolithic figure found in northern Europe (fig. 11.1a,b) was typical of the population as a whole.

On the other hand, such images are unlikely to be purely imaginary: the artists had probably seen at least a few people with this body shape. Furthermore, the conspicuous and sometimes exaggerated representations of the adipose masses suggest that such body conformations had social significance to the population in which they occurred and may indicate that they were venerated or desired. Shattock (1909) described similar enlargements of relatively small, but visually conspicuous adipose depots that were attributed to sexual selection. How do these selectively enlarged depots compare to the distribution of adipose in other mammals? How did the human pattern of fatness evolve? Are any of the exceptional features of humans part of our primate heritage?

Mammals

The distribution and thus probably the functions of adipose tissue in mammals (and birds, which also control their body temperature internally) differ from those of the ectothermic vertebrates (i.e., fish, amphibians, and reptiles). Adipose tissue occurs in a dozen or more discrete depots and are associated with several viscera and skeletal muscle (fig. 11.2). Contrary to the impression established in textbooks, there usually is no continuous layer of "subcutaneous" adipose tissue in most species. Site-specific differences in relative adipocyte volume, biochemical properties, and anatomical relations to other tissues characterize distinct depots (Pond and Mattacks 1985a; Pond 1986; Pond et al. 1992b, 1993a,b, 1994, 1995).

Examination of more than 200 specimens from more than 50 species showed consistent features of the anatomy of mammalian adipose tissue (fig. 11.2) (Pond and Mattacks 1985a, b, 1988, 1989; Pond 1986, 1987). We found that the principal difference between species is the relative abundance of adipocytes in the various depots. Homologous depots can be

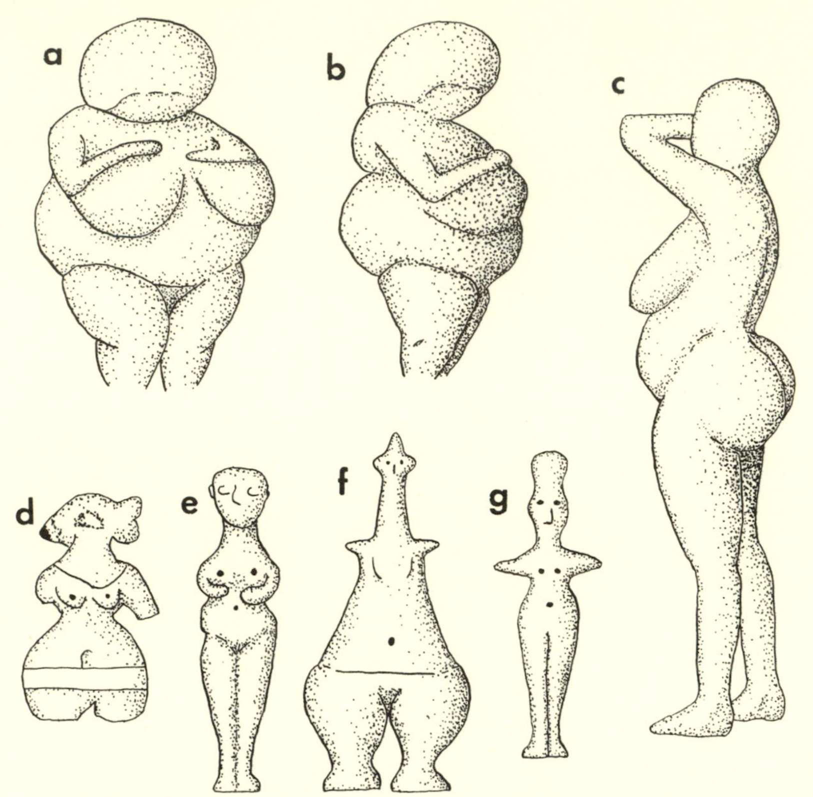

FIGURE 11.1. Representations of the female body: *(a)* front view and *(b)* side view of a small female figurine of Palaeolithic age from northwestern Europe; *(c)* Hottentot woman; *(d)* figurine from Bronze Age India; *(e–g)* various figures from Late Bronze Age of Eastern Europe.

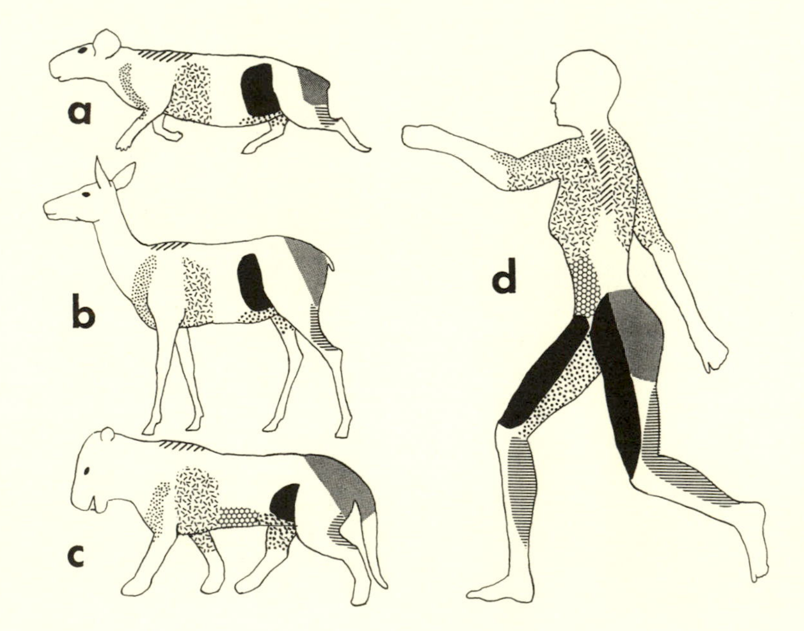

FIGURE 11.2. The principal superficial adipose depots in *(a)* guinea pig, *(b)* deer, *(c)* tiger, and *(d)* human. The specimens are drawn in postures that show as much of the body area as possible. Homologous depots have the same style of shading. Shading indicates the relative area, not the relative volume of the depots.

identified in different species, regardless of the body composition of the individual species examined. Furthermore, the ratios of volumes of adipocytes in recognized depots remain remarkably constant over a wide range of species (Pond and Mattacks 1989).

Studies of metabolic activity (e.g., in vivo measurements of glycolytic enzymes and of lipid turnover) reveal site-specific differences in the physiological properties of different depots. These differences are found in homologous depots of all mammals and are characteristics of each anatomical location of the adipose depot, rather than of the size of the adipocyte (Mattacks et al. 1987; Pond and Mattacks 1987a,b, Pond et al. 1992a; Pond 1992b). Thus, although relative adipocyte volume provides a convenient means of identifying the depots, we do not yet understand its functional significance.

Primates

This general mammalian plan is also present in primates. Our investigations show that macaque monkeys (*Macaca*) have the usual mammalian depots, and their cellular dimensions and biochemical properties are similar to those of other mammals (Lewis et al. 1983; Pond and Mattacks 1987b).

Marmosets, macaques (Pond and Mattacks 1987b), lemurs (Pereira and Pond 1994), humans (Pond et al. 1993a), and probably all other primates, however, have an additional "paunch" depot on the outer surface of the abdominal wall (fig. 11.2c, d). In lean monkeys and lemurs, it forms a shallow narrow band along the midline, but it becomes quite massive in obese specimens, extending to the sternum and to the crest of the hip bone on the side. This depot seems to accumulate fat selectively. It may be minimal (less than 5% of the total adipose tissue) in lean but not emaciated specimens. In contrast, in obese humans and captive primates, it may be large enough to overlie and obscure partially the typical depots while the other depots expand more slowly with increasing fatness. That this paunch depot is often massive in lemurs (Pereira and Pond 1994), comprising up to half of all adipose tissue in obese specimens, suggests that it is an early and basic feature of primate anatomy that humans and other higher primates have inherited almost unchanged (Pond 1991, 1992a). Many Carnivora have a small amount of adipose tissue (less than 5% of the total) at the homologous site, but there is no evidence for its selective enlargement, even in species that naturally become massively obese in the wild (Pond et al. 1992a, 1994, 1995).

In all species for which there are sufficient data, we always found a fair amount of inter-individual variation in the relative masses of depots that cannot be attributed to age, sex, or lean body mass, even in genetically homogenous wild populations (Pond et al. 1992b, 1993b, 1994, 1995, Pond 1994) and in primates bred and maintained under carefully controlled laboratory conditions (e.g., Pond and Mattacks 1987b; Kemnitz et al. 1989; Jayo et al. 1993). Such "fat patterning" in humans (Björntorp et al. 1988, Bouchard 1990) and captive monkeys (Shively et al. 1987; Kemnitz et al. 1989) has been studied with almost obsessive enthusiasm for many years. In spite of the vast quantities of detailed data obtained, attempts to relate body conformation to normal and pathological metabolic properties have, on the whole, been disappointing.

As a comparative biologist, I am not surprised by this failure: the partitioning of adipose tissue between depots may be inherently variable in humans and other primates, as it seems to be in naturally obese wild animals (Pond 1994). Naturally obese arctic mammals live in a tough environment where natural selection is stringent: 90% of arctic foxes on Svalbard die in their first year of life, before they are old enough to breed, but variability in the distribution of adipose tissue persists. Like the enlargements or reductions of superficial adipose depots that evolved among some human populations (Shattock 1909), such features have a negligible impact on Darwinian fitness. So far as we can tell, wild arctic foxes, reindeer, or wolverines are not disadvantaged by having a bit more adipose tissue here and a bit less there, so why should it matter in humans?

The incidence of obesity has been thoroughly investigated in captive colonies of primates and can be compared to other mammals. Only 5 of the 873 adult female pig-tailed macaques examined by Walike and colleagues (1977) were obese; all the obese animals were female and at least 8 years old. About 10% of the captive macaque monkeys (*Macaca fascicularis*) examined by Pond and Mattacks (1987b) and by Laber-Laird and coworkers (1991) were obese. The fatness of half of the captive lemurs examined by Pereira and Pond (1994) was also within the range of values measured in wild mammals. Some carnivores also become obese in captivity (Pond and Ramsay 1992), and arctic species are particularly prone to obesity for a large part of the year (Pond et al. 1992a, 1994, 1995). Looking at the "big picture" of comparative obesity, the most one can say is that primates might be slightly more susceptible to obesity than any other group of mammals when living on an artificial regime of diet and exercise.

The distribution of adipose tissue in obese monkeys apparently is more variable than in obese individuals of other species. Both Pond and Mattacks (1987b) and Laber-Laird and colleagues (1991) reported selective accumulation of intra-abdominal fat and massive growth of superficial depots maintained under similar conditions. Unfortunately, there is no detailed information about the anatomy and distribution of adipose tissue in ape species that are more closely related to humans. So the question of how and when the exceptional features of hominid and, later, human adipose tissue evolved remains open to speculation (Pond 1991).

Anatomical Organization of Adipose Tissue

Site-specific growth of adipose tissue has been measured thoroughly by anthropologists and physicians (Harrison 1985; Bouchard and Johnston 1988). However, functions of different depots and the biology of sex and populational differences are not well understood. Since humans cannot be investigated experimentally, the first step to elucidating the evolutionary origins and physiological implications of adipose tissue is to compare data from humans with information from other species.

Figure 11.2 shows that the distribution of fat depots is similar in many mammals, including humans. Species differ in the relative development of the fat depots, and the extent of each depot shown here is typical for well-nourished, but not obese, individuals. Both the thickness and the area covered by adipose tissue increase with increasing fatness. The bipedal posture and associated limb proportions affect the arrangement of the adipose tissue. Thus, humans have extended hips and long thigh segments, so the groin depots of typical mammals cover much of the hip and thigh to form a subcutaneous layer of adipose tissue that extends from the pelvis to the knee. In fat individuals, the depots become massive and terminate abruptly just above the knee, as is especially noticeable in obese women (fig. 11.1 a, b, f).

There is a bilaterally symmetrical depot on the anterior lateral thorax extending along the upper arm to the back of the elbow. It is slight in species such as deer (fig. 11.2b), but extensive in rodents (fig. 11.2a) and carnivores (fig. 11.2c). The basic anatomy is the same in primates, but, in humans, the upper arm segment is relatively long and the trunk is flattened from front to back. Consequently, this fat depot extends over the front, side, and back surfaces of the chest and, in addition, over the back and inside aspects of the upper arm (fig. 11.2d). It forms the subscapular site on the upper back (i.e., below the shoulder blade) and is associated with the mammary glands on the front of the chest. Thus, in quadrupedal mammals, this depot forms a condensed mass on the side of the trunk but in humans appears as an extensive subcutaneous layer of adipose tissue over the chest and upper arm. In men, the parts of this fat depot are approximately similar in thickness. In women, the upper chest portion of this depot may be enlarged and forms the adipose tissue component of the breasts.

The adipose tissue on the back side of the upper arm and at the subscapular site are among the easiest places to measure fat as part of skinfold thickness. Anthropologists traditionally record these sites as independent measures of fatness (e.g., Garn et al. 1987). However, comparative anatomy shows that these sample sites are parts of the same fat depot. It is no wonder that measures of their thickness correlate so much more highly with each other than with other depots (Bouchard 1987; Garn et al. 1987).

There are more examples of homologies between fat depots in humans and other mammals that are obscured by differences in the arrangement of the muscles and skeleton. For instance, the pair of depots that form large human buttocks are homologous with the usually small depots around the back part of the pelvis and the tail in other mammals. In humans, the anatomy of the fat depot behind the knee, called the popliteal depot, is also the same as in other mammals. But, again, because of differences in the arrangement of the joints and muscles, the adipose depots appear to be different in bipedal humans and quadrupedal mammals. In four-legged, running mammals, unlike humans, the knee remains flexed. The three flexor muscles insert along the leg and cover the calf muscle along most of its length. The popliteal depot is bounded by the major locomotory muscles of the hindlimb. It almost always is enclosed between the muscles; only a small area contacts the skin (fig. 11.2 a,b,c). In humans, on the other hand, thigh muscles that flex the knee attach just below the knee joint. This change in the site of insertion of flexor muscles means that the fat depot behind the knee extends from the back of the thigh to the ankle. Because the calf muscles that move the ankle and foot are covered with skin, part of the popliteal depot appears to be subcutaneous although its anatomical relations are still like those of other mammals.

Like other intermuscular adipose tissue, the popliteal depot may serve as a ready source of energy for the adjacent musculature. The relatively large fat cells have an unusually high capacity for taking up glucose and for the breakdown and resynthesis of triacylglycoids, properties that increase with exercise (Mattacks et al. 1987; Pond and Mattacks 1987a; Mattacks and Pond 1988; Pond et al. 1993a).

The basic features of human adipose tissue anatomy, therefore, are similar to those of other primates and mammals. Changes in posture and thus in trunk and limb proportions in humans make it appear that adipose tissue covers a greater part of the body than is the case in most other mammals. Because adjoining depots overlap each other in moderately obese people, the majority of individuals studied in modern, Western populations appear to have an almost continuous layer of subcutaneous adipose tissue. In fact, in fat individuals, it is difficult to distinguish the depots both by palpation or medical imaging technology in living subjects and during cadaver dissection.

Abundance of Adipocytes in Adipose Tissue

Severe obesity is common among modern Western humans. But, are the adipocytes too numerous, or is each one too large? How many adipocytes is too many adipocytes? How big an adipocyte is too big? Such questions are very difficult to answer especially since the health of most individuals of the population may not be physiologically optimal, and some may have recognizable disorders of energy metabolism.

To answer these questions, measurements from wild animals are combined to generate allometric equations that relate adipocyte number to body mass (Pond and Mattacks 1985b). Using such equations, we can calculate the expected number of adipocytes for any animal of known body size and compare our predictions with observations of real animals. This comparison shows that "normal" humans of both sexes have 3 to 30 times more adipocytes than expected for a wild mammal of the same size. This increase in adipocyte complement is much greater than that of other mammals in even

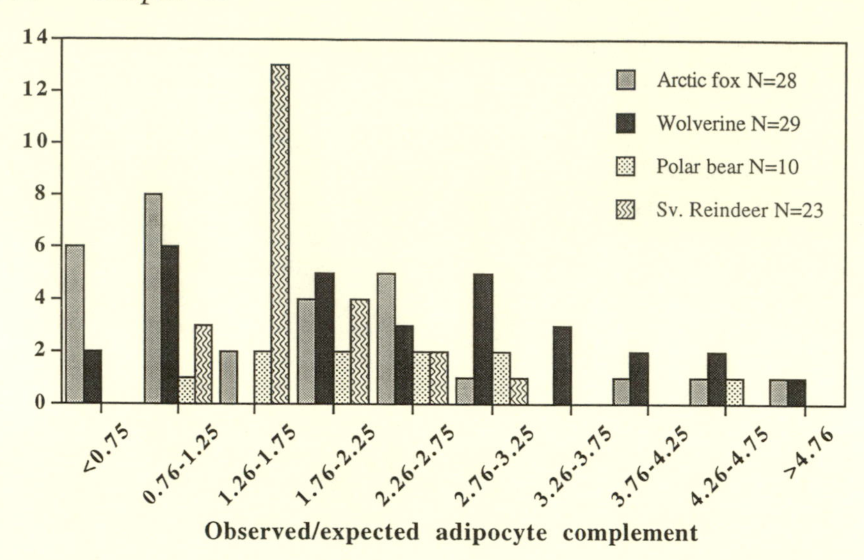

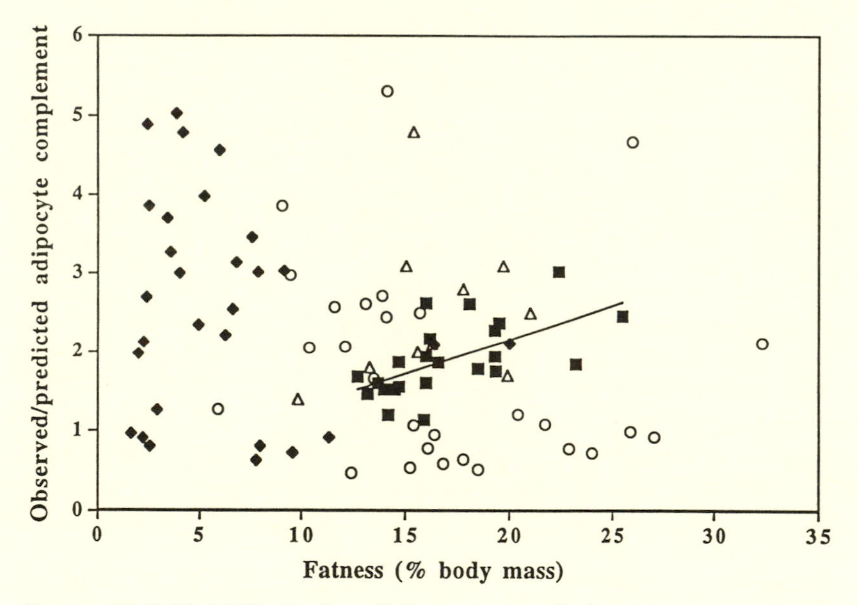

FIGURE 11.3. Variability in the cellular structure of adipose tissue in naturally obese wild mammals. *Top*, Histogram of the frequency of different ratios of observed adipocyte complement to that calculated from equations derived from temperate-zone and tropical mammals (Pond and Mattacks 1985b) for 90 arctic mammals. *Bottom*, The relationship between fatness (mass of all dissectible adipose tissue as a percentage of total body mass) and the ratio of observed adipocyte complement to that calculated from equations derived from temperate-zone and tropical mammals for the same specimens. *Circles*: arctic fox *(Alopex lagopus)*, $N = 28$, N.S. Triangles: polar bears *(Ursus maritimus)*, $N = 10$, N.S. *Closed diamonds*: wolverines *(Gulo gulo)*, $N = 29$, N.S. *Closed squares and regression line*: Svalbard reindeer *(Rangifer tarandus platyrhynchus)*, $N = 23$, $r^2 = 0.375$, $P < 0.01$. Data from Pond et al. (1992, 1993b, 1994, 1995).

naturally obese species such as pigs, camels, whales, and bears. The adipocyte complement of Western people is extremely variable and does not correlate with fatness (Sjöström and Björntorp 1974) or the tendency to become obese in middle age (Sjöström 1981). This variability among humans was long thought to be a degenerate, possibly pathological consequence of modern lifestyle, but, in fact, comparable, equally unexplainable variabilities are seen in naturally obese mammals (Pond 1994).

Some data for four naturally obese arctic species (arctic fox, wolverine, polar bear, and Svalbard reindeer) are shown in fig. 11.3. The reindeer have only two to three times more adipocytes than expected from the data from temperate-zone and tropical mammals (fig. 11.3, top), although all specimens were about as fat as young humans (fig. 11.3, bottom). The data for the three species of Carnivora are much more variable. Some arctic foxes and wolverines have fewer than the expected number of adipocytes, whereas a few others have five times as many. At the time they were collected, there was no consistent relationship between adipocyte complement and fatness (fig. 11.3, bottom). Thus, in these naturally obese wild animals, whether adipose tissue consists of numerous small cells or fewer, larger ones does not seem to be among the major influences on appetite, hunting ability, or metabolic efficiency. Reversible "fattening," such as that of pregnant female polar bears before they enter the breeding den, is largely due to enlargement of adipocytes. Where both lean and fat tissues are growing at similar rates, however, increases in the number of cells seem to predominate so that average cell size remains approximately constant (Ramsay et al. 1992).

Compared to those of wild animals, human adipocytes are smaller than expected (and coincidentally about the same size as those of laboratory rats). One reason for the exceptional abundance of relatively small adipocytes in modern humans may be that the mode of growth of primate adipose tissue is atypical. Adipocyte proliferation seems to contribute much more to expansion of adipose tissue in captive macaque monkeys than is the case in laboratory rodents. Growth of adipose tissue in adult humans may be similar.

As in humans (Sjöström 1988), the correlation between total mass of adipose tissue and mean adipocyte volume in monkeys is weak or statistically insignificant (Pond and Mattacks 1987b). In other words, the microanatomy and mechanisms of growth of human adipose tissue are much more like that of other primates and wild mammals than that of rats raised under uniform conditions. Studies of laboratory rodents can elucidate the basic biochemical mechanisms, but they are unhelpful, even misleading, for explaining the cellular structure and gross anatomy.

Functions of Adipose Tissue

The primary function of adipose tissue is as a long-term energy store. In this role, its mass varies greatly, sometimes over a short time. In a swimming, walking or running, or flying animal, an added mass causes minimum disruption to balance if it is located at or near the center of the body. The lipid stores of many fishes, amphibians, and reptiles, for example, are concentrated in a few localized depots that usually are situated inside the abdomen and near the center of the body mass (Pond 1978).

In contrast, adipose tissue in mammals (and birds) occurs in discrete depots in association with several different organs, including viscera, skeletal muscle, and the skin. Because of its distribution, adipose tissue is thought to have acquired additional functions of insulation or support of other organs. These functions would require wide distribution over the body surface as well as around "delicate" organs. In fact, both these roles are to some extent incompatible with the original role of adipose tissue as an energy store.

Insulation

The idea that superficial adipose tissue insulates the body against heat loss and protects underlying tissues from mechanical damage is widely stated as fact in biological literature, almost always without supporting evidence (e.g., Wood and Bladon 1985; Brown and Konner 1987; Williams 1990). However, in spite of its widespread acceptance, there are few data that support the insulation theory, even in the case of semiaquatic mammals.

Trapped air and stagnant water are effective thermal insulators, and the formation of a stagnant layer of air or water around the body through the development of a thick fur pelt is the principal mechanism of heat retention in many animals. In contrast, tests of adipose tissue from human cadavers show that it has only a little less than half the specific conductivity of muscle or stagnant water (Hatfield and Pugh 1951), and numerous observations on mammals native to cold climates suggest that adipose tissue's contribution to insulation may be minimal compared to that of fur.

Insulative capacity of fat is proportional to its thickness. The adipose tissue should be thickest over the parts of the body most vulnerable to cooling, and the superficial layer should be selectively spared during fasting. But Ashwell et al. (1986) found that, in obese women on a severely restricted diet, the superficial and intra-abdominal depots were depleted at the same rate.

The relative thicknesses of the superficial depots in humans are not consistent with a role as insulation in air or in water. The back, head, and neck are much more exposed to the elements during both walking upright and swimming than are the front of the trunk or the inner surface of the thigh, yet superficial adipose tissue is thickest over the abdominal wall (paunch) and, in women, as part of the breast adipose tissue and on the upper parts of the limbs (fig. 11.2d). There is no need to postulate some special adaptive process to account for this aspect of the distribution of adipose tissue.

Studies of human populations living in arctic and temperate zones show that the greatest differences are in the proportions of the skeleton and musculature, not in the distribution or abundance of adipose tissue (Johnston et al. 1982). Eskimos pursing a traditional lifestyle and diet, in fact, have less subcutaneous adipose tissue than do Canadians of European descent. Laboratory studies of young men also show that heat loss in cold water depends more on body proportions and muscle development than on skin-fold thickness (Toner et al. 1986). Because infants are smaller and more sedentary, one might expect adipose tissue to be more important as an insulator. But the ability of neonates to maintain a constant body temperature correlates more closely with the lean body mass than with the thickness of their superficial adipose tissue (F. E. Johnston et al. 1985).

Allometric Scaling of Adipose Tissue

There is an alternative explanation for the presence of thick superficial adipose tissue in large semiaquatic carnivores. I compared the masses of superficial and intra-abdominal adipose tissue in various wild and zoo-bred carnivores that I have dissected over the last 10 years with those from some wild polar bears, shot by native Eskimo hunters, that we examined in 1988 (Pond and Ramsay 1992). The superficial adipose tissue increased isometrically with with body mass, but the mass of the intra-abdominal depots scale regularly as lean body mass$^{0.74}$, that is, it becomes proportionately smaller in larger mammals. Figure 11.4 shows a similar allometric comparison integrating Pond and Ramsay's (1992) data with many similar measurements from arctic foxes (Pond et al. 1995) and wolverines (Pond et al. 1994). Once again, there is a fair amount of interindividual variation in the partitioning of adipose tissue between internal and superficial depots even in these genetically uniform populations, but overall the mass of superficial depots increases in simple proportion to body mass. The coefficient of allometry relating mass of the intra-abdominal depots to lean body mass becomes slightly higher than that fitted to the smaller data set (Pond and Ramsay 1992) but still very significantly less than unity.

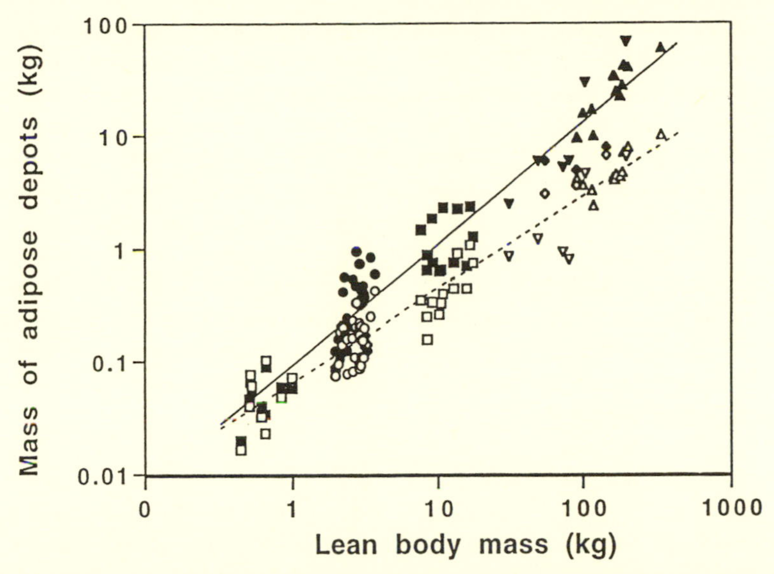

FIGURE 11.4. The mass of all superficial (closed symbols and solid line) and all intra-abdominal (open symbols and dotted line) adipose tissue as a function of total body mass in 70 adult and subadult Carnivora over 8% fatness. Squares: Mustelidae: (9 ferrets, 6 European badgers, 6 wolverines); circles: Canidae (22 arctic foxes); diamonds: Felidae (5 domestic cats; 1 jaguar; 1 tiger; 1 lion); upright triangles: 13 polar bears; inverted triangles: 6 brown bears. The regression lines shown are fitted to all data except those from polar bears. Superficial depots α (lean BM)$^{1.05}$, $r^2 = 0.885$. Intra-abdominal depots α (lean BM)$^{0.807}$, $r^2 = 0.875$. Redrawn from data in Pond and Ramsay (1992) and Pond et al. (1994, 1995).

In small mammals such as ferrets, about half the adipose tissue is internal and half superficial, although in all species studied, the proportion in superficial depots increases with increasing fatness. The proportion inside the abdomen decreases with increasing body size (independent of fatness) so that in large bears, less than 17% of all adipose tissue is internal. The body surface area scales as body mass $^{0.66}$, that is, it becomes proportionately smaller with increasing body mass. With a small area to cover, the superficial adipose depots are thicker in larger animals (Pond 1992b), even if the specimens are not fatter (i.e., if the ratio of the masses of lean and fat tissues is constant). The superficial adipose tissue is 10 times thicker in a 500-kg bear than in a 0.5-kg rat from this effect alone, quite apart from the fact that a greater proportion of the adipose tissue is superficial, and bears are often fatter than rats.

The measurements from the wild polar bears match the regression lines fitted to the data from tropical and temperate-zone carnivores, so these data provide no evidence for any change in the distribution of adipose tissue in arctic polar bears compared to those from warmer climates: the partitioning of adipose tissue between internal and superficial sites arises entirely from the fact that polar bears are very large and become very obese. It was not necessary to invoke a net shift in the partitioning of adipose tissue between internal and superficial depots to explain the thick layers of subcutaneous fat in polar bears.

This effect is particularly clear-cut among the Carnivora—possibly because the abdominal cavity is relatively small in these mammals, with the liver and guts each amounting to only about 2% of the body mass in large species such as polar bears (Pond and Ramsay 1992)—but it also happens in wild arctic reindeer (Pond et al. 1993b). In such species, an apparent change in the distribution of adipose tissue arises as a direct consequence of an increase in its *abundance*, and in the change in body proportions that arises from increasing body mass, and probably has nothing to do with thermal insulation. In humans, the preponderance of superficial adipose tissue may have arisen in the same way as it did among these large carnivores. Anthropologists have reached similar conclusions using different information. For instance, skin-fold thickness of traditional and recently urbanized Canadian Eskimos suggests that "normal" adipose tissue is mainly internal but "additional" fat is deposited in superficial depots in young adults of both sexes (Schaefer 1977).

We can summarize these findings by concluding that, in general, the extra fat goes into superficial depots especially in large mammals. As already mentioned, humans are relatively large primates and certainly the most obese of the living species. The preponderance of superficial adipose tissue in our own species may have arisen as a simple consequence of our large body mass, in the same way as it has done among large obese carnivores such as bears. Apparent change in the distribution of adipose tissue arises as a direct consequence of increase in abundance and in change in body proportions that result from an increase in body mass. This change probably has nothing to do with thermal insulation.

Biochemical data also suggest that insulation is not a primary function of the superficial depots. Homologies between adipose depots (fig. 11.2) enable us to extrapolate to humans from small animals that can be investigated experimentally. The dwarf hamster, *Phodopus sungorus*, is a small-bodied rodent native to the severe climates of Mongolia and Siberia where it remains active throughout the year. To determine whether fat depots serve primarily as insulators or as energy stores, we measured rates of lipid turnover in the major depots of these animals at rest and after an hour of moderate exercise (Mattacks and Pond 1988). Although adipose tissue on the back, neck, and shoulder is properly positioned for insulation, the lipids in these depots are most readily mobilized. There are large site-specific differences in the rates. Furthermore, the intra-abdominal depots have the metabolic properties that would be expected of adipose tissue with a principal function of insulation. Thus, the distribution of functional properties of the depots are not matched by their appropriate biochemical processes. If the homologous depots in humans have similar properties, then the insulating function of adipose tissue is clearly not well integrated with its role as an energy reserve.

Organ Protection

Another explanation for the distribution of adipose tissue is its role in the protection of vital organs (Pond 1978; Wood and Bladon 1985). In many harem-forming mammals such as red deer and elephant seals, the males become fatter before the breeding season, and they eat little and lose weight during the mating season. In a few species, some of the additional adipose tissue accumulates on visually conspicuous sites making the animal appear more massive. In a few such species, notably elephant seals, such prominent superficial adipose depots also sustain many injuries inflicted by rival males (Halliday 1980).

As with arguments for insulative properties of fat depots, the evidence does not support this hypothesis. Adipose tissue is not distributed in a way that would protect delicate and vulnerable parts of the body, because it is minimal over exposed vital organs such as the head and neck. Neither the site-specific differences in collagen content of adipose tissue, nor its scaling to body mass are consistent with the idea that adipose tissue is *adapted* to the protection of vital organs (Pond and Mattacks 1989; Pond 1992b; Pond and Ramsay 1992).

The Possible Role of Adipose Tissue in Reproduction by Women

Sex differences in the distribution of adipose tissue are among the most distinctive features of our species. Just as many of the supposed contrasts between humans and other mammals in the distribution of adipose tissue reflect a tendency toward human fatness, some sex differences are due primarily to the fact that women generally are fatter than men and, therefore, some superficial depots are exaggerated. To account for these differences we have to explain why humans are so fat, and when in our evolutionary history we became so fat. Why are women and children fatter than men? Are these properties adaptive? If so, what causal or historical relationship does increased fatness have to other human life-history attributes? Finally, what is the significance of sex and individual differences in the shape of certain superficial depots?

Sex Differences in the Abundance of Adipose Tissue

Researchers frequently attribute sex differences in the distribution and abundance of adipose tissue to energy storage during pregnancy and breast-feeding. Some authors, notably Frisch and McArthur (1974) and Frisch (1988, 1990b) observed that, at least among American women raised on nutrient-rich, Western diets, menstruation and ovulation are inhibited when adipose tissue falls below 17% of the body mass, a phenomenon they believe is adaptive (also see McFarland chap. 12; Vitzthum chap. 18). Other anthropologists argue that variables in addition to fat are involved, such as nutrition and work effort (Ellison 1990). Studies of non-Western populations show that pregnancy and lactation are physiologically and socially demanding (Galloway chap. 10; Vitzthum chap. 18; Panter-Brick chap. 17), although possibly not more so than for most wild mammals (Pond 1977). In many mammals, endocrine mechanisms may prevent conception or produce spontaneous abortion if maternal energy stores are insufficient.

In some wild mammals, including most nonhuman primates, adipose tissue is normally present in such small quantities that its contribution to the total energetic cost of reproduction may be small. Furthermore, the extra nutritional requirements for pregnancy and lactation in primates, particularly the large species, are very low compared to those of other mammals. Compared to other mammals, the milk of higher primates, and women in particular, is low in fats but rich in lactose (Ben Shaul 1962). The milk must be synthesized mainly from carbohydrates or protein precursors in the diet, not from stored lipids. In humans, increased energy expenditure during pregnancy and lactation is so low that it is difficult to measure (Prentice and Prentice 1988). Among women in The Gambia, West Africa who are engaged in farming, the correlation between milk production and skin-fold thickness is weak, even among lean women eating a barely adequate diet (Prentice et al. 1981). In humans, as in most other mammals, appetite, intestinal area, and digestive efficiency increase during late pregnancy and in lactation (Pond 1977). These changes indicate that much of the additional requirements of breast-feeding are normally met by increased intake and utilization of dietary nutrients. Large quantities of adipose tissue, however, could be critical to successful reproduction in uncertain or fluctuating seasonal environments with possibilities of severe famine (see also Panter-Brick chap. 17).

Frisch's ideas are not consistent with other basic aspects of human ecology. All the fossil and comparative evidence indicates that hominids lived in tropical or warm-temperate terrestrial habitats, exploited many different food sources, hunted and gathered food cooperatively, migrated over long distances, and lacked seasonal breeding. All these features suggest that, far from being dependent for successful breeding on a transient glut of food, or vulnerable to the failure of a single food source, the omnivorous, cooperative, nomadic habits of humans and their hominid ancestors buffer them against severe, prolonged starvation.

Since many other mammals reproduce successfully, often with much less adipose tissue, it is difficult for me to see why an omnivorous, nonhibernating, social species should require so much adipose tissue to produce such slow-growing offspring. When the ecology and physiology of humans are compared to those of other large mammals, the conclusion that the high adipose tissue content of women is an adaptation to the energetic cost of maternity seems unlikely.

In such discussions, it is important to remember that "fitness" and "success" have slightly different meanings to sociologists and psychologists than to biologists. Many of the traits that sociologists measure are much more relevant to economic advancement than to evolutionary "success." A rich diet may produce taller, longer-lived offspring, but these changes may have little bearing on net fecundity. Although acute malnutrition decreases fertility, chronic undernutrition has little impact on lifetime fecundity (Bongaarts 1980). Adequately nourished people may be more effective parents, hence contributing more to the next generation, than overnourished people.

Sex Differences in the Distribution of Adipose Tissue

Sex differences in the distribution of adipose tissue may seem to be a peculiar, unique feature of humans (but see McFarland chap. 12). In humans, unlike most mammals, the distribution of adipose tissue is influenced by sex hormones throughout adult life (Krotkiewski et al. 1983). These sex differences are visually conspicuous and important for many forms of social and sexual behavior. However, when corrected for differences in body composition, stature and skeletal proportions, they actually are relatively minor. They are most marked in small depots—particularly those on the breast, buttock, thigh, and calf (Bailey and Katch 1981)—and in younger, leaner individuals (Schaefer 1977).

Sex differences, however "minor," are among the most thoroughly studied of all secondary sexual characters of humans (Harrison 1985; Marshall and Tanner 1986). The sex-specific superficial adipose depots have been exhaustively measured and discussed. However, with only data from contemporary humans and some fragmentary historical information, the question of their origins is far from resolved.

Few issues have been the focus for a wider range of speculation based on fewer facts than the evolutionary origin and physiological function of women's breasts. Morris (1967) suggested that the upper chest swellings mimic the protruding buttocks and, thereby, promoted face-to-face intercourse. Morgan (1982) believed that the enlargement of the breasts occurred during a possible aquatic phase of human ancestry as an adaptation to the transport of infants in deep water. Gallup (1982), much influenced by the writings of Frisch and McArthur (1974), concluded that prominent breasts are an indicator of sexual status that became necessary following the elimination of an overt estrus phase. Cant (1981) also was persuaded by the critical-energy-store hypothesis and suggested that both the breast and the buttock adipose depots are indications of a woman's nutritional status and hence of her ability to raise children successfully.

Detectable sex differences in the distribution of adipose tissue are minimal in the great majority of mammals, including most primates: an intensive search revealed only a few minor sex differences in the distribution of adipose tissue in macaque monkeys; all differences were more conspicuous in older, fatter specimens (Coelho 1985). Even prolonged, intensive energy storage that is integral to reproduction, seems to have little impact on adipose tissue distribution. For example, there are sex differences in the chemical composition and microanatomy of adipose tissue of adult polar bears (Ramsay et al. 1992), but no differences in its anatomical distribution (Cattet 1990). Because the human situation has few parallels in other mammals, interspecific comparisons and experimentation with laboratory species contribute little to elucidate its evolutionary origin and physiological implications.

However, comparative anatomy can help to interpret the special features of adipose tissue in humans. The mammary glands in most mammals are associated with the adipose tissue in the groin and the abdominal wall as the mother nurses while standing or lying down (fig. 11.2c). In all higher primates, the nipples are located on the chest, and nursing takes place while the mother is sitting or carrying the infant under her belly. Mammary adipose tissue forms from sections of the bilaterally symmetrical depot on the chest (fig. 11.2d) that is present, albeit in smaller quantities, in other mammals. In spite of the importance attributed to them, the breasts are normally a relatively small depot (Campaigne et al. 1979) about 0.5 liters or roughly 4% of the total adipose tissue in young women. Rebuffé-Scrive (1987) expressed surprise that, in lactating women, lipids seem to be more readily mobilized from the "femoral" (groin) depots than from the adipose tissue in the breast. But, the human situation, in which adipose tissue located in the breast is not that most readily mobilized for milk production, is to be expected from our examination of comparative anatomy.

The reduction of body hair in modern humans makes the enlargement of the breast before and during breast-feeding more conspicuous than in furred mammals. Furthermore, the mammary glands do not mature until toward the end of pregnancy in most mammals. But in humans, growth of the mammary adipose tissue in girls is among the earliest anatomical changes in puberty, preceding menarche by about 2 years. The breasts may be almost full size before normal fertility and adult sexual and maternal behavior have developed. Breast adipose tissue apparently has no special physiological relation with the mammary gland (Rebuffé-Scrive 1987). It does not necessarily enlarge disproportionately with the number of children produced (Lanska et al. 1985). These facts suggest that mammary adipose tissue in women and girls is not solely, or even primarily, related to its role in the energetics of lactation. It simulates and exaggerates the form of the lactating breast, generating the appearance of fertility in girls long before they are actually capable of successful reproduction.

From early childhood, girls are significantly, and often conspicuously, fatter than boys on a similar diet. In addition, adolescence begins earlier and is completed sooner in fatter, taller girls (Marshall and Tanner 1986). Sex differences in adipose tissue distribution arise in part from differences in the timing of its relative growth and in average body composition. Expansion of muscle and adipose tissue on the hip and thigh also begins early in adolescence in girls. The femoral and buttock depots in women have properties that may enable the lipid in them to be mobilized selectively in lactation (Rebuffé-Scrive 1987), but this process has never been demonstrated in other mammals. Furthermore, pelvic skeletal growth does not reach adult dimensions until approximately 5 years after menarche (Moerman 1982). Thus, in spite of its importance to successful reproduction, maturation of the pelvis is not complete until long after the development of the conspicuous secondary sexual characters.

Far from being an accurate indicator of total body composition, the thickness of buttock adipose tissue correlates very weakly with that of other depots (Harrison 1985). Breasts, buttocks, and other anatomically minor but visually conspicuous adipose depots are much more convincingly interpreted as sexual signals that evolved in connection with permanent estrus (Szalay and Costello 1991) than as physiologically indispensable energy stores. Szalay and Costello's theory may account for the evolution of sex difference in distribution of adipose tissue but it does not completely explain why the tissue is so much more abundant in women than in men on a similar diet and exercise regime and why women are much more susceptible to severe obesity.

Obesity may be rarer in boys and young men than in girls and women but is more strongly associated with genetic and endocrinological abnormalities in men (Krotkiewski et al. 1983). As in women of similar body composition, much of the adipose tissue in young males is superficial and it becomes obvious in only a small minority of very fat boys. The rate of growth of adipose tissue during male adolescence lags behind that of the skeleton and musculature, so that boys typically become leaner in their teens. They accumulate adipose tissue relatively slowly during early manhood (Marshall and Tanner 1986).

Mammary adipose tissue in women regresses after fertility has declined, although it is not necessarily accompanied by a change in body composition (Lanska et al. 1985). However, in many postmenopausal women and in middle-aged and elderly men, adipose tissue often accumulates in the intra-abdominal depots. These depots together with the paunch on the outer wall of the abdomen produces the characteristic potbelly and spindly limbs (Borkan and Norris 1977) that seem to be unique to primates. This body shape is widespread among older people of both sexes (Sjöström 1988) and occurs in some captive primates (Pereira and Pond 1994), but it is rare in other mammals, even when they have been kept in captivity for many years. Such body conformation may be a symptom or a cause of physiological malfunction or other changes with age (Björntorp 1987; Bray 1988). Such characters that appear only in the elderly, of course, are not subject to natural selection in the same way as those that develop before or during reproduction and parenthood.

Summary

The sparse data on the "natural" distribution and abundance of adipose tissue in primates suggest that the basic anatomy of the human tissue is similar to that of terrestrial monkeys and thus probably was inherited directly from our primate ancestors. Superficial adipose tissue appears to extend over a greater area of the body in humans than in other terrestrial mammals because of changes in the proportions of the limbs and in the shape of the girdles, dorso-ventral flattening of the thorax and abdomen, and the bipedal posture of the hip, knee, and shoulder. Many of the differences between humans and other primates can be attributed to the fact that modern people are much fatter than wild monkeys. Many of the exceptional features of the distribution of adipose tissue in humans arise from its greater abundance. In several different lineages of naturally obese mammals, superficial adipose tissue becomes proportionately more massive with increasing fatness. Anatomical, ecological, and biochemical information provide no evidence that the distribution of adipose tissue in modern humans has evolved primarily as an adaptation to thermal insulation or as protection from mechanical damage.

Adipose is important as an energy store, but comparison of the energetics of reproduction, anatomy of adipose tissue, and composition of milk in humans and other mammals shows that the special features of the distribution of adipose tissue in young women are probably not adaptations to energy storage for reproduction.

Some of the sex differences in the gross anatomy of adipose tissue arise because women are much fatter than men of the same age following similar lifestyles. Other sex differences that consist of selective expansion of visually conspicuous but anatomically minor depots may have arisen under sexual selection. The accumulation of mesenteric and omental adipose tissue in men and older women has no parallels in wild mammals and may not be physiologically adaptive.

12 Female Primates: Fat or Fit?

Robin McFarland

OVER THE past century, we have refined Darwin's ideas about individual survival and reproduction. Over the past decades, long-term studies of mammals—especially primates—in which individuals are followed throughout life (e.g., Altmann et al. 1977, 1988; Cheney et al. 1988; Goodall 1986; Fedigan chap. 2; Pavelka chap. 7) have revealed the subtleties of interaction between individuals and environments. Most recently, the life-history approach, emphasized in this volume, has helped us to understand selective forces that influence survival and fertility (Zihlman et al. 1990; Morbeck chap. 1). Such an approach provides the opportunity to understand selective pressures that influence individuals from captive and free-ranging groups living in a range of environments and to appreciate the complexity of the interaction between the individual and the environment as organisms move through the life stages from youth to old age.

Conception, pregnancy, and lactation require energy. All mammalian females must balance the demands of reproduction with survival and maintenance activities such as predator avoidance and foraging. Primate females share the ability to lactate with all other mammals but have elaborated a distinctive pattern of reproduction. Primate females experience longer gestation, lactation, and offspring dependency relative to other mammals of comparable body size (Harvey 1986). Primate life-history characteristics have evolved in tandem with complex social interactions that enhance the ability of the offspring to survive to maturity. Strong social ties mean that the mother, even after weaning, often provides warmth and protection and transfers knowledge to the growing offspring (Altmann 1986a). From the point of view of the female, the evolution of long-lasting parent-offspring interactions means that much of adult life is spent pregnant, lactating, and caring for an infant or growing juvenile (Altmann 1980, 1983).

The life-history framework provides the context for my own research on the relationship between body fat and reproductive outcome in captive female pigtail macaques. Energy intake and output influence reproductive outcome. For example, food availability (Altmann et al. 1988; Cheney et al. 1988), physical condition, and activity (Altmann 1983) influence a female's ability to produce viable offspring. I directed my research toward the question of how body composition (i.e., relative proportions of lean body mass and body fat), a reflection of both activity and access to nutrition, is associated with female reproduction. Specifically, I examined the relationship between storing and drawing on fat with reproductive outcome.

In order to situate historically the debate about the relationship between body fat and reproduction, I first provide some background information about human studies. Second, I discuss the relationships among nutrition, activity, and body composition and nonhuman primate reproduction. Third, I discuss the results of my own study of reproduction and body fat in female monkeys. Finally, I place the information in an evolutionary context and discuss the implications for human evolution.

Nutrition, Body Composition, and Reproduction: Human Studies

Historical Background: Frisch and the Critical-Fat Hypothesis

In the 1970s Frisch and colleagues suggested that a particular level of fatness may be an important determinant of human reproductive ability. In their view, specific amounts of adipose tissue relative to body weight determine the onset and maintenance of menstrual cycles by affecting ovarian function (Frisch and Revelle 1970; Frisch and McArthur 1974; Frisch 1978a; 1990a).

Frisch and colleagues developed the critical-fat hypothesis. Based on a study of 181 adolescent girls, they calculated that girls need about 17% body fat in order to begin sexual cycling and that mature women need about 22% fat in order to maintain regular, ovulatory cycles or to restore cycling after weight loss (Frisch and McArthur 1974; Frisch 1978a). This information was supplemented by studies of female athletes and anorectic women with low percentages of body fat who no longer had normal sexual cycles (Frisch et al. 1981).

Frisch's work directed attention to an important issue: the connection between body fat and reproduction. Until the 1970s, the functional significance of adipose tissue was virtually ignored (Pond 1984; chap. 11). Perhaps Frisch's greatest contribution is that she shifted the emphasis from description of body composition to a discussion of the biological function of fat within an evolutionary framework. Her research paved the way for debate that continues to focus our attention on the evolutionary significance of body fat in men and women.

Fat and Reproduction

Recent studies indicate that the relationship between fat levels and ovarian function is not as simple as Frisch initially proposed. There is no doubt that extremely thin women or women who have experienced severe weight loss may not have regular menstrual cycles (Ellison 1990). The proposal of a "critical" level of fat has been questioned, however, because even women with less fat seem to be able to conceive (Prentice and Whitehead 1987).

We now know that moderate thinness and moderate weight loss associated with fluctuations in food availability may be related to reduced ovulatory frequency and luteal suppression, even though menses may be regular (Lager and Ellison 1990). For example, among Lese horticulturalist women of the Ituri Forest, those with the poorest nutritional status (i.e., based on weight for height) or those who lost significant amounts of weight over a period of 4 months had lower levels of progesterone during their regular cycles. This indicates that they ovulated less frequently than did better-nourished women (Ellison et al. 1989). Even if they did ovulate, they were less likely to produce a viable corpus luteum. Ovarian suppression was most severe just after the peak of a period of seasonal food shortage (Ellison et al. 1989).

The effect of food availability, a feature of the environment, on reproduction is mediated through a complex interaction among the brain, hormones, and reproductive organs. Access to food may affect the basic organs of the reproductive process: the hypothalamus, the anterior pituitary gland, and the ovaries (Bronson 1989). Neuropeptides synthesized in the hypothalamus control the release of gonadotropins from the pituitary. Two gonadotropins are of central importance for reproductive ability: FSH (follicle stimulating hormone) and LH (luteinizing hormone). They stimulate the maturation of eggs and luteinization of the follicles to form the corpora lutea. Synthesis of the neuropeptides is, in turn, regulated by pathways in the hypothalamus and other parts of the brain. Ultimately, environmental input, including availability of food, may influence the synthesis of neuropeptides that control the release of gonadotropins (Bronson 1989).

Different researchers emphasize slightly different proximate mechanisms for the effect of nutrition on the hypothalamic-pituitary-ovarian axis. Frisch (1978a, 1990a) emphasized the direct role of body fat. She drew on evidence that adipose tissue is an extragonadal source of estrogen because conversion of androgen to estrogen takes place in the abdomen, omentum, and fatty marrow of the long bones (Frisch 1990a; Simpson and Mendelson 1990). This has led to the suggestion that levels of circulating estrogen (correlated with amounts of adipose tissue) may act as a signal to the hypothalamus (Frisch 1990a). Fat may influence the direction of estrogen metabolism to more or less potent forms. Obese women, for example, have more potent forms of estrogen (Lustig et al. 1990). A decrease in the concentration and potency of circulatory estrogen associated with a lack of fat storage might be a signal of abnormal metabolic rate to the hypothalamus and may reflect a potentially harmful reduction in environmental resources (Frisch 1990a).

Frisch (1990a) suggested that either weight loss (i.e., a change in energy balance) or a change in body composition resulting from a loss of fat (i.e., potential energy) may influence the hypothalamic-pituitary-ovarian axis and, thereby, regulate ovarian function. Adipose tissue stores could link the environment to reproductive function if the hypothalamus can somehow sense the amount of energy that is available in fat stores. It is, however, not known if the brain can monitor the status of fat stores (Van Itallie and Kissileff 1990). Ellison (1990), in contrast to Frisch, emphasized the association between energy balance and hypothalamic function, rather than body composition. He suggested that weight loss may be a signal to the brain that food sources are inadequate and that this may lead to suppression of ovarian function.

Dietary Composition, Body Weight, and Reproduction

Other recent studies examine the relationship between dietary composition or body weight and reproductive outcome, that is, successful ovulation, gestation, and lactation resulting in surviving offspring. The impact of body fat is not assessed directly in these studies, but they do explore the association between nutrition and reproduction.

Studies of the relationship between dietary composition and menstrual cycles have produced conflicting results. For example, women who ate high-fat diets (46% of calories) over two complete menstrual cycles did not show significant differences in levels of plasma luteinizing hormone, progesterone, or estradiol when compared to women on a low-fat diet (25% of calories) (Hagerty et al. 1988). There was no evidence that increasing short-term dietary fat affects levels of circulating hormones. Vegetarians with diets high in fiber and low in fat were, however, more likely than nonvegetarians to experience irregular menses or none at all (Pedersen et al. 1991). Surveys of human populations indicate that the age of menarche is later in those areas where intake of dietary fiber is high and intake of fat correspondingly low (Hughes and Jones 1985).

Diet influences not only ovarian function but also another aspect of reproduction, the birth interval. Studies of people in Guatemala, Mexico, and The Gambia indicate that the length of time between births among women whose diets are supplemented is shorter than the interval among women who do not receive nutritional supplements (Chavez and Martinez 1973; Delgado et al. 1978; Prentice et al. 1986).

Likewise, body weight is associated with the length of the birth interval. A study in Bangladesh indicates that women who weigh more at parturition have a shorter period of postpartum amenorrhea. Women who weighed more than 44 kg had a mean period of postpartum amenorrhea of 13.6 months versus 15.9 months for those women who weighed 38 to 43.9 kg; those who weighed less than 38 kg had a mean birth interval of 17.6 months (Ford et al. 1989). Similarly, a study in Belize shows that women who weigh more relative to height have shorter periods of postpartum amenorrhea (Fink et al. 1992).

Although the authors of these reports suggest that dietary composition and body weight influence reproductive function, it is difficult to distinguish the particular influence of dietary components and related nutritional status on the hypothalamic-gonadal pathway because diet, hormones, and the amount of body fat are not consistently assessed.

Sex Differences in Body Composition

Humans exhibit sex differences in both the distribution and proportions of body fat, a feature that has been noted for decades. In men, muscle represents 43% of body weight, whereas it represents 36% (Forbes 1987) in women. In percentage of body fat, women are consistently fatter than men (e.g., Bailey 1982, skin-fold thicknesses and total body potassium; Clarys et al. [1984], dissection; McCance and Widdowson [1951], urea dilution; Segal et al. [1985], bioelectrical impedence analysis and hydrostatic weighing). Between the ages of 16 and 18, the body of a well-nourished woman is 26–28% fat, whereas that of a man is about 14% fat (Forbes 1987).

In humans, there are sex differences in distribution of fat as well as in proportions of fat and lean mass. Androgens influence the deposition of fat in the trunk and abdomen, whereas estrogens promote the deposition of fat on the hip and thigh (Forbes 1987), thus contributing to the distinctive human female shape. Fat in the hip and thigh is more metabolically active during lactation than is fat in other depots (Rebuffé-Scrive et al. 1985; Pond chap. 11) and is most resistant to weight loss, suggesting that this depot may be a useful reserve for reproduction.

The combination of Frisch's proposal of the critical-fat hypothesis, ecological studies that highlight the link between nutrition and reproductive function, and recent physiological studies that emphasize the functional significance of body fat for normal ovarian function and lactation, all provide a context for understanding the functional, evolutionary importance of sex differences in body composition.

Significance of Human Studies

Two points emerge from studies of humans. The first is that the exact impact of nutrition on the hypothalamus is not understood. It seems clear that in all mammals food either directly (through diet) or indirectly (through fat storage) influences the hypothalamic-pituitary-gonadal axis. It is possible that availability of food, composition of the diet, and the amount of energy that is stored synergistically influence reproductive capacity (Lustig et al. 1990). A second point is that we need more precise estimates of body composition if we are to define any influence it might have on reproduction. The attempt to assess directly the impact of nutrition on physical condition and ovarian function sheds light on the interplay between the environment (e.g., quality and quantity of food), the individual, and reproductive outcome.

Bridging the Gap between Humans and Other Mammals: The Contribution of the Comparative Perspective

The predisposition of mammals to store fat is particularly important for females. Adipose tissue provides a ready source of energy for reproduction. This is especially true during lactation because the metabolic costs of reproduction are greatest then (e.g., Pond 1977, 1984; Stini 1982; Worthington-Roberts and Rodwell Williams 1989; also see Pond chap. 11; Zihlman chap. 13).

The triggers of ovarian function may not be the same in all mammalian species. Fat stores probably are more important in larger mammals. Large-bodied mammals can store proportionately more energy and, consequently, draw on more reserves (Lindstedt and Boyce 1985) than can small mammals such as rats, which have different life-history characteristics and different patterns of energy use.

In mammals, there appears to be a continuum of dependency on fat stores for reproduction. In the extreme case of the northern elephant seal, the female does not forage at all while lactating but relies solely on fat stores to maintain both herself and her infant (Widdowson 1976; Costa et al. 1986; Reiter chap. 4). The female may lose up to 58% of her initial body fat during the approximately 4 weeks that she is transferring energy to her pup (Costa et al. 1986).

Animals have many demands on their lives for both survival and reproduction, and there may be circumstances that constrain how much energy they can store in the form of fat. Some species have little fat buildup and must meet the demands of reproduction by eating more food. Wild rabbits, for example, store little fat (less than 4% of body weight) and are at the opposite end of the continuum from seals (V. G. Thomas 1990). Rabbits have the ability to store more fat in captivity but apparently, in the wild, pressure from predation keeps fat storage low. Whereas elephant seal mothers are relatively immobile during parturition and lactation, rabbits need to be light and agile and fast to escape predators. Too much fat interferes with locomotion; heavier, slower rabbits are subject to higher rates of predation.

Primate females seem to fall between seals and rabbits. Primates have both the ability to store fat and to increase food intake. Since most catarrhine primates are large-bodied, they can carry some body fat and still remain mobile and active, walking and foraging for long hours with a group, possibly carrying one offspring while pregnant with the next infant (Zihlman 1992). Patterns of fat storage in females are influenced by survival life-history features such as the need for mobility to escape predators and forage for food and, in addition, by energetic requirements of reproduction.

Nutrition, Body Fat, and Reproduction: Nonhuman Primate Studies

Anatomical studies suggest that features of nonhuman primate adiposity are similar to those of humans. For example, the distribution of fat depots is similar in monkeys and humans (Pond and Mattacks 1987b; Pond chap. 11). Like humans, nonhuman primate females and males have different relative amounts of body fat and also different propensities to accumulate fat during growth and development.

In contrast to the few anatomical studies, a plethora of ecological studies suggest that nutrition and related body weight play a role in a female primate's ability to reproduce successfully. Most studies provide indirect evidence for the relationship between nutrition and fertility because, particularly in free-ranging studies, the effect of diet on physical condition is rarely assessed. Clues about energetic demands and female response to them are provided by behavioral studies suggesting that females adjust their activity levels and diets in response to reproductive demands.

Anatomical, physiological, and ecological studies together suggest that primate females have the ability to respond to the energetic demands of reproduction both by storing and drawing on fat reserves and by adjusting feeding behavior and activity patterns. Several aspects of female life have been shaped by the need to gestate, feed, and care for offspring.

Sex Differences in Body Composition

Only in recent years have a significant number of studies concerning sex differences in nonhuman primates begun to emerge. Although sex differences in the amount of fat in nonhuman primates are not as dramatic as in humans, other female primates consistently have relatively more body fat than do males. Males, on the other hand, have more lean body mass relative to total body weight. For example, 5-year-old baboon males have 6% body fat, whereas females have 16% (Rutenberg et al. 1987). In a study of captive adult pigtail macaques, Walike et al. (1977) found that males have 8.7% body fat, whereas nonobese females average 12.7%. Obese females average 40.5% body fat (Walike et al. 1977). That female orangutans have more fat relative to body weight than do males is suggested by dissection (Morbeck and Zihlman 1988).

Differences observed in adult males and females arise in early stages of life. Eighteen-week-old male baboons have significantly more lean body mass than do females (Rutenberg et al. 1987). When infant baboons were fed a low-calorie diet, females gained substantially more fat than did males (Lewis et al. 1984). Males and females fed a high-calorie diet gained about the same fat mass, but females had significantly less lean body mass (Lewis et al. 1984).

Skin-fold thickness at several sites (neck, subscapular, suprailiac, and triceps) was greater at every age from birth to 8 years in female baboons, even though the mean weight, crown-rump length, and triceps circumference of males exceeded those of females throughout that 8-year period (Coelho et al. 1984; also see Pond chap. 11).

In summary, much evidence suggests that, developmentally, female primates are more disposed to accumulate body fat than are males. Although sexual dimorphism in this feature is most pronounced in humans, sex differences in body composition may have a long evolutionary history within the primate order. Sexual dimorphism in body composition may have evolved because females need more energy reserves to offset the demands of pregnancy and especially lactation (Coelho 1974; Grand 1983).

Feeding Behavior and Other Activities

Female primates have been observed to adjust their diet according to their reproductive state, either spending more time feeding during pregnancy and lactation or increasing intake of specific foods. For example, lactating howler monkeys spend significantly more time feeding than do other adult females (Smith 1977). Baboons spend more time feeding as their pregnancies progress (Silk 1986). Lactating female baboons in different populations increase feeding time, sacrificing either resting or social time to elevate their food intake (Altmann 1980; Dunbar and Dunbar 1988).

Captive galagos increase caloric intake and consumption of protein during lactation (Sauther and Nash 1987). Similarly, female guenons shift to foods with high-protein content during the part of the year when they are pregnant and lactating (Gautier-Hion 1980).

Another indication of extra energy requirements is that females may eat more or different foods than males of the same species (Waser 1977; Gautier-Hion 1980; Cords 1986). In some prosimian species, for example, females are dominant to males in feeding and, therefore, have access to the best foods (Pollock 1979; Jolly 1984). Feeding priority also plays a role in some species of small South American monkeys. Among marmosets and tamarins, the female leads the group, is the first to arrive at feeding trees, and has her choice of insects and fruit. In contrast to the typical primate pattern, males carry and groom infants; virtually the only contact the mother has is during nursing. This allows the female extra time and mobility to feed on insects, which are rich in protein (Wright 1984). Behavioral adjustments by both males and females allow the females to obtain more energy than males and to conserve energy because they do not carry offspring.

Male and female dietary differences have also been observed. Female chimpanzees eat more termites in East Africa (McGrew 1981) and more nuts in West Africa (Boesch and Boesch 1984); both foods are rich in protein, and the nuts are rich in lipids. Among mangabeys, adult females spend more time feeding than do males, and they forage more extensively for insects (Waser 1977). Similarly, among two species of guenons, adult females are more insectivorous and/or folivorous than males, which are more frugivorous (Cords 1986). Increased proportions of insects and leaves in the diet enhance protein intake (Clutton-Brock 1977).

In summary, adjustments in feeding behavior appear to be one way that females increase energy intake, particularly when they are expending relatively more energy, such as during pregnancy and lactation. Activity adjustments offset the demands of reproduction.

Ecological Studies: Nutrition, Body Weight, and Reproduction

Studies of both captive and free-ranging primates suggest that availability of food, dietary composition, and physical condition related to nutrition influence aspects of reproduction, including the age of sexual maturity and first reproduction, the length of intervals between births, and infant survival. These ecologically oriented studies of nonhuman primates have demonstrated that, in at least some populations, nutrition is related to reproduction (Bronson 1989). Although the studies do not directly assess the connection between nutrition and physiological processes underlying reproduction, they provide information about the relationship between the environment and reproductive outcome. They complement the studies of direct effects of nutrition and related body composition on reproductive function.

Sexual Maturity and Age of First Reproduction

Comparison between free-ranging and captive primates, as well as provisioned and nonprovisioned populations, reveals the dramatic impact of nutrition on reproductive timing. Captive primates with access to abundant food consistently achieve sexual maturity at a younger age and give birth earlier and more often than their free-ranging cousins. In captive populations of baboons, for example, females reach first estrus at 3 to 3½ years of age and conceive their first infant within the next year. In contrast, free-ranging baboons at Amboseli, Kenya attain first estrus at 4½ to 5 years, and conception first occurs at 6 years of age (Altmann et al. 1977; Altmann and Alberts 1987). Captive chimpanzee females undergo menarche two to three years earlier than animals in the wild and give birth at ages 8.9 to 11.3 years versus the 13 to 15 years typical of some free-ranging animals (Coe et al. 1973; Goodall 1986).

Similarly, there are differences in the age of reproductive maturity among free-ranging groups of the same species with access to different food resources. One baboon troop at Amboseli supplements its diet by feeding at a garbage pit near a tourist lodge. Individuals within the "garbage group" fed for shorter periods of time per day than individuals in a nonprovisioned troop (22% of time versus more than 60%), yet the females in the group with access to human leftovers experience first estrus at 3.3 years of age, similar to captive females (Altmann and Muruthi 1988).

Similarly, social groups of vervet monkeys in Kenya were observed in adjacent territories that differed in availability of food. One group had access to ample mature acacia trees and water holes, one group had access to a low-quality habitat, and one group had access to an intermediate territory. Females in the group with the highest-quality resources gave birth for the first time an average of more than one year earlier than females living in the lower-quality environment. The group with access to the most abundant food had a mean age at first birth of 4.4 ± 0.6 years; the group that had access to the territory of intermediate quality gave birth for the first time at 5.1 ± 0.6 years; the group with access to the low-quality habitat had a mean age at first birth of 5.7 ± 1.1 years (Cheney et al. 1988).

The effect of access to nutrition on reproductive parameters within a population is evident in studies of Japanese macaques for which comparative data have been collected during periods of provisioning and nonprovisioning. Two sites in Japan show similar differences in the age of first reproduction when food was provided. At Mt. Ryozen, during provisioning, the age of primiparity was 5.2 years; without provisioning, first birth occurred at 6.7 years, a difference of 1.5 years (Sugiyama and Ohsawa 1982). At Koshima, the age at first birth during provisioning was 6.2 years; in the nonprovisioned period, it rose to 6.8 years (Mori 1979).

These studies (and others, e.g., Strum and Western 1982) show that variation in the age at attainment of sexual maturity and first reproduction of primate females is associated in some populations with access to food. The timing of these life-history events is affected by quality of the environment.

Birth Interval

Another aspect of fecundity that is influenced by the environment is the length of the birth interval. It consists of (1) postpartum amenorrhea (associated with lactation), (2) the waiting period to conception, and (3) gestation (Bongaarts and Potter 1983). Gestation is far less flexible than the period following birth and accounts for little of the variation in the time to next conception. The factor that most influences the average length of the birth interval is the duration of postpartum amenorrhea, which is influenced by lactation and possibly by the availability of food for both mother and offspring.

Studies of nonhuman primates indicate that nutrition and the length of the birth interval are associated. Among vervet monkeys at Amboseli, Kenya, birth interval varied between groups with access to different nutritional resources. Groups with access to high- and medium-quality diets experienced average birth intervals of 13.7 and 13.5 months, respectively, whereas members of the group with the low-quality diet had an average birth interval of 17.1 months (Lee 1987). Among baboons, females that weighed more (and presumably were in a better nutritional state) had shorter birth intervals than lighter females (Bercovitch 1987). At another site in Kenya, birth intervals in a population of baboons increased as access to food decreased over a 5-year period (Strum and Western 1982). In 1977, the average interval was 17.4 months; by 1981, the average had increased to 30.3 months.

Infant Survival

Whether offspring survive is one of the most important determinants of completed lifetime reproduction. After the infant is born, the primate mother promotes its growth and survival by providing milk, transportation, warmth, and social knowledge.

Adequate nutrition may affect the mother's ability to care effectively for her offspring, including the amount of milk she produces during lactation. Although the amount and quality of milk is crucial to infant survival, they are seldom measured. At least one study demonstrated the link between nutrition and lactation. Among captive female baboons, three groups were fed different amounts of food. Group 1 was fed ad libitum, group 2 was fed 80% of the ad libitum amount, and group 3 was fed 60% of the ad libitum amount (Roberts and Coward 1985). All females were followed until 10 weeks postpartum, when the infants were weaned. Groups 1 and 2 (fed ad libitum and 80% ad libitum) exhibited no significant differences in the amount of milk produced. However, by 9 to 10 weeks, the amount of milk produced by mothers in group 3 (fed 60% of the ad libitum amount) dropped to 63% of that produced by the ad libitum group, indicating that the underfed mothers did not produce as much milk (Roberts and Coward 1985).

Survivorship of infants in some free-ranging populations, measured as rate of mortality, increases with improved nutrition. This could be due to the improved physical condition of the mother who can lactate more efficiently or the increased availability of supplemental foods for the infants. Among Japanese macaques, the frequency of infant survival differs between periods of provisioning and nonprovisioning. For example, at Mt. Ryozen, the survival rate in animals less than 2 years old dropped from 85.4% during the period of artificial feeding to 72.3% under natural conditions (Sugiyama and Ohsawa 1982). At Koshima, survivorship of young within 1 year of birth declined from 85.1% in the period of artificial feeding to 31.2% in the period of natural feeding (Mori 1979; Watanabe et al. 1992).

Infant mortality among vervet monkeys in Amboseli, Kenya was especially low in the group occupying the home range that contained food resources of the highest quality (Lee 1987; Cheney et al. 1988). The percentages surviving the first year were 61% in the group with access to the best resources, 42% in groups inhabiting ranges of medium-quality, and 41% in low-quality ranges.

Infant mortality is influenced by the mother's condition even before conception. Rhesus females who weighed less at conception, for example, tended to have babies of lower birth weight that were less likely to survive (Riopelle et al. 1976).

Altmann has suggested that the diet of female baboons at a young age (30 to 70 weeks) influences the number of their offspring that survive to one year (Altmann 1991). He observed the amount and type of food that females ate when they were being weaned. More than 10 years later, he assessed the number of surviving offspring that they produced. Those that had higher energy intakes while they were infants were more likely to produce more surviving offspring than females with less optimal intake (Altmann 1991). Altmann's findings suggest that diet, even at a very young age, may influence lifetime reproductive outcome.

In summary, long-term studies of fertility and survival in captive and free-ranging populations have contributed to our understanding of flexibility in life-history characters among individuals. Variation in timing of reproductive events such as age at sexual maturation and first birth, length of birth interval, as well as survival of offspring are influenced by nutrition. Food availability is a key environmental feature that affects reproduction.

Case Study: Body Composition and Reproduction in Female Pigtail Macaques

I studied captive female pigtail macaques (*Macaca nemestrina*) housed in the Regional Primate Research Center at the University of Washington. My intent was to study directly the association between body composition and conception and/or successful pregnancy outcome in a nonhuman primate species.

The purpose of the study was to (1) identify factors that influence conception, (2) identify variables that are associated with pregnancy outcome, and (3) explore the interactions between body composition and reproduction through a longitudinal study of females of different parities from preconception, through pregnancy, and into the postpartum period. Other variables that may influence reproduction also were taken into account. These include age, parity, health status, and social rank (McFarland 1992).

Interval to Conception

Female reproduction over a lifetime is shaped by three major features of life history: (1) the length of the reproductive life span, (2) fertility (the actual number of infants per reproductive year), and (3) survivorship of offspring (Fedigan et al. 1986; Fedigan 1991). The first question I addressed concerns fecundity, because frequency of female reproduction is related to the interval to conception. I explored factors that influence the time it takes to become pregnant once females have the opportunity to mate.

I used a new methodology for the determination of body composition: bioelectrical impedance analysis. Bioelectrical impedance analysis has been successfully used to determine body composition in humans but has rarely been applied to the study of nonhuman primates. In essence, it measures the transmission of electrical currents in the body and, since resistance varies in different bodily tissues, calculates the amount of fat present in the body. In addition to bioelectrical impedance analysis, standard measurements, including skin-fold thickness, circumferences, weight, trunk height, and crown-rump length were taken to assess fatness and size (see McFarland [1992] for discussion of methods).

I used survival analysis (the Cox proportional-hazards model) to examine the relationship between one aspect of reproduction—the interval to conception—and fatness and size, age and parity, and social rank. The duration of the interval to conception is the time that a female was exposed to a male either until conception or the end of the study (i.e., the period during which she could conceive).

I found that the interval to conception is negatively correlated with percentage of fat (as calculated from bioelectrical resistance) and anthropometric variables (e.g., weight, skinfold thickness). Fatter females were with males for a shorter period of time than thinner ones before they conceived. The interval is not significantly associated with age, parity, or social rank. It appears that, even in this captive sample where females have access to abundant food, individual variation in body composition is one factor that influences the interval to conception. This result corresponds to other studies that show that physical condition is related to the length of time between births. (e.g., baboons, Bercovitch 1987; vervet monkeys, Whitten 1983; macaques, Mori 1979).

Pregnancy Outcome

In addition to intervals between births, I examined the relationship between body composition and survival of offspring. I found that maternal nutritional status, as reflected by anthropometric and bioelectrical measurements, is associated with pregnancy outcome among the pigtail macaques. Females that produced surviving offspring weighed significantly more and had larger arm circumferences before they conceived than did females that had spontaneous abortions, stillbirths, or infants that died shortly after birth. The majority of females that had surviving offspring experienced increases in skin-fold thickness during pregnancy. Abdominal skinfold thickness, and presumably fat, during pregnancy was significantly greater in the females that bore surviving infants (McFarland 1992).

Some of the females did not nurse their offspring, because the mothers rejected or abused the infants, failed to produce milk, or the infant failed to cling and suckle properly. Infants that were separated from their mothers were raised by humans. Among females producing offspring that survived for at least 6 months, those that lactated were significantly fatter than those that did not. Infants produced by lactating mothers were significantly larger at birth than those produced by nonnursing mothers (in both weight and femoral length). Perhaps the smaller babies produced by the leaner mothers were not able to effectively cling, suckle, and maintain the visual and vocal contact necessary to evoke proper maternal responses, including milk production.

Many factors that influence successful pregnancy outcome have been identified, including maternal experience (Altmann 1980), social rank (Altmann 1980; Turner et al. 1987), and the mother's weight (Riopelle and Hale 1975; Prentice et al. 1986). Body composition also may influence pregnancy outcome because mothers without abundant energy reserves may be more likely to miscarry or have stillbirths, or produce infants of low birth weight. Females that are in better physical condition may be more likely to produce babies that thrive. Healthy mothers and infants may more easily develop strong social attachments that aid in ensuring survival of offspring to independence (Smuts chap. 5).

Interaction between Body Composition and Reproduction

I found that short-term changes in the amount of fat during and after pregnancy in pigtail macaques are similar to those experienced by females of other species, including humans. The pigtails gain weight and fatness during pregnancy and lose fatness in the postpartum period (McFarland 1992).

Short-term fluctuations in body composition during one complete reproductive cycle correspond to energetic demands placed on females by pregnancy and lactation. Weight gain during pregnancy is a general mammalian characteristic (Widdowson 1976) and is partly due to fat deposition (Thomson and Hytten 1977). Women lose fat in the postpartum period, whether or not they are lactating (Forsum et al. 1988; Brewer et al. 1989; Sadurskis et al. 1989). Some researchers have suggested that females gain weight in addition to the weight of the growing fetus during pregnancy in order to prepare for the energetic demands that they will experience during lactation (Widdowson 1976; Pond 1984).

Short-term changes in the relative amounts

of fat and lean tissue associated with pregnancy and lactation may be cumulative and translate into long-term changes. Pregnancy and lactation place energetic needs on females above what is required to maintain themselves when they are sexually cycling (Altmann 1983; Dunbar 1988). Usually, as I have pointed out, females can offset these demands by increasing caloric intake, adjusting activity, or drawing on fat reserves.

Studies of people living in diverse environments suggest, however, that, in some circumstances, women may suffer long-term deficits in their own energy reserves because they do not completely recover from reproductive losses between successive births. Although women who have access to adequate nutrition may actually accrue fat throughout their reproductive lives (e.g., Texan women; Butte and Garza 1986), other women who consistently have limited access to food may suffer long-term deficits in the amount of fat (e.g., Au of Papua New Guinea; Tracer 1991). The effects may be compounded if women also engage in strenuous activities such as agricultural work or long-distance travel by foot, as they do in many societies (Ellison 1990; Panter-Brick 1991, chap. 17; Peacock 1991).

Data from the captive pigtail macaques suggest no clear long-term effects on body composition as a result of successive births. In fact, females appear to accrue fat and weight through their reproductive lives. These females have access to adequate nutrition and are relatively sedentary. They may be compared with women who have access to plenty of food and who do not perform hard physical labor (Butte and Garza 1986). Such women may gain fat with each pregnancy because they do not lactate for great lengths of time and do not lose all of the weight that they gained during pregnancy (Sadurskis et al. 1989).

In summary, the study of the relationship between fat and fertility in female pigtail macaques suggests that, even in a population of monkeys with access to ample food, females that have more body fat tend to have (1) a shorter interval to conception once they have the opportunity to conceive, and (2) are more likely to have surviving offspring and to lactate for them. In short, they have an improved reproductive outcome compared to females with less body fat. In addition, the female monkeys have a pattern of fat deposition during pregnancy and mobilization after birth that is similar to the human pattern.

Implications for Understanding Human Reproduction

Studies of mammals, including human and nonhuman primates, suggest that fertility is related to energy availability and use: food, fat stores, and activity. All mammals provide food for their offspring after birth. This relates to the ability to store fat during pregnancy and draw on reserves during lactation (Pond 1977, 1984). Natural selection has enhanced the propensity to store fat as a way to maintain an energy balance that ensures one's own survival and, at the same time, in females, enhances the survival and development of offspring.

Fat is not the only tissue that responds to reproductive demands. Bone also remodels itself during pregnancy and lactation (Galloway chap. 10). During pregnancy, bone mineral is laid down, whereas during lactation, reserves in the skeleton are drawn upon. This mirrors the pattern of fat deposition and use where fat is laid down during pregnancy and drawn upon during lactation. Body composition is only one aspect of female bodies that has been modified to accommodate the complex demands of activities that women undertake throughout life.

Studies of women in different environments provide clues about selective forces that have shaped women's bodies through human evolution. Through most of our evolutionary history, humans made a living as nomads who gathered and hunted over a large home range. Recent studies of women in hunting-gathering and other nonindustrialized societies have broadened our knowledge of the range of human reproductive patterns (Prentice and Whitehead 1987; Ellison et al. 1989; Panter-Brick 1991).

One of the best-studied hunting-gathering populations is the !Kung of southern Africa (see Draper chap. 16). Major features of !Kung reproduction include production of 4.7 children over the average woman's life (Howell 1979). The average length of the birth interval is 44.1 months (Lee 1979). Suckling of infants is frequent and intensive (Konner and Worthman 1980), and breast milk is an essential part of a child's diet until about age 3, because the bush diet lacks commonly preferred weaning foods such as cow's milk or grain (Bentley 1985).

The long birth intervals are probably the result of several related factors. First, there is a long period of lactational amenorrhea due to hormonal suppression of ovulation (Konner and Worthman 1980). Second, birth intervals may be prolonged because of nutritional inadequacy. Females may not resume sexual cycling as early as they might if they had access to abundant food (Howell 1979). Third, !Kung women vigorously engage in subsistence activities and travel long distances carrying heavy loads, including dependent children (Bentley 1985). Their strenuous physical effort may contribute to suppression of ovarian function (Ellison 1990).

!Kung females' lives differ from those of urban, industrialized women in that the !Kung experience later menarche, many years of lactation, longer birth intervals, earlier menopause, and fewer menstrual cycles throughout life (Howell 1979). They experience about 15 years of lactational amenorrhea and just under 4 years each of pregnancy and menstrual cycling. Women in industrialized societies, on the other hand, reach menarche at an early age and menopause relatively late in life, do not lactate for long, and experience short periods of lactational amenorrhea. As a consequence, the average urban woman may experience 35 years of menstrual cycling (Lancaster 1985).

What patterns of human reproduction would explain the elaboration of fat deposition in women? We share major features of reproduction with other primate females. For example, all catarrhines normally produce a single infant. Both nonhuman primates and women who live in nonindustrialized societies expend energy on subsistence during pregnancy and lactation (Altmann 1980; Panter-Brick 1991; chap. 17; Peacock 1991). Humans in hunting-gathering societies have slightly shorter birth intervals than do the great apes (Galdikas and Wood 1990). Milk composition is similar in primate species; primate milk has low protein content compared to carnivores or artiodactyls (including cows) and has more fat relative to protein (Oftedal 1984). The similarities between humans and nonhuman primates suggest that differences in the length and quality of lactation cannot explain the propensity of women to accumulate fat.

Explanations of fatness in women focus on aspects of reproduction, including mating opportunities. Some researchers invoke sexual selection to explain the evolution of sexual dimorphism in body composition. One recurring theme is that permanent breasts and buttocks replaced the cyclical sexual swellings of apes and monkeys and served to attract men who then provided food for women and their offspring (Morris 1967). In this scenario, which has been updated with different twists (Lancaster 1985; Szalay and Costello 1991, Pond chap. 11; also see Zihlman chap. 13), the long period of human juvenile dependency increased the need for bonds between males and females. Paternal investment became increasingly important in humans (e.g., Lovejoy 1981, 1993), and males chose to mate with females who had "attractive" fat depositions. Fat served as an advertisement of ability to reproduce and raise young (Huss-Ashmore 1980; Lancaster 1985).

The invocation of sexual selection to explain human sexual dimorphism in body composition is less satisfactory in light of research that has been conducted over recent years. Comparative information from mammals, including primates, suggests that fat is important for the maintenance of a balance of energy. Physiologically, fat depots are mobilized in response to reproductive state (e.g., Rebuffé-Scrive et al. 1985). Furthermore, there is evidence that adiposity is associated with ovarian function (Ellison 1990; Frisch 1990a).

The pattern of deposition of fat in breast and hip regions may have more to do with the ability of women to travel and forage than with the ability to form long-term social bonds with provisioning males (Zihlman chap. 13). Distribution of fat in the trunk may be related to the development of bipedality. Fat is distributed near the center of gravity so that it does not interfere with mobility (Zihlman 1992).

The long period of juvenile dependency, accompanied by rapid brain growth, in addition to the evolution of a new mode of locomotion (bipedality), and new foraging patterns that accompanied the transition of early hominids to a savanna mosaic environment may all have been factors that contributed to the human female adaptation (Zihlman and Tanner 1978).

Conclusion

The ultimate goal of anthropologists is to understand human evolution. A marked sex difference in relative amounts of fat and lean tissue characterizes our species. Within the framework of comparative studies, however, the distribution and amount of fat in humans is consistent with our mammalian and primate heritage. Reproductive outcome and body composition are related. This indicates that the ability to store energy in the form of fat has evolved, at least in part, in response to reproductive demands.

Part V

Women in Human Societies

What It Means to Be a Human

HUMANS share biobehavioral survival and reproductive life-history features with other catarrhines, primates, and mammals. We can be viewed as a variation on these themes, as shown by the chapters in this book. The use of the comparative, functional, and evolutionary approach to human studies elicits new insights into our standing within the natural and cultural worlds.

Humans exhibit a set of distinctive species-defined characteristics. Many of these are anatomically linked to the evolution of bipedalism and increased brain size and complexity (Zihlman chap. 13). These characters, however, represent only a small portion of the features that distinguish us and show how humans have capitalized on the "mammalian primate model."

Humans are long-lived, with expansion of the life cycle and blurring of the life stage boundaries as we grow to biological and social maturity. A relatively longer time during childhood invites more chances for factors in individual life stories to influence survival and eventual reproduction. Throughout our lives, we develop and maintain long and complex social relationships (see Rodseth et al. [1991] for discussion). Within our kin group, this complexity is enhanced, necessitating both devotion of time and effort to maintaining social networks. Reproduction involves the birth and rearing of offspring to weaning as defined for many mammals as well as the continued provisioning of children throughout much of their lives. The result of this intricate group life is that we can anticipate the survival of our family members. These biologically grounded behaviors increase opportunities for our immediate survival and help to ensure our future survival and reproduction.

HUMAN FEATURES

Extended Life Stages

The period of growth and development in humans is longer than that of other catarrhines, expanding the time during which the mother-infant relationship can develop and become intensified. One of the primary features of this expansion of the life stages is that the offspring is provisioned by the parents as both an infant and a juvenile. This responsibility usually rests with the mother. Apes and Old World monkeys also exhibit long-term relationships between mothers and their juvenile, subadult, and even adult offspring. The mother provides a protective environment and enhances the chances that the young will avoid predation and social aggression from within the group. Mothers also facilitate finding and processing of food and promote the establishment of their own as well as their offspring's social relationships. Mother-infant relationships in nonhuman catarrhines do not, however, generally involve exchange of food. After weaning, juveniles are nutritionally self-sufficient.

Humans are not unique among the catarrhines in maintaining associations with their offspring throughout the life stages. Provisioning of children and the proximity and guidance of the parents and other kin both diminishes the risks of starvation and protects from predation, accidents, and social aggression. This may greatly reduce the possiblity of juvenile mortality, a feature often emphasized in current life-history theoretical studies (Harvey 1990; Janson and Van Shaik 1993).

At birth the human infant demands active caretaking as she or he is unable to "grip and go" as do other primate infants. Touch, smell, and sight of the mother are important to the infant not only for food, thermoregulation, and protection, but also for psychological well-being and normal growth. Parental investment among humans is particularly intensive, using the primate reliance on the sense of sight and touch for individual recognition. As with other primates, older human infants tend to accompany the parent, usually the mother, wherever she needs to go. As the child grows, however, the mother often increasingly relies on the assistance of others, including grandmothers, aunts, and older siblings of the child.

Young children learn to interact and function within their social worlds. Brain growth is accelerated so that the brain reaches adult size well before full body size is achieved. Culture and language, acquired during early childhood, necessitate the recognition, for instance, of distinctive aspects of sound or actions. Children learn culture-appropriate responses and their consequences. Brain growth and anatomical organization appear critical to the development of full human communicative skills. It is also important for the establishment of a cultural world view by which to organize and explain our surroundings.

The expansion of the early life stages in humans also allows additional time for interaction of the body with the physical environment in which it must continue to survive and eventually reproduce. Physiological reactions to environmental stimuli allow for modulation of growth and development to exploit or cope with factors present inside and outside the body. Flexibility in the pace of growth also allows for slowing or "catch-up" periods rather than a more strict adherence to life-cycle timing via a genetic template. We literally are the product of what we experience, a mammalian feature that is exaggerated in humans (Vitzthum chap. 18).

Information Exchange

Human reproduction has been called the exchange of information through two channels: the genetic and the cultural (Boyd and Richerson 1985; Worthman 1993). We cannot divorce ourselves from culture, and our childhoods are steeped in the passage of information between generations and within peer groups. The rapid growth of the brain and its relatively large size in adults has made possible, and may be simultaneously necessitated by, the evolution of the complex social organization, traditions, and relationships that we consider culture.

The nutritionally dependent juvenile, seen above, can also be depicted as informationally dependent (King 1991; Worthman 1993). As the child grows, the sources from which she or he can draw this information expands to wider social networks. Gradually, the child is also able to test and develop new information. In order to accomplish this, however, a protective environment in which she or he can explore, play, and learn is established. To become a viable human adult, development of both subsistence and social skills is essential. The acquisition of social abilities is often of greater importance to future reproductive success than physical size or age at sexual maturity as predicted under conventional life-history theory.

Involvement of children in the lives of their younger siblings also forms an important channel for information exchange. This two-way interaction allows the infant exposure to appropriate and inappropriate behaviors for a different age and developmental stage while under the protection of the caregiving sibling. The caregiver gains experience in meeting the demands of a younger child, experience that may greatly enhance the survival of the sibling's own offspring.

The role of siblings in the rearing of subsequent offspring is found in other mammals, including other primates (Nicolson 1987). Humans, however, as usual, have expanded this role and in many cultures, require work of children as part of the economic unit of the family. Here, many of the gender-differentiated behaviors become apparent (Morelli chap. 15). Parental expectations of the roles of their daughters and sons in the future and the traditions within each culture help guide the assignment of tasks and the level of interaction seen between adults and children.

The Predictable World

Humans do not live in a closed system. Energy supplies can be increased or modified in form, as is readily apparent in the manner in which humans interact with their surrounding environment. For instance, our close relatives among the primates appear to rely on a diverse diet and maximize the possible alternative food sources that can be exploited in times of shortage. In humans, the transition from primate foragers has increased the predictablity of food so that individual and group survival is less frequently in jeopardy.

Mammals are individual foragers. Each member of the group is responsible for acquiring, processing, and consuming his or her own food. The mother-infant dyad, which functions as shared foragers with the mother obtaining and distributing food to her offspring, is the primary exception. In most cases, the offspring will learn to do the same for itself from observing the mother. In some primates, such as chimpanzees, however, the mother may actively teach her youngster to process foods that are more difficult to obtain (Boesch 1991b; Morbeck 1994). A second exception to the individual forager model among the primates occurs with the hunting and distributing of meat. Studies of chimpanzee hunting activities demonstrate that this cooperative effort also involves distributing this highly favored food to other members of the group (Goodall 1986; Boesch and Boesch 1989).

Awareness of the ability to manipulate the environment so that it will yield additional energy supplies is already present among the primates. No doubt, it was important to our own evolution. By 4 to 6 million years ago, hominids, members of the lineage that has produced modern humans, had evolved bipedal locomotion (Lovejoy 1988). While freeing the hands for gathering, transporting, and preparing food items, bipedalism allowed both a potentially greater diversity and, thus, reliability of nutritional energy sources. Restructuring of the primate body plan in terms of body proportions and movement and weight-bearing abilities (e.g., heavy, long hindlimbs), in turn, also must have been expressed as changes in pattern of growth and development of youngsters. Feet "evolved" for humanlike bipedal walking and hands only for manipulation, not habitual locomotion, differ from strong, grasping feet and hands of nonhuman primates. In addition, infants and juveniles most likely had shorter hindlimbs than typical of their parents. Mothers were confronted with offspring that could not travel the distance or at the pace of the social group and needed to be carried, either by their hands or some other means.

Hand use clearly is associated with the making of tools, and stone tools are preserved in the archaeological record of 2 million years ago (Schick and Toth 1993; Kingdon 1993; also see McGrew 1992). These probably developed from a concept of tools based on wood, bone, and fiber. In most cases, stone tools appear as used cobbles or simple flakes of uncertain use. It is likely that they served a multitude of functions including the processing of food items. Our ancestors, like modern chimpanzees, apparently understood form-function relationships and the nature of their behaviors for the short-term manipulation of environmental resources. Again, humans engineered a way to obtain foods that were otherwise not very accessible and thus made their own survival more predictable.

The effects of the ability to control fire by humans were dramatic. The exact time of the appearance of the control of fire in the course of human evolution is hotly debated. Less disputed are the changes that it brought (Clark and Harris 1985). These include (1) the ability to cook, denaturing proteins and making them more readily digestible; (2) the use of fire to catch and, possibly, kill prey; (3) the use of fire to defend against predators and aggression during competition for resources; and (4) the use of fire to help fashion tools. Furthermore, we see fire as critically important because it could transform night into day, enabling humans to devote even greater portions of their lives almost exclusively to the enhancement of culture. This last effect of the use of fire, while often not recognized, allowed humans to convert from the realm of biological time to that of cultural time.

With prehistoric, as well as contemporary, gatherer-hunter groups we see evidence of further development of the concept of time. Planned seasonal mobility became the norm as groups followed both plant and animals throughout the year. People facilitated the growth of plants by clearing surrounding competitive plants or redirecting resources so as to enhance the growth of desired foods. Communal gathering was used to collect foods that had a short harvest life. No longer were the hunters forced to follow the herds of game animals or wait for the appearance of the animals in the humans' home territory. Groups could move to specific areas where animals were predicted to congregate. Hunts could be planned cooperatively to situate the game in the ideal location for the kill.

The benefit of this planning was that, again, the food resources became more predictable. In such a long-lived species, with heavy parental investment as well as investment of time and effort in social relations within the group, the knowledge that there is adequate food to guarantee survivorship becomes important. Although life may still be difficult, it is less subject to the oscillations imparted by the environment.

Predictability in food resources appears to become paramount with the introduction of more extensive cultivation, agriculture, and animal husbandry. In most skeletal series that track the transition of human populations from a gathering to an agricultural lifeway, there is an overall decline in the average health. This is, however, accompanied by an increase in population size. Life may be hard, but more people survive because there is a greater chance that at least minimal food will be available. Most traditional agricultural groups make extensive use of a number of crops as well as continuing to use wild foods. This mixed dependence also helps protect against the catastrophe that can ensue if there is general crop failure. However, people cannot automatically shift from agricultural foods to wild foods in years of a poor harvest as, at these times, even most wild foods are extremely scarce. Trade provides a mechanism through which food and other resources can be acquired, and virtually all human groups do negotiate some form of trading arrangements as a hedge against inclement times or as a means to obtain desirable items (Morelli chap. 15; Draper chap. 16).

As with the shifts in gathering and hunting patterns noted earlier, agriculture reinforces the concept of time. Seasonal activities must maximize the exploitation of natural conditions. Soil preparation must occur before planting, which corresponds to the rains. Harvesting must occur before the onset of adverse weather conditions. Cultural mechanisms that assist in the calculation of these times are found among agriculturalists.

With the Industrial Revolution, there has been a shift to the exploitation of fossil fuels and technology to help avoid any shortages in resources. We accomplish this by greatly increasing the energy that we incorporate into the human consumption and expenditure network. This is a primary disrupting force of energy flow and allocation of energy among humans. The available resources in modern societies, including non-Western and developing nations, is not a fixed amount. Trade-offs as proposed by conventional life-history theory are less calculable if there is the possiblity for additional resources occurring in the future.

Our drive to establish predictability has brought us to our current situation as a species. By failing to understand the complexity of the ecological situation, we frequently undermine the delicate balance in our attempts to increase production of material that we desire or limit what we find undesirable. Our determination to meet the needs of our immediate social groups over the short term blinds us to the long-term and even global consequences of our activities.

The Concept of Time

The human understanding of time has grown out of the biological, social, and cultural changes in our evolutionary past. First, we have long lives and maintain relationships with each other throughout our lives. During this time, we develop the complex neural mechanisms and accumulate memories by which we can store and retrieve this information. Second, like other catarrhine primates, we depend on the recognition and identification of individuals and the associated long-term social relationships. The complexity of our intelligence is, most likely, linked to our social life (Jolly chap. 19). Humans, for example, frequently remember events in the social context even if the event is "nonsocial." For example, the dates of political campaigns may be remembered as coinciding with the birth of a child or a death in the family.

Third, we have used culture and language to organize and explain the world and transmit this understanding to others. Time is viewed as part of the natural world and, as such, subject to a similar form of organization and categorization. Although the depiction of time varies throughout the world, some recognition of seasonality and human generational time is universal.

To some extent, our concept of time has given us unwarranted confidence in our abilities to plan. While formulating plans for the future, including such "decisions" as reproduction, even humans frequently fail to comprehend fully the long-term consequences. Planning for the future, for contemporary humans, is a concerted effort—it does not "come naturally." Our evolutionary history promotes immediate survival needs and our awareness of the past, present, and future does not mean that we are willing to forgo these needs and desires.

The Cultural Overlay

Zihlman (chap. 13) follows the comparative, functional, and evolutionary perspective set out in this volume. She builds on her studies of the biology and behavior of apes (Zihlman chap. 8) combined with information from modern humans, fossil, archaeological, and paleoecology records to provide a glimpse of our evolutionary past as hominids and humans. Special emphasis is placed on the similarities and differences of females and males.

Borgognini Tarli and Repetto follow Zihlman's approach taken in the previous chapter but focus on what we can read from bones and teeth about the nature of variation associated with sex. Comparisons are made cross-culturally, as defined by archaeological and historical data, and through time in populations believed to be biologically continuous. Their functional perspective links much of the variation as seen in quantitative sex differences with effects of socioeconomic and labor-related behaviors as well as flexibility in growth and development that results in adult form. Their work exemplifies the extent of plasticity as preserved in skeletons (e.g., in body size and proportions) as a response to environmental influences. We can also assume flexibility in reproductive physiology. This key aspect of human life is the interplay between biology and culture, a theme illustrated in the remaining chapters and emphasized by Vitzthum for its evolutionary significance.

Field studies of our own species quantify what modern people do and confirm the importance of these studies of natural history in the cultural context. Morelli highlights the cultural component of the biocultural interplay. Like Zihlman and Borgognini Tarli, however, she is interested in female-male similarities and differences. Her comparison of the socialization of girls and boys into culturally appropriate gender roles in two societies demonstrates the impact of differences in nutritional base and predictability of resources.

Draper also illustrates the contrast in gender roles in two kinds of economic systems but in the same society over time. Individuals are linked directly to their populations. Her chapter concludes with a discussion of changes in gender roles associated with sedentary lifestyle, socioeconomic interactions, and related demographic changes. This demographic shift may, ultimately, have possible evolutionary implications.

Panter-Brick also uses a comparative and functional approach with two human populations dependent on agriculture. Instead of focusing on how children are socialized into their societies, she concentrates on the other end of the spectrum—how the mother copes with all phases of reproduction while simultaneously working to support her family.

Vitzthum places her study of women into an evolutionary construct. Using detailed information on one biobehavioral aspect of maternal investment, she develops a critical assessment of the flexible response system so important in human reproduction. She integrates her observational data with an understanding of the physiological processes that underlie the expressed behavior. This physiology and its flexible development reflect our shared ancestry with primates and mammals. Her discussion allows us to understand how evolution works in contemporary populations.

13 Women's Bodies, Women's Lives: An Evolutionary Perspective

Adrienne L. Zihlman

FOR MANY years I have been writing about the role of women in evolution, an interest that grew out of my early research on the origin of human locomotion and functional interpretations of fossil pelvic and limb bones. Hominid (i.e., the human family) locomotion allows individuals to walk long distances and to carry objects. Early in my academic life, I proposed that survival features such as locomotion evolved to promote male activities such as hunting. Women had been largely invisible in reconstructions of early human evolution, and their activities were passively portrayed. However, by the 1970s, I began to rethink women's roles in evolution. As a consequence of this new perspective, Nancy Tanner and I wrote several articles on women and human origins (Tanner and Zihlman 1976a; Zihlman and Tanner 1978). We questioned traditional assumptions such as: hunting is the center piece of human origins; the sexual division of labor is a defining feature of the human way of life; and a pair bond is the primary social and reproductive unit. Our writings, by emphasizing the female contribution to subsistence, focused on putting women into the evolutionary picture as active participants in mate choice and in the wider social world.

Ongoing debates in anthropology focus on sex versus species differences in early hominids. For example, the basis for drawing conclusions about sex differences in behavior has derived from the supposed ability to attribute a male or female sex to fragmentary or isolated fossil bones and teeth. But little thought was given to how physical features affect the behavior of females and males. In this chapter several of my interests converge: examining the relationship between anatomy and behavior; sorting variation in fossil anatomy (e.g., sex vs. species characters), and portraying women's activities in evolutionary reconstructions. Going beyond the earlier theories on women's activities in prehistory—which have not been based on morphology—I explore how women's bodies and lives reflect continuity with an evolutionary past.

The comparative perspective of evolutionary "layers" illustrates how the physical, behavioral, and social dimensions of women's lives and bodies result in the adaptability that is an important life-history feature of the human species. An evolutionary perspective incorporates two dimensions, process and time. The process of evolutionary change through natural selection operates on individuals during each stage of life from infancy through adulthood. The cumulative effect of change is expressed over generational time in populations. A longer time frame encompasses recent human history, more distant hominid origins, and extends back to nonhuman primate and earlier mammalian ancestors. A life-history perspective brings time and process together, for example, by taking into account growth and development as individuals move through life, as well as long-term adaptations for both survival and reproduction (Morbeck 1991b; chaps. 1, 9; Zihlman et al.1990).

Life-history theory allows the exploration of females and males in terms of what each sex has to do to survive and reproduce. Although females and males of the same species share species characters, at the same time each sex diverges to a greater or lesser degree (Zihlman chap. 8).

Many reconstructions of hominid evolution, instead of examining females in their own terms, interpret either women's bodies or lives relative to those of men. In addition, the many facets of women's lives often are ignored in preference to a narrower focus on one aspect, such as child care or labor. I offer alternative ways to view what usually is presented narrowly as the sexual division of labor and a monogamous mating pattern. This chapter discusses how women's bodies and lives are dynamically intertwined throughout life as they have been throughout evolutionary time.

The Role of Locomotion

One basis for the evolutionary success and behavioral flexibility of women and other female primates is their continued mobility during late pregnancy and lactation. Locomotion is a thread that demonstrates the connections among structure, function, and behavioral capabilities. Movement and weight bearing affect the whole animal and are important aspects of time and energy budgets. As a survival life-history feature, the mode of locomotion affects individual well-being through all stages of life. The structural basis lies within the body, in muscles, bones, and nervous system, and the behavioral expression occurs in the context of the social group and physical environment. Locomotion, then, illustrates the dynamic interaction among evolutionary history, reproduction, and individual experiences within an environmental context (Zihlman 1992). For the human lineage, bipedal locomotion is a distinguishing character. It appears early in hominid evolution and offers a means to connect life-history characters with our evolutionary past.

Exploration of the connection between behavior and morphology sets forth key features that emerged during successive stages in mammalian, primate, and hominid evolution. Primates share with other mammals the basis for social life in mother-infant interaction. By deepening and lengthening the bond, the primate stage opened up new possibilities for that interaction. Then, by shifting to a new locomotor pattern, hominids further elaborated on the communication system and laid the foundation for the greater behavioral flexibility that is characteristic of the human species. With this sequence in mind, a way to integrate, rather than to separate, women's biological, social, and cultural dimensions is feasible as I demonstrate in the next sections of this paper.

Mammalian Revolution: A New Kind of Social Life

The foundation for a radically new kind of social life emerged during the mammalian revolution some 180 million years ago. In contrast to the ancestral reptiles, mammals are characterized by a complex of structures and interactive behaviors: (1) female mobility while supporting a fetus or newborn, (2) infant-maternal contact through lactation and suckling with maternal care, and (3) audio-vocal communication.

For females, the mammalian locomotor system enables foraging and traveling even during the awkward, later stages of pregnancy and during lactation (Pond 1977). This mobility is possible because of a well-developed muscular system, flexible and mobile joints, as well as the development of motor skills and a high metabolism.

Gestation, lactation, and suckling result in defined stages of growth and maturation. Fetuses and newborns receive from their mothers warmth, nourishment, and protection from environmental perturbations. For infants, the ability to suckle is made possible by the appearance of new neuromuscular structures in the oral-pharynx region not found in reptiles (K. K. Smith 1992). In addition to suckling, this region is associated with all other major functions of mastication, respiration, swallowing, and vocalizations. The appearance of a set of teeth at weaning promotes independence in feeding, allowing the weaned infant to adopt the adult diet (Pond 1977). During juvenescence, body and bone growth continue, and finally sexual maturity and reproductive capacity are reached (Morbeck chap. 1).

The close association of young and adults, however minimal, is unlike the pattern for most reptiles (i.e., probable ancestors of mammals), in which the young must survive without adult assistance or risk being eaten by them. In mammals, contact between a lactating female and her suckling young is facilitated by the infant's keen sense of smell and hearing and the ability to emit vocalizations and to perceive these high-frequency sounds (MacLean 1990). Using smell, females, their young, and litter mates identify each other, and using vocalizations and hearing, communicate their locations, pleasures, and distress. In mammals, the distress call of young may be a universal vocalization (MacLean 1985). Therefore, the survival of young mammals depends on maintaining contact with the mother, promoted by olfactory and audiovocal communication.

Among mammals, the intensity of the bond between female and young varies by species; it may be minimal or extensive. Social interaction and communication begin at birth and are linked to and facilitated by the emotions. Active social interaction is fundamental and pervasive in mammals, initially between mother and offspring and extending to interactions between the young and their litter mates or older siblings. Social play probably originated among litter mates (MacLean 1990), becoming a major activity during juvenescence; it is associated with a period of growth, when learning and practicing of motor and social skills is stressed. The bonds developed during this period become the foundation for social bonds with other individuals later in life.

The social interactions and communications associated with the period of infancy and juvenile development are facilitated by sensory developments as part of major changes in the structure and function of the brain. The limbic system comprises a new part of the forebrain that has no counterpart in birds or reptiles. Damage to the limbic area of the brain also interferes with maternal care (MacLean 1985), and the limbic system is the seat of the emotions. In other words, it is part of being a mammal (see Smuts [chap. 5] for its importance in primates).

Anatomy and Physiology

The mammalian system of reproduction and infant care imposes greater energetic demands on females than do other vertebrate reproductive systems; it involves physiological responses in bone, fat, and hormones. The system has shaped female anatomy and physiology through the appearance of new structures: mammary glands and fat deposits. Milk production, in conjunction with maternal care, imposes increased energetic demands on females. The mobilization of fat stores and calcium from bones promotes effective lactation and thus ensures both the survival of a female's offspring and the maintenance of her own health (McFarland chap. 12; Galloway chap. 10). In addition, acquisition of sufficient fat reserves promotes a female mammal's own growth and maturation, the conception and development of a fetus, and later the support of a neonate and infant (Pond 1984; Bronson 1989; McFarland 1992).

Fossil Record

There is no direct fossil evidence of mammary glands, fat stores, or brain structures in the earliest mammalian fossils, but all living mammals—including the egg-laying prototheres (platypus and echidna) and marsupials—produce some type of milk to feed the young. The behavioral occurrence of lactation suggests that it is a fundamental mammalian feature that emerged in the earliest stages of mammalian evolution (Pond 1977).

Although fossils do not preserve soft tissue, they do provide evidence for mobility and communication. Complete skeletons in the fossil record indicate a flexible locomotor system for foraging, finding mates, and avoiding predators (Crompton and Jenkins 1973). The masticatory system also ensures the processing of the large quantities of food necessary for maintaining the energetic demands associated with locomotor independence and reproductive activities. Fossils indicate that the middle ear, emerged from two ear ossicles homologous with bones from reptilian jaw bones, increased acuity for reception of higher sound frequencies (Davis 1961; Crompton and Parker 1978). The implications of the development of the mammalian vocalization capacity is that such features made possible new ways for individuals to communicate with each other.

In sum, young mammals, as contrasted to other young vertebrates, require more time and energy to survive to reproductive maturity; consequently, mammals produce fewer offspring than other vertebrates. The mammalian caretaking system—with a potentially long and intense association between young and mother and among siblings (i.e., as in primates, including humans)—made it possible for social learning to extend over a long period during development. This system laid the foundation for the behavioral adaptability that characterizes a number of mammalian groups. The variety of mammals illustrates the diverse ways that female mammals meet the demands of survival, mating, and caring for offspring. The primate and later human way of life build on the mammalian foundation of mobility, lactation, and communication.

Primate Revolution: Behavioral Flexibility

Primates, one of many orders of mammals, expanded into new realms of behavior. The mammalian features—mobility during lactation and of communication—were enhanced by extending the life stages and increasing the time and energy females invest in each offspring. Longer periods of gestation, infancy dependency, and especially childhood were the result.

Just as lactation and suckling underlie the mammalian way of life, female behavioral flexibility and the ability of infants to cling to their mother's hair underlie the primate way of life. Primate locomotion is based on the ability to climb by grasping. Hands and feet with long, mobile digits are equipped with nails and sensory pads and have an opposable thumb and big toe (Washburn 1951).

An infant primate (except humans) is able to hold on to its mother from birth. The hands and feet, equipped with considerable muscle and joint flexibility, function effectively in grasping. Newborn catarrhine primates' hands and feet, for example, are relatively twice as heavy as those of adults. While suckling and getting a free ride, the infant manipulates and explores the environment and, at the same time, develops motor coordination and social skills for later independent survival. As it grows, its body proportions change in response to the acquisition of childhood locomotion and manipulative skills (Grand 1983; Morbeck chap. 1).

From the female's point of view, her primate locomotor system equips her for carrying young on her body through all reproductive stages. In contrast to the marine mammals that suckle their infants on the beach (Ono chap. 3; Reiter chap. 4), or to those mammals that keep their young in dens or nests, most primate females, regardless of the presence of helpless infants, remain in social groups that travel around the home range all the year around (Hiraiwa-Hasegawa chap. 6; Pavelka chap. 7). This mobility underlies female behavioral flexibility and plays a key role in the ability to adjust her behavior to meet the demands of reproduction.

Such mobility also affects the infant. Because the infant moves with its mother from birth, it acquires an instant social network through its mother. This social group consists of both sexes and all ages categories. From the bonds established initially with its mother and siblings, the infant expands its social repertoire by forming bonds with other group members (Pavelka chap. 7). Thus, the close physical and emotional attachment of primate infants to their mothers provides the experience for establishing position in the social order (Smuts chap. 5).

Primate infants are socially responsive from birth. They are born with large brains and are precocial in their sensory development (Portman 1990). Like other mammals, they perceive the world through olfaction and touch and also actively grip their mothers. But, unlike other mammals, primates rely on vision, and monkeys, apes, and humans have the ability to perceive color, depth, and details of faces. Continual close physical contact between an infant and its mother facilitates the development of face-to-face communication. From early social experience, primate infants learn the meaning of the nuances of facial expressions, body postures, and gestures that form much of the communication system and promote the development and maintenance of complex social networks.

In many catarrhine species, the infant primate—by learning from its mother's actions and support—acquires a social rank similar to that of its mother (Pavelka chap. 7). Generations of females, or matrilines, contribute to the long-term stability of social groups in most species. Social connections during life and across generations require "brain power," and primates have large brains relative to body size. The primate ability to anticipate and manipulate the behavior of other group members promotes the exchange of information within and between groups (Cheney and Seyfarth 1990). This "social intelligence," as Jolly (1966; chap. 19) labels it, promotes linkages within and between generations and is the basis for behavioral adaptability. Consequently, primates have increased potential to respond to environmental perturbations or to adjust during their lifetime to specific social situations. This ability is acquired during the prolonged experience of growing up. It is related to the long period of lactation and maturation of the brain that continues into adolescence and is acquired before sexual maturation (Lancaster 1985).

Meeting the Demands of Reproduction

Perhaps the best illustration of primate adaptability and individual behavioral flexibility comes from analysis of adult female lives, the major portion of which are spent in pregnancy or nurturing young. When they are not pregnant or lactating, females experience sexual cycles of ovulation and menstruation (Richard 1985). Sexual activity persists until conception takes place and may continue in some species into the early stages of pregnancy. At about the weaning stage, adult females become pregnant again within a few months of weaning, beginning a new cycle of reproduction.

Females and their infants adjust in a variety of physiological and behavioral ways in order to meet the combined energetic demands of surviving, caring for offspring, and maintaining social contacts. Responses combine alteration of travel time and mode, foraging time, and dietary composition, social activity, and infant carrying (see table 13.1).

Anatomy and Physiology

The physiological capabilities of female primates are like those of other mammals. They can add body weight consisting of fat deposits before conception and when pregnant. This added weight is then available for later lactation. Females gain weight and are at their heaviest during pregnancy and at their lightest in the later stages of lactation (Bercovitch 1987; McFarland 1992; chap. 12). During lactation, calcium is drawn from a female's long bones and is transferred via the breast milk to the growing infant for bone and tooth formation. After the infant is weaned, the mother's bone calcium is replenished (see Galloway chap. 10).

Individual Activity

Time-allocation studies on focal animals help estimate energy budgets during different reproductive phases (Coehlo 1986). To alter their energy budget during different reproductive phases, females adjust daily or seasonal activities in a variety of ways. For example, they may increase foraging time and food intake, consume higher-quality foods by entering feeding areas first, taking priority in feeding, or having access to particular feeding places (Dunbar 1977; Waser 1977; Gautier-Hion

TABLE 13.1. Female Accommodation to Reproductive Demands

Anatomy/Physiology
Body weight
Body composition and fat stores
Calcium mobility
Individual Activity of Female
Travel
Time spent
Substrate and height in trees
First access to feeding areas
Feeding
Dietary composition
Quantity of food
Time spent
Mother-Infant: Lactation and Weaning
Time and distance infant carried
Variation in lactation
Early weaning of infant
Late weaning of infant
Social Dynamics
Decrease in social interaction during lactation
Preferential access to feeding sites
Sharing of caretaking of infant with others

1980; Cords 1986; Kano 1992). Additionally, during pregnancy and lactation, females may decrease the distance they travel or spend less time traveling (Altmann 1980).

Mother-Infant Interaction

The species pattern of female biology and infant development affects the duration and nature of mother-infant interactions. For example, among baboons, mothers accommodate by altering the time and distance they carry suckling, growing infants (Altmann and Samuels 1992). The youngest infants, which are completely dependent nutritionally, are carried by their mothers during all travel and foraging, for a total of 8 to 10 km per day. By 8 months of age, both carrying time and distance declines to almost zero (Altmann and Samuels 1992). The physiology of lactation also can vary even within a single species (e.g., Japanese macaques [Tanaka 1992]).

In addition, the relationship between growth and weaning may vary among and within species. Fast-growing infants may wean earlier, as in patas monkeys, whereas slower-growing baboon infants wean later (Altmann 1980; Chism et al. 1984). Even after weaning, most catarrhine females continue to interact with juvenile offspring that remain emotionally dependent, and this social activity contributes to a female's energy output (Morbeck chap. 1; Smuts chap. 5; Pavelka chap. 7).

Social Group

As noted previously, primate reproduction and caretaking take place in the context of social life and affects all group members. Females and their offspring, even if adults, form the core of social groups of many catarrhine species. Consequently, older primate females generally contribute to generational continuity, to passing on traditions, and to the stability of social groups over the long term.

Social living provides many benefits to a mother. For example, it offers protection for the lactating females, their infants, and especially vulnerable juveniles. Because social interaction requires that individuals devote time and energy at all life stages to establishing and maintaining social connections, the time spent in social interactions and relationships varies during females' lives. From an energetic perspective, for instance, the time a mother has available for feeding or resting may be increased by her infant's time spent with a group member "baby sitter." Or, a mother may decrease the time spent grooming, as do savanna and gelada baboons when they are particularly stressed during lactation. When the infant is weaned and is no longer being carried, the mother may again engage in more grooming (Altmann 1980; Dunbar and Dunbar 1988). This very flexible reproductive system thus allows a variety of ways to promote both survival of a mother and investment in the biosocial future of her infant.

Fossil Record

Fossil evidence for grasping hands and feet appear among early primates about 50 million years ago. Catarrhine primates are found in the fossil record some 35 million years ago. Forward-facing bony eye orbits and an expanded occipital (visual) region of the cranium suggest the development of stereoscopic and, probably, color vision and day living. Later, less than 20 million years ago, large incisors and square low-crowned molars suggest fruit eating, and limb bones indicate generalized quadrupedal locomotion. Although there is little direct fossil evidence for interpreting social behavior, it is possible that at least small social groups (10–15) existed.

In sum, the combination of features—locomotion, long gestation and lactation, and complex communication—contribute to a reproductive system with considerable investment by adult females. Moving from reptiles to mammals to primates entails an increasing commitment of female time and energy to producing and rearing offspring. At the same time, the entire group contributes to the welfare of its members and allows the context for young to be protected during their long period of dependency and provides time for learning. A fundamental part of the behavioral flexibility and adaptability of primates is the ability to adjust behavior to meet the competing demands of survival and reproduction.

Hominid Revolution: Bipedal Walking and Communicating

The human family, the Hominidae, builds on the combined mammalian and primate foundations of infant care, female mobility, and social communication. The origin of hominids from ape ancestors came about during the acquisition of a new locomotor pattern based on upright posture, weight bearing, and movement. The new way of life was based on habitual bipedal walking that, in turn, profoundly modified body shape and proportions.

For all hominids, the new locomotor pattern enabled long-distance travel and carrying. This pattern probably emerged when early humans came to rely heavily on ground resources dispersed seasonally and spatially, as they moved away from the forests into the more mixed vegetation of the savanna mosaics (Laporte and Zihlman 1983). Individuals then could forage over a wide area, collect and carry a variety of foods, and scan the landscape for resources and possible danger. Upright posture makes for more effective displays against predators by permitting vocalizations, jumping, arm waving, or throwing objects. Their hands and upper limbs could be used for manipulation and gesturing rather than for weight bearing during locomotion.

The distinctly human body shape and proportions can be attributed to this locomotor adaptation. In body proportions, humans have massive lower limbs and light upper limbs relative to total body weight (Zihlman 1992). The size and shape of the human pelvis and associated musculature play a major role in locomotor function. The short and broadly curved human pelvis provides a link above with the upper trunk and below to the thigh, leg, and foot. The curves of the pelvis and thigh accommodate the attachment of muscles for hip extension and rotation and reflect the underlying massive musculature and large joint sizes.

There are physiological and mechanical costs of this upright posture. Increased biomechanical stresses on hip, knee, and ankle joints and in the vertebral column, especially in the lower back, result from the vertical body alignment and intense compressive forces generated during bipedal locomotion. The brain now lies above the heart and sufficient blood flow is maintained by overcoming gravity (Falk 1990). This locomotor system, a "makeover" from that of a quadrupedal ape, must have offered advantages that outweighed these difficulties (Zihlman and Brunker 1979).

Bipedal locomotion and consequent body reorganization affected hominid females even more than males. For women, the pelvis serves multiple functions and possible competing selection pressures: the basis for bipedal locomotor function; the pathway to conception, exit for birth, and a framework for positioning the genitals; and a means of internally carrying a fetus and externally carrying a nonclinging, suckling infant, and possibly food and implements. For these reasons, looking at the hominid way of life in greater detail from a life-history perspective—from the point of view of females, infants, and juveniles—offers a different perspective on the hominid adaptation in contrast to a traditional, usually male, point of view.

Women's Bodies: Mobility, Work and Reproduction

As is becoming clear from studies of women cross-culturally and of women athletes, the physical capabilities of women are indeed amazing. During several million years of human prehistory, until food production some 10,000 years ago, women routinely traveled long distances to collect and carry food, and nursed and transported their infants during their early years. They had to adjust to competing demands of ovulation, gestation, lactation, work, and child care (Shostak 1981; Lee 1979). Women's bodies have evolved in response to the range of activities throughout the life course needed to survive and reproduce. Their anatomy and physiology accommodate an extended period of breast-feeding an infant with a great deal of brain growth after birth, while maintaining their own mobility and survival.

The human species is unusual among mammals and primates in having abundant body fat, and women have significantly more fat than men and children (Pond this volume). Relative to body weight, women, on average, have more than 25% fat, whereas men have only 14% (Forbes 1987). Other primates—female macaques, for instance—have about 13% body fat (9% in males), which is less than half of the percentage for women (McFarland chap. 12). Therefore, this lesser amount of body fat may be more typical of nonhuman primates.

Several issues center on this nature of tissue composition. First, why do women have so much body fat? Second, how does the amount of fat affect the proportion of other bodily tissues, especially muscle? Third, what can account for the specifically human pattern of fat distribution (Pond chap. 11)? In order to address these issues, we must look at not just fat, but the whole body and its function.

There is not complete agreement about the function of human body fat. Laboratory and field research suggest that, from an evolutionary perspective, women meet the competing demands of lactation and work effort by using fat reserves as a cushion against nutritional variation. In the long evolutionary time frame, such reserves probably contribute to successful reproductive outcome at all levels: in ovulation, conception, successful completion of a pregnancy, lactation, and survival of infants and children.

The distribution of human and primate body fat follows a mammalian pattern: storage in the pectoral region, thigh, and base of tail (or buttocks in the absence of a tail) (Pond and Mattacks 1987b; Pond chap. 11). These locations lie near the center of gravity, a placement that does not interfere with efficient locomotion. Locomotor function is vital to all aspects of survival and reproduction, and the ability to walk long distances, to carry food and infants, and to use the upper limbs and hands cannot be compromised.

The greater percentage of body weight devoted to fat in women, compared to other primates and to men, alters the proportion of other body tissues. For example, in macaques, 40 to 45% of body weight is devoted to muscle, a figure that is typical for many other quadrupeds (Grand 1977a). In women, 36% of body weight is muscle. Men, who have about 43% muscle, have departed less from this primate pattern than have women.

With lesser amounts of muscle tissue, which muscles groups have been modified in women? Muscles of the hip, thigh, and leg provide propulsion, braking, and balance functions during bipedal walking, and these muscle groups seem least affected in women. Women are capable of endurance in long-distance walking and running, and physiological research indicates thigh muscle strength is similar in women and men (Dyer 1982).

Humans share with other apes (hominoids) a large clavicle, broad and flat chest (Schultz 1969a), and well-developed shoulder and arm musculature (see Zihlman chaps. 8, 13). The upper trunk of women departs more noticeably from this pattern than does that of men. Women are less muscular in their upper trunk and upper limbs. Nonetheless, women are capable of heavy work, and they do such work throughout their lives.

Women's bodies, then, build upon the mammalian and primate features of fat storage and mobile calcium, but during hominid evolution, a greater proportion of body fat was added, reshaping tissue composition. The modified proportions of fat to muscle tissue does not diminish women's locomotor abilities, endurance, and strength. Women's bodies not only perform many tasks but also do so during their reproductive lives. In their postreproductive lives, women continue to participate in economic activities, in caretaking, and in the social activities of the group.

Women's Lives: Mobility, Work and Reproduction

For most of human history, humans had a nomadic way of life. It is under these conditions, rather than those of our more familiar village or urban life, that human lives and bodies evolved.

The interrelationship between locomotion and mobility, work effort, and the several dimensions of reproduction in women's lives have been highlighted by anthropological studies (e.g., Marshall 1976; Shostak 1976; Lee 1979). Studies on the gatherer-hunter !Kung in Botswana provide a perspective on how nomadic life affects women. The findings challenge the traditional and static image of immobile, stay-at-home, and physically weak women (Lee 1968, 1969; Marshall 1976; Shostak 1976). Data on the contributions of women—from before through after the reproductive life phases—to the economic and social life of such societies replace the old stereotypes. The subsistence pattern requires that people traverse a large home range to carry implements, find water, collect, hunt, and transport food and implements. Women travel long distances, carry heavy loads, find and collect food for their families, take care of others, all the while carrying and nursing their infants.

Women accommodate to frequent travel, to the effort of finding, picking, digging, and carrying food and firewood back to camp during pregnancy, infant-carrying, and lactation (Lee 1972, 1979). A !Kung woman nurses and carries each child for more than 3 years; consequently, the birth interval is over 4 years. Early studies recognized that lactation acted as a contraceptive by suppressing ovulation and contributed to wide spacing of births (Konner and Worthman 1980). The birth interval perhaps also is compounded by nutritional stress and disease and by the effort of collecting food while carrying infants; together these factors contribute to amenorrhea (Howell 1979).

Research during the past few years has focused on the variables that influence the outcome of a woman's reproduction during her lifetime (Ellison 1990). Ovarian function, time spent in activities (an indirect measure of energetics, see Panter-Brick chap. 17), and the types of activities women perform have been measured (Ellison et al. 1989; Peacock 1985) under a variety of conditions. Together these studies demonstrate the complex interrelationship between individual physiology—including body weight, nutritional status, ovarian function, and milk production—and external, ecological variables—such as food availability and caloric intake and disease—on reproductive outcome (Prentice and Whitehead 1987). In addition, the behavioral dimension of individuals, such as the distances they travel and the amount of work done each day, clearly intersect with cultural practices relating to subsistence patterns and child care (Panter-Brick chap. 17; Vitzthum chap. 18).

A cross-cultural perspective on child care, according to time-allocation studies, indicates that women add child care to their other tasks (e.g., Efe foragers in Zaire [Peacock 1991]). In other words, women with and without children spend a similar amount of time engaged in multitask work.

In foraging and horticultural societies, women can and do perform the same range of tasks as men. For example, work patterns are influenced by the seasons, and among the Tamang in rural Nepal, increased efforts coincide with rice planting during the monsoon. All agricultural labor is valued equally, and everyone works. Mothers and non-childbearing women engage equally in agricultural tasks (Panter-Brick 1992a; chap. 17). Among the Agta in the Philippines, women with and without children work hard, take risks, travel many miles over rugged forest slopes, and hunt with bow and arrow (Estioko-Griffin and Griffin 1981; Estioko-Griffin 1985). Every able-bodied Agta is capable of accomplishing any task necessary for the group's subsistence. Among the Efe in Zaire, task assignments are not determined by physical strength, but rather by the energetic constraints imposed by pregnancy and lactation in a food-limited environment (Peacock 1991). Women also cooperate in child care. For instance, women may suckle each other's infants so that Efe women who are lactating may continue their work in the fields. In addition, women may travel shorter distances than men, in order to return home by evening to prepare the meal.

Cultural practices affect the relationship between women's work and their reproductive outcome. For example, among the Tamang, children are weaned after the age of 2 years because they become too heavy to carry. These juvenile children are not taken to the distant fields but are left alone or with an older sibling. They are at risk because their health is compromised by inadequate nutrition and sanitation. In fact, there is a high mortality rate for children under 5 years (Panter-Brick 1989). Because their child-care practices and work patterns differ, children of the Hindu and Tibetan ethnic groups in the same region do not have high mortality rates. Children in these families either travel with their mothers or stay home with senior women, often the grandmother.

In each culture, the particular combinations of work and child-care practices may affect survival of children and therefore also influence women's lifetime reproductive effort. It is evident that when all dimensions of women's lives are looked at in a variety of cultures, women around the world "do it all." Women are physically capable of performing a full range of tasks including hard labor; child care is simply added to the other tasks that women already perform.

The Fossil Record

During the course of human evolution, two anatomical changes that profoundly influenced the hominid way of life are documented in the fossil record: bipedal posture and locomotion and the development of a large brain. The earliest evidence for the adoption of upright posture and two-legged locomotion dates to more than 3.5 million years at Laetoli, Tanzania. There, tracks of hominid footprints, along with teeth and jaws in the same fossil deposits, indicate the appearance of bipedal hominids (Leakey and Hay 1979). At this early stage of evolution between 3.5 and 2 mya, fossil evidence provides a picture of early hominids. They had small brains (450 cc)—little larger than those of modern common chimpanzees—small canine teeth, and large molars and premolars. Body proportions indicate short rather than long lower limbs—proportions similar to those of apes—and somewhat shorter upper limbs, suggesting a lower center of gravity than in living apes (Zihlman 1992). Between 1.5 to 2 million years ago, the genus *Homo* developed a larger brain size (850–1000 cc) and smaller molar teeth. A nearly complete skeleton of *Homo erectus* from Kenya shows body size and proportions that appear to be similar to those of modern humans (Walker 1991; Walker and Leakey 1993).

The fossil record provides little evidence of soft tissue and, therefore, few clues about when an increase in body fat might have occurred. Except for teeth, jaws, and crania, which are more often preserved as fossils, remains of the rest of the skeleton are rare and incomplete. Sorting females from males always is problematic, even for fossil pelves. In modern humans, this part of the anatomy distinguishes between women and men, because the female pelvis changes during growth and development to create a birth canal large enough to accommodate the passage of a large-brained neonate. But brain size in early hominids is small, so that the pelvis is probably similar in females and males, much as it is in modern apes (Hager 1991).

The fossil record, however, does provide evidence for rates of development and offers opportunities for reading some of the life-history characters in the bones and teeth of past hominids (Morbeck chap. 9; B. H. Smith 1992). Studies of hominid dentition demonstrate that the rate of development, although distinct in early hominids, is more similar to that of chimpanzees rather than to that of modern humans (Bromage and Dean 1985; B. H. Smith 1986). By the time *Homo* appears, overall brain size is larger and the life stages devoted to growth and development of the brain are presumably longer. Probably, at this time, the length of gestation increased over that of the apes, and infants required greater care and a longer period of lactation. Therefore, the length of the juvenile stage also may have increased, which in turn, required mothers and other caretakers in the social group to adjust to this longer social dependency.

Social Communication and Dependent Infants

Looking at human evolution from the perspective of mothers and infants also permits interpretation of the evolution of the large brain and the origin of human language. The two obvious anatomical changes that occurred during human evolution—bipedal locomotion and enlarged brain size—had profound effects on infants, both in their growth and development as well as in their relationships, first to their mothers and, later, to other members of the social group. Both changes required that, compared to other primate infants, hominid infants receive longer and more intensive care after birth.

The reorganization of the body that occurred with hominid bipedal locomotion affected infants' ability to cling and their mothers' needs in assisting them. In hominids, the hands are not involved in weight bearing during locomotion and are less massive in humans (1% body weight) than in quadrupedal chimpanzees (2% body weight [Zihlman 1992]). Furthermore, modifications in the feet for greater stability during walking on two limbs suggest the feet are less effective for clinging. Also interfering with a human baby's ability to cling to its mother might have been the reduction of human body hair. This probably occurred early in human evolution in order to reduce excess heat generated from activity in the open savanna (Zihlman and Cohn 1988). For a variety of reasons, therefore, early hominid mothers had to assume a more active role in supporting infants. Infants, in turn, assumed a more active role in maintaining contact with caretakers through enhanced communication channels (Borchert 1985; Borchert and Zihlman 1990).

The hominid communication system builds on the mammalian ability to give and receive vocalizations for mother-infant interaction (MacLean 1985; Zihlman chap. 8). Added to this is the primate emphasis on visual, especially face to-face, communication, which is facilitated by the active holding of infants and the close physical association of infants and mothers during nursing and carrying (Tanner and Zihlman 1976b). During hominid evolution, both infants and mothers actively maintained mutual contact by elaborating both the visual—gestures and facial expressions—and vocal modes. For example, the hands no longer used in clinging to a caretaker, now became available for more effective gesturing. The development of smiling, reaching and pointing, cooing and crying helped to ensure contact with mothers and other caretakers and thus contributed to the infant's survival.

Language may have emerged from this infant-mother interaction as part of the changes accompanying the adoption of bipedal locomotion. From the point of view of the infant, babies may have first made use of soft vocalizations in order to coordinate activities with their mothers. This need for babies to maintain a more extensive vocal dialogue with their mothers may have produced the initial development and later evolution of sensorimotor intelligence (Borchert 1985; Borchert and Zihlman 1990).

Furthermore, infant and mother hominids, like other primates, functioned in the larger social context and not as an isolated unit. Therefore, vocalizations also could have facilitated interactions between the infant and other group members who also contributed to its survival. Viewing language origins as enhancing the survival of infants and juveniles is consistent with the pattern of evolution of mammalian communication. Vocalizations that are part of infant-mother interactions characteristically show up in adult mammals' greeting displays and sexual behavior (Gould 1983; MacLean 1985).

After weaning during the long juvenile period, young hominids are vulnerable; their body size has not yet reached the adult state; they are practicing their social skills and learning about the environment. A juvenile hominid, like other juvenile primates, is still dependent—especially on its mother and siblings—for food sharing, protection, and for emotional well-being. However, in the context of the social group, young hominids also develop relationships with playmates and other adults. Vocalizations, gestures, and facial communication would have facilitated interaction with others and would have enhanced survival at this life stage also.

The increase in brain size and change in pelvis size during the last 2 million years of hominid evolution indicate that, at some point, infants were born with relatively larger brains compared to other catarrhine primates. Like other primates, human infants have well-developed special senses at birth (tactile, auditory, taste, olfactory, visual), but are very immature in motor function and are, therefore, quite helpless. The human brain is the anatomical basis of language, memory, and social intelligence. It develops rapidly postnatally, from 25% of adult weight at birth to about 95% by 10 years of age (Tanner 1990), and as it grows, so does the learning rate of children. Aside from learning language, juveniles acquire social roles early in life—by the age of three (Morelli chap. 15; Draper 1975b). If the evolution of large brain size is interpreted from the point of view of infants and children, then it relates to the ability of the young to learn a lot quickly so that they survive to sexual maturity. Once language developed, however, it could be applied in all realms of life.

The origin of language might relate more to this early stage of life, rather than to the traditionally emphasized model of language as a product of hunting and therefore a male activity. Language origins are often held up as marking the transition to modern humans, which implies that language was a survival feature, rather than a more specific way to communicate.

Summary: Lessons for Reconstructing Early Human Social Life

Through the discussion of women's bodies and lives as dynamic, variable, and adaptable, this chapter offers another perspective for looking at women during prehistory in terms of their physical attributes, workload, and reproductive system. When placed in an evolutionary framework, the abundance and distribution of women's body fat is usually interpreted as a sexual attraction for men. Interpretations about body fat should consider the overall function of women's bodies over evolutionary time. With this perspective, major changes in women's bodies—including the abundance of fat—emerged in order to accommodate many kinds of activities, such as collecting and preparing food, traveling, nursing, and carrying dependent but growing infants, as well as caring for children after weaning.

Reconstructions of early hominid social life, almost without exception, postulate monogamy as a hallmark of humanity, that is, monogamy supposedly originated some 5 million years ago and has remained unchanged until the present time. The basis for this conclusion is the supposed high parental care by males. This supposition ignores three facts. First, in most cultures, men have little direct involvement in child rearing. Second, women add child care to a wide range of other tasks and often cooperate with each other to handle child care. Finally, it also omits the fact that children are not just part of an isolated mother-infant unit, but are part of a social group, all of whom contribute to a greater or lesser degree to the well being of children. Children, in turn, begin contributing to the society, by practicing skills or by actively participating in caretaking of siblings or assisting in herding, food preparation, and other activities (Morelli chap. 15).

The sexual division of economic labor often is viewed as a defining hominid feature. The implication usually is that women do the "easier" tasks and men do the "heavy," risky work, like hunting. This assumption obscures the fact that cross-culturally women engage in vigorous activities, ones that involve risky behavior, skill, long-distance travel, and the carrying of heavy loads. When the full range of women's activities are acknowledged, it is difficult to view early hominid women's activities as rigidly limited.

In conclusion, women's bodies and lives were shaped over evolutionary time as mammals, primates, and hominids and did not arise merely from conditions evident in the recent historical past. Once that is acknowledged, then it is possible to shift from defining women's bodies and lives only in terms of anatomical and physiological traits or only in the stage of adult life. The cross-cultural variation in women's lives highlights their adaptability and exemplifies the behavioral flexibility established as part of this evolutionary background.

Acknowledgments

The ideas presented in this paper have emerged over a number of years and have benefited from discussions with many colleagues. On this draft, comments from R. McFarland, B. McLeod, M. E. Morbeck, and K. Nichols were particularly helpful. Financial support from the University of California, Division of Social Sciences and the Academic Senate Committee on Research is gratefully acknowledged.

14 Sex Differences in Human Populations: Change through Time

Silvana M. Borgognini Tarli and
Elena Repetto

SEX DIFFERENCES in humans and other species are evolutionary adaptations. Females and males are variations on the species' life-history attributes. Sexual dimorphism, or the variations of sizes and structures in females and males, reflects differences in how each sex survives, mates, and rears offspring.

The human skeleton allows us to study sex differences in present and past populations. Physical features that distinguish women and men include a mosaic of characters that are revealed at different levels of biobehavioral and ecological organization (Morbeck chap. 1).

Sex differences, in part, result from species-specific growth, developmental, reproductive, and aging patterns in women and men. These variations are products of genetic instructions, that is, species characters and X or Y chromosomes that carry DNA or genes. Genetically guided programs produce general size, shape, internal architecture, and composition of human teeth and bones. "Overlays" of female or male features distinguish the sexes, for example, the broad pelvis of adult women associated with giving birth to big-brained infants.

Environmental interactions throughout an individual's life provide another influence on skeletal features. These include, for instance, quantity and quality of food, activity types, their durations, and intensities, social and biological health, and reproductive efforts in females (Morbeck chap. 9; Galloway chap. 10). The social/cultural and economic/technological conditions of humans especially pattern the ways in which environmental factors operate during an individual's life and, in addition, when and how these factors are recorded in teeth and bones.

Skeletons preserved after death document the historic and prehistoric records of past human individuals and their groups. Historical and archaeological records, including written descriptions and material objects, help to place skeletal series and their presumed biological populations in their respective environmental contexts. Teeth and bones give information about individuals, sex differences, and variation within and among populations that can be related to social/cultural and economic/technological attributes.

In this chapter, we measure skeletal features and compare female and male values to describe the nature of sex differences in four sets of human groups that lived at different times, places, or social/economic environments. By quantifying the nature of sex differences in modern *Homo sapiens*, we illustrate the many ways that females and males, within one species, adjust to environmental conditions during their lives. The varying degrees of differences between the sexes as expressed in a mosaic of skeletal features highlight the inherited range of flexible responses to various living conditions throughout their lives (Morbeck chap. 1, 9; Vitzthum chap. 18).

Sex differences in humans and nonhuman primates have intrigued human biologists, paleoanthropologists, and primatologists. Our interest in explaining sex differences in human populations begins with research on the skeletal biology of individuals who lived in the past and their groups. Our aim is to reconstruct subsistence patterns, regime-related activities, and, more generally, the way of life.

The life-history perspective emphasized in this volume includes processes of maintenance, growth, and reproduction. Sexual dimorphism results from differential growth of females and males and, in many ways, is tied to reproductive traits. However, in addition to this evolutionary perspective, our comparative, functional approach has resulted in recognition of several patterns of sex differences in skeletal series that represent people living under different conditions. We can show that the diverse nature of physical differences between female and male skeletons also reflects influences of a variety of environmental, social, technological, and cultural factors (Borgognini Tarli and Masali 1983; Masali and Borgognini Tarli 1985; Borgognini Tarli and Mazzotta 1986; Borgognini Tarli and Repetto 1986a, b, 1989; Borgonini Tarli and Marini 1990).

Relationships among various degrees of sexual dimorphism—for example, differences associated with inadequate nutrition or with economic division of labor resulting in different activities—have been shown for specific populations, although with apparently conflicting results (Frayer 1980; Larsen 1981; Stini 1982; Wolfe and Gray 1982a, b). Therefore, we systematically examine skeletal series from groups that experienced economic transitions as documented by historical and archaeological records. Such situations may have involved changes in activities or sex roles, and we were interested in finding out if these are expressed in female and male skeletons as sexual dimorphism.

Four sets of comparisons of sex differences in skeletal series representing human populations are explored (table 14.1). Case studies are drawn from Europe and the Mediterranean region. These cultures vary in terms of time, place, or social, economic, and technological conditions. Comparisons of sex differences in skeletal series include: (1) Upper Paleolithic peoples and Mesolithic groups (circum-Mediterranean populations during Mesolithic Transition); (2) Etruscans and other Italian Iron Age populations; (3) Iron Age and Roman Age populations in the Marche Region of Italy; and (4) Egyptians during the Predynastic and Dynastic Periods.

Sex Differences: Theory and Methods

Our emphasis is on the nature of sexual dimorphism as revealed by skeletons of individuals as indicators of changes in social/economic structure in defined historic or prehistoric populations. We measure teeth and bones of individuals but analyze data at the level of skeletal series that represent populations. Before discussing the details of sexual dimorphism in these case studies, we must consider the theoretical background and primary methodological problems of quantifying and interpreting similarities and differences of females and males in human skeletal series.

Theoretical Background

Three processes, each working inside the boundaries of human species-specific life-history characters, can produce observable, measurable sex differences. These include: (1) reduced time or rate of growth of females or males; (2) extra time or rate of growth of females or males; and (3) divergent growth of both sexes. The same or different degrees of sexual dimorphism in different groups can be obtained by any one or a combination of these mechanisms.

Among nonhuman primates, physical features of males generally are larger than those of females (i.e., positive dimorphism). However, features that are similar in size or shape also occur (i.e., monomorphism) and, in some cases, female features are larger (i.e., negative dimorphism). Recent human populations

TABLE 14.1. Sex Differences in Human Skeletal Series taken from literature, except for Uzzo Mesolithic, Matera Neolithic, and some Arene Candide material and Sovana Etruscans, which were measured personally the authors. The number of individuals (males and females) in each skeletal series is indicated in the sample list

Population	Time	Cultural and Economic Profile	Samples			Sex Differences
				M	F	
Circum-Mediterranean populations during Mesolithic transition	~10,000 years ago	Hunting/fishing and gathering vs. farming and animal husbandry	*Upper Palaeolithic*			Trend toward overall reduction of sex differences through time
			Afalou-bou-Rhummel	21	14	
			Arene Candide	5	2	
			Balzi Rossi	4	2	
			Brno	2	1	
			Cro-Magnon	2	1	
			Dolni Vestonice	2	1	
			Laugerie Basse	1	1	
			Mladec	2	2	
			Oberkassel	1	1	
			Ofnet	5	7	
			Prdmosti	4	2	
			Taforalt	39	31	
			Mesolithic			
			Columnata	20	16	
			Hoedic	4	5	
			Lepenski Vir	3	1	
			Muge-Moita	34	24	
			Teviec	7	8	
			Uzzo	5	2	
			Vlasac	23	18	
			Neolithic			
			Ligurian caves	22	8	
			Matera	9	4	
Iron Age populations and Etruscans in Italy	3,000–2,500 years ago	Increased urbanization and social complexity; men higher ranked than women vs. men and women similar in social status	*Iron Age*			Sex differences in face but not in brain case and limb bones
			14 series	470	329	
			Etruscans			
			Certosa	4	4	
			Monterozzi (Tarquinia)	16	13	
			Populonia	3	2	
			Sovana	14	11	
Iron Age populations and Romans in Marche region, Italy	3,000–2,500 years ago vs. 2,500–1,500 years ago	Farming and animal husbandry vs. increased sexual division and specialization of labor; men higher ranked than women in both periods	*Iron Age*			Sex differences increase in all the skeletal regions, face excepted
			Camerano	43	34	
			Camerino	8	6	
			Mossa di Fermo	7	6	
			Novilara	19	11	
			Numana	33	42	
			Pieve Torina	8	8	
			Roman Age			
			Civitanova Marche	62	57	
			Fano	6	6	
			Potenzia	67	34	
			Urbino	34	35	

TABLE 14.1 (*cont*)

Population	*Time*	*Cultural and Economic Profile*	*Samples*	*M*	*F*	*Sex Differences*
Predynastic and Dynastic Egyptian populations	7,000–5,100 years ago vs. 5,000–2,300 years ago	Agriculture, hunting, fishing vs. strong central political power with men apparently not higher ranked than women	*Predynastic* Gebelen	15	19	Sex differences in face are greater than brain case; women show little change through time; men become taller and more slender
			Dynastic Gebelen and Asyut	126	118	

generally show positive dimorphism in most skeletal parts. On the basis of "informed speculation," this relationship has been extended to include fossil hominids (Wood 1985; but see Zihlman chaps. 8, 13).

Many anthropologists assume that, in spite of an overall increase in body size, the degree of sexual dimorphism in modern *Homo sapiens* is reduced compared to that of earlier hominids (e.g., Frayer and Wolpoff 1985). However, this generalization about the size relationships of females and males may not be valid. Modern human populations show varying degrees of average size differences between females and males. Our data for past human populations confirm and illustrate this potential for fluctuations in the degree of sexual dimorphism in *Homo sapiens.*

Researchers suggest a variety of possible reasons for variation among populations in the degree of physical differences between women and men. These include, for instance: (1) the effect of mating system, which implies greater dimorphism in polygynous societies (Alexander et al. 1979; contradicted by Wolfe and Gray 1982a,b); (2) the effect of parental investment, which implies greater dimorphism in societies with reduced contributions by the father to child care or well-being (inconclusive results by Wolfe and Gray 1982a,b); (3) the effect of stresses caused by inappropriate nutrition, which implies reduced dimorphism in ecologically or economically marginal social strata or societies (Hiernaux 1968; Stini 1969; but see Eveleth 1975; Lopez-Contreras et al. 1983); and (4) the effect of labor division, which implies reduced dimorphism in societies with less differentiated sex roles (Brace 1973; Frayer 1977; Larsen 1981; Borgognini Tarli and Masali 1983; but see Armelagos and Van Gerven 1980).

These interpretative models have not always been verified in skeletal or living populations. In fact, simplistic causative models are very seldom validated in biology, especially in human biology, since they also must integrate the effects of complex social/cultural and economic/technological conditions. Humans are said to occupy a "cultural niche" that buffers the influences of the natural environment on many physical traits. Careful consideration of sexual dimorphism in terms of these cultural environmental factors, however, allows us to test hypotheses on the observed patterns of sex differences.

Methodological Problems

We confront several methodological problems in evaluating the degree of sexual dimorphism and in assessing the range of its fluctuation as determined from human skeletal series. A whole organism–whole life perspective, for example, shows that it would be impossible to separate the genetic and environmental components of sexual dimorphism. But, we must do the best we can. This involves confronting problems of data collection and analysis. For instance, inadequate data bases and methodological problems related to statistical analyses include: (1) an inherent circularity in evaluating the degree of sexual dimorphism in skeletal samples in which identification of females and males requires previous knowledge of the range of skeletal sex differences; (2) small samples with unequal numbers of females and males and the related problem of determining whether a sample represents its population; (3) the influence of different kinds of features on the interpretation of the "typical" degree of sexual dimorphism for a given population; and (4) the mathematical effects of different measured variables and methods of quantitative analyses.

The biological problem (i.e., the genetic basis for the expression of sex differences) cannot be faced directly. But, difficulties with the data base and statistical analyses may be solved both by careful selection of measured features and by simultaneous use of many techniques for determining sex in skeletons (Borgognini Tarli and Repetto 1986a,b). Our studies on prehistoric and early historic European and circum-Mediterranean skeletal series illustrate how to deal with these problems.

Through our work, we demonstrate the importance of evaluating the effects of the choice of metric traits (the size dimensions of the bones), confronting the issues related to sample composition, employing different types of indices used to measure the degree of dimorphism, and, finally, applying an allometric correction (i.e., accounting for a change in shape resulting from change in size). These data and results of our analyses are part of the case studies presented in this chapter.

What to measure in the skeleton? We measure a variety of skeletal features that include, of course, traditional measurements that can be compared to those published in the literature. In addition, we incorporate functional indicators of overall size, robusticity, and muscularity, and measurements of bone diameters influenced by thickening and remodeling under load-bearing conditions (table 14.2).

How to assess the degree of sexual dimorphism in humans? Anthropologists evaluate the extent of sex differences in skeletal series in different ways. Each method includes a certain number of variants that are linked to each other by arithmetical relations. Since the same data are used, they often give rise to equivalent results (Borgognini Tarli and Repetto 1986a).

We prefer two types of indices. One is the percent difference between male and female means weighted on the male mean. This is based on the calculation of the difference between male and female averages (crude or size-corrected). In order to allow comparisons between traits with a different order of magnitude, the difference is weighed on one of the two average values or on the average between the two sexes (Thieme and Schull 1957; Zihlman 1976; Garn et al. 1972; Brace and Ryan 1980; Frayer 1980; Hamilton 1982; Borgognini Tarli and Masali 1983; Larsen 1984; Perzigian et al. 1984; Rose et al. 1984). The other important index used in our work is based on univariate studies of the distributions of the metric traits in the two sexes, and we evaluate the area of overlapping or of nonoverlapping of the two curves (Bennett 1981; Chakraborty and Majumder 1982; Borgognini Tarli and Masali 1983).

If only a small number of individuals make up a skeletal series, another methodological issue arises. Several skeletal series from different places must be combined to construct an adequate sample for statistical analyses. The choice of the pooling procedure for these composite samples may affect results.

Most researchers report data based on crude female and male average values within given time periods or geographical regions (Wolpoff 1976; Frayer 1977, 1980; Gray and Wolfe 1980; Wolfe and Gray 1982a,b). However, this procedure is not appropriate for both biological and statistical reasons (Borgognini Tarli and Repetto 1986a). From the biological viewpoint, the degree of sexual dimorphism is peculiar to each population, which has its own sources of variation. The commonly used pooling procedure obscures the contribution of each subsample as a result of the blending produced by the calculation of grand means. Furthermore, the pooling procedure based on general aver-

TABLE 14.2. Skeletal Measurements

Functional Region	*Skeletal Part*	*Measurement of Size and Shape*
Head	Braincase	Length Breadth Forehead (minimal frontal) breadth Height
	Face	Cheek (bizygomatic) breadth Upper face height Nose breadth Nose height
Trunk	Backbone (vertebral column) Shoulder and hip joints	Height of the vertebral bodies Clavicular length Pelvic breadth
Limbs		Maximum lengths, diameters, and circumferences
Upper limb	Humerus (upper arm) Radius (forearm)	
Lower limb	Femur (thigh) Tibia (leg)	

age values does not allow the distinction of "between"- versus "within"-group variability. From the statistical viewpoint, it is better to weigh the contribution of each subsample to the average value in terms of sample size because small samples can have exceptionally low or high degrees of dimorphism. Therefore, we calculate the degree of sexual dimorphism in composite samples as a weighted average among the degrees of all the subsamples.

The influence of allometry also poses a methodological problem. Changes in size with resulting changes in shape can affect anatomy in different ways. Two kinds of allometric factors discussed by anthropologists include: first, so-called ontogenetic scaling (i.e., with males as scaled larger versions of females in the case of positive dimorphism); and, second, a departure from ontogenetic scaling with dissociation between variations in size and variations in shape. When the effects of allometry are not the object of the research, the metric data to be analyzed should be adjusted in order to neutralize the influence of the allometric component. The mathematical description of allometry is given by Huxley (1932); the geometric representation of the allometric function is a straight line on a logarithmic scale. The most direct correction factor for allometry, therefore, is a log-transformation of the raw data (Corruccini 1972). We have done this for female and male sample means.

To assess the extent of sex differences, male/female mean values for each variable in the different groups are evaluated by a variety of statistical techniques and tests of significance. These include, for example, one-way analysis of variance, Mann and Whitney-Wilcoxon test, and Spearman's rank correlation coefficient (since data that are represented by indices may not be normally distributed).

SEX DIFFERENCES IN PAST POPULATIONS: FOUR COMPARISONS

The Upper Paleolithic-Mesolithic Transition

Cultural and Economic Profile

The first example centers on the shift from hunting/fishing and gathering to food production (mainly farming and animal husbandry). This gradual economic transition took place in the Mediterranean region after the last period of glaciation about 10,000 years ago. It represents substantial changes in human environmental conditions, economic systems and population structure.

Among the more important economic/technological innovations of this period are: new techniques for working flint, bone, and antler, and new types of tools; improvements in transportation, including navigation; and intensification of plant and animal use, including use of domesticated dogs and a greater reliance on fishing. An increase in population density apparently accompanied these changes in subsistence patterns. The archaeological record reveals a greater number of sites per unit area, larger settlements, more regular settlement systems, and more permanent house structures. This indicates increased sedentism, that is, people were staying in one place for longer periods of time (Newell 1984). Archaeological inferences also suggest that the marriage system may have shifted from a bilocal (i.e., two-place) pattern to a patrilocal (i.e., husband's place) pattern (Constandse-Westermann and Newell 1988).

Sex Differences

Sexual dimorphism, as determined from skeletons, undergoes a general reduction from Upper Paleolithic people to those with Mesolithic tool traditions. Except for facial and upper limb measurements, the overall sex differences in these skeletal series are statistically significant.

Since a reduction of the degree of sexual dimorphism can be due to reduced growth of males, extra growth of females, or convergent reduced growth of both sexes, we compared the weighted average values for the most size-related variables (e.g., braincase variables and trunk and limb lengths and circumferences) in the two groups (Borgognini Tarli and Repetto 1989). Both sexes undergo a size reduction that is, on the whole, more marked in males. The difference between male and female reduction is significant in length and circumference measurements.

The reduction of sexual dimorphism during the Mesolithic transition has been considered a result of a major change in the environmental conditions and in the related change in economic system, technology, and subsistence patterns. The main factors affecting this change are said to be labor division and change in hunting strategies (Frayer 1980, 1981). According to Frayer, stature, cranial, and dental measurements point to an overall size reduction of males and a general stability of females. Our analyses show that size reduction, although less marked, also occurs in females. In particular, an appreciable reduction of long bone circumference measurements in females indicates that a gracilization process took place in both sexes. This would suggest that factors other than labor division are involved.

It is very difficult to isolate each factor that affects size and shape during an individual's life, but other information about population density and social structure during the Mesolithic transition in Europe suggests a reduction of gene flow accompanied the settling of groups (Constandse-Westermann and Newell 1982; Newell 1984; Newell and Constandse-Westermann 1986). Reduced gene flow could lead to decrease in body size and, as a consequence, to the reduction in dimorphism. Furthermore, modification in food procuring, processing, and storage techniques could have been accompanied by reduced labor division or role specialization (Borgognini Tarli and Repetto 1986b). Other potentially influential environmental factors, such as nutritional stress, affecting males more than females, remain unproven in the absence of specific skeletal indications.

Iron Age Populations and Etruscans in Italy

Cultural and Economic Profile

During the Iron Age (i.e., about 3,000–2,500 B.P.), the Italian peninsula was inhabited by a variety of peoples with different language and cultural identities. Direct archaeological data as well as indirect historical data found in the writings of Roman authors who report on the origins of Rome suggest that the non-Indo-European peoples probably represented the indigenous population, whereas the Indo-Europeans were recent invaders from Eastern Europe (Bietti Sestieri 1985).

Signs of reorganization of the settlement system suggest a culmination of urbanization at the beginning of the Iron Age (about 2,800 B.P.). The rise of an aristocracy, the formation of the Thyrrenian "city-states" and of the Adriatic "territory-states," and increased social complexity characterize this period (Salmon 1967; Peroni 1969; Bietti Sestieri et al. 1988; Macchiarelli et al. 1988).

The presence of two funeral traditions, one that burned the dead and the other with burial, shows social stratification (Bietti Sestieri 1985). Funerary goods in tombs with male skeletons are differentiated by rank. Women, regardless of their social position, are represented both in archaeological and historical sources as spinners and weavers (Bartoloni 1988).

In contrast, evidence for the social position of Etruscan women does not fit this pattern. Their status is documented directly by grave goods and by figurative art (mainly sculptures and paintings from monumental tombs) and is described in historical accounts. Etruscan women of the upper social class enjoyed formal education and participated actively in public life. Children apparently were identified by both parents since a mother's name usually appeared alongside that of the father (Cateni n.d.; Durant 1957; Pallottino 1984; Bartoloni 1988). Unless they were Roman women, designated only by their family name, Etruscan women were indicated by first name and family name, which they maintained after marriage (Cateni n.d.). Moreover, monogamous marriage rules were applied to men and women without discrimination. In addition, the laws on divorce were not exclusively in favor of the husband but also considered the rights of the wife (Aziz 1977; Cateni n.d.).

Sex Differences

Sexual dimorphism of Etruscan skeletal samples as compared to other Italian samples from the same period shows appreciable differences only in the facial measurements. All the other sets of measurements show small and fluctuating differences. On the basis of these data, no difference in sexual division of labor between Etruscans and the other Italian populations of the Iron Age can be deduced. The difference in social position of the Etruscan women as compared to the women of other Italian populations, therefore, is not reflected in the metric traits examined here.

Iron Age and Roman Age Populations in the Marche Region of Italy

Cultural and Economic Profile

The skeletal series and archaeological/historical data document populations living in the Marche Region during an earlier Iron Age (i.e., 3,000–2,500 B.P.) and a later Roman Age (i.e., 2,500–1,500 B.P.). The most important changes in these populations through time were associated with economic activities, especially, sexual division of labor.

The Iron Age economy at its beginning apparently was based primarily on domestic products obtained by agricultural and husbandry activities combined with hunting, fishing, and gathering. Agriculture, based on wheat and barley, was rather poor and technically unsophisticated. Animal husbandry was limited to swine, sheep, and goats. Cattle were used for transportation and field work. Subsistence activities involved all the members of the communities according to age and sex. Increased labor specialization probably was related to metallurgy. Funerary goods, in addition, suggest that all of the women were spinners, but only a few of them also were weavers (Bietti Sestieri 1985, 1986).

The social position of women during the early Roman period was very similar to that of Iron Age Italian women. A woman's primary commitment was to her family and household. The *paterfamilias* even had the power of life and death over his wife and daughters. The only difference from the early period was that

Roman law sanctioned the differentiation of sex roles. Thus, this status hierarchy was the legalized norm.

The transition to the Roman period involved major changes in economy and social structure. Land exploitation and husbandry became primarily male activities. Agriculture developed strongly with the introduction of new vegetables and fruits, olives, and grape vines and with techniques of soil fertilization and crop rotation. Breeding stock included poultry.

Although initially a family activity, land use progressively transformed into large estates in which labor was performed by members of the lower class or by slaves. In the Marche Region about 2,300 B.P. a strong "Romanization" cultural process began with the foundation of numerous cities. Urbanization and increasing social stratification were accompanied by growing labor specialization. New working categories appeared, for instance, artisans (masons, goldsmiths, coppersmiths, workers in bronze and iron, ceramists, carpenters, weavers, rope makers, tanners, shoemakers, cooks), merchants, traders, administrators, masters and preceptors, writers, actors, and musicians. These jobs were predominantly male activities.

Historical accounts describe the position of women in the Roman State before the early Empire epoch in terms of social, legal, and political inferiority, lacking freedoms enjoyed by males. Sexual discrimination began at birth since the Roman law established that female newborns could be killed at the discretion of the father. Women were considered essentially on the basis of their capability of being mothers, as was the case in many ancient European societies (Y. Thomas 1990).

Sex Differences

Roman Age skeletal series show a systematic and appreciable increase in the degree of sexual dimorphism in all the functional regions examined with the exception of facial measurements. This increase cannot be related to changes in body size since, in both sexes, all of the metric traits considered are substantially similar in the two periods (Borgognini Tarli and Mazzotta 1986). The same increase in sexual dimorphism from Iron to Roman Age is shown by all the other Italian skeletal series studied (Borgognini Tarli and Mazzotta 1986). Therefore, we suggest that there is a relationship between the above-described changes in socioeconomic structure and an increased degree of sexual dimorphism. Other possible interpretations, such as the immigration of new people into this region, remain historically unproven.

Predynastic and Dynastic Egyptians

Cultural and Economic Profile

The final case study compares predynastic (7,000–5,100 B.P.) and the dynastic (5,100–2,300 B.P.) populations in Egypt. This transition involved changes in subsistence patterns and in social structure, many of which are comparable to those already described for the Iron-Roman Age transition in Italy.

Predynastic economy was based on primitive agriculture supplemented by fishing and hunting. Handicrafts were mainly represented by pottery, weaving, and boat manufacture. Trade and metallurgy were very poorly developed.

The population unit was the *nomos*, a settlement governed by a local chief and comprising primarily members of the same family or tribe. Women's roles are not documented in the archaeological record but can be inferred from the importance of female divinities in later dynastic religion. For example, the great goddess Isis represents the original priority and independence of the female element in procreation and inheritance as well as the leading function of women in plant domestication.

During the Dynastic period, formation of a strong, central political power with religious attributes was accompanied by social stratification and labor specialization. The domestic organization of agriculture was replaced by a large estate system, with most land being owned by the Pharaoh and the upper social classes. Peasants worked as forced labor and, in addition, were impoverished by taxes. Numerous categories of artisans were formed, most of which were similar to those described above for the Roman period. Most workers were free men, and a worker's job was passed from father to son.

Upper-class women apparently were not subordinate to men. They participated in public life, were involved in business, and could even become queen. Property and condition followed inheritance through the mother's family. This privileged condition of women ended, however, with the Ptolemaic period and a strong Greek influence (Durant 1956).

Sex Differences

The availability of particularly well preserved skeletons allows us to evaluate the degree of sexual dimorphism not only in terms of sets of skeletal measurements, but also in terms of reconstructed body build (Borgognini Tarli and Masali 1983; Masali and Borgognini Tarli 1985; Borgognini Tarli et al. 1987). Our technique is based on procedures developed by Correnti (1960) for the analysis of body build in children and in athletes.

Results illustrate that women in Predynastic and Dynastic periods show little variation through time, whereas men become taller and more slender. In the Egyptian samples, facial measurements are more dimorphic than are braincase measurements. In contrast, increase in the degree of sexual dimorphism observed in Italian samples during the Iron-Roman transition is not observed in ancient Egyptians in spite of similar socioeconomic change. It seems reasonable to suggest, then, that different environmental conditions, social status, and/or genetic background may account for the different patterns.

Discussion and Conclusions

The variables that show the greatest degree of sex differences are functional measurements that are sensitive to environmental conditions. For example, the long bone shafts of the limbs reflect different load-bearing activities of females and males. In terms of the degree of sexual dimorphism, these "eco-sensitive" variables are followed, in decreasing order, by limb bone length, then measurements of the face and braincase. This pattern characterizes all of our skeletal series in spite of the different sample sizes and compositions (Borgognini Tarli and Repetto 1986a,b).

A second conclusion drawn from our work is that general statements cannot be made about trends of reduction or increase in the degree of sexual dimorphism in skeletal samples. The various skeletal regions, measured in a variety of ways, contribute to a mosaic of features that combine both genetic and environmental factors. Our studies show the importance of the effects of social and physical stress, subsistence patterns and diet, and economic division of labor in the degree of sexual dimophism in skeletal samples.

The overall reduction of sexual dimorphism shown in the skeletal series from Upper Paleolithic to Mesolithic peoples in Europe, in part, may represent reduction in overall body size. However, when data are corrected for allometric effects, significant differences between the degrees of sexual dimorphism for many variables are still present. Explanations about the nature of the environmental pressures during the Paleolithic–Mesolithic and Neolithic transition as the basis for a changing pattern of sexual dimorphism require much larger samples, especially from Upper Paleolithic populations. The complexity of the interpretation of a trend toward size reduction is compounded by the broad geographic range and huge time interval that exist among the skeletal samples. Consequently, multiple factors, differentiated in time and space and with changing interactions, may contribute to the apparent trend.

The increase in sexual differences shown by the skeletal samples from the Marche Region of Italy during the Iron and Roman Ages illustrates the usefulness of the degree of sexual dimorphism as an indicator of both socioeconomic change and of division of roles by sex in a specific regional context. Similar cases of regular diachronic trends also are observed at the regional level in different ecological conditions and with different types of social and economic transitions (Larsen 1981, 1984; Goodman et al. 1984; Rose et al. 1984).

The Etruscan samples indicate that not every difference in sex roles and social status is reflected by differences in the degree of sexual dimorphism in the skeletal traits examined here. In spite of the well-known difference in the status of women, significant sex differences are not apparent from the skeletons. These results may be compromised, however, by uneven comparisons between regional and national composite samples. More work with these samples also needs to be completed.

The ancient Egyptians, when compared with other situations with a socioeconomic transition, illustrate greater stability of body measurements in women. Moreover, the variation of the pattern of sexual dimorphism from that observed in analogous social and economic transitions stresses the importance of understanding the social and environmental conditions in *particular* times, places, and socioeconomic contexts.

In summary, sexual dimorphism is not a unidimensional phenomenon (Zihlman 1976, 1982; Armelagos and Van Gerven 1980). The extent of differences between the sexes needs to be analyzed from many perspectives. We must use many kinds of measurements, carefully choose statistical techniques, and, of greatest significance in studies of human populations, consider in detail the social, cultural, economic, and technological conditions of a given time and place as an integrated part of our analyses of sexual dimorphism.

Acknowledgements

We thank Alessandro Canci for useful discussion during the preparation of this manuscript and organizers and participants in the conference, "Female Biology, Life History, and Evolution" held in Santa Cruz, California in September 1990, for substantial help in improving this chapter. Research is supported by the Italian National Research Council (C.N.R.), grant 89.05182.CT15 and by grants from the Ministry of University and Scientific Research (M.U.R.S.T.).

15 Growing Up Female in a Farmer Community and a Forager Community

Gilda A. Morelli

In a Farmer Settlement, February 23

Emakabwana, a young farmer woman, breathes heavily as she carries a food-laden wicker basket up a steep hill made treacherous by the morning rain. Secured to her hip by an old cloth is her 3-month-old son whose body bumps rhythmically in time to her steps. Behind her is Uese, her 4-year-old daughter, who precariously balances a small pot of water on her head, careful not to slip in the mud. As Emakabwana enters the village, she pauses to greet her husband and his friends who are sitting under a leaf-roofed, open-sided structure, protecting them from the mid-afternoon equatorial sun. Scanning the village, she notices that the other villager women are sitting next to their cooking fires preparing the day's meal. Young girls help their mothers pound *mahogo*; young boys scramble around the cooking fires in a wild game of tag. As Emakabwana leaves the relative coolness of the men's sitting place, her husband calls after her "Emakabwana! How long will it take you to prepare my afternoon meal? And warm me some water so I can wash my hands before eating."

In a Foraging Camp, February 25

Mau, an adolescent forager boy, sits in camp with his brother's 15-month-old daughter draped across his lap; lulled to sleep by the not-so-distant music of a finger piano. Mau reaches over to stir his pot of *sombe* as a group of young boys and girls play "shoot the fruit" using child-sized bows and arrows. The children come dangerously close to Mau's cooking fire, and he utters a disapproving "aa-ooh!" Nearby, a woman face-paints a young man. Keeping his eye on the children, Mau looks for someone to fetch him some more firewood. As he scans the camp, he notices a group of women preparing for a fishing trip, while others lounge, smoking tobacco along with the men.

It was scenes like these that kindled the curiosity of a young researcher on her first field trip to the Ituri rain forest. They helped reinforce my growing sense that the social arrangement of everyday life differed for Efe foragers and Lese farmers living in northeastern Zaire. It seemed clear that the roles and responsibilities of men and women, and the relationships between them, were not the same for the Efe and the Lese. I wondered whether these differences meant that the social experiences of forager and farmer children were also different. And if any differences existed, what role did they play in shaping the development of gender-appropriate ways of thinking, feeling, and behaving? The many questions that surfaced during my first months in the field helped to crystallize the issues surrounding the life of children that became the focus of my work for the next 2 years.

It is fitting to emphasize children in the context of understanding women's role in human societies. Research on the life course of human females is concerned primarily with adults, often focusing on the strategies they use (no conscious intent implied) to help insure their survival and reproductive success. However, the strategies developed by women constitute a lifelong process, beginning at birth. To understand the life course of females, we must include in our endeavors the study of children. By observing girls and boys as they actively participate in the routines of their community, we can begin to understand what growing up female means to them.

Conceptual Framework

Most scholars agree that human social roles and relationships are shaped by the biological differences between males and females, even though they do not agree as to the way in which these differences influenced the course of human evolution (Dahlberg 1981). But, how is knowledge about gender roles appropriated by each generation of individuals? Are gender roles an invariant aspect of human behavior relying solely on maturational processes for their expression? Or do they develop in the course of participating in the activities of one's cultural community? My thinking about the development of gendered behavior is strongly influenced by two related approaches that bring together development and culture by studying the child in context. The first, sociohistorical approach was advanced by Vygotsky, a Soviet psychologist writing in the 1930s (see Vygotsky 1978) and expanded on by cultural psychologists who were dissatisfied with prevailing developmental theories that ignored culture (Cole 1985; Wertsch 1985; Rogoff and Morelli 1989; Rogoff 1990). Champions of this approach share the view that studying the developing child separate from context is a futile undertaking. They believe, instead, that the child's construction of reality emerges from his or her participation in the day-to-day practices of the community. It is in the context of daily routines with more competent community members that the child's knowledge about him- or herself develops, as do the skills needed to engage in culturally appropriate activities. In keeping with the major tenet of this tradition, the unit of analysis is the individual in sociocultural activity (Wertsch 1985; Bruner and Haste 1987; Rogoff 1990).

The second approach revises the sociohistorical tradition by bringing into focus the role of the child in activity (Rogoff 1990). Rogoff introduced the concept of *guided participation* to highlight children's contribution to directing the course of the activities in which they participate. She suggested that cultural knowledge is *appropriated* by the child in the course of participating in activity. It is not internalized as Vygotsky and others argue. Valsiner (1988) extended the idea of the active child by maintaining that the child and his or her social partner act together in co-constructing the meaning of the child's actions.

Gender, Children's Activities, and Social System

Social systems play an important role in structuring sex-related differences in human behavior, an observation first made by Margaret Mead (1935, 1949). Research on the development of gendered behavior shows that the way children construct sex-appropriate ways of thinking, feeling, and behaving is shaped by features of community life such as economic production, male/female roles and responsibilities, male/female relationships, and beliefs about children and their development (Barry et al. 1957, 1959; Brown 1973; Bolton et al. 1976; Erchak 1976; Welch 1978; Cone 1979; Hendrix 1985; Whiting and Edwards 1988; Rogoff 1990). Indeed, the word "sex" has been replaced by the word "gender" to reflect the social and cultural constructiveness of this biological category (Lloyd 1987; Schlegel 1989).

Research on Setting and Children's Companions

One perspective that informs our thinking about the relation between the social system and gender stresses the importance of the setting in structuring children's activities with community members (e.g., Whiting and Edwards 1988). A prominent feature of setting that shapes development is the "category of person" with whom children are involved. Whiting and Edwards (1988) believed that by interacting with different "categories" of people (grouped on the basis of gender, age, and kinship) children practice distinct patterns of behavior. People of a particular "category" share certain behavioral qualities and, as a result, model and teach specific behaviors to and elicit specific behaviors from children. For example, children given the responsibility of caring for infants are likely to practice nurturant behaviors because infants elicit nurturance from those with whom they interact. Mothers, by comparison, elicit dependent behaviors from infants and children.

The regularity with which children encounter different categories of people and the opportunities made available to them is related, therefore, to the settings they frequent. In many communities, for example, girls work harder than boys and, in addition, often begin to work at an earlier age (Whiting and Edwards 1973; Rogoff et al. 1975; Wenger 1983; Munroe et al. 1984). Their chores often consist of doing household errands and caring for infants and toddlers. Childcare responsibilities provide girls with the opportunity to practice nurturant behaviors as well as to anticipate the needs of others (Wiesner and Gallimore 1977). In the course of being cared for predominately by girls, infants also receive culturally important messages about who provides care. Household and childcare responsibilities are often done near the homestead under the watchful eyes of mother and other adult women. Thus, girls' spatial range and, therefore, the extent to which they experience varied activities may be restricted. Further, girls are more likely to be exposed to and model the activities of adults and to be supervised by them.

Boys, on the other hand, generally are assigned fewer work-related responsibilities. More of their time is spent either playing and socializing with peers (particularly other boys) or being involved in solitary ventures. In addition, boys are allowed greater freedom of movement such that by 8 to 11 years of age boys spend most of their time outside of the homestead. As a result, boys may experience a more diverse range of activities than girls. In addition, boys are likely to be exposed to and model the behavior of same-gender peers rather than adults. Furthermore, their activities are often unsupervised. Play and other peer-oriented activities provide boys with the opportunity to practice skills associated with negotiating interpersonal conflicts and contribute to their understanding of egalitarian relationships (Hartup 1979).

When boys and girls from different cultural communities are assigned to similar settings, similarities in their behaviors should be expected. When settings differ, so should behavior. Support for this view is found in the work of Whiting (Whiting and Whiting 1975) and others (Ember 1973; Whiting and Edwards 1988). In Orchard Town, USA, for example, girls had few opportunities to interact with infants and toddlers. They scored higher on masculine-type behaviors (e.g., egoistic dominance) and lower on feminine-type behaviors (e.g., prosocial dominance) than girls who had more experience caring for young children (Whiting and Edwards 1973; Whiting and Whiting 1975). In Kenya, Luo boys assigned "gender-inappropriate" tasks were more likely to behave in traditional female ways than similar-aged boys assigned more "gender-appropriate" tasks (Ember 1973). Interestingly, however, this was true only if performance of the task took place within the confines of the home, suggesting to Ember that "the context of the task situation may be more critical than the task itself" (Ember 1973:437).

Research on Social Representations of Gender and Children's Activities

What factors help shape the social and physical arrangement of children's everyday life? Community members structure the activities of boys and girls in ways consistent with their representation of gender, a representation that is culturally based (Archer and Lloyd 1985; Bruner and Haste 1987; Fivush 1989). In the course of participating in cultural routines, community members both provide children with the opportunity to practice skills considered appropriate to their gender and guide children's construction of their feelings about and understanding of their actions.

Research conducted in the United States and Britain suggests a relation between people's representations of gender-appropriate behavior and their interpretation and structuring of children's activities. In a rather ingenious study, Condry and Condry (1976) manipulated college students' perception of gender by showing them a videotape of a 9-month-old infant in gender-neutral clothing playing with several different toys. Half of the male and female students were informed that they were watching a girl, half a boy. When asked to rate the infant's behavior, the students watching the "boy" described him as being angry, those watching the "girl" described her as being afraid in situations where the infant's behavior was ambiguous. Condry and Condry concluded that gender stereotypes influence adult perceptions of infant behavior. But do gender-based perceptions of infant behavior actually influence the way in which people respond to infants? There is evidence to suggest that they do.

Smith and Lloyd (1978) asked mothers to play with infants that were similar in age to their own. Half of the male and female infant actors were dressed in gender-appropriate apparel, half were dressed in cross-gender apparel. When the baby was presented as a boy, mothers responded with physical actions; when presented as a girl, mothers responded with soothing and comforting actions. In another laboratory-based experiment, unsuspecting students were asked to wait in a room before participating in a study. Before leaving the students alone, the researcher told them that there was a sleeping baby in the adjoining room, and asked them to get the researcher if the baby cried (actually it was a recording of an infant). Half of the males and females participating in the study were told that the baby was a boy, half a girl. Females were slower to contact the researcher on hearing the baby cry if they thought the baby was a boy; males were slower than females irrespective of the infant's perceived gender. Thus, perceived gender influences people's behavior toward infants; in this study, the behavior of females was more strongly affected than males.

Studies examining adult-child interactions in laboratory situations demonstrate a relation between the child's gender and nature of interaction. The differences observed may differentially foster the development of emotional and cognitive capabilities in boys and girls (Rothbart and Rothbart 1976; Carpenter and Huston-Stein 1980; Archer and Lloyd 1982; Carter and McCloskey 1983; Weitzman et al. 1985; Deloache et al. 1987; Lloyd 1987; Hron-Stewart 1988; Fivush 1989). For example, one of the ways in which community members guide children's knowledge about themselves is to provide an evaluative framework from which past (and present) events can be interpreted. Evaluative frameworks help children make sense of their experiences and communicate culturally appropriate reactions to events, including appropriate ways of feeling.

Insofar as the frameworks made available to boys and girls differ, so might the way boys and girls remember their past and construct their present. To investigate the possibility that boys' and girls' memories about the past are organized differently, Fivush (1989) recorded mothers' conversations with their 30- to 35-month-old children, coding the emotional terms used to talk about children's past experiences. Although mothers were recorded using the same number of emotional terms in conversation with sons and daughters, the content of the terms differed. In conversations with children, mothers focused more on positive emotions when talking with daughters. They tended not to attribute negative emotions to daughters, but did so to sons. And they were more likely to talk about the emotional state itself with daughters, but the causes and consequences of emotions with sons. Thus, boys and girls are learning to evaluate their past differently and these differences may play an important role in shaping their understanding of gender-appropriate emotional reactions.

Not only do mothers differ in the way they talk to their sons and daughters about emotional reactions to events, they also differ in the opportunities they provide that are associated with the development of certain cognitive skills. Hron-Stewart (1988) observed the strategies mothers used to help their 2-year-olds in a problem-solving task. Mothers were more likely to provide their daughters with specific hints, thus preventing them from exploring the task on their own. The results are consistent with those obtained in an earlier study by Vaughn et al. (1981).

The studies described represent some of the diversity of research on the development of gendered behavior. The approaches differ with respect to the communities studied, issues addressed, and theoretical orientations adopted. However, the studies converge in advancing the view that gender-appropriate ways of thinking, feeling, and behaving emerge from activity with community members who structure children's experiences in ways consistent with their culturally based representation of gender.

Taken collectively, the research demonstrates that by studying children whose lives are differently arranged, we can begin to understand how features of a sociocultural system shape children's activities and, thus, their appropriation of cultural knowledge, including knowledge of themselves as boys or girls. My work on young boys and girls from two cultural communities extends our understanding of the relation among social systems, children's activities, and gender role development.

The People

The communities with whom I worked allowed for a unique study of gender role development in young children. First, the Efe foragers and Lese farmers are sympatric communities involved in an exchange relationship that has been going on for generations. They not only play an important role in the lives of one another but inhabit and exploit a similar ecosystem (Wilkie 1988; Bailey and DeVore 1989).

Second, the Efe subsist primarily by gathering and hunting, a mode of subsistence that characterized more than 90% of our species' evolutionary history. Study of Efe children may therefore provide important clues to the sociocultural and physical context in which our ancestors lived. Although our understanding of the conditions that shaped evolution of human behavior may be enhanced by studying people like the Efe, for obvious reasons, statements based on extant forager populations must be made cautiously and interpreted carefully.

Another reason why the knowledge gained by studying the Efe must be used judiciously is because there is no prototypical forager group. Although foraging communities share a method of subsistence, they may differ in other important respects. Draper and Cashdan (1988), for example, noted important differences between children of the !Kung and of the Hadza, a forager group living in Tanzania. Hadza toddlers are more likely to be left in the camp while women forage for food, even though the number of available caregivers in camp do not appear to differ between the two communities. In addition, Hadza children are more productive gatherers, supplying approximately half of their calories (Blurton Jones et al. 1989). Differences in young children's activities also are observed between two forest-living foraging communities, the Aka of the Central African Republic and the Efe of Zaire (Tronick et al. 1987, 1989; Hewlett 1989; Winn et al. 1989). Thus, detailing aspects of the lives of Efe forager girls and boys adds to our knowledge of forager children's experiences and heightens our awareness of the diversity of child-rearing practices among forager communities.

Third, the study of forager and farmer children living in the same ecological setting allows one to speculate on how shifts in subsistence practices influence the daily arrangements of children's activities and, thus, the development of gender role behavior. The move from a technologically simple to a more technologically complex mode of production has been associated with increased labor requirements, more hierarchical authority patterns, increased sexual division of labor, greater use of child labor, and greater differentiation among individuals (Draper and Cashdan 1988).

These changes have important implications for children's daily experiences as suggested by the work of Draper and Cashdan in their study of 4- to 14-year-old sedentary and bush (i.e., mobile) !Kung children. For example, sedentary children spent more time working than bush children, a pattern observed in the adults of their community. Further, sedentary children spent more time out of camp and, therefore, spent less time under the watchful eyes of adults. With respect to interpersonal relations, sedentary children received fewer requests from adults, including mothers, and adults received fewer requests from children. This last finding suggests that !Kung adults living in a sedentary community play a less prominent role in the lives of their children, at least with respect to requests, than !Kung adults living in a nomadic community.

One of the contributions of the work by Draper and Cashdan is its focus on the role of technology in structuring the activities of children. A second contribution is its focus on the experiences of boys and girls living in technologically different communities. In the bush community, few gender differences were found in the behaviors measured, a finding related to the fact that bush children do little or no work. By comparison, differences were observed between sedentary boys and girls. Sedentary boys spent more time out of camp, an activity likely to be associated with a range of experiences not available to girls. The more restricted movement of sedentary girls might help explain the finding that they were slightly more likely to receive requests from adults than sedentary boys. Finally, although similarities exist in the amount of work performed by sedentary boys and girls, differences were found in the type of work they do. Sedentary girls engaged in work that required them to remain near their homes and in the vicinity of adult women. Boys engaged in work that required them to leave the confines of their homesteads.

It is clear from the work of Draper and Cashdan on the !Kung that a shift from a foraging to a more horticultural means of production is associated with a wide range of changes for both adults and children. The resulting changes have important implications for the role children play in the life of the community and, therefore, for the development of gendered behavior.

Research on Children from a Forager and Farmer Community

On the basis of my initial observations of Efe foragers and Lese farmers, I became intrigued with the idea of exploring in detail the social arrangement of children's everyday lives. It appeared that community activities differed in ways that had implications for the development of gendered behavior. The perspective emphasizing the relation between children's participation in community-structured activities and psychological development provided me with a way to conceptualize my research. In the next section I describe some of this work, illustrating the importance of studying the child in cultural context.

Children's Activities

To answer questions concerning the development of gendered behavior, I observed the day-to-day activities of 15 foraging and 14 farming 22- to 39-month olds using a focal-subject sampling technique (Altmann 1974). I watched each child for 6 hours, recording in sequence the occurrence of preselected behaviors. Observations were evenly distributed over the daylight hours.

I was interested in knowing if the activities of forager boys and girls, and farmer boys and girls, were arranged differently. To explore this question, I examined the amount of time (measured in 1-minute intervals) boys and girls from each community spent in household economic routines. I chose to look at economic routines because of the work of Whiting and other cross-cultural researchers who place importance on the relation between children's work activities and gender role development (Ember 1973; Whiting and Whiting 1975; Bolton et al. 1976; Whiting and Edwards 1988). Economic activity was defined in two ways. The first included errands and chores, and the second emphasized watching and mimicking the work activities of others.

Not surprisingly, five of the seven farmer girls spent more time doing errands and chores than did all seven farmer boys. Furthermore, six of the seven farmer girls spent more time watching and mimicking the work activities of others than did five of the seven farmer boys. Thus, farmer girls participated more often in daily economic activities, both as doers and observers. A different pattern of results emerge when the activities of forager children were analyzed. Forager boys and girls *were remarkably similar* for each behavior examined.

These findings suggest that the activities in which farmer boys and girls participate begin to diverge at an early age. Two- to 3-year-old farmer girls are more likely to spend their time involved in economic household routines than are farmer boys. This pattern of findings is not replicated for forager boys and girls. Why do gender differences exist in young children of one community, but not another? Supporters of the sociocultural perspective suggest that the role of community members in guiding children's participation in activity should be examined to help answer this question.

Children's Economically Related Directives

To study community members' guidance of children's participation in activity, I decided to focus on what children are told to do. By telling children what to do, community members communicate to children expectations concerning appropriate behavior as well as provide them with the opportunity to realize these expectations. Further, children practice certain skills in the course of participating in the activity. I examined how often children were asked to engage in household economic routines such as errands and chores for the same reasons I chose to examine children's economically related activities.

The study of economically related directives to children suggests that members of the farming community channel the behavior of boys and girls in ways different from members of the foraging community. Compared to farmer boys, farmer girls were asked to engage in economic household routines more often and the proportion of economic related to other directives was greater for them. Thus, economically related directives figured prominently in what farmer girls are asked to do. Foragers are less likely to differentiate between boys and girls in requests to participate in economic routines, and the proportion of economically related directives to other directives is similar. It is therefore reasonable to assume that members of the two communities differ in their expectations of boys and girls and that these differences are reflected in economically based directives to them. But who are the community members responsible for guiding children's participation in work routines?

Children's Partners in Directing Work Routines

Who assumes the responsibility of guiding children in different types of activities depends on a variety of factors, not the least of which includes what they perceive as their social role(s), and their responsibility to the child, as well as the extent to which they are available to the child. Each of these factors is related to features of the child's cultural community.

I explore the question of who channels the behavior of young farming and foraging children by examining the role of women and men in requesting children to participate in economically related activities. Women appear to be the major players in asking farmer boys and girls and, also, forager girls to carry out errands and chores. By comparison, women and men appear to be equally active in asking forager boys to engage in economic routines.

Summary of Findings

The research shows quite clearly that the activities of farming boys and girls begin to diverge early in life with girls spending more time in economic routines. Community members' role in fostering the development of gendered behavior appear to be quite remarkable in that they were more likely to request a farmer girl to participate in economic household routines than they were a farmer boy. Requests to engage in economic activity, no doubt, take on an added significance for farmer girls as they made up the bulk of the requests received by them. Not all community members, however, assume the responsibility of guiding children in work routines; women play the most important role for both boys and girls.

In contrast, the activities of foraging boys and girls, at least at this age, do not show the pattern of differences observed in farming boys and girls. Forager boys and girls are equally likely to engage in household economic activities. Community members also treat boys and girls similarly in that they are as likely to direct girls' involvement in economic routines as they are to direct boys' activities. Although girls received the majority of economically related directives from women, boys receive them from both women and men.

The extent to which farmer girls and boys participate in economic activities clearly differs, and community members play a role in fostering the differences observed. Forager boys and girls, however, do not differ in the time spent in household economic routines; nor do community members differentially encourage boys and girls to participate in these activities. Since children's knowledge about themselves as boys and girls emerges from engaging in the day-to-day activities of the community, and economic activities have been singled out as important in shaping gender role development, it is reasonable to assume that farmer boys and girls differ in their construction of gender and in the competencies that they are acquiring. In contrast, forager boys' and girls' knowledge about gender, as well as the competencies acquired, may be less divergent at this age.

Discussion

The sociohistorical approach and the concept of guided participation advance the view that the day-to-day routines in which boys and girls participate, and the nature of their participation, are shaped by features of the sociocultural system. Thus, we need to consider the broader cultural context in which boys and girls develop in order to understand more fully the social development of gendered behavior. In the following discussion, I highlight the important aspects of community life that guide the activities of farmer and forager boys and girls and, thus, their appropriation of knowledge about gender.

Lese Farmers

The Lese are slash-and-burn horticulturalists whose work cycle is entrained to seasonally appropriate tasks such as clearing, burning, planting, weeding, and harvesting. Lese farmers live in roadside villages that rarely change location. Each village generally contains several families, but each family's living quarters often is spatially and sometimes visually isolated from the others. Descent is traced though the paternal line, and residency is with the husband's family, with some exceptions.

Lese farmers spend most of the daylight hours out-of-doors; houses are used for sleeping, storage, and protection from inclement weather. Women's and children's village activities that include cooking, eating, and socializing take place primarily near their houses in kitchen areas called *mafika*. By comparison, farmer men's village activities take place primarily in a semi-enclosed sitting area called the *baraza*, where it is customary for them to wash, eat, socialize, and relax. The men's *baraza* often is located near the road and is separate from the house and *mafika*.

The social and physical separation of men and women when in the village is promoted by beliefs concerning the relationship between them. Some of these beliefs were made clear to me during my first days in the field. After the customary greeting of the village men, who were sitting in their roadside *baraza*, I was told that I would be much happier if I sat with the women in their kitchens. Infants and young children are also discouraged from joining men in the *baraza* and often do so only after receiving an invitation.

When not in the village, women are generally found in their fields or *shamba*, as fieldwork is essentially a woman's responsibility. During active agricultural periods, they spend most of their day in the *shamba*, leaving only men and elderly people in the village. Fieldwork is not a communal activity, except during short periods around planting and harvesting. Further, women are not able to socialize easily with one another when involved in fieldwork because of the distance separating fields. Thus, women are regularly isolated from each other's company, and from the company of men, when laboring in the field.

Children accompany their mothers to the fields because women are also responsible for providing child care. This means that children, too, are often isolated from the company of men. Rarely do farmer men engage in gender-inappropriate activities like child care, especially when such activities are available for public scrutiny. One farmer woman told me that "when I work in the fields, my husband sometimes helps me in child care by carrying our 1-year-old. But as soon as another person, especially a man, approaches, he is quick to return our son to me because he wants to avoid the possibility of being teased." Other activities not considered men's work include preparing food, cutting or carrying firewood, and fetching water.

Lese men work hard 3 months of the year when land is cleared for planting. During this 3-month period, women may visit relatives or spend time in the village making or repairing baskets, sleeping mats, and the like. Men work very little for the remaining portion of the year and only occasionally help their wives with fieldwork. More often than not, men are found relaxing in their *baraza*, or visiting friends in a nearby village.

The roles and responsibilities of men and women, their physical and social arrangement when in the village, and the scheduling of work routines shape young children's social experiences. This includes messages about what it means to be a male and female in their community. For example, community life among the Lese farmers restricts the amount of time children spend with men. When children are in the company of men, they are likely to observe and participate in a rather constrained range of activities. Instead, young children spend the majority of their day (and night) in contexts dominated by females and their activities. Thus, the social opportunities available to young boys and girls on a regular basis are of a particular type. What might this mean for the developing farmer girl and boy?

Young farmer girls are provided with the opportunity to watch and participate in activities that foster the development of gender-appropriate competencies and are encouraged to do so by community members. Their degree of involvement in productive work supports Brown's thesis that girls growing up in cultures where women are the primary food producers are trained in subsistence activities at an early age (Brown 1973). Since engaging in productive work promotes responsible behavior (Erchak 1976), it is likely that farmer girls are being socialized (among other things) to behave responsibly. It is a common Lese belief that girls learn *kufuata njia ya banabake* (to follow the path of women) by participating in adult women's work routines.

Young farmer boys, on the other hand, do not have gender-appropriate activities readily available to them, in part, because most of the work done by adolescent and adult males takes place outside the village. When village men are in the village, they prefer to relax. As a result, farmer boys neither engage in the work surrounding them (which falls primarily under the domain of women's work), nor are they encouraged to do so by community members. Just as farming girls are behaving in ways sanctioned by the community, so are farmer boys.

The finding that women are active in directing children in economic activity whereas men are not is not surprising given the organization of life in this farming community. Children spend little time in the company of men, and when they are in each other's company, men may not feel that it is their responsibility to organize children's economic routines.

Young Lese farming children are growing up in a community where the roles and responsibilities of men and women differ. These differences are communicated to children both in the behavior of men and women and in the way children's activities are guided by them.

The Efe Forager

The Efe[1] are a short statured people who acquire forest foods by gathering and hunting, mainly with metal-tipped arrows. Cultivated foods are also important in the Efe diet and are obtained from the Lese in exchange for forest products and/or services.

The majority of Efe live in transient camps established in small, forested areas cleared of vegetation. As with the Lese, descent is traced through the male line, and resident patterns are with the husband's locality. Each family consists of brothers and their wives, children, unmarried sisters, and parents. The Efe build leaf huts that are used primarily for sleeping, food storage, and protection from inclement weather. The huts are typically arranged around the camp's perimeter creating a large, open, communal space. Since most day-to-day activities take place outside, within this communal space, they are in clear view of other camp members.

Camp members do not regularly synchronize their day-to-day activities. During the mid-morning and early afternoon when most out-of-camp activities take place, one or several individuals are likely to be found in the camp resting, taking care of children, preparing food, or socializing. The almost continual presence of people in the camp provides mothers with an opportunity to leave their children in the camp while, for example, foraging for food; this opportunity is often taken.

Cultural practices concerning the social and physical arrangement of community members make it easy for men and women to work together, usually side by side, in clear view of children. This is in marked contrast to the social separation observed between men and women in the farming community. Cultural practices also make it more acceptable for fathers to eat with their families, even though each family is never far away from any other family. Indeed, it is quite common for families to carry on conversations with one another while sitting in front of their respective huts.

Many camp activities are shared by men and women, although some activities are considered more appropriate for members of one gender. It is not unusual, for example, to see a man involved in some economic household routine such as preparing food for consumption. However, women rarely accompany men on hunts and, when they do accompany them, they do not kill forest game with bows and arrows.

Since the amount of time young Efe forager children spend in the camp is considerable, ranging from 80% to 95% of daytime hours (Morelli 1987), they are developing in a community where men, women, boys, and girls are physically and visually available to them. They have the opportunity to be involved in a variety of camp activities and to practice a variety of skills. Although Efe children can participate in the economic activities more commonly engaged in by males or females, community members do not strongly encourage one activity over another.

Given the roles and responsibilities of forager men and women and the arrangement of people and activities in camp, it is not surprising that forager boys are as likely to engage in economic pursuits as are girls, and that community members share in the responsibility of directing young boys' work routines. What is surprising is that women are the primary director of forager girls' work routines. One explanation requires that we bring into focus the role of children in shaping their joint involvement with community members. Forager girls may choose to spend more of their time in proximity to women, thus placing themselves in situations where their activities are more likely to be under the direction of females. Boys, by comparison, may choose to distribute more evenly the time they spend close to men and women; thus, placing themselves in situations where their activities are likely to be under the direction of both men and women.

Young Efe forager children are growing up in a community where the relationship between men and women is far more egalitarian than is the relationship between farmer men and women. This is communicated to children both in the behavior of men and women and in the way children's activities are guided by them.

Summary

The sociocultural communities in which forager and farming children develop differ. The differences are reflected in children's social activities, in culturally structured opportunities, and in the lessons appropriated by children in the context of everyday life. The differences in the arrangement of farmer boys' and girls' participation in the routines of their community undoubtedly foster differences in their understanding of gendered behavior and in the skills acquired by them. Forager boys' and girls' participation in cultural activity is not as differently arranged and, in all likelihood, their understanding of gendered behavior, including the roles and responsibilities of males and females, may be more similar than different.

Acknowledgments

This research was supported by grants from the National Science Foundation (BNS-8609013), the National Institute of Child Health and Development (1-R01-HD22431), the Spencer Foundation, and Faculty Research Funds from Boston College. I would like to thank David S. Wilkie for his critical reading of this manuscript, and Cathy Angelillo for assisting with its preparation.

Note

1. The Efe are commonly referred to in the literature as pygmies. We have chosen to use this term sparingly. Although the term pygmy may be informative to the reader, it is considered pejorative by the Efe.

16 Institutional, Evolutionary, and Demographic Contexts of Gender Roles: A Case Study of !Kung Bushmen

Patricia Draper

In a Foraging Camp, 1968

Sa//gai is a 10-year-old foraging girl. It is 9:00 A.M. The sun is already hot; even the sand is already uncomfortably warm under her bare feet. She scoots back into the partial shade of a nearby bush. She looks across the camp, her quick glance taking in the 10 grass huts that are the residences of the several families, some with young children, that form the hunting-and-gathering band she and her family live with. She looks about for something to do. Her mother and infant brother have already left for the day with some of the other women from the camp; some of the other women are also carrying nursing children. They will be gone for the day gathering bush foods. She sees her father a few feet away. He sits in the shade of a storage platform near her parents' hut and is sharpening his arrows, preparing for the next day's hunt. Several children of the camp play near him. When they come within an arm's reach, he shoos them away. The poison on the arrows is lethal; he wants them at a safe distance.

Sa//gai's glance comes to rest on an old couple, //Oka and /=Oma, her grandparents, sitting together at their own fire outside their grass hut. Two other adults, also members of the band, sit with them. She wanders over and sits cross-legged between her grandparents. /=Oma, her grandfather, is pulling the hot *mongongo* nuts out of the hot ash where he has been roasting them. As they cool, his wife, //Oka, cracks them expertly between two stones. Wordlessly, Sa//gai reaches out her opened palm to her grandmother. The old woman drops several warm nut meats into Sa//gai's palm and resumes cracking. Still chewing the nuts, Sa//gai stands and moves off to a clearing at the edge of the camp. Here some of the older children are playing a throwing game. She joins in without changing the pace of the game.

In a Settled Village, 1988

//N is a 50-year-old !Kung woman. She and her husband, N//au, live in a settled village about 10 kilometers from !Angwa, the regional administrative center of about 100 people. It is 8:00 A.M. This morning she stands outside her mud hut, thinking about how there is too much to do. Her husband, N//au, has already milked the few cows they have that are giving milk. He left the milk, in two brimming pails, with her for churning. Then, with the help of //N's grown son, and two nephews, N//au drove the cattle and goats to the well. N//au will be away most of the day at !Angwa, seeing the local headman about a business matter.

//N calls her grown daughter, N!ai, to take the pails of milk and prepare them for churning. This is a task that N!ai usually performs, often with the help of Di//au, //N's own mother. Di//au has been widowed for years and, since widowhood, has lived with her daughter and son-in-law. Di//au, though old, is an important help to her daughter's village in making the homestead run.

N!ai appears in answer to her mother's call. She straps her 3-year-old to her back as she approaches. //N looks at her daughter with a mixture of pride and dissatisfaction. N!ai is strong and beautiful. She finished 6 years of schooling in !Angwa and can speak and write in SeTswana, the language of the country. This is an unusual accomplishment for a !Kung of any age. N!ai is a good daughter and willing to work. Yet N!ai did not marry her child's father. Her lover, a !Kung man her own age, had a good job in a distant town and did not want to marry her. N!ai's first child was "a child of the bush," as the !Kung say. Now her belly is swelling again; this time the baby is by a Tswana (non-!Kung) man. "Well," //N thinks, "At least she is here, helping us to live."

//N calls out to three men of her village who are saddling donkeys for a trip to !Angwa. She wants them to carry sour milk with them to give to her son, her youngest child, a boy of 13 years. He attends school at !Angwa and boards with a Bantu family during the week. On weekends, he comes back to his parents' village to help with the work.

Knowing that the morning milk is being churned and that other milk is going to her boy at school, //N sets off for the garden. She picks up a bucket on her way out of the village so she can bring water from the well on her return trip. Yesterday she saw a break in the garden fence. It must be repaired before the goats get in and ruin the crops.

!Kung Ethnography

These scenes describe observations in !Kung villages made 20 years apart at the beginning of my research among !Kung San in 1968 and, more recently, in 1987–1988. In the past 20 years, I have made several trips to the western Kalahari in Botswana where !Kung (also known as Zhun/wasi of !Kung San) live. Over this time, I have observed marked cultural changes as the people have moved from a tradition of hunting and gathering to living as full time, settled food producers. The !Kung are one of several indigenous southern African groups known, collectively, as Khoisan. They contrast with the predominant Bantu-speaking populations of southern Africa in being smaller in stature, lighter in skin color, having cultural traditions of foraging rather than pastoralism, and having distinctive non-Bantu languages characterized by click phonemes. In recent decades, the !Kung have had increased contact with cattle-keeping peoples (such as the Tswana and Herero) who have been moving into regions of the Kalahari that were previously occupied primarily by !Kung.

The !Kung economic transformation has come about gradually over several decades. This process had begun well before I started my studies. However, in the late 1960s when my research began, I was able to work with two populations, one living as hunter-gatherers, another as settled, food producers and clients to Bantu patrons. In the late 1980s, the !Kung of western Botswana were all settled and deriving the bulk of their subsistence from stock raising, gardening, and government-distributed surplus foods. The people obtained some foods from the bush but only on a sporadic basis.

In this chapter I have two objectives. One is to detail the consequences of changes in social structure for gender roles and to show how economic change affects the lives of males and females. The second objective is to explain how my perspective on the relationship between behavior and social structure has changed over time. More particularly, I have been influenced by ideas in evolutionary biology. Although I still see human beings as embedded in a cultural context, I also see that other insights come by viewing humans as another mammalian and primate species. These ideas will be elaborated more fully in a later section of the paper.

My early research focused on the ways in which social and economic practices affect the socialization of children for gender roles, the relations between women and men, and the organization of family relationships. As will be detailed below, child and adult gender roles are affected by social and economic practices. However, the description of correlation does not speak to causation. *Why* did the !Kung who lived by food-producing adopt more sex-segregated and gender-asymmetric customs? *Why* was the interaction between the sexes among the foragers more egalitarian? A brief ethnographic account of foraging and settled !Kung is provided below.

The lifestyle of foraging !Kung has been well described in numerous monographs and papers (Lee 1969, 1979; Lee and DeVore 1976; Marshall 1976; Howell 1979; Solway and Lee 1990; Yellen 1990). The foragers lived in small, kinship-based, mobile bands, moving many times during the year in pursuit of game or vegetables. The environmental limitations were severe for the hunter-gatherers. Standing water, outside of the rainy season, was available at a few hand-dug wells. Most times of the year, group size was limited to 30 to 40 people, because a larger population exhausted available plant and animal foods, even with frequent moves during the year.

Twenty years ago, some !Kung were still living primarily by hunting and gathering. They alternated bush life with occasional visits to other !Kung who lived at cattle posts as clients and hangers-on of cattle-keeping groups, chiefly of the Tswana or Herero tribes. The contrasts between the foragers and the settled people as they lived in the 1960s will be described below. By the late 1980s, the !Kung of western Botswana all were settled around permanent water and attempting to make a living by a mixture of stock-raising, gardening, work for Bantu, government dole, and occasional foraging. The contemporary !Kung will be described in a later section of the paper.

The lifestyles of the !Kung who, even 20 years ago, were settled, have been less well described. The settled !Kung lived more or less permanently near year-round water sources. They subsisted by foraging, hunting, and gardening, and through ties of clientship to Bantu cattle keepers (Wiessner 1982; Wilmsen 1982, 1988; Lee 1984; Biesele et al. 1989). Living conditions were not as severe for settled !Kung as they were for the foragers. Settled people were spared the frequent moves and having to live away from water for several weeks at a time. It was common for young !Kung men to work as cattle herders for Bantu cattle owners; many !Kung families lived on the fringes of Bantu villages as clients and sometimes servants of the pastoralists. While their lifestyle was impoverished in comparison with their Bantu patrons, many !Kung were attracted by the prospect of regular food and the opportunity to learn the economic techniques of the wealthier cattle people (Draper 1975b, 1991).

Child Behaviors

My chief interest was to understand how child behavior varied according to whether the children lived in foraging groups or in sedentary groups. At that time, a number of scholars, influenced originally by the culture and personality school, were devising more systematic ways to explore the processes whereby culture was transmitted via the socialization process to children. These researchers focused on empirical studies of overt behavior (in place of the more elusive concept of "personality") and on social institutions and physical environment (in place of the more global concept of "culture"). Works by Barry et al. (1959), Whiting and Whiting (1975), Whiting (1965), Ember (1973), Whiting and Edwards (1973), Wiesner and Gallimore (1977), and Morelli (chap. 15) are examples of this approach.

Over a period of about 18 months, during different seasons and different times of the day, I collected a series of randomized, timed, behavior observations on about 75 children, divided about equally between the two groups. More-detailed descriptions of the methods employed can be found in published articles (Draper 1973, 1975a, 1976, 1978; Draper and Cashdan 1988). I also was concerned with adult work roles, for I knew that adult activities would shape the lives of children. I collected daily work-activity diaries on men and women of the foraging groups, as well as weekly censuses and diagrams of the many band encampments.

The behavior of the foraging children shows little sex differentiation. The most obvious explanation for the similarities in girls and boys is that children are not trained to do economically useful tasks and, as a result, escape early pressure for sex-differentiated behavior. In many societies, particularly agricultural or pastoral ones in which much adult time is spent in subsistence work, children are pressed into service at early ages to help adults with their work (Barry et al. 1957). Girls typically are assigned more tasks than boys, and at earlier ages than boys, because their mothers want help and reason that the girls will be learning gender-appropriate tasks. Boys spend more of their early years free of pressure for responsibility and obedience. Adult men and older boys eventually train them in their work roles, but training begins at later ages for boys than for girls.

!Kung children of the foraging groups do virtually no useful work until very late ages, well over 12 years. This is an unusually permissive regimen for children in comparison with the expectations for child labor in many other traditional societies where data on children's work have been collected (Whiting and Whiting 1975). The foraging children are freed from work because of the nature of the work their parents do and because of the environment in which they live. Men and women in their hunting and gathering rounds can walk long distances in search of food, and they realize that children cannot keep up. Children become tired, slow the work party down, and may need to be carried, thus reducing the efficiency of one or more adults. Furthermore, for about 7 months of the year, there is no rainfall in the Kalahari. Much of the terrain in which the adults forage is dry and hot. If children accompany the workers, extra water will have to be brought for them, also increasing the work load of the adults.

The solution !Kung parents have devised is to leave children behind at the main camp. Other adults can be counted on to be there to supervise children. In this society, most adults actively hunt or gather only every other day, or less frequently. Both girls and boys have equal amounts of leisure during the day. Even child care does not seem to fall heavily on the girls. Mothers look after most of the needs of infants and toddlers themselves, not using children on a regular basis as mother's helpers.

Sedentary children, however, show more sex differences, mirroring the greater sex differentiation in the roles and behaviors of the adults. Sedentary children also do more economically useful tasks, a fact explained by the more numerous responsibilities of their parents and the beginning attempts by parents to put their children to work. Girls stay closer to home base and helped their mothers; boys are more frequently out of the home village, often doing chores associated with domestic stock. From early ages, boys range freely with other boys, have more contact with Bantu and, relative to girls, have a head start in learning non-!Kung languages (Whiting and Whiting 1973).

On the other hand, even in the foraging groups, the behavior profiles of girls and boys are not identical. In fact, they differ along many of the same lines that female and male juveniles differ in nonhuman primate species (Hiraiwa-Hasegawa chap. 6). Girls are physically less active, more likely to be in physical contact with others. They are spatially less far-ranging, more likely to receive interruption and redirection by adults (Draper 1975a; Draper and Cashdan 1988). Girls are more likely to spend time with adults where their behavior is closely supervised. Boys more often play with peers and are less likely to be close to adults and adult supervision. Even though the requirements of daily living among the foragers do not facilitate gender-segregated treatment of children, it appears that girls and boys gravitate to different microenvironments. I could see that "choices" made by girls early in life could be implicated in a common pattern of sex differentiation in which girls receive, at earlier ages, stronger pressures for compliance and obedience. Staying closer to adults, particularly women, they have less time to explore on their own and in the company of peers.

Adult Behaviors

Adult behaviors in the foraging and sedentary groups also differ. Men and women of the foraging groups are egalitarian in their dealings with each other. They are typically found in mixed-sex groups in the camp settings, although their work is usually done in same-sex groups. Women do not show deference to men. Living in small bands without well-developed leadership roles, they arrive at decisions by a consensus in which women participate along with men. Women as well as men are involved in primary production. Women retain control over the foods they gather and to whom it is distributed. Women as well as men leave the camp during a day's work. As a result, both men and women are well acquainted with the terrain through which the group moves over months and years.

Sedentary !Kung, on the other hand, show clearer gender segregation. Certain tasks are more strongly identified with one sex or the other. Men and women do not mix as easily and informally in everyday settings, and women avoid and defer to men in many situations (Draper 1975b, 1991).

These characterizations of gender relations among !Kung in the two settings are based on many observations of the interactive styles between men and women and the amount and types of work done by both sexes. Social structural features also promote gender egalitarianism. Many customs among the foragers make it difficult for men to coerce women by virtue of their superior strength (Draper 1975b). For example, the prevailing bilateral residence arrangements insure that women among the foragers are rarely, if ever, in the position of living away from their own kin. As a result, a woman cannot be isolated by her husband and mistreated by him, nor can young women be easily intimidated by their husbands' kin because their own relatives are typically close by. The close proximity of individual households in the foraging bands and the people's custom of spending nearly all of their waking and sleeping time out-of-doors in full view of other villagers mean that women benefit from the informal protection of public surveillance, which acts as a deterrent to most acts of male aggression.

Among the sedentary !Kung, many features of the social ecology (which among the foraging !Kung seem to have guaranteed a kind of informal gender egalitarianism) are being dismantled. With greater property accumulation comes greater male control over property, particularly the more prestigious forms of property such as livestock and cash. The work of women is increasingly domestic and confined to the village locale. Perhaps even more important to male ascendancy in the settled camps are men's interactions with the Bantu pastoralists who were neighbors of the settled !Kung. !Kung men act as intermediaries for women and children vis-à-vis the Bantu. Not surprisingly, it is the men who have mastered the Bantu languages and not the women. Sedentary women, relative to foraging women, have lost ground to men. They have become confined and relegated to a back seat in many public settings. Men have moved to the forefront, acquiring cash, language, and livestock. But *why*? What is there in accumulating material property and adopting a more complex economy that accelerates and intensifies gender asymmetry?

My thinking about the relations between women and men parallels my thinking about the sex differences in the behavior of girls and boys. The behaviors changed concomitantly with contextual change, but any statements I make about these correlations is not an explanation. Cultural practices among the sedentary !Kung account for substantial gender segregation, whereas a different set of socioecological factors among the foraging !Kung account for a relative absence of gender-segregated behaviors. Yet, the behavioral profiles of the girls and boys in the foraging groups are not the same, despite the absence of cultural pressure for gender differences. There is another reality beneath performance and context that shapes the behaviors of girls and boys.

I continued to think that the socioecological approach had value; culture and economy play a major role in structuring gender relations. But, culture, institutions, and economy do not tell the whole story. I began to think harder about what it is about the particular form of society among the nomadic !Kung that promotes gender egalitarianism and how that acts as a constraint on the more familiar pattern of gender asymmetry that appeared to "lurk beneath the surface" of social interactions and social institutions.

At this stage in my thinking and writing (mid-1970s) I was concerned with several issues. In the !Kung case, property accumulation and economic change were associated with gender segregation and increasing female subordination. But *why* should things work this way? Furthermore, why should the more "primitive" economic conditions of the foraging !Kung promote gender equality? There are other descriptions of hunter-gatherer groups in which sex-role symmetry is well established (Hart and Piling 1960; Rose 1960), so the mere fact of a hunter-gatherer subsistence base does not guarantee equality between the sexes. Socioecological factors enter into the structuring of gender relations, but social structure and cultural ideology do not provide good theoretical reasons for why gender asymmetry develops in the first place. It takes the sociobiological perspective to fit additional pieces of the puzzle, as will be discussed below.

Sociobiology

In the 1970s I began to read the developing sociobiological literature (DeVore 1965; Trivers 1972; Dawkins 1975; Williams 1975; Wilson 1975; Alexander et al. 1979; Dickemann 1979; Zihlman 1981; Lancaster 1985). My graduate training in anthropology had stressed the "natural history" approach to the study of "man." However, I began to think more seriously about what it means to be mammal and primate, as well as "human." I did not abandon a commitment to empirical measures of human behavior, grounded in an understanding of social structure and ecology, but I began to recognize that humans are like other organisms with the same problems to solve, such as staying alive, mating, and parenting.

Reading the sociobiological literature made me appreciate fully the fact that men and women, like the sexes in other mammalian species, have important reproductive asymmetries. In nonsocial species in which males play no parental role in provisioning the mother or in defending the young, the reproductive specialization of the sexes and the dimorphism in size do not have necessary implications for the ability of one sex to coerce the other. In other species that live in multimale or heterosexual groups, male-female relations assume the familiar male superior/female inferior hierarchy as a consequence of males competing among themselves for access to females. On the other hand, in mammalian species in which the parental role is critical to the survival of young, male and female interests more nearly coincide in the welfare of their young. In such species, it is usual to see some forms of mate guarding on the part of both male and female members of the pair. Either mate can be disadvantaged if the other deserts the pair for another mate. The displays of jealousy are often more intense in the male than in the female. This is explained on the grounds that the male member of a monogamous pair can be cuckolded (tricked into rearing the young sired by another male), whereas the female cannot, under natural circumstances, be tricked into bearing offspring that are not genetically her own.

Humans carry the mammalian specialization to an extreme by producing only a few offspring who mature slowly and who require large amounts of parental care in order to survive. Women are committed to a disproportionate amount of this parental work since, unlike males, they cannot recoup one or a few infant or child deaths by finding another mate (Lancaster 1985). A woman who loses a child has lost not only that individual with whom she has personal ties, but she has lost irreplaceable reproductive time. A man who loses even all his children may experience an acute sense of personal loss, but he can replace them by establishing one or more additional mating relationships with other women. The reproductive inequality between the sexes gives rise, of course, both to the behavioral and somatic dimorphisms between the sexes and to the differing reproductive potentials of the sexes.

Following this line of reasoning, one sees the human female is encumbered to an extent not seen in other species. Because of the extremely dependent state in which young are born and because of their slow development, any roles that conflict with a woman's reproductive roles are generally avoided by her as an individual or denied her by other interested parties, especially her kin and her mates. The human mother must continue to invest high levels of parental care in several young simultaneously. Unlike other primate females who greatly reduce care of the next oldest offspring (by no longer nursing or carrying the subadult juvenile) when a new infant is born, a woman maintains not one but several dependent offspring, albeit at different stages of dependence (Lancaster 1985). With each new child, she adds to her encumberment and goes farther and farther into "debt" in the sense that her dependents multiply, but her physical reservoir of energies remains the same. In order to rear offspring, a woman must have help. Some aid comes from her kin, but nearly all human groups attempt to regulate access to the reproductive capabilities of women by designating a mate (husband) and making him and his kin share in the work of rearing or defending the children.

Humans live in groups that include numerous other individuals who are eligible as mates. As a result, the sexual contract (paternity certainty in exchange for protection and economic resources) has consequences for other male-female relationships besides marital ones. Daughters and sisters of men, for example, have a vested interest in maintaining alliances with their male consanguines, not only because they benefit from men's labor but because these men protect them from other men. A woman's mate, if she has children by him, is also more likely to benefit her and her children than are foreign, unrelated men.

It is probably true to say that in the past environments of evolutionary adaptation in which human social, psychological, and sexual behavior have been molded by natural selection, a woman has had few degrees of freedom. A woman who mistakenly judges her economic or social resources risks the survival of her offspring, who, in technologically "simple" societies, are wholly dependent on her nurturing, especially during the first few years of life. This fundamental and rather dismal picture (from the point of view of modern individualistic and humanitarian values) needs to be kept in mind when thinking about behavioral and institutional inequalities. I do not suggest that women are without choices or strategies, nor that they have been selected for passivity. I do argue that women relative to men must be extremely cautious in their economic and reproductive careers because, unlike men, they have only limited ability to recoup their losses.

As a cultural anthropologist, I have found that an extremely interesting aspect of cultural variability is the extent to which the underlying reproductive asymmetry of men and women is institutionalized. Any circumstance in which fertility is low and monogamy is imposed, either as a result of ecological constraints or social conventions, is a good place to look for restraints on the ability of males to coerce females. The reason is that, in this situation, the reproductive interests of the sexes are the same (Alexander et al. 1979; Alexander 1987) or more nearly identical than in other human groups. Under the ecological conditions in which some contemporary hunter-gatherers, such as the !Kung, have lived, the requirements for male labor are sufficiently high that most men can only support one mate and her children at a time (Hewlett 1988). In technologically more advanced societies that have surpluses in the form of stored grains or herds of domestic animals, the fitness interests of the sexes are not the same. Men can advance their own fitness at the expense of other men by competing for access to more than one mate. Where resources are potentially abundant and can be disproportionately controlled by a single man or alliances of men, male-male competition has direct consequences for male fitness (Dickemann 1979; van den Berghe 1979). In such social systems, the variance in male reproductive success is high, and men have much to gain by winning in competitive engagements with other men and, thereby, gaining access to women.

The sociobiological perspective provides another piece of the puzzle about sex, gender asymmetry, and culture. The substantial gender equality that is seen among the foraging !Kung must be related to the ecologically imposed low ceiling on male-male competition. In the hunting-and-gathering setting, the best a given man can do, reproductively speaking, is the best a woman can do. Not surprisingly, monogamy is the prevailing marriage form; polygyny is allowed but rarely practiced. Because of the "simple" technology, the requirements of mobility, the inability to store surpluses, and the unpredictable nature of both vegetable and animal resources, men and women must work equally hard to maintain themselves and their dependents (Lee 1968). Ecological uncertainty makes sharing, especially the sharing of animal protein, a necessity, and a consequence of the sharing is minimized male-male competition.

By settling at permanent water and beginning to adopt the food-producing practices of the Bantu, !Kung raise the ceiling on male-male competition. In the late 1960s, the shift to sedentism was not far advanced, for the major incursion of Bantu (primarily of the Herero tribe) took place in the late 1950s. Few !Kung in these early years owned stock. A few men managed small herds on a contract basis for Bantu owners, but most of the settled !Kung were clients or hangers-on to the Bantu, whose numbers were increasing steadily. The men who worked for the Bantu had definite advantages. Their families drank milk and received occasional payments of meat and agricultural produce. In comparison with men living in the nomadic groups, they had few obligations to share their food.

There also were disadvantages to settling with Bantu. !Kung complained that their compensation for work was not sufficient and, in truth, the !Kung servants lived in much poorer circumstances than their Bantu patrons. The main complaint of !Kung who worked for Bantu in the 1960s (as well as now) was that they were not paid well enough to get ahead, to become independent stock owners in their own right.

Another disadvantage to !Kung men working for Bantu is that some Bantu men wanted to take !Kung women as wives. More often, rather than marrying them, Bantu men enter into informal liaisons with !Kung women, not marrying them, but fathering children by them and not contributing to their support. This put !Kung men in the unenviable position of having fewer !Kung women to marry themselves and seeing their daughters begin reproductive careers without husbands who could help their wives and in-laws. Most men who become servants or clients to the Bantu manage to find !Kung wives, but not a few of them rear at least one child born to their wives by a Bantu lover.

In the context of the competition from Bantu men over access to women of their group, it is not too surprising to find changes in the structure of gender roles among the sedentary !Kung in the 1960s. If !Kung men have moved into more prominent "brokering" roles vis-à-vis Bantu, and if they encourage their women to remain in the background, these moves are understandable as strategies to reduce the sexual competition with men of a wealthier group. If !Kung women "cooperate" by reducing their spatial mobility and by minimizing their contact with the pastoralists (especially the men of these groups), this also is understandable if they want to maintain their claims on support and protection from men of their own groups. !Kung women can be expected to avoid contact with non-!Kung men as long as men of their own groups are willing and able to assist them.

In this discussion I hope to have made clear the different perspectives on behavior that can be gained by adopting, on the one hand, a social structural perspective, and on the other, a sociobiological perspective. The two models are in no way antithetical; in fact, they provide complementary kinds of insight. The changes in subsistence practices from a life based on foraging and economic egalitarianism to a life based on a settled, food-producing economy and increased potential for economic differentiation, led to a multitude of changes in !Kung social practices. Some of these changes are visible in the mundane organization of adult work roles and the socialization of children. Other changes are visible in the realm of sexual, reproductive, and gender relations between men and women.

New Directions, New Data, and Old Interests

In 1987–88 I spent 15 months in the Kalahari studying aspects of aging and life-cycle development among the !Kung.[1] As part of the study, I collected demographic data, the core of which came from reproductive interviews with about 330 adults ranging in age from 18 years to more than 75 years. These data speak directly to matters of sex, reproduction, and gender differences, although, in this context, the questions are not about gender asymmetry and sexual egalitarianism but about sex differences in reproductive strategy.

In brief, the acculturative changes that were already apparent in the late 1960s have continued. All !Kung in western Botswana are now settled, and the Bantu presence has increased. !Kung now subsist on their stock (mostly goats with a few cattle), sporadic hunting and gathering, and a periodic government dole of corn meal and beans. Some !Kung men work as cattle herders for Bantu families. A recent 7-year drought has made gardening nonproductive. Many !Kung customs remain in place: village residence remains largely bilateral, and marriages are almost entirely monogamous. The biggest change is in the timing of marriage and in the frequency of out-mating by !Kung women with non-!Kung men. In traditional !Kung practice, women are married in their early teens. This custom was continued by !Kung in the 1950s and 1960s, both among those settled and those living in the bush. Today women marry at later ages. Most significantly, a rising number of women do not marry, or if they do, marry after having had one or more children out of wedlock, often by non-!Kung men.

What this trend will mean in the long run for gender relations is not clear. In the short run, it seems advantageous to women. In the old days !Kung girls were married young and to men an average of 5 to 10 years older than they (Howell 1979). As girls, they had little power to refuse the matches arranged by their parents. Though many of these early marriages ultimately failed, it was often after a long period of unhappiness and resentfulness on the part of the young brides. The reasons for these early marriages were economic. Parents wanted their daughters to marry in order to secure their welfare and so that the family of the bride could gain access to the hunting skills of the groom. In former times, the hunting prowess of men was highly esteemed by people living in the bush as well as by the settled !Kung. By all accounts, the game was more abundant formerly, before fences disrupted the movements of the game herds and before the introduction of domestic stock and mounted hunting.

Today, many !Kung women do not marry until their middle or late teens. Others do not marry at all but establish informal liaisons with men by whom they have children. These unions often do not lead to marriage; instead, the girl stays at home with her own parents and kin. Her lover visits her in her village but does not have rights in her as a husband.

At least two aspects of cultural change facilitate the rise in multigenerational extended families and female-headed households. First, middle-aged men and women (and in some cases the elderly parents of middle-aged people) have sufficient wealth to keep their daughters at home *whether or not* they have husbands. Sedentary people work harder than do foragers, and there are many tasks to be completed. The grown daughters are welcome as working members of the villages as are the children they bear. Second, young !Kung men in the modern acculturated setting have less to offer young women (and the parents of young women) in comparison with what they could offer as hunters when people lived as foragers. As mentioned above, game is scarce, largely because of the increased numbers of cattle. Alcohol is now available in the areas of !Kung settlement, and the frequent drinking and resulting knife fights involving !Kung men do not enhance their reputations or prestige. It appears that, at present, husbands and sons-in-law are less important than they once were. The labor that men provide is still needed, and the social positions that men occupy as heads of families are still important. However, under the new economy, one man, as head of an extended family compound, can provide reasonably well for his wife and children. Women still need such men, but it appears that they do not need them as husbands. If they have fathers, brothers, sons, or uncles to provide an umbrella of protection, it is apparently sufficient.

Some data I collected in recent field work with !Kung show some of the effects of cultural change on the reproductive differences among different cohorts of men. These data come from reproductive, employment, and acculturation interviews collected from about 110 men over the age of 20 years. Analysis of these data shows the effects of cultural change on the completed fertility of men of different age cohorts. Older men (over about 60 years of age) had about 3.5 children and a small variance. These older men began their reproductive careers well before the arrival of the Bantu in large numbers. Younger men, roughly 45 to 62 years of age had a higher average number of children (about 4.0) and a greater variance. Men of this age group spent the bulk of their reproductive years during a time of increased exposure to Bantu influence and presence. Inspection of the individual cases who make up the middle cohort shows that men who had high reproductive success differed from those with low reproductive success in having long-term residential association with Bantu and more years of employment with the cattle owners (Kranichfeld and Draper 1990; Kranichfeld (1991).

A comparison of the reproductive success of men and women of different ages makes a similar point (Draper and Buchanan 1992). These data are based on reproductive interviews with about 330 adults, ranging in age from 18 to over 80 years. The total numbers of children (3.6) born to old people (60 years and over) are equal for the two sexes, whereas for a younger age cohort, (45–59 years of age), the sexes do not have the same fertility. !Kung women of this 45–59 year cohort average about 4.6 children. Men of the same age group have produced an average of about 3.5 children. The discrepancy may be explained by the common pattern whereby men marry younger women and continue fathering children at later ages than women. In this case, the age-hypergynous factor does not account for the difference.

Figure 16.1 "Children of Mixed Ethnicity as a Percentage of All Live Births" shows the percentage of mixed-ethnicity children born to women of the age groups 18–29, 30–44, 45–59, 60+. As can be seen, older women produced few children by mating with Bantu men. Younger cohorts of women have produced a steadily increasing percentage of children fathered by non-!Kung men. The men of younger cohorts, the bulk of whose adult lives have been spent during the period of maximum acculturative pressure from Bantu, are losing out to Bantu in reproductive competition for !Kung women.

Another surprising piece of this newly emerging pattern in family organization and reproductive strategy is the finding that the survivorship of the children of "mixed matings" is

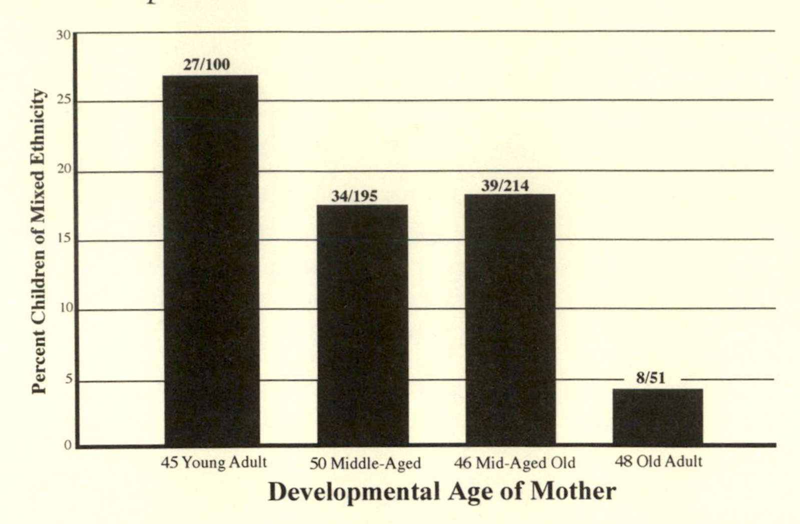

FIGURE 16.1. Children of mixed ethnicity as a percentage of all live births; percentage of children of mixed ethnicity born to women of the age groups 18–29 years, 30–44 years, 45–59 years, and over 60 years.

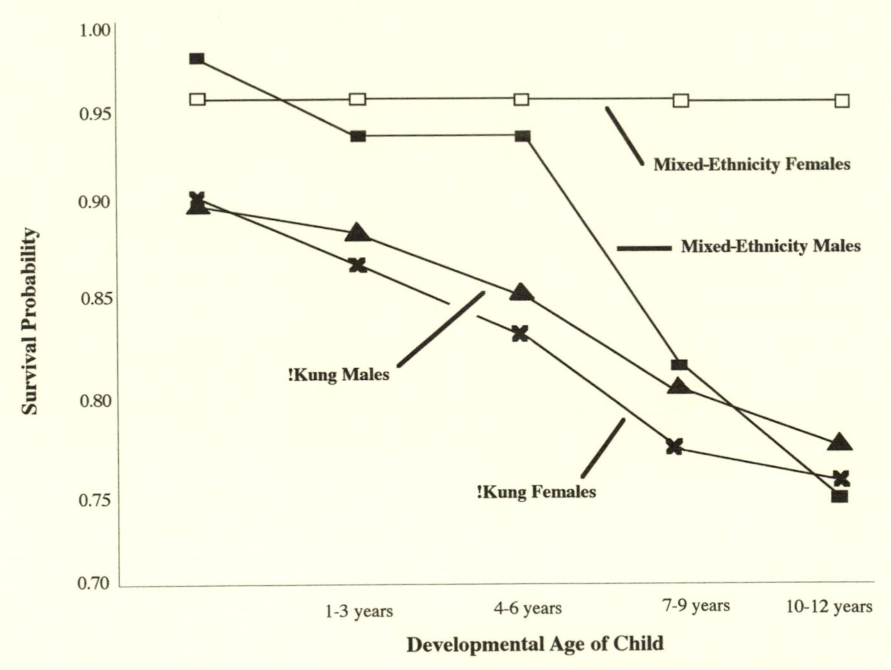

FIGURE 16.2. Survival probabilities for children of interviewed women, by ethnicity and sex.

higher than that of children both of whose parents are !Kung. These findings, and those discussed below, are based on data on about 700 children born to interviewed parents. At all ages, the children of mixed parentage have a higher probability of surviving than do the full !Kung children (Kranichfeld 1991; Draper and Kranichfeld in progress).

A more interesting picture emerges as shown in figure 16.2 "Survival Probabilities for Children of Interviewed Women, by Ethnicity and Sex." In this display, it is evident that primarily the girls of mixed ethnicity survive at the highest levels, whereas, after about age six, mixed boys lose ground and at age 10–12 years are surviving at lower probabilities than the !Kung males. Given that the modern !Kung are living in a sharply stratified system, and given that they occupy the lowest rung on the ladder, it is tempting to speculate that the sex differences in survivorship, particularly among the children of mixed ethnicity are related to the better marriage or mating possibilities for girls in comparison with boys.

What does all of this mean for any future structuring of gender roles among the !Kung? The "data" presented in this last section are of a different type and pertain to reproductive events, not to everyday economic and social behaviors of men and women. Nevertheless, the demographic data indicate that new developments are taking place and, reproductively speaking, the sexes are not sharing in them equally. !Kung are experiencing profound cultural changes not only in the economic changes but in the increased exposure to non-!Kung ethnic groups whose wealth and levels of education greatly exceed those of the !Kung. Often in considering culture change, one thinks about the impact of new conditions on the recipient group as a whole. In the !Kung case, the entire ethnic group is affected by unilateral changes, yet the responses to those changes are not uniform across the population. Different age cohorts are affected differently as are the two sex groups.

Most !Kung men spoke volubly about the increase in mixed-parentage births. In private conversation, they heaped ridicule and scorn on women who go with Bantu men. They say (as do the mothers of children of mixed ethnicity), "The Bantu do nothing for these children. They just drop them." (But if this is true, why do the children of mixed ethnicity have higher probabilities of surviving to later ages?) One particular man spoke, telling of the male view, and was especially vehement about the liaisons between women of his group and non-!Kung men. He said bitterly, "Besides, where do you see a Bantu woman sleeping with a !Kung man? These women are big and important, and they have possessions. The Bantu women say that a !Kung man is a little thing of no account. They don't notice us."

These demographic developments would not have taken place if other changes were not already in place. Coming in from the bush, of course, exposed !Kung to economic and sexual competition from a wealthier and technologically superior group (Draper and Kranichfeld 1990). As !Kung left the bush, gradually and experimentally, at first, later in larger numbers and with a more permanent commitment to settled life, they found themselves less and less able to exploit the bush for wild foods. The men's hunting role was more severely affected than the women's gathering role, and these effects have accelerated in recent decades. Even by the 1960s, the abundance of game was reduced, according to informants. The majority of !Kung men were not able to compensate quickly enough by replacing their lost value as hunters with acquired value as workers and managers in the new sedentary economy. Meanwhile, what game was left was being disproportionately bagged by Bantu who hunted with firearms and were mounted on the horses and donkeys that the !Kung were not yet able to afford.

Increasing numbers of !Kung women have chosen to mate (not, at present, to marry) with non-!Kung men. This indicates an inability or unwillingness (or both) on the part of the guardians of women to control their reproductive behavior. Does this represent "gains" for women, or does it indicate cultural breakdown? What do the apparent "facts" of the demographic changes I have described have to say about the structuring of gender relations among the !Kung? Are women better off or worse off? In comparison with whom? On the basis of what measures? The question about the future structuring of gender relations among !Kung is not easy to answer. It is evident, however, that it is important to take into account the social structural context of people's lives, the historical epoch in which they live, the ages of specific individuals at the time they experience certain events of cultural change, and the underlying reproductive biology of the population undergoing the change. In the instance of contemporary !Kung, the reproductive changes leading to changes in family structure have been at least as influential in guiding the future course of !Kung modernization as have been changes in !Kung social structure.

Acknowledgments

Research was supported by a grant Number AG03110 from the National Institute of Health. Principal Investigators were Christine Fry of Loyola University and Jennie Keith of Swarthmore College.

17 Women's Work and Energetics: A Case Study from Nepal

Catherine Panter-Brick

Profile of a Tamang Woman: One Observation Day in September 1983

Nangsye Tamang, 21 years of age, spent the night at her father's house, having left her husband with the cattle penned on the mountain side. At dawn, her first household tasks were to fetch water at the village fountain, then walk over to the water mill where she nursed her child, and slept a little, while her sack of maize flour was being ground for the midmorning meal. Around 10 A.M., she left for the fields to weed millet with a group of 14 people, who had arranged to work on each other's fields in reciprocal exchange. She carried her baby girl on her back during the day, interrupting her work to breast-feed when the baby cried. She also gave her a little maize flour and millet beer when the workers stopped for lunch. The labor group worked hard, as a line of men hoed ahead and a line of women progressed steadily through the terraced fields, bending over to stick millet shoots into the ground with thumb and forefinger. At times, they would break into a song and a dance, improvising witty verses on a familiar melody known as the "Love Song." The work ended somewhat early, at 6 o'clock, and all returned home. Nangsye carried grass for the cattle in a big bamboo basket, with her child on top, and reached her father's home in time to fetch water and cook the evening meal.

An Ecological Perspective of Energetics

In most areas of the world a woman must maintain her own health, contribute economically to her family, and, at the same time, reproduce, and nurture her children. Women's work and reproductive efforts in a subsistence economy are examined here from the viewpoints of energetics, ecological anthropology, and human adaptability. Following a discussion of the theoretical basis for this approach, I use a case study from Nepal to illustrate how major ecological constraints structure a range of behavioral choices that have important, measurable biological consequences for survival and reproduction.

Why study energetics? The literature points to intriguing relationships among food intake, workloads in subsistence tasks, economic survival, mother-child health, and associated reproductive performance. Consider the following passage from Rajalakshmi (1980:187–88): "The poor woman in India and in many similar countries lives on a diet providing [few] kcal/day . . . and does not appreciably increase her intake during pregnancy and lactation. . . . In spite of this she achieves a satisfactory weight gain . . . during pregnancy . . . , produces a reasonably healthy infant . . . and nurses with remarkable success. . . . She often manages to achieve all this through several successive parities without a decline in quality of her performance and without losing weight in the process. . . . How does the poor woman maintain her body weight and overall health status?" In other words, how does a woman with little energy to spare survive and cope with repeated reproductive demands?

More than a decade ago, Durnin (1978:11) called for research on the "supposed extra energy needs of pregnancy and lactation," given that "very little evidence exists in practice to support these extra needs. We have some evidence that there is a very considerable compensatory reduction in physical activity by both the pregnant and the lactating woman, which almost may negate the supposed increased energy requirements." To date, a number of physiological and behavioral coping responses have been identified, such as the use of body mass and fat reserves, a smooth performance of tasks, and reduced physical activity or basal metabolic rate (Torun et al. 1982; Lawrence et al. 1985; Prentice and Prentice 1988; Durnin and Drummond 1988; Ferro-Luzzi 1988).

It also has been shown that mothers decrease their "optional" activities, while maintaining levels of "obligatory" work. Californian women, who only do light physical work, for example, can reduce daily energy expenditure by only 6% during pregnancy, modifying leisure behavior rather than occupational work and child care (Blackburn and Calloway 1979). Rural women in The Gambia, who have heavy agricultural workloads, can lower activity levels by 25% in the month before delivery, but achieve this through decreases in housework and leisure, not time devoted to farming (Roberts et al. 1982). And Papua New Guinean mothers travel less than other women and spend more time sitting, mostly to nurse (Greenfield and Clark 1975).

What is the situation of pregnant and lactating mothers in other communities where women assume a heavy workload and cannot afford to discontinue subsistence activities? Many ethnographies report that women continue to work throughout pregnancy, often giving birth at the work site (Jimenez and Newton 1979). For example, Shostak describes the !Kung of Botswana as follows: "Once the pregnancy is known to the group, others offer help. . . . Still, the woman is not viewed as in need of protection, nor is she expected to cease her normal activities. She continues to travel the usual distances to gather food and is apt to return with her usual load . . . many women maintain their normal work routines until the day they give birth. Pregnancy is thought of as a given; it is 'women's work.'" (1990:178)

Although such reports do not offer quantitative data, they do highlight a fundamental problem for women who work outside the home and who face important time and energy constraints. How do such women combine motherhood with their subsistence responsibilities? Is there a conflict between childbearing and food-producing activities, and where do women's priorities lie?

The compatibility—or incompatibility—between subsistence work and child care is a fundamental issue that has prompted conflicting views about women. Feminist theories of gender inequalities have stressed that responsibilities of motherhood keep women from achieving equal social status with men (Rosaldo and Lamphere 1974). A once-classic model of hominid evolution proposed that women's reproductive role led to a sexual division of labor whereby women gathered food and nurtured children while men hunted and provided defense. This view was redressed by evidence that suggests an active role for females in food-getting and processing (Zihlman 1981; chap. 13). A wide variation of sex roles among nonhuman and human primates also suggests that childbearing did not preclude women from hunting in hunter-gatherer groups (Goodman et al 1985). Yet, children can limit a woman's economic activities (Hurtado et al. 1985) and employment opportunities (Leslie 1988). In addition, women's work has consequences for maternal and child health, for example, as demonstrated in The Gambia (Prentice and Prentice 1988). These "costs" raise the question of the adaptiveness or success of the existing range of productive and reproductive behavioral strategies.

The topic of motherhood, thus, has generated a great deal of attention. Debate has centered on vital issues such as how poor women manage to reproduce "against the odds" on tight energy budgets, the degree to which they continue to work in food production when childbearing, and the consequences or tradeoffs among possible behavioral alternatives. In the 1980s, "the basic data available [about women were] inadequate and insecure" (Nestle Report 1982). In particular, a careful evaluation of Third World rural communities is still required in order to document women's actual working behaviors, the factors that constrain or facilitate economic and reproductive activities, the range of possible alternative choices, and the consequences of existing strategies.

A focus on energetics is justified at both descriptive and theoretical levels (Harrison 1983). Time and energy are factors that influence an individual's behavior in many different environments; they serve as "currencies" that lend themselves to quantification. Measures include behavioral data from time-allocation studies and physiological data to estimate energy expenditure. The concept of energetic efficiency (allocation of time for particular activities and expended energy) has proved a useful explanatory tool when discussing human subsistence strategies, especially for communities at high altitude where productivity is limited (Thomas 1976), and for hunter-gatherers whose lifestyle is closely tied to the environment (Winterhalder and Smith 1981). Reproductive behaviors also have been described as the product of a trade-off between the allocation of time and energy to food acquisition and parenting (Hill and Kaplan 1988).

Although the process of acquiring energy is basic to the survival of all individuals, the degree to which energetics will "explain" observed behaviors in human communities depends on the importance of this constraint for the local population. A wide range of features of the biological, social, technological, and physical environments can influence human behaviors. These include shortages of time, of nutrients and energy, of sexual partners, of labor, land or capital, culturally defined scarcity, and ecological uncertainty or unpredictability of resources (Winterhalder 1980; Jochim 1981).

Behavioral strategies aim to alleviate the most pressing of these constraints. Some behaviors will be *effective* at overcoming exisiting limitations of time and energy. These are adaptive in allowing individuals to pursue other goals relevant for reproductive success. Other behaviors will be *efficient* in the use of time and energy, representing the best possible choice given available resources (E. A. Smith 1979, 1983). These are significant in promoting, relative to other competing strategies, greater economic security and reproductive success. Thus, it is necessary to measure the relative costs and benefits of several behavioral alternatives. Their outcomes, in terms of survival and reproduction, provide an ultimate test of their success in coping with existing limitations.

A Case Study from Nepal

An example from Nepal shows why considerations of the use of time and energy are important. I spent the last 12 years studying the farmer-pastoralist Tamang people and the low-caste Kami blacksmiths of Salme village (1870 meters [6135 feet], northwest Nepal). The foothills of the Himalayas are rugged, with marked seasonal and altitudinal variation. High altitude, low temperature, seasonal rains, and limited technology severely limit food production.

The Tamang operate a diversified economy based on agriculture, animal husbandry, and the use of common forest and pastures. Fortunately, the community has access to a vast expanse of land, an entire mountainside of 30.7 square kilometers rising from 1350 to 3800 meters (4429 to 12467 feet). Five staple crops (i.e., paddy rice, maize, millet, wheat and barley) are grown at successive altitudes, on terraces carved out of the mountainside, and harvested at regular intervals throughout the year. Plots are dispersed and small, because families aim to cultivate crops at different altitudes in order to stagger the harvests, and because land is divided equally between sons at each generation. Such extensive cultivation ensures self-sufficiency in food, but also requires a sustained and substantial workload. It generates a shortage of labor among Tamang families, who live predominantly in nuclear households. The Kami, by contrast, derive income from the smithy and have fewer agricultural and animal resources. Compared to the Tamang, Kami women spend comparatively less time away from home.

Women's Work: Dual Roles of Food Production and Reproduction

Work and Food Production

How do Tamang women organize their work? Are their behaviors directed toward making an efficient use of time and effort? In order to describe and evaluate their activities, a detailed time-allocation study and measurements of energy expenditure were conducted for a representative sample of 87 villagers. Individuals were followed continuously by two village assistants on several observation days per season, and their activities recorded from 6 A.M. to 7 P.M., yielding a total of 7678 hours of minute-by-minute observations during the year. The energy cost of specific activities was also measured by the total volume and oxygen content of expired air during the performance of the task (Panter-Brick 1993a).

This question proves particularly relevant in regard to travel on the mountainside, a prerequisite for all outdoor subsistence work. The Tamang journey over an extensive area, each day on foot, with or without loads, up and down slopes with 20% to 40% incline. Not surprisingly, they have devised means of saving themselves considerable effort. This is shown by their allocation of time and extent of household labor devoted to different activities and by their use of mobile housing.

A Tamang family builds a temporary structure, known as a *goth* (Nepali), on the mountainside near the place of work. It provides overnight shelter for both humans and cattle. The shelter is built of wooden posts and bamboo mats; it can be dismantled, carried, and set up again whenever needed. Mobility is a key feature. The goth is moved every 2 to 3 days on a circuit that averages 19 kilometers (11.8 miles) over the year. Dung and urine from livestock penned on a terrace during the night dramatically improve soil productivity, and the goth is located on fields that must be manured and ploughed before planting. The goth also is set close to sources of fodder, either pastures where animal graze during the day, or forests and fields where foliage and grasses are cut.

Husbands or fathers are usually in charge of a goth. When convenient, their wives or daughters sleep there overnight and, therefore, make shorter journeys to the fields or forest than they would have made had they traveled from the village home. The extent to which women save travel time by sleeping overnight in a goth depends on its variable location on the mountainside. In the winter, for instance, shelters are placed nearby the village at mid-altitude on maize and millet fields. Tamang women covered the "home-goth" journey (24 minutes one way) quite frequently, both residences being conveniently close to the fields. (Journeys from the goth averaged 10 minutes, those from home 19 minutes). They spent an average 4.7 kilocalories/minute walking on more or less level ground, and 3.8 kilocalories/minute carrying (more slowly) loads of up to 55 kilograms (121 pounds). They saved only 14 minutes of travel time and little energy by sleeping in a goth instead of at home.

By contrast, in the spring, goths are placed at high altitude on wheat and barley fields (61 minutes away from home), and women spent several nights in a row in one or the other residences. Journeys from a goth averaged 16 minutes to the fields and 34 minutes to the forest, as compared to 27 minutes and 65 minutes from home. Women spent an average 5.5 kilocalories/minute climbing uphill for 56 minutes, and 3.1 kilocalories/minute carrying their loads downhill for 54 minutes. In the spring, the use of a goth represents a substantial economy of time and effort: without it, women would have traveled on average, an extra 110 minutes (1.8 hours) per day and expended 475 kilocalories to reach their place of work.

In addition, the goth obviates the need to carry manure to dispersed fields and fodder to animals stabled at home. The time and energy required for such tasks would greatly exceed the effort of moving the goth to a new location, which amounts to carrying three or four loads of 40 kilograms (88 pounds) every 2 to 3 days. An alternative residence also gives family members greater flexibility to meet multiple responsibilities for tending crops and animals. In all, the use of a mobile shelter is an adaptive behavior responding to the very real need to spend considerable time and energy to manure and cultivate widely dispersed land holdings. It appears to be a successful strategy given specific resources in land, animals, and labor, and Tamang goals for self sufficiency.

Maintenance of a goth has agricultural importance. Agronomists have measured a 10-fold difference in the yields of nonirrigated staple crops. This correlates highly with the extent of fertilization and the number of nights the cattle are penned in goths on a particular field ($r = 0.73$; Pierret-Risoud and Risoud 1985: 160). Use of a mobile goth raises soil productivity and, at the same time, enables the Tamang to cultivate the most distant terraces, thereby extending the areas of cultivation.

Goths, so useful in the spring, are dismantled in the monsoon. This is the time when men and women both transplant paddy rice in irrigated fields, which need no manure. They also move millet shoots in rain-fed fields, which are mixed with maize and cannot accommodate cattle. In neighboring villages at lower altitudes, where land and forest resources are scarce, the Tamang have relinquished goths in favor of fixed shelters and the practice of carrying loads of manure to the fields. Thus, the practice of keeping mobile shelters is abandoned whenever it fails to make good ecological sense.

In addition to its role in subsistence economy (and survival), use of the goth is important from the viewpoint of health and reproductive outcome. Residence in the goth has a favorable effect on nutritional status. Goth-tenders have greater access to milk from cattle, averaging per capita intakes of 68 grams/day as compared with 19 grams/day for village residents (Koppert 1988). Milk is particularly important for young children, who average 30 grams/day, as it provides otherwise scarce animal protein. Moreover, the goth has a positive effect on daily energy balance by reducing the amount of energetically expensive travel, a significant consideration in a community where per-capita energy intake is only 1847 kilocalories/day. Use of the goth also may dampen fertility rates since Tamang spouses frequently are separated while meeting simultaneous household and agropastoral tasks in diverse locations. This may partly explain their long interbirth intervals (37.7 months). At the same time, it facilitates the integration of maternal work and child care by reducing the amount of time that working mothers spend away from children who are past weaning age and left behind in the goth. They can give their children a better start in life and, thus, promote their own reproductive success.

The Tamang calculate for themselves the relative costs and benefits of goth management, modifying their behavior accordingly. Although goths save travel time and effort, not all households can afford them (owning at least three animals and mobilizing a herdsman full time is required). Therefore, some families "rent" a goth for a specified number of nights on certain fields. Others pool animal and human resources to share one. Alternative behaviors, therefore, are sensitive to various land/cattle and land/labor ratios (Pierret-Risoud and Risoud 1985) and, furthermore, to changes in the environment. Thus, individual households make a number of critical choices regarding work organization to address very specific ecological constraints. To date, goth management is modulated from outright ownership to reciprocal partnership; ownership emerges as the best strategy (an "optimal strategy") for self-sufficient households with a given threshold of cattle and labor in relation to land resources. A careful examination of the relative costs and benefits attached to alternative behaviors helps to explain why certain choices are made from the range of alternatives.

Work and Child Care

In addition to subsistence work, women organize child-care activities. How do Tamang women combine both responsibilities? Because of labor shortages within most households, Tamang women continue to work during the childbearing period and to travel extensively on the mountainside. Behavioral choices again are guided by ecological imperatives. Seasonal time-allocation data comparing the work actually done by pregnant or lactating mothers and nonpregnant or nonlactating women show that Tamang women adopt different priorities during different times of the year.

Tamang women take babies with them to the work place, carrying them on their backs in a bamboo basket supported by a head strap. Toddlers, too heavy to carry and still too small to walk, are left behind. In late winter, women who are pregnant or lactating spend less time outdoors, devoting 39% less time to agriculture, husbandry, forest work, and travel, and spending more time on indoor household tasks. They averaged a total 3.3 hours/day outdoors (working 2.5 hours and resting 0.8 hours), as compared to other women who average 5.4 hours/day (working 4.1 hours and resting 1.4 hours). Winter is a time of lower workloads, when the Tamang need not work long hours, so long as they fulfill a variety of tasks (planting one crop, harvesting another, tending cattle, processing grain) without delays of schedule.

A different situation prevails in the monsoon (mid-June to September), when the Tamang assume much heavier workloads and focus on urgent agricultural tasks. The beginning of the rains (June-July) signals intense work effort, because paddy and millet shoots must be planted early in order to catch the best of the growing season. In the monsoon season, women spend 8.2 hours/day outdoors, working 5.1 hours/day excluding time for rest (in fact, they work 7.7 hours in the late June to early July, 6.2 hours in late July, and 3.0 hours in September, excluding rest). All women, including third-term pregnant women and newly delivered mothers, follow this pattern (Panter-Brick 1989). At this time of year, men and women of all ages and childbearing status form large labor groups, dividing tasks among themselves and pooling resources to speed up work. Men plough and irrigate fields while women weed and plant them. This "economy of scale" underscores the need for efficient labor organization in response to a seasonal time constraint (Panter-Brick 1993b).

An example will illustrate the consequences of time mismanagement. One Tamang family was 2 weeks late transplanting rice, having insufficient time and manpower to plant both paddy and millet crops in early July; as a result, it expected its rice yields to be reduced by half (Smadga 1986). In the monsoon, Tamang households simply cannot afford to release childbearing women from their responsibilities in food production. Therefore, how much subsistence work Tamang pregnant or lactating women perform is a consequence of the costs incurred as a result of a loss of their labor, the magnitude of which varies in different seasons of the year.

In contrast, low-caste Kami households in Salme village depend primarily on male craft work in the smithy. Women have lighter outdoor work responsibilities with less seasonality. Whereas all adult Tamang women show similar work profiles in outdoor tasks, individual variation is more evident among Kami women. Young Kami daughters-in-law are sent to the fields or are hired as wage laborers for the Tamang, whereas senior mothers-in-law tend to stay at home. Neither pregnancy nor lactation changes this situation (young Kami mothers take their babies to the fields, or leave them behind with older women in the village). In this case, work patterns are an expression of status in the household, which is elevated with age and the birth of children. These distinctions are made possible for the Kami by the lower demand for female labor in their community.

Biological Consequences for the Mother and Child: Reproductive Outcome

What are the consequences of female work for the adult woman who sustains an important workload to achieve economic security while also providing for her child?

In Salme village, the seasonal nature of working behavior and the contrast between two socioeconomic groups or castes facilitates the evaluation of a number of different strategies. Two findings are of interest. First, childhood mortality is high, and growth rates are slow for the Tamang, particularly in the monsoon. Overall death rates are 175/1,000 in the first year of life, for the Tamang, and 158/1,000 for the Kami. In the first five years, death rates are 271/1,000 for the Tamang and 279/1,000 for the Kami. Thus, Kami infant mortality is somewhat lower. Second, the castes differ in both the total number of births and in birth spacing. The Tamang average birth intervals (37.7 months) that are 8 months longer than those of the Kami (26.7 months). Therefore, it is important to examine the consequences of women's behaviors for both child survival and reproductive performance.

Child Health and Survival

Does the priority given to agricultural tasks in the monsoon prejudice the health of Tamang children by restricting time available for child care? Without doubt, the monsoon is the most difficult time of year for all children. It is difficult to determine how much this predicament results from an unsanitary environment or inadequate maternal care. However, different factors will influence survival (or death) of infants versus that of young children.

Differences exist between infant-care and child-care strategies as seen in the degree to which babies and toddlers have access to the mother. Babies are carried to the work place while older children often are left behind. Tamang mothers seem to integrate subsistence work and infant care rather well, even during peak agricultural activity in the monsoon. For example, there is no evidence of seasonality in lactation times for 1- and 2-year-olds, who are nursed for 6 minutes at 85-minute intervals during the working day throughout the winter, spring, and monsoon (Panter-Brick 1991). Lactating mothers respond to both infant demands and nursing opportunity. They structure feeds in such a way as to nurse when other women are simply resting, eating, or standing, especially when the labor group takes meal breaks in the fields.

In contrast, mothers cease to nurse 3-year-olds during the monsoon, leaving them behind to accelerate the process of weaning, or to facilitate journeys to the fields, which are made more difficult by the rains. These children stay by themselves from dawn to dusk until adults return from the fields. They eat leftover food, which is easily contaminated by bacteria under conditions of high temperatures and humidity. In terms of nutritional status, 3- to 6-year-olds are the most vulnerable age group (Koppert 1988). Thus, women's work may not adversely affect the health of babies, since work schedules allow for nursing at the work place. However, older children are deprived of adequate care. In the hill regions of Nepal, overall death rates of infants less than 1 year old are comparable between women who work at home and those who work on farms. Death rates of children 1- to 5-years old, however, are considerably higher for the farming women (Gubhaju 1985).

Nevertheless, the decision of Tamang women to concentrate on infant care at the risk of neglecting older children is in a sense an appropriate choice. Survival is the key issue for Tamang infants (two-thirds of childhood deaths occur before age 1), whereas for older children the concern becomes one of nutritional well-being (Panter-Brick 1992b). This is quite obviously a behavioral "survival strategy" that makes the most (but not the best) of a bad situation. Improvements on this strategy would include better supervision of the children left in the village, keeping them away from dirt, and feeding them more frequently. The nature of subsistence responsibilities, which includes walking up and down slopes and carrying loads, requires that older children, too heavy to carry, are left behind. Infant mortality rates, already high, would certainly increase if women worked long hours and left their babies at home. For instance, in another Nepali community at higher altitude, women deem the mountain environment too difficult and dangerous to take children to the fields. They leave them with other caretakers. Mortality is extremely high, 226/1,000 for less than 1-year-olds and 317/1,000 for 1- to 5-year-olds (Levine 1987). In Salme, because of the shortage of available caretakers, Tamang mothers take chief responsibility for their children and give priority to the youngest in their care (Panter-Brick 1992b).

Tamang women space births more than 3 years apart. This facilitates the integration of the roles of food producer and child care. Since mothers carry young children with them, breast-feeding of 1- and 2-year-olds at the workplace helps to delay another pregnancy (Vitzthum chap. 18). Long interbirth intervals minimize both sibling competition for parental attention and depletion of maternal energy reserves.

More-detailed calculations of the cost and benefits associated with subsistence and child-bearing activities are required to demonstrate whether the observed birth interval is a result of selective adaptation (see Blurton Jones et al. 1989). Thus far, results only give a tentative indication of the magnitude of selective pressures involved.

Mortality shows a direct relationship with birth interval in Salme (see also Gubhaju 1986). Death rates in the first 5 years of life increase dramatically by 160% with shorter than average birth spacing (296/1,000 for intervals less than 36 months, relative to 185/1,000 for longer intervals; Koppert 1988). After exclusion of cases in which an older sibling died (an event likely to shorten birth intervals), it is apparent that the increase in mortality is due to infant deaths (before age 1 year) which increase by 238% when birth intervals are less than 36 months. Child mortality (between 1- and five-years old) is relatively unaffected, possibly because children have already survived infancy, the most critical period of life. Even when birth intervals are short (<36 months), surviving older siblings average 27 months of age, old enough to be fully weaned. When birth intervals are long (>36 months), they average 52 months, old enough to withstand a loss of maternal care, repeated infections, and other vicissitudes of life. To summarize, in addition to the ecological, social, and demographic constraints identified above, a woman's child-care strategy is related to the priority given to her work. Her decision to concentrate attention on one particular child is guided by the latter's age and chances for survival at different stages of the life cycle.

Reproductive Performance

How does sustained work effort affect a woman's reproductive ability? This issue long has been the subject of debate and a strong impetus for new field studies. Now there is increasing evidence that reproductive function is sensitive not just to maternal energy reserves ("critical levels of fat for weight" reviewed in Frisch 1990a; see Pond chap. 11; McFarland chap. 12; Zihlman chap. 13) but to changes in maternal energy balance (i.e., degree of weight loss and exercise; Ellison et al. 1993b). A flexible reproductive strategy in response to energetic constraints would decrease mortality risks for newborns and mothers, and would be favored within the population by natural selection (Ellison 1990). It is interesting to explore this hypothesis for the Tamang and the Kami, since the former experience higher workloads, longer birth intervals, and, in addition, birth seasonality.

Socioeconomic group differences in fertility could be due to a combination of social practices, nursing schedules, and nutritional and energetic factors. As discussed above, Tamang women, in contrast to the Kami, sometimes live apart from their spouses, residing in cattle shelters on the mountainside. Tamang women breast-feed their infants for 3 years, whereas Kami women breast-feed for only 2 years. Subtle differences in nursing behavior could hasten a new pregnancy (Vitzthum chap. 18). For instance, Tamang women nurse at increasingly longer and regular intervals as children grow older and heavier and are left behind. Kami housewives nurse at irregular intervals irrespective of the child's age, as they are more likely to combine breast-feeding with extra supplementary food prepared at home (Panter-Brick 1991). During my study, lactating mothers were persuaded to give small amounts of blood after breast-feeding, which permitted the analysis of hormonal responses to suckling a baby. The Tamang show significantly sustained levels of prolactin hormone. This is consistent with their more intensive nursing behavior and longer birth intervals (Stallings et al. 1994).

Finally, work patterns show obvious differences between the socioeconomic castes. Tamang women lose up to 3% of their body weight (2.8 kilograms) during the monsoon, which they regain in the winter. This is a rather modest seasonal change, but individual women may experience considerable weight loss and gain. Moderate weight loss is known to be associated with changes in reproductive function (Ellison et al. 1993b). Ellison and I have demonstrated the consequences of seasonal weight loss for ovulation frequency and reproductive ability. Hormonal profiles analyzed for the Tamang show that they experience disturbances in their menstrual cycles as a result of weight loss in response to increased work effort in the monsoon (Panter-Brick et al. 1993).

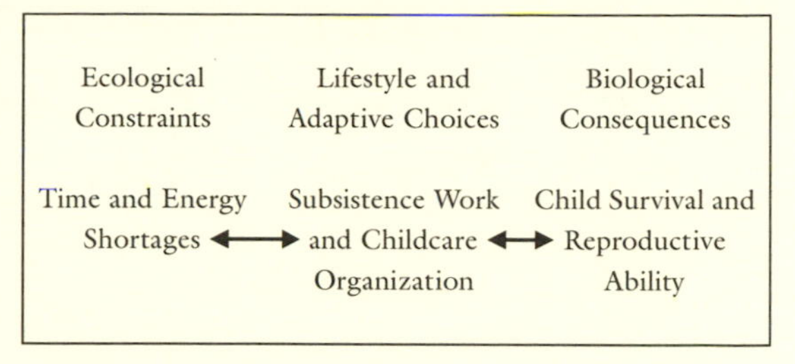

FIGURE 17.1. Links between ecological constraints, behaviors, and biological consequences.

CONCLUSION

Rural women in Nepal hold important and varied roles in their families and communities. I have described their individual and group life-history strategies in terms of ecological conditions and possible biological consequences for maternal and child health. The focus on time and energetic efficiency, subsistence, and child-care strategies, fertility, and child mortality (fig. 17.1) is particularly helpful for evaluating women's travel and subsistence activities and infant-and-child care behaviors. Women's behaviors are strategic choices, selected from a range of alternatives, to cope with, for instance, seasonally changing circumstances in the environment, or a conflict of priorities associated with childbearing, or a shift of priorities for rearing children at different ages postpartum. Combining an ecological approach with a life-history perspective serves to highlight the critical nature of such choices for an individual and for the community. Women's actual behaviors in production and reproduction are true survival strategies with significant consequences for subsistence, childbearing, and well-being.

18 Flexibility and Paradox: The Nature of Adaptation in Human Reproduction

Virginia J. Vitzthum

THE STUDY of the life history of human females involves many of the same questions as the study of any mammalian species. Yet, as culture-bearing organisms, humans are unique. Because human reproduction is shaped by a dynamic interplay of biology, culture, physical environment, and personal decisions, the challenge is to incorporate these dimensions into the investigation of human female life history. Clearly, there are many avenues by which to approach the study of human reproduction, and several fields—particularly, medicine and demography—have their differing perspectives (e.g., Bongaarts and Potter 1983). However, all acknowledge that ovarian function plays a key role in having offspring. Considering this role then, it is remarkable how little we know about variation in ovarian function and the factors responsible for it. In fact, many assume such variation is insignificant, although a growing body of evidence suggests otherwise (see Ellison 1990; Ellison et al. 1993a; Bentley 1994; Vitzthum et al. 1994).

I specifically am interested in three biocultural determinants of variation in ovarian function: 1) breast-feeding (or lactation), 2) nutrition, and 3) workload (or energetics). My aim is to describe how human biology and its interaction with culture can explain the variable responsiveness of the reproductive system to these behavioral and environmental influences.

The study of individuals through careful observations and measurements serves as my research focus to provide insights into the physiological mechanisms underlying variation in ovarian function. However, in developing these ideas, I find myself shifting between different levels of inquiry and interpretation. From the individual, I move up as well as down (also see Morbeck chap. 1). The individual woman communicates what is needed to understand the culture in which she operates; variation among these individuals generates the demographic characteristics of the population, which is the evolving unit (i.e., moving up in the analysis).

I then shift down again, for insights into the evolution of reproductive function derived from a knowledge of individual variation in turn enhance an understanding of individual function. Although I am brought to the individual—my starting point—again, I now have more than description (what), even more than mechanism (how); I have the beginnings of explanation (why). The impetus for my own fieldwork was a concern with learning how everyday life affects reproductive physiology. If we are to understand the evolution of the reproductive system, we must gather more information, more carefully and more thoroughly than ever before, and, most importantly, in the contexts in which women spend their lives.

There are striking similarities in the rhythms and patterns of life for women in widely divergent cultures and locales (see Morelli chap. 15; Panter-Brick chap. 17; Draper chap. 16). For most human females, childhood is short and nearly all of life is a constant balancing act of production and reproduction. Throughout infancy, childhood, and adolescence, a girl's mind develops in accordance with the cultural environment while her body does so in response to the physical surroundings (Morelli chap. 15). Both processes are complexly interdependent and go largely unnoticed. And, both are critical to her ability to function effectively as an adult. As a girl child, she will learn when and how and what to feed her siblings, and hence her own infants; such practices will directly affect ovarian functioning. Intricate hormonal processes will respond to food intake and workload levels, adjusting her growth rate, her age at menarche, and, eventually, her lifetime reproduction. The flexibility and lengthy developmental period characterizing the human species presents us with an extraordinary challenge. We must go beyond documenting variation in adult women to a full consideration of their stories as individuals developing and responding in a variety of cultural and ecological settings.

The People of Nuñoa

At an altitude of more than 13,000 feet, Nuñoa is among the more remote communities of the southern Peruvian Andes. My first view of the locale where I lived and conducted fieldwork during the summer months of 1985 was from the back of a fully loaded slat-sided truck, lumbering over the single dirt road that links the 4,500 townspeople to the nearest rail station. In subsequent trips, I would hunker down for warmth like everyone else, but this time I stood tall to take in the harsh beauty of the altiplano, a terrain too high, cold, and dry for trees or much else.

The starkness of this environment marks its inhabitants and is woven into the fabric of Andean culture. Lamenting tunes—"El Condor Pasa" sung by Simon and Garfunkel probably being the best known to North Americans—reflect the difficult realities of high mortality and arduous labor. But, beneath the rigors, there is a richness to life. Old traditions that link the people to the land and to each other serve to strengthen community and familial ties. Like flowers on a brown landscape, animals are decorated with bright woolen tassels and women's shawls defy color theorists. Festivals are commonly multiday feasts with special costumes and gay dancing.

Like women worldwide, Nuñoa mothers engage in a daily juggling of numerous demands to meet the needs of themselves, their families, and their children. Both immediate and long-term goals must be met; strategies must be developed. The choices made by women can have a profound effect on reproductive patterns and the well-being of mothers and their children. To understand these consequences accurately, I knew that I would have to observe and record their daily behaviors and experiences.

The majority of the population (and the focus of much of my own work) consists of its poorer members, Quechua Indians distinguishable by several cultural attributes. Their first—and for many, only—language is Quechua; especially among women, Spanish is rarely spoken or understood. Women typically wear non-Western dress. This includes as many as seven layers of short full skirts; a couple of blouses and perhaps a cardigan; and the ubiquitous *manta*, a brilliantly colored, finely handwoven, heavy shawl folded and tied across the shoulders to carry goods or a sleeping infant. Their dark hair always is arranged into a pair of long thick braids, often ending in bright woolen tassels and topped by a bowler hat.

Most homes consist of three or four low, small independent rooms with dirt floors, each opening into a center courtyard. They are constructed of the large stones that litter the landscape and make farming so difficult; roofs are usually of straw. The kitchen roof commonly has a small opening that generally is insufficient to allow the escape of the dark smoke given off by burning llama chips in the stone hearth.

Most Quechua in Nuñoa rely upon wage labor and agropastoralism for subsistence, but employment is scarce, and low-quality soil, intense solar radiation, and reduced atmospheric pressure make for poor agricultural output (Thomas and Winterhalder 1976). The main crops are potatoes and native cereals (*quinoa* and *cañihua*). Some may herd a few llama, alpaca, or sheep. Many try to augment their incomes by selling small items such as gum or safety pins at the local Sunday market. The more successful are able to offer llama wool, handicrafts, tape cassettes, and other items for sale.

Next on the socioeconomic scale are the *cholos*, a transitional class retaining some traditional[1] lifeways but marked by an adoption of Western dress and access to relatively steady cash income as, for example, storekeepers or truck owners. Many successfully sell a wide variety of goods in the Sunday market. They usually live in well-constructed adobe homes, some with tin roofs and kerosene stoves for cooking and heating. Spanish is commonly spoken by *cholos*, even if it is not their first language.

The Spanish-fluent *mestizos* are the upper class of the community. They live in attractively plastered adobe homes with tin roofs and wooden floors. Many have second homes in more modern towns. Some have relatively large herds of cattle or sheep and are usually able to hire others to tend their herds and work their fields.

Membership in any of the three described socioeconomic classes is not fixed but situationally dependent, and the economic resources of the groups overlap to some extent (Mishkin 1946; Escobar 1976). Although a very few are relatively well-off, life still is not very comfortable in Nuñoa. The town lacks modern sanitation facilities. Water commonly is hauled from a shallow river or community spigots since only a few homes have their own courtyard faucet. Electricity is available only sporadically and almost exclusively to public buildings. Kerosene is the only alternative fuel to llama chips for heating and cooking.

Since the mid-sixties, socioeconomic differentiation within Nuñoa has been magnified by a substantial increase in the landless and unemployed sectors of the population (Leatherman et al. 1986; Luerssen and Markowitz 1986; Carey 1988). These changes are partly a consequence of the 1968 agrarian and 1972 land reforms, neither very successful in more equitably redistributing land resources (Ferroni 1980; Aramburu and Ponce 1983; Painter 1983; Orlove 1987). The concomitant growth of the weekly market, boasting several varieties of fresh fruits and vegetables, and the increased importation of commercial foods in local stores adds further to the widening socioeconomic disparity. Although there is markedly greater diversity in the goods and foods available (Escobar 1976; Luerssen and Markowitz 1986), most are relatively expensive and usually require cash payment. Although herds are common, nearly all slaughtered animals are sold outside the town, where a better price is obtained, and thus contribute little to the diets of most Nuñoans. As a result of these economic conditions, most Nuñoans do not have sufficient means to meet their basic dietary needs (Leonard 1987). Many have neither access to adequate land for farming and herding nor opportunities for steady wage employment. Simply put, food scarcity is common.

In this social and economic context, women are reproducing, nursing, and raising their children. Although early contributors to the family's labor pool, children benefit from the considerable affection of both parents and other family members. I was interested in investigating the relationship between suckling patterns and female fecundity. Because breast-feeding in Nuñoa, as in other nonindustrialized populations, is an integral part of daily life, I was able to observe nursing behavior directly and avoid the methodological limitations of laboratory settings or the unreliability of self-reports (Quandt 1987; Vitzthum 1992b, 1994b).

How Does Breast-Feeding Affect Reproduction?

Not long ago, Western medical science generally considered a claim for a relationship between breast-feeding and the prevention of a new pregnancy to be an unfounded "old wives' tale." During the last three decades, there has been a dramatic reversal in opinion. Today it is well established that breast-feeding (or lacta-

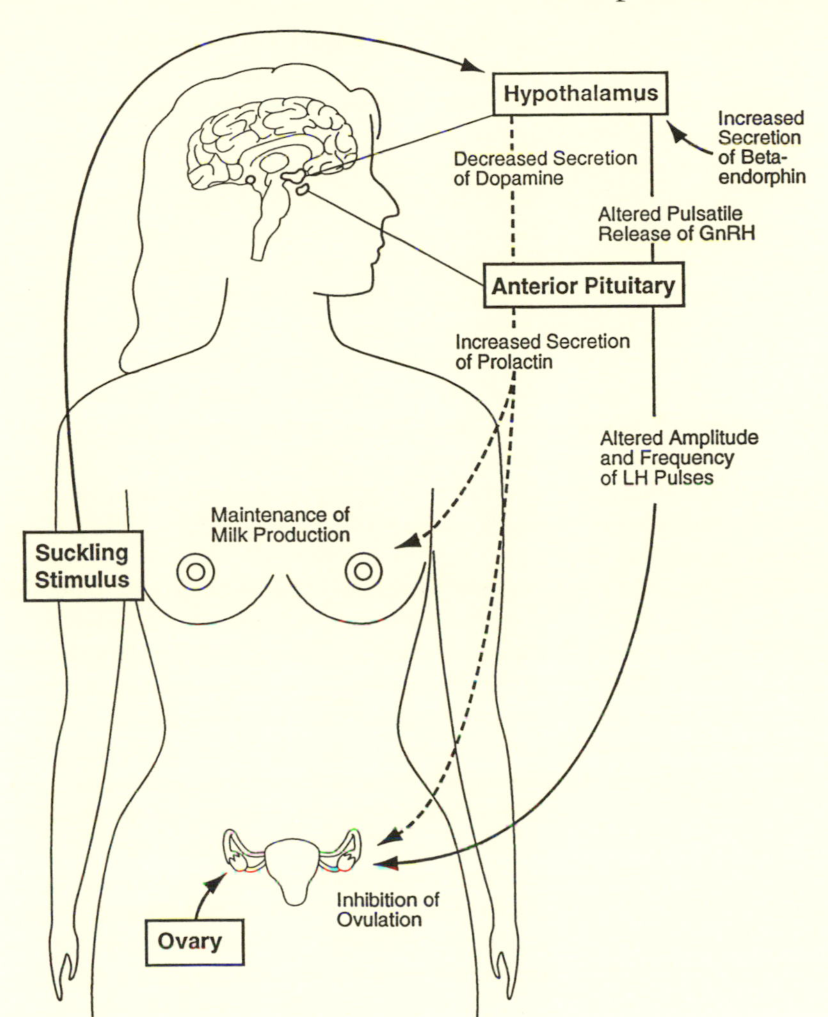

FIGURE 18.1. Physiology of ovarian function.

tion) affects ovarian function. In fact, some estimates of the relative contribution of various determinants of ovarian function conclude that breast-feeding is the most important factor influencing fertility levels in noncontracepting populations.

The debate surrounding the neuroendocrinological mechanisms underlying the relationship between breast-feeding and ovarian function remains unresolved (see fig. 18.1). Laboratory studies have demonstrated that the onset of nursing activity is associated with a substantial and rapid rise in serum prolactin levels (Tyson 1977; Howie et al. 1982; McNeilly et al. 1982, 1985; Robyn et al. 1985). Prolactin peaks 15 times higher than prenursing basal levels can occur within 30 minutes of the beginning of suckling. Once nursing ceases, a return to basal levels is reached in about 3 hours, barring the resumption of nursing (Noel et al. 1974). Evidence indicates that the action of suckling sends a signal to the hypothalamus (a small region at the base of the brain) to decrease secretion of dopamine, a reduction that triggers the anterior pituitary (also called hypophysis, a small gland near the hypothalamus) to increase its production of prolactin. It appears well established that this increase in prolactin induces the production of milk in breast tissue. However, any suppressive effect of prolactin on human ovarian function remains unproven (McNeilly 1993; McNeilly et al. 1994). Although there are at least two hypothetical pathways by which prolactin may act (directly on the ovaries [the gonads] and/or inhibition of pituitary release of lutenizing hormone [LH], a gonadotrophin responsible for normal menstrual function), there is strong evidence that prolactin is not principally responsible for lactational amenorrhea (Schallenberger et al. 1981; Short 1984; Glasier et al. 1986; Polson et al. 1986; Franceschini et al. 1989; Glasier and McNeilly 1990; Tay et al. 1992). Rather, suckling appears to disrupt the normal hypothalamic pulsatile release of gonadotrophin-releasing hormone and hence inhibits the pituitary release of LH. The suckling stimulus may mediate hypothalmic activity by stimulating the release of beta-endorphin, but the role of opiates is questionable (McNeilly 1993). Collectively, these hormonal circuits often are referred to as the hypothalmo-hypophyseal-gonadal axis. They are likely to be the important pathways by which the environment can influence reproductive functions.

Whatever the specific mechanism, a considerable body of evidence verifies that a reduction in suckling is primarily responsible for the return of ovulation some time after giving birth (Bonte and van Balen 1969; Perez et al. 1971; Berman et al. 1972; Jelliffe 1976; van Ginnekan 1977; Delgado et al. 1978; Konner and Worthman 1980; Gray 1981; Andersen and Schioler 1982; Howie and McNeilly 1982; DaVanzo et al. 1983; Gross and Eastman 1983; Knauer 1984; Short 1984; Dobbing 1985; Habicht et al. 1985; Lunn 1985, 1992; Wood et al. 1985; Campbell 1987; Eslami 1988; Diaz 1989; Vitzthum 1989, 1994a,c; Gray et al. 1990; Tay et al. 1992; Gray 1993; McNeilly 1993; Worthman et al. 1993). As the infant grows, the decreasing reliance on the breast for nutritional needs and the concomitant physiological changes in the mother lead eventually to the resumption of ovulation. What remains uncertain is which characteristics of breast-feeding structure are primarily responsible for postponing ovulation and preventing another pregnancy. Based on clinical evidence, it appears that suckling duration is important when frequency is low (i.e., the intervals are long) but less critical when suckling occurs 10 to 20 times per day (McNeilly et al. 1985). Yet, it is still unclear how suckling effort (strength), suckling duration, and the length of the interval between feeds (equivalent to suckling frequency) interact to inhibit ovulation. Must women breast-feed every hour in order to suppress ovarian function, as has been suggested? Or are longer intervals permissible if coupled with a greater suckling duration? What are the consequences for ovarian function of variation in breast-feeding structure?

Aside from the theoretical issues, there are important practical reasons for teasing apart the relationship between suckling patterns and lack of ovulation. It has been suggested that women be encouraged to breast-feed as a means of contraception. To succeed, reproductive health centers must make appropriate recommendations to their clients. If, for example, ovarian suppression really does require very frequent, if very short feeds, efforts to encourage breast-feeding under labor conditions in which the mother is separated from her infant for long periods of time would have little effectiveness in reducing fertility, even if the child is exclusively fed on the breast. In fact, such a situation could prove deleterious to infant health.

Methods

I observed mothers and their infants during the postharvest season (late June through early September), when nearly all of a woman's daily activities take place in the confines of the courtyard and rooms of her home. Since they do not need to work in the fields at this time, women engage in a wide variety of other tasks related to food preparation (cooking, processing *chuño*—a freeze-dried potato—for storage), textile production (wool cleaning, spinning, and weaving), minor wage labor (laundry), and household maintenance (cleaning, sewing, shopping). These daily activities are sporadically punctuated by the arrival of visitors, who may stay to chat and graciously share coca leaves or tea, and the demands of children, especially the nursing infant.

The 10 observed women ranged in age from 20 to 44 years, their infants from 3 to 21 months. During an observation period (lasting from about 4 to 6 hours), the activities of every person present in the home were recorded to the nearest minute. In particular, I maintained a constant awareness of a mother and her infant, and breast-feeding activity was recorded to the nearest second. In other words, every movement of the infant's mouth on and off the nipple was documented.

This precise timing of suckling revealed several previously unrecognized characteristics of breast-feeding behavior and highlighted some critical methodological limitations in earlier studies (Vitzthum 1989, 1992b, 1994b,c). For example, there is a lack of consistent definitions of nursing activities, making it impossible to compare the results of different studies. In the Nuñoa study, nursing behavior was analyzed as suckling periods separated by nonsuckling intervals. Suckling periods were defined at three levels: (1) nursing event: the period of suckling when the nipple is in the infant's mouth, that is, the exact duration in seconds of suckling activity; (2) nursing episode: one or more events separated by less than 5 seconds (Wood et al. 1985); and (3) nursing session: one or more episodes separated by less than 1 minute. A simple analogy for event, episode, and session is bite, course, and meal.

In addition to timed observations, these 10 women and an additional 20, each currently with an infant less than 3 years old, were interviewed on their reproductive histories and infant-feeding practices. The information gathered in these interviews must be evaluated judiciously since some types of data are not accurate, yet they are widely solicited in a number of studies. Of particular importance is that a comparison of interview and observational data for the same women revealed little concordance between timed data and a mother's memory of her daily breast-feeding behavior (Vitzthum 1992b, 1994b). For example, mothers typically report suckling duration in multiples of 5 minutes, a practice that obscures considerable variation. Clearly, a correct representation of daily suckling patterns requires more-detailed and accurate data than that obtainable in surveys. Careful observations in a natural context are a critical component in evaluating the effect of daily breast-feeding structure on reproductive function (Vitzthum 1994c).

Results

Clustering

Nursing episodes are clustered into nursing sessions, each session separated by a lengthy intersession interval. This demonstration of clustering is critical to employing the appropriate quantitative methods in analyzing variation in breast-feeding structure. Part of the current ambiguity regarding which aspects of the behavioral repertoire of nursing are responsible for a lack of ovulation is likely an artifact of the statistical techniques employed, some of which assume a nonclustered distribution (Vitzthum 1994c).

Diurnal Variation

Total breast-feeding time and intensity varies with the time of day and generally is greater in the morning than in the afternoon. This pattern is consistent with a hiatus in feeding during the night. Because of the very cold climate, infants in this population are swaddled and unable to nurse at will. Hence, night nursing is of low frequency, perhaps only two or three times a night. The observed diurnal variation in nursing also may reflect simple availability of breast milk. In a study of a different population (Stini et al. 1980), milk production was shown to be greatest in the morning. Because nursing does vary diurnally, analyses of the morning and afternoon suckling patterns are kept separate in my analysis.

The Role of Suckling Duration

The Nuñoa study provides the first evidence from a nonclinical field setting that the duration of individual feeding sessions can play a determining role in maintaining postpartum anovulation. With increasing time since birth, the probability of the resumption of ovulation increases and the mean duration of morning suckling sessions decreases. Very simply, and not surprisingly, as an infant ages, she/he nurses less. The expected average session duration for a newborn is about 9 minutes and is reduced to about 1 minute in a 2-year-old. In contrast, intersession interval durations do not vary significantly with the infant's age. The expected morning intersession duration, irrespective of the infant's age, is about 45 minutes.

In the afternoon, suckling duration is not related to the age of the infant. Rather, a significant positive relationship is found between mean session duration and the mother's age, independent of the infant's age. Expected mean session duration is about three times greater in a 40-year-old mother than in a 20-year-old. This finding may be related to a reduction in the replenishment capabilities of older women or to the presence of surrogate mothering.

There is some evidence from other studies that older women are not able to produce as much milk as younger mothers (Prentice et al. 1986). If true, an infant of any age would have to suckle longer to acquire the same—or perhaps even less—milk. However, if the association seen here is a reflection of differences in milk-production capability, one expects a similar relationship between session duration and mother's age in the morning as well as in the afternoon. But, my analyses did not reveal such a relationship. This may mean that differences in women's milk-production capabilities are not as critical in the morning after a long night's rest from feeding and a concomitant build-up in milk stores. It may be that differences in replenishment capabilities become apparent later in the day, after several feeding sessions.

It also may be that there is longer session duration in older women because these women almost always have a daughter of sufficient age to act as a surrogate mother or babysitter. This daughter's labor frees the mother to attend to other chores. During the time these data were collected, this babysitter generally was unavailable in the morning hours as she was attending school. However, her presence was of particular importance in the late afternoon when the mother was preparing the evening meal. In such a situation, a greater number of episodes might be clustered into each session that does occur whenever the mother breaks away from her household activities. In contrast, a younger mother often keeps the child with her while preparing a meal, wrapped in a shawl and slung on her back, available for shorter but slightly more frequent sessions.

Interval Duration

In contrast to the role of session duration, the duration of intervals between sessions does not appear to be contributing to variation in postpartum fecundity in this population. This is a very surprising finding as most previous surveys and laboratory studies had concluded that interval rather than suckling duration is the variable of consequence. This investigation demonstrated that there are at least three methodological limitations that may explain this discrepancy: (1) self-reported surveys of nursing behavior are usually unreliable; (2) definitions of nursing variables vary substantially among studies; and (3) no previous analyses had recognized that nursing episodes are clustered as parts of nursing sessions. Aside from methodological issues, I would argue that an either/or approach to analyzing suckling patterns is inappropriate. Rather, suckling and interval durations probably interact to suppress ovarian function, but it is likely that this dynamic varies among individuals and populations because of cultural and behavioral differences (Vitzthum 1989, 1994c).

The Effect of Nutrition on Reproduction

Nuñoa proved to be a setting in which I could initially explore the impact of nutrition on ovarian function because of the growing disparity in nutritional status among its inhabitants in the last 20 years. Relatively isolated and traditional communities like Nuñoa are often misperceived as socioeconomically homogeneous. In fact, within the Quechua population of Nuñoa, there is marked variation in socioeconomic status that is associated with substantial differences in maternal fecundity, caloric intake, and infant feeding practices.

The role of nutrition has been part of a larger debate surrounding the analysis of human fertility, first ignited by the "critical-fat hypothesis" developed by Rose Frisch (1976, 1978b, 1988; Frisch and Revelle 1971; Frisch and McArthur 1974). Frisch argued that a female would cease to ovulate below a particular body-fat threshold, a proposal that produced criticism of her methodology, analyses, and interpretations (Johnston et al. 1975; Billewicz et al. 1976; Trussel 1978, 1980; Reeves 1979; Bongaarts 1980; Menken et al. 1981; Ellison 1982). However, whatever the shortcomings of Frisch's work, it stimulated considerable discussion and research that continues today. Clearly, it would be extraordinary if a mammalian reproductive system did not respond to nutritional factors (see Pond chap. 11; Galloway chap. 10; Bronson 1989). There are two pathways—neither of which has been studied thoroughly—by which nutrition could affect ovarian function: (1) indirectly, via infant-feeding practices that encourage a reduction in suckling and, as described above, would lead to the resumption of ovulation, and (2) directly, through maternal nutritional status. Through both routes, socioeconomic variation and cultural attitudes and perceptions can exert an influence on whether a woman has a child and whether she can successfully support the growth of that child.

As part of my exploration of variation in maternal fecundity, caloric intake, and infant-feeding practices, I partitioned my interviews with 30 women into relatively poorer and better-off samples of women (low socioeconomic status [SES] and mid-SES) on the basis of several criteria, including the education and occupation of the household head, dress, household construction and household goods, and evaluations by community members. In these subsamples, maternal age range and mean age were nearly identical (low SES: 19–44 years, 29.8 years; mid-SES: 18–40 years, 29.3 years). Mid-SES infants were slightly, but nonsignificantly younger than low SES infants (12.4 vs. 16.5 months). Sixty-seven percent of the infants were female; quantitative analyses indicate that infants of both sexes are nursed and fed in similar fashion in Nuñoa (Vitzthum 1988, 1992a). All 30 interviewed women reported an absence of either traditional or modern contraceptive use.

Before considering socioeconomic differences in infant-feeding practices, it is useful to describe those that are shared among nearly all Nuñoa Quechua women (Vitzthum 1988, 1992a). Generally, infants sleep with their mothers and are nursed on demand from shortly after birth. Day nursing is prolonged, and night nursing may continue for as long as a year after the full cessation of day nursing. Bottle-feeding, if used, is initiated at about 5 months of age. Solid foods are introduced at about 1 year of age; infants are not given any special foods, and, as they mature, they are given "the foods of the house." Food sources other than breast milk are treated as complements, rather than as alternatives, to nursing. During the second year of life, an infant may receive substantial nutrition from breast-feeding, bottle-feeding, and solid foods. Mothers reported that adequate infant development is the most common reason for the cessation of breast-feeding. The onset of menses or another pregnancy is the second most important reason. Thus, as the infant matures, the relative contributions of the various food sources will gradually and continually change in response to a number of factors, including the infant's perceived needs and the resources of the household. However, nursing may be curtailed by some mothers before the infant is of an "adequate age" if another pregnancy occurs.

Socioeconomic Variation in Maternal Fecundity

Based on time since giving birth and current menstrual status, probit analyses (Finney 1971) allowed a comparison of the median durations of postpartum amenorrhea in the socioeconomic samples (Vitzthum 1989). The difference is fairly dramatic: about 9 months for better-off women versus 22 months for poorer women. At 12 months postpartum, this translates into a probability of menses nearly seven times greater among the better-off women. As noted above, this substantial difference in fecundity can be brought about either directly, as a consequence of maternal nutrition, or indirectly, via infant-feeding practices.

Socioeconomic Variation in Maternal Nutrition

To evaluate variation in nutrition, individual food intake was weighed at every meal on several days throughout 1985 for 175 persons from 30 Nuñoa households (Leonard 1987) that were partitioned into socioeconomic samples comparable to those in my study and comprised many of the same individuals. Analyses revealed substantial socioeconomic differences in caloric intake and dietary composition. Members of relatively better-off households consume about 200 calories/person/day more than poorer individuals. When seasonally partitioned, during the preharvest "lean season" this difference increases to about 400 calories/person/day (on average, about 1200 vs. 1600 calories/person/day). Better-off families consume more nonlocal food items, for example, rice, oatmeal, evaporated milk, and oil. In contrast, poorer families eat more potatoes, *chuño*, and traditional grains, and substantial amounts of cheap processed flour. An analysis of nutrient composition revealed that fat consumption is four times greater in better-off families. These nutritional differences are the result of differential access to store-bought food items because of the greater cash income of the relatively better-off families. Along with demonstrably improved health (Carey 1988) and child growth patterns (Leonard 1987), greater nutritional resources may be directly contributing to the higher fecundity of relatively better-off Nuñoa women.

Socioeconomic Variation in Infant-Feeding Practices

Infant-feeding style (Quandt 1985, 1986; Van Esterik and Elliott 1986) encompasses those culturally meaningful assumptions that guide infant-feeding decisions. Shared aspects of infant-feeding behavior persisting across socioeconomic classes may be thought to reflect a culturally specific style of infant feeding, whereas class differences can be conceived of as transformations of this underlying framework.

Despite a shared style, as described above, there are striking differences in practices between the socioeconomic samples. Most notably, the time from birth to completed weaning is 7 months shorter for daytime nursing and 14 months shorter for night nursing in the better-off sample relative to their poorer counterparts (16.6 vs. 23.6 months; 28.8 vs. 43.2 months).

An earlier termination of breast-feeding often is attributed to the adoption of bottle-feeding (Bonte et al. 1974; Wenlock 1977; Popkin et al. 1982), but, in fact, similar proportions (about 40%) of each sample claim to use some bottle-feeding. Further, the infant age at introduction of the bottle among these women is also very similar (about 5 months) and long before the end of breast-feeding in either sample. Clearly, bottle use does not mean an end to nursing. It is used as a complement rather than a substitute in both samples. Earlier solid food supplementation is also thought to be responsible for earlier breast-feeding termination, but again the age of solid food introduction is very similar in the two samples (about 13 months). As with bottles, the feeding of solids does not signal an end to nursing. The question remains, Why is the time to weaning shorter for infants in the better-off sample?

I think that, in addition to the mechanics and timing of food supplementation, we need to consider its nutritional quality. The practice of giving "the foods of the house" translates into very different diets among Nuñoa infants. As discussed in the previous section, there is marked household variation in dietary resources. Poorer mothers reported that various teas and herbal infusions are the most common liquids bottle-fed to their infants. In contrast, better-off mothers frequently fed canned milk and fruit juice, available because of their access to nonlocal store items. Bread, potatoes, and chuño are common first foods for all infants, but better-off women include eggs, meat broths, and chicken in their infants' diets. Poorer women never reported the inclusion of meat or dairy foods (these foods are rarely a part of any poorer household's resources). Better-off families are also consuming greater quantities of food. This suggests that these infants may be receiving more as well as higher-quality foods. Feeding more-nutritious supplemental foods is a critical avenue through which economic factors influence infant-feeding routines.

The Effect of Workload on Reproduction

In the time since my return from Nuñoa, I have been struck by an apparent contradiction. Clinical studies of American women indicate a sensitivity of ovarian function to even moderate amounts of stress (Pirke et al. 1985, 1989; Schweiger et al. 1987; Lager and Ellison 1990). In particular, athletic women often do not menstruate each month (generally referred to as exercise-associated amenorrhea) and may have to stop training to become pregnant. Yet, Nuñoa women, in arduous conditions characterized by marginal nutrition and heavy workloads, average more than 6 and may have as many as 12 pregnancies (Vitzthum 1988). On the one hand, in American athletes, the ovarian system is responding negatively to workload; on the other, in Nuñoa women it appears relatively less sensitive.

To examine this paradox, a review of the literature shows that the nature of workload outside the laboratory is very understudied (Vitzthum and Smith 1989). Instead, virtually all information concerning the relationship between ovarian function and activity is found in the sports medicine literature. Furthermore, surprisingly little is actually known because of a variety of problems with research designs in different disciplines.[2] Equally problematic is the general lack of a research framework or theoretical paradigm within which to conduct investigations. The medical sciences persist in viewing all cases of amenorrhea as pathological. An evolutionary perspective, while not ignoring pathology, considers the possibility that a change in ovarian function may be an appropriate response to environmental conditions (Frisch 1978b; Wasser and Isenberg 1986; Leslie and Fry 1989; Ellison 1990, 1994; Vitzthum 1990, 1992c, 1995; Worthman 1990). Whatever one's perspective, much remains to be learned about the relationship between activity and ovarian function. In the course of a literature search, important questions that require investigation came to light. Although continuing research since our review has contributed much to an understanding of exercise-associated subfecundity (Rosetta 1993; Cumming et al. 1994), the issues outlined below remain far from settled.

First, what are the activity-associated variables that trigger physiological responses? There are a number of characteristics associated with exercise or workload that may lead to changes in ovarian function, most of which can be grouped into six categories: (1) body composition, (2) weight and/or fat loss, (3) dietary intake, (4) activity characteristics, (5) reproductive history, and (6) psychological stress. For each variable, there is at least some evidence that it plays a role in exercise-associated amenorrhea. For instance, data from Ellison's lab at Harvard provide support for the role of weight loss (Lager and Ellison 1990). The argument is that the organism seeks to maintain energy balance. In the face of even moderate weight loss, there is a reduction in reproductive activity, which requires energy expenditure, until a balance is reachieved. The investigations of Deuster and her colleagues (1986) are among the few studies to document adequately a role for diet in excercise-associated amenorrhea. Relative to athletes with normal menstrual cycles, they found that the diet of amenorrheics was 33% lower in fat, markedly higher in vitamin A, and lower in zinc. They entertain possible roles for zinc and vitamin A in ovarian function, but as yet the actual mechanisms, if any, are unknown. The role of dietary fat in ovarian function has generated considerable speculation, but specific mechanisms are yet to be demonstrated. There is also compelling evidence that mental stress may be as important as physical stress in producing hormonal changes. More than 80% of West Point female entrants experience amenorrhea during their freshman year. In this sample, such a high percentage of menstrual dysfunction did not appear to be explained solely by physical stress, and percent body fat, body weight, and weight loss were not contributory factors.

Second, what are the physiological pathways that link activity and ovarian function? There is much work to be done in this area, although it is clear that activity profoundly affects the body's hormonal environment (Cumming et al. 1994). It is intriguing that beta-endorphin has been implicated in ovarian suppression by suckling during nursing (see fig. 18.1) and may also play a role in exercise-associated subfecundity. Laboratory studies clearly have demonstrated a rise in plasma beta-endorphins with exercise. What remains unexplained, however, are the dramatic differences in response by individuals during the same exercise regime. Additionally, it is unclear how short-term hormonal changes observed during exercise could lead to the chronic hormonal changes associated with menstrual dysfunction.

From what we know thus far, ovarian function is not an all-or-nothing proposition (Prior 1985; Ellison 1989, 1990, 1994; Vitzthum and Smith 1989; Vitzthum 1990, 1995). There is no one level of activity that will shift an individual from a fecund state to one of full ovarian dysfunction. Rather, there appears to be a gradation in responses, any of which may reduce the probability of reproduction. In addition to investigating the demonstrable variation in response among American women (Shangold and Levine 1982; Boyden et al. 1983; Howlett et al. 1984), we need to consider variation among women in different populations. Is the effort of walking great distances with heavy loads or hoeing fields the same as a daily sprint? Are the consequences of each activity for reproductive function comparable? We really don't know. The next step in our investigations must involve comparative field studies of women's activity levels in different populations (see Bentley 1985; Panter-Brick chap. 17).

Biology, Culture, and Female Reproduction

The Evolution of a Flexible Response System

The framework I have developed (Vitzthum 1990, 1992c, 1995) brings together the sometimes paradoxical information regarding human ovarian function. In agreement with others (Prior 1985; Cameron 1989a,b; Ellison 1990; Parfitt et al. 1991), I propose that ovarian function is gradual (rather than all-or-nothing) and that the postponement of reproduction under poor conditions may be evolutionarily advantageous (Ellison 1990; Peacock 1990). In addition, I propose that ovarian function in adult women depends on conditions experienced during adolescent development such that some women are better able to tolerate environmental stressors because of prior exposure to these conditions. This flexibility has been shaped by natural selection.

In developing an adaptive explanation of variable responsiveness among women to seemingly identical conditions, several points about natural selection and how it shapes variation over time are important. First, natural selection operates at the level of the individual relative to the other individuals in the population and relative to the specific environmental conditions of that population. Second, the outcome of individual reproduction is over a lifetime. The environment changes constantly; thus, every moment between birth and death can provide an opportunity for one individual to outscore the other. It is the cumulative relative fitness at the end of life that determines the most adapted genotype.[3] Third, natural selection operates on any morphological, physiological, or behavioral trait as long as the populational variation in the trait is correlated with a variation in genotypes. Although particularly complex, natural selection operates even if the trait is not determined solely by genes and even for traits capable of changing during the course of one's lifetime.

Virtually all physiological processes are continually responding to environmental conditions. Observations and innumerable studies have documented often considerable variation in various physiological processes among individuals and populations. In a few cases, the evidence is good that physiological functioning in the adult reflects environmental conditions experienced during preadult stages. For example, relative to individuals raised at sea level, respiratory physiology among individuals raised at high altitudes is quantitatively superior, thus allowing them to tolerate low-oxygen conditions more effectively (Baker and Little 1976). In other words, much of the variation we see in physiological processes may be due to variation in environmental conditions. Finally, because physiological processes are controlled by proteins produced by genes, it is also likely that at least some of the physiological variation reflects genetic variation and, hence, is subject to natural selection. In other words, individuals whose physiological responses to environmental conditions allow them to have the most offspring, relative to other individuals, are the best adapted in their population.

Four generalizations regarding human reproductive physiology emerge from this background discussion and provide a foundation for understanding adult ovarian function: (1) variation in ovarian function is known to exist among women and populations; (2) human physiology, in general, is flexible and responsive to environmental conditions; (3) human adult physiology reflects a particularly lengthy developmental period; and (4) human reproductive systems are subject to natural selection.

Let us now consider a hypothetical population comprising two types of women: those that ovulate no matter what sorts of environmental conditions exist, and those whose ovaries will adjust their functioning according to the quality of the environment. Which physiological response (i.e., which phenotype) is the best adapted? In other words, which response will lead to the greatest number of total offspring in a lifetime?

The answer depends on two factors. First, in a particular set of environmental conditions, what is the probability of successful reproduction (i.e., conceiving and giving birth to a healthy infant) relative to the probability under the best conditions that ever exist in that environment for that population? If the current conditions are relatively poor, the phenotype that continues to function (ovulate) as if things are fine will be wasting effort. The phenotype that waits out the poor conditions (i.e., temporarily stops ovulating) will be saving energy that can be used when things get better and the probability of successful reproduction is higher. Over a lifetime, the latter phenotype is expected to have more children, on average, than the former. The outcome explains, in part, why it is adaptive to stop ovulating under poor conditions.

But what if conditions stay poor? Over a lifetime, a phenotype that never ovulates is definitely at a disadvantage relative to one that tries to reproduce, even if the chances for success are not very high. This brings us to the second factor: What is the probability that the current conditions will get better or worse before the next opportunity to conceive? If conditions are going to get better, the adaptive choice is to wait. If conditions remain poor for a long time, it is best to try to reproduce before all chances are lost. The process of adjusting to new conditions and the resumption of normal functioning is known to occur for many physiological processes and is called acclimatization.

Clearly then, given these two factors, the best-adapted phenotype is the one that can stop reproduction in the face of less than good conditions but resumes ovulation if the poor conditions persist. In other words, a flexible system responsive to environmental conditions and capable of adjusting that response over time will be favored by natural selection.

Now we are brought to trying to explain how a woman's body can judge the quality of the environmental conditions. In fact, our bodies constantly receive and process information about our environment. In the case of reproduction, a critical environmental component is the availability of sufficient food (energy) to bring a pregnancy to successful completion. This is the basis of my interest in the consequences of nutrition, workload, and breast-feeding for reproduction. Our bodies monitor our energy intake (nutrition) and output (workload and lactation), and the reproductive system responds accordingly. Changes in these variables are physiologically interpreted (most likely, by the hypothalmo-hypophyseal-gonadal axis described in fig. 18.1) as messages about the environment and about the likelihood of successful reproduction.

At this point, we need to add one more piece to this theoretical framework. Recall that ovarian function in some women appears to be relatively insensitive to negative environmental conditions (e.g., Nuñoa women), whereas others are known to experience a cessation of normal ovarian function under seemingly less difficult conditions (e.g., American women in controlled clinical studies). This apparent paradox of variable response to identical conditions (to state the same phenomenon from another perspective) is explainable. Each woman has an individual life story—a long developmental period—that has shaped her adult physiological responses to environmental conditions.

Two critical points are important. First, the conditions experienced during the developmental period are not identical for all women across all populations. Second, the best environmental conditions found in one population and, hence, the "optimal" conditions in that case, may be conditions that are uncommon and relatively poor in a different locale. Thus, reproduction under conditions that appear arduous is to be expected if those are the conditions experienced during the individual's developmental period. Arduous they may be, but these conditions are also normal in that population. Natural selection dictates that populations must have some reproductive output or become extinct. In contrast, for those who grow up under good conditions (i.e., high-energy availability and low-energy output), a relatively mild reduction in those good conditions may cause the reproductive system to cease functioning temporarily. The system is used to the good life and has every reason to expect things to get better shortly. Should conditions continue to be less than the best known by that population, it is to be expected that eventually ovarian function will resume. There is some reproductive output, and acclimatization takes place.

Collectively, these ideas explain the flexibility of the reproductive system, its ability to respond to environmental conditions, and the apparent paradox of different sensitivities to difficult conditions. The explanation relies on two key premises. First, the physiological responses of a woman at a moment in time are a reflection of her entire previous life story. Second, the human reproductive system reflects our special evolutionary history. It is a system that has been shaped by natural selection, and its functioning reflects the process of adaptation.

The Role of Culture

The human reproductive system is subject to the same selective pressures as that of other mammals and primates. Many of the same life-history issues are pertinent. At the same time, humans depart from the mammalian system because people make personal decisions in a cultural context. How then might culture interact with an evolutionarily adapted flexible reproductive system?

Put simply, culture complicates matters considerably. Energy availability is among the most important environmental variables determining successful reproduction. Cultural beliefs, however, can create new and equally important factors that influence physiological processes and, thus, modify reproductive patterns. Many examples come to mind, including beliefs that copulation during menstruation has the highest probability of conception, that colostrum is harmful, that intercourse uses "life energy" and should be avoided, that breast-feeding and intercourse are incompatible. These and many other practices and beliefs, most with relatively obvious consequences for reproduction, have been documented by numerous anthropologists.

Davis and Blake (1956) identified a list of "proximate determinants," a relatively few number of biological or behavioral variables by which all other factors, including the vast array of cultural practices, affect reproduction. Reformulations since then (cf. Bongaarts and Potter 1983; Wood 1990) have proven valuable. However, the intricate mechanisms underlying the dynamic interactions between biology and culture remain obscure and unappreciated. The consequences of an abstinence taboo, for example, are relatively straightforward. It is much more difficult to measure and predict the effect of mother's perceptions about infant development on infant-feeding decisions and, hence, on ovarian function.

A False Dichotomy: Suckling Duration versus Interval Duration

The observational data from Nuñoa provided the first evidence from a natural setting that suckling duration plays a role in determining the duration of postpartum amenorrhea. This finding differs from the only two other precisely timed observational studies of nursing up to that time, both having concluded that interval duration was of singular consequence (Konner and Worthman 1980; Wood et al. 1985). Aside from methodological differences that may or may not have contributed to this discrepancy (see Vitzthum [1989, 1994c] for a fuller discussion), I would argue that, from what we know of the physiology that underlies lactational amenorrhea (of particular importance is the role of the gonadotrophin pulse generator; see fig. 18.1), suckling and interval durations should interact to suppress ovarian function, but this interaction can vary among populations.

Seemingly comparable magnitudes of daily nursing can result from widely divergent schedules in the duration and frequency of suckling sessions. Variation in suckling effort (strength and rate of suckling) adds to this complexity in breast-feeding structure (Vitzthum 1994c). Thus, it is likely that different populations can achieve the same effect on ovarian function by different combinations of means. Conversely, among different women, similar total daily time spent nursing may not have comparable effects because of differences in the underlying structure of breast-feeding. Furthermore, if there are factors—biological or cultural—that affect interval duration but play no role in regulating suckling duration, then it is to be expected that variation in interval duration will correlate with variation in ovarian function. Thus, it will be seen as the variable of consequence. Conversely, if there are factors that affect only suckling duration, this component of nursing behavior will be held responsible.

There are likely to be several aspects of biology and culture, which vary among populations, that impinge on both components of breast-feeding structure (Vitzthum 1989, 1994c). For example, different populations are typified by different activity patterns. During Wood's data collection, the Gainj women in his study spent their entire day engaged in the singular task of working their gardens. There was little, if anything, to structure the time between nursing sessions, hence, interval durations were free to vary. In contrast, Nuñoa women are constrained in their activities by regular school and work schedules that may contribute to structuring their days and, thus, interval durations. This may explain the relative lack of variation in interval duration and the clear importance of variation in suckling duration in contributing to variation in ovarian function in Nuñoa women. Clearly, the task cannot be limited to an investigation of variation in suckling and interval durations. We must also determine how that variation is influenced by the cultural and physical environment if we are to unravel the linkages between culture and biology in human reproductive functioning.

Style and Substance: The Role of Food Quality in Ovarian Function

Now let us broaden our scope beyond suckling patterns to the wider array of infant-feeding practices. Figure 18.2 depicts the dynamic interplay among the factors that influence infant-feeding behavior, breast-feeding duration, and female reproduction in Nuñoa. "Infant-feeding style" encompasses those culturally meaningful assumptions that inform maternal decisions regarding infant feeding and weaning. Information, including physiological status (infant development and maternal fecundity), is mediated by this system. The decision to complement nursing with other foods, begun in part as a response to the growing infant's needs, will reduce infant suckling, enhance maternal fecundity, and may prompt breast-feeding cessation. Bottle use probably augments this process because larger amounts can be fed more easily. Finally, food choice, dictated by the "foods of the house," can shorten breast-feeding duration through at least two pathways. First, given sufficient nutritious foods, the infant is genuinely satiated and will suckle less than an infant receiving teas. Reduced suckling also will enhance maternal fecundity and prompt nursing termination. Second, sufficient high-quality foods will promote infant development, a consequence known to influence breast-feeding duration in this population (Vitzthum 1988, 1992a). Even though supplementary foods and bottles are not perceived as alternatives to nursing, their use hastens the onset of culturally recognized cues to terminate breast-feeding.

The expanding market economy in Nuñoa can influence these relationships in several ways. First, by introducing previously unfamiliar "Western" styles such as bottle-feeding. However, in Nuñoa this method does not appear to have been markedly differentially incorporated by those of greater affluence. Second, improved maternal nutrition may be directly contributing to enhanced maternal fecundity. Finally, higher-quality infant foods will influence infant suckling. Thus, greater access to the resources of the market economy acts to decrease breast-feeding duration and influence ovarian function without changing the essential framework by which mothers are guided in the feeding of their children. In sum, this intricate interaction between biology and culture is neither unidirectional nor obvious, yet its influence on ovarian function would go unrecognized if we had failed to incorporate cultural context into this study.

Closing Thoughts: Variable Culture Meets Flexible Biology

The net consequence of cultural variation is to increase the environmental variation among women and their populations. Discovering all the determinants of human reproduction and the mechanisms by which they act is difficult, time-consuming, and ultimately exhilarating. This is no small task we have set ourselves, but it is rewarding. Paradoxes will unfold and prove themselves consistent if we remember that each woman has an individual life history embedded in a cultural and environmental context. Similarly, our species has an evolu-

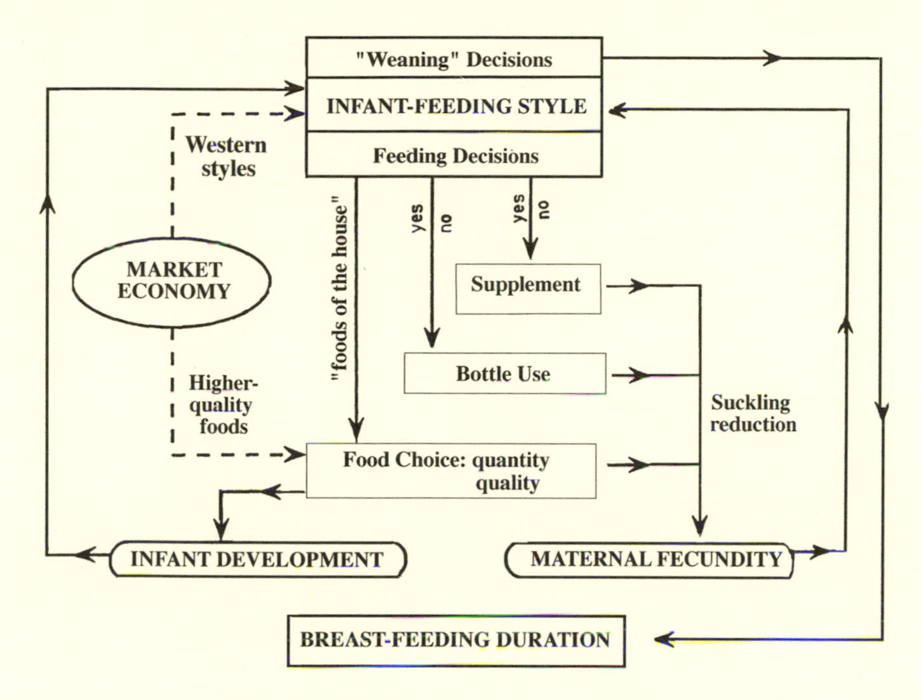

FIGURE 18.2. A model of the dynamic interplay among the factors that contribute to variation in ovarian function within Nuñoa. Infant feeding style encompasses those culturally meaningful assumptions that inform maternal decisions regarding infant feeding and weaning; information, including physiological status (infant development and maternal fecundity), is mediated by this system (from Vitzthum 1994a).

tionary history. The functioning of a woman's reproductive system reflects the environmental conditions she experienced as she grew to maturity. Taken together, the inevitable, and somewhat daunting, conclusion is that we can expect to find considerable variation in human reproduction. This variation is neither pathological nor paradoxical. Rather, it is an extraordinary adaptation to the varying world in which our species evolved.

The women in the far-away community of Nuñoa welcomed me into their homes, allowed me to record their daily struggles, successes, and the nuances of their behavior, and gave me their medicines when I was sick. In graciously sharing their lives with me, they taught me much about women everywhere and about myself. I have never met a good anthropologist who did not appreciate the privilege of these experiences. Much more work in the settings where women live out their lives needs to be done. Much will be gained.

Notes

1. As used in this paper, "*traditional* refers to practices not significantly altered by modernization or westernization . . . [and] does not imply stasis or homogeneity in practices" (Conton 1985).

2. These include (1) the use of markedly different methodologies; (2) a frequent absence of hormonal assays; (3) statistically prohibitive small sample sizes; (4) substantial intra- and interindividual variation; (5) a lack of control for individual subject characteristics such as age, age at menarche, and parity; (6) a lack of control for different activity characteristics such as training regimen, intensity, and type of sport; and (7) multiple definitions of menstrual dysfunction.

3. Lifetime reproductive success (LRS) is neither the only nor, perhaps, the best measure of fitness (see Sober

1984; Endler 1986; Burns 1992; Chisholm 1993). This common definition is used here as a starting point for exploring these hypotheses regarding reproductive function. However, alternative fitness definitions may prove more appropriate and will be considered as this work progresses. For example, if selection favors traits that minimize the intergenerational variance in offspring number (see Seger and Brockman 1987; Chisholm 1993), then delayed reproduction may have selective advantages that are not apparent when only LRS is considered. Interestingly, and somewhat counterintuitively, the best-adapted genotype is not always the most common in the population, nor will it necessarily ever be. Refer to Cavalli-Sforza and Bodmer (1971) for a complete explanation.

Part VI

Life History, Females, and Evolution

Life History, Females, and Evolution

JOLLY pioneered the importance of social intelligence, especially in primates. Using the broad perspective of life history, as in this volume, we see social intelligence as a survival life-history feature. Jolly also was one of the earliest field researchers to identify the role of female dominance in survival and reproductive outcome. Here, she builds on a continuing interest in sex and intelligence, especially in primates, including humans.

Social intelligence, especially in long-growing, long-lived, big-brained, group-living primates, also promotes shuffling of information. It can be viewed at different levels: (1) where it fits in the whole organism–whole life scheme (survival and reproductive "decision-making" of individuals in context of ***bauplan*** abilities and mediated by many factors in the environment); and (2) how it fits in the overlays from a phylogenetic-historical perspective (e.g., language exists in humans, who have a relatively recent, unique evolutionary history in geologic time, but social intelligence characterizes most primates and many mammals.

Information exchange provides stability of social groups through time. Females are important. At least among most primates, females pass along the social networks and traditions of the society as they raise their offspring to maturity (i.e., through long periods of learning during infancy and childhood) (Jolly chap. 19; Pavelka chap. 7; Zihlman chaps. 8, 13). Here, Jolly extends her interest to artificial intelligence. As she suggests, computer programmers are "females"; they control "resources," that is, the ideas/traditions that keep society functioning. It may be better to say that primate females, because they are mothers and provide the biological, social, and physical environment for their infants, "guide" the development of biological maturation and social behaviors that are genetically programmed by species-level life-history characters.

Using evolutionary thinking, Jolly explores the parallels between sex and intelligence. Just as intelligence probably evolved because of the advantages presented by the free flow of ideas and the recombination of concepts, the point/counterpoint of levels and theoretical approaches with and between the chapters of this volume aim to produce a more complete understanding of the complexities of life.

McLeod complements the perspective of evolutionary biology by viewing the issues presented in this volume in terms of her own training in the social sciences. She reflects on what it is like to be a human female with newfound interest in her evolutionary ancestors. Her comments are enlightening for, as biologists ourselves, we were fortunate to expand our horizons by learning more about the cultural overlay.

19 Social Intelligence and Sexual Reproduction: Evolutionary Strategies

Alison Jolly

> In short, sex has evolved so that your children are not condemned to be just like you—and intelligence means that you are not condemned to remain just like yourself.

MANY evolutionary biologists have drawn parallels between physical evolution and cultural history. Many have pointed out the fundamental differences. Some have even tried to quantify their interactions. Darwin himself, Julian Huxley, J. B. S. Haldane, Kenneth Boulding, E. O. Wilson, Richard Dawkins, and John Maynard Smith come to mind, as well as Teilhard de Chardin's vision of our planet radiating thought. Their common theme is consideration of human thought and culture as a biological innovation, the product of natural selection, which in turn is changing the evolution of our species and of the biosphere.

This essay makes a more modest comparison. It pursues an analogy between sexual reproduction and social intelligence. Sexual reproduction and social intelligence are systems for transferring information between individuals. Sex and intelligence also allow information from different sources to be combined and used by one individual. This gives rise to some formal similarities between the two processes. Both innovations have allowed complex symbioses and multidimensional cooperation that, in turn, transform the biosphere. The processes can be contrasted, respectively, to asexual reproduction and to innate behavior, where a circumscribed set of instructions is passed from a single blueprint. In short, sex has evolved so that your children are not condemned to be just like you—and intelligence means that you are not condemned to remain just like yourself.

I offer this article for a book on female life histories and evolution, because there seems to be a fundamental question as to why females should put up with males at all. What is the function of a class of organisms—that is, males—whose contribution to the next generation is to move information around? Some males, including humans, also accept the female roles of biologically and socially nourishing as well as defending their offspring, but this is true in only a minority of species.

In the argument that follows, old interests and new ones come together. I have always been fascinated by the evolution of intelligence (Jolly 1966). I argued that primate society has been the major stimulus for the evolution of intelligence, far more important than hunting, tools, or even mental foraging maps. Lemurs (and other mammals) have complex social behavior, but they do not have anything like a monkey's interest and understanding of objects. Monkeys, in turn, have even more complex societies, but still surprising gaps in understanding their environment (see Cheney and Seyfarth 1990). I credit myself (and the lemurs' example) with seeing the importance of socially evolved intelligence, but the idea crystallized in the course of trying to explain it or argue it through with my husband Richard Jolly. (This is yet another example of the social matrix of ideas.) Humphrey (1976) took this concept further, pointing out how animistic humans are, attributing intentionality to thunderstorms and ill-luck (as well as husbands), perhaps because our minds evolved to deal primarily with other minds. Now, primate society and intelligence are widely recognized as major influences on each others' evolution (Goodall 1986; Byrne and Whiten 1988; Cheney and Seyfarth 1990).

This chapter emerged from a course I recently taught on the biology of sex and sex roles, which builds bridges among biology, anthropology, and women's studies. This has forced me to go back and read basic biological literature and then to justify it to amazed majors in literature. The costs and benefits of sexual reproduction versus asexual reproduction are endlessly discussed among evolutionary biologists. I have quoted current, authoritative works, adding nothing new.

What is new (at least to me), is the step-by-step analogy between the two modes of information transfer. The literature on the advantages of sexual reproduction concentrates on the costs and benefits to the individual. Looked at from that angle, exchanging ideas and modes of behavior with other individuals seems as risky and as rewarding as adopting sex in much the same ways.

In the context of this edited volume, *The Evolving Female: A Life-History Perspective*, the nature of sexual reproduction in most animals and—especially in mammals and long-lived primates—the sharing of information about social environments are life-history attributes important to survival as well as reproductive outcome (see Morbeck chap. 1).

Costs and Benefits of Sexual Reproduction

"Nor do we know why nature should thus strive after the intercrossing of distinct individuals. We do not even in the least know the final cause of sexuality; why new beings should be produced by the union of the sexual elements, instead of by a process of parthenogenesis. . . . The whole subject is as yet hidden in darkness," wrote Charles Darwin in 1861 (p. 61). A hundred and twenty years later, Bell (1982:19) could still say "Sex is the queen of problems in evolutionary biology. Perhaps no other natural phenomenon has aroused so much interest; certainly none has sowed so much confusion." A large part of that confusion stemmed from the general assumption that sex could be explained by the advantage to the species of more flexible possibilities and more rapid evolution. When Williams (1975) and Maynard Smith (1978) elaborated Darwin's original insight that sex, like most other biological phenomena, is selected for or against on the basis of its advantage to the individual parents and their offspring, then it seemed even more mysterious why individuals should go through sexual fuss and bother instead of copying themselves as a nice, safe clone.

The fundamental cost of sex, as well as its advantage (and indeed, for some, its definition), is recombination of genes. The complex of genes that sufficed to produce a parent who survived to reproduce are broken up and reshuffled with genes from a stranger. This seems a terrible risk, and if it is so common, it must then have equally dramatic advantages.

Table 19.1 (from Lewis 1987) lists the generally considered costs of sex. They bear a rough similarity to the costs of depending on learning rather than instinct—the breaking up of behavior that was good enough for the parent, to start over again from an unsure base, and the time, energy, physical risk, and social demands in doing it the complicated way. Even the "cellular-mechanical costs" have a parallel in that meiosis in a given organism may take 10 times as long as mitosis. Similarly, development of behavior from learning takes longer than instinctual activities.

One major class of costs of sex, however, is not apparently relevant: those derived from anisogamy. Anisogamy is the union of unequal gametes. We do not have two obligately

TABLE 19.1. Costs of Sex (from Lewis 1987)

Costs not derived from anisogamy
Recombination
Cellular-mechanical costs
Meiosis: cell division resulting in reduction of chromosomes to one-half the normal content
Syngamy: sexual reproduction with fusing of gametes
Karyogamy: fusion of gametic nuclei with interchange of nuclear contents
Fertilization
Exposure to risk
Minimal density for reproduction
Costs derived from anisogamy
Genome dilution (cost of males)
Sexual selection
Sexual competition (conflict, exposure)
Dual phenotype specialization

TABLE 19.2. Benefits of Sex (after Stearns 1987a)

Stability
DNA repair
Muller's ratchet
Variability
Tangled Bank
Environmental
Sibling diversification
Red Queen
Versus evolving competitors, predators, and prey
Versus evolving parasites
Selection arenas

TABLE 19.3. Historical Sequence of the Evolution of Sexual Reproduction (from Stearns 1987a)

1. Asexual reproduction
2. Limited recombination (bacterial)
3. Meiosis
4. Mating types
5. Anisogamy
6. Gender (products of sexual selection)
7. Incompatibility types

distinct intellectual sexes. The basic property of anisogamy is that one sex puts more intellectual capital into each of its reproductive products and therefore becomes a scarce resource for the other sex. For example, on the level of individual human lives, it is true that most women put more personal thought and energy into each of their children than the fathers do. However, on the level of ideas, there do not seem to be a class of heavyweight, propertied ideas, which are sought out by another class of lightweight ideas.

This means that there is no close parallel to the "two-fold cost" of genome dilution (Maynard Smith 1971, 1978). If a female with a limited supply of eggs is to blend her precious genetic material in such a way that each child has only half of her genome, sexual reproduction must be at least twice as advantageous to her if it is to persist in competition with asexual reproduction. This also is called "the cost of males." A female who provides the cytoplasm and energy for her egg to become another individual, in this view is "parasitized" by the male who dilutes by half her contribution to each offspring. In isogamy, where union occurs between gametes of similar size and genetic contribution, as for ideas, if there seems to be no energetic limit on how different genes are assorted among the offspring. Any significant advantage of the mixed forms over the single-parent forms selects for mixing.

Current suggestions of the benefits of sexual reproduction are summarized in table 19.2. A wealth of mathematical models and some experiments attempt to quantify the magnitude of the various effects. One crucial insight of the past decade is that the chief advantages of sex may have been different during early evolution, and in bacteria today, from the advantages of sex for ordinary eukaryote organisms. In other words, the origins and maintenance of sex may have different explanations. Table 19.3 (Stearns 1987a) lists the stages that probably led to fully developed sexual reproduction. We shall see that the costs and benefits of social intelligence resemble the more chaotic early stages of sexual reproduction, rather than the codified, later form.

Stability as a Benefit of Sex and Intelligence

Sex

At the origin of sexual reproduction, among simple bacteria and protozoa, the transfer of DNA may have been primarily a means to promote stability, rather than change. This is Bremerman's (1979, 1987) hypothesis of sex as DNA repair. Conjugating bacteria transfer variable lengths of their ring of DNA. The receiver lines up the donated segment with the corresponding stretch of its own ring, snips out the old, and splices in the new, donated piece. A single gene may control the tendency to give or to receive. If that gene happens to be transferred, the receiver becomes a donor when it next engages in conjugation.

The hit-or-miss process should not perhaps be dignified by the name of sex. It is only one of the ways bacteria transfer genes, another being via viruses. Margulis and Sagan (1986) pointed out, however, that such transfer functions as a "massive health-care system" for bacteria, because it allows a healthy section of DNA to replace one that has been damaged. This must have been particularly important under Earth's early atmosphere, before the ozone shield developed, when Earth was bathed by high-intensity, mutagenic, ultraviolet light. "The splicing and polymerase enzymes for repair became the enzymes of sexuality. DNA recombination is far older than it is in the laboratory. Thus, in the evolutionary sense, *ultraviolet repair preadapted bacteria to sexuality*. . . . When the source of new DNA was different from that of the damaged DNA, the repair process became a form of prokaryote sexuality: new DNA was formed from more than a single potential source." (Margulis and Sagan 1986:49.)

Margulis (1981) has shown that the eukaryotic cell evolved from a symbiosis of several bacterial types. A large bacterium became the "host," eventually forming a nuclear membrane to shield its own genetic material both from disruptive oxygen and from all its guests. Oxygen-using forms became mitochondria; photosynthetic cyanobacteria became plastids.

Although it is controversial, Margulis (1981) and Margulis and Sagan (1986) believed that threadlike spirochete bacteria joined this precellular community, giving rise to the microtubules that form the mitotic and meiotic spindles as well as to sperm tails. They pointed out that there is only one form of microtubule organizing center, or basal body, for the microtubular structures within any one cell. Cells that use it for flagellae, including sperm with their motile tails, are not able to divide by mitosis or meiosis. In protozoa, the flagellum can be reabsorbed and afterward converted into the mitotic spindle. In higher organisms, once a cell develops a flagellum, it is then fixed for that cell's life, and the cell will die without further division. Sperm depend on the egg for the apparatus of division.

Meiosis, with its reduction division of paired chromosomes, is much less widespread than mitosis. To progress to true meiosis, the chromosomes must first be diploid (i.e., having two homologous sets of chromosomes). How could diploid cells themselves have originated? Margulis and Sagan (1986:152) suggested they came either from incomplete cell division, or by cannibalism and incorporation of the nucleus of a conspecific. "At this point, sex, rescue from death, cannibalism, and incomplete digestion were much the same thing."

This scenario of the early world implies that DNA repair was the crucial advantage to gain from sharing DNA with a stranger. Diploidy itself, with its potential to make up for a defective gene by another, functional allele, and the transfer of chunks of DNA, produced at least some healthy, surviving individuals. It also shows a world where sex is not, or was not, fully codified. Part, or even a fairly randomly chosen part, of the genetic information could be shared—not necessarily an orderly half as in the higher plants and animals.

Intelligence

Learning is a life-history feature that can play a similar buffering role in the behavior of individuals. In situations where the end behavior should be fairly uniform to be adaptive, but the initial circumstances are often subject to change or to environmental insult, an innate lock-and-key behavior would often fail. One example is mother-infant bonding in humans. There is probably a period of an hour or two following birth, when the mother's hormones prime her to accept the newborn child (Trevathan 1987). This seems to be true of all other mammals that have been studied. Some, however, such as goats normally reject an infant that is presented after that period (Klopfer and Klopfer 1968). Human mothers, and at least some other primates, will come to love a baby a day later after cesarean, or a month later as it emerges from an incubator, or years later, through adoption. The child's reciprocal love is similarly multiply determined. Of course, mother-infant bonds do not always form, but learning can repair failures in earlier experience.

A second type of situation in which learning produces a similar response to varied inputs is imprinting. Imprinting occurs in animals as a life-history attribute when it is important to know the characteristics of the individual parent, as in precocial ducks that swim after their own particular mother around the pond. The variability of input—the different mothers—is not an "insult," but it is not possible to be programmed by innate cues. Again, our own human infantile attachments have often been compared to imprinting (Bowlby 1969; Lorenz 1970). Given our dependent and vulnerable childhoods, and the vagaries of human growth, courtship, and parenthood, such stabilizing learning is presumably central to our survival and reproduction.

Muller's Ratchet (Muller 1964) is an aspect of the danger of continued asexual reproduction, which can be redressed by sexual reproduction. Muller pointed out that in an asexual clone, mutations accumulate through time. There is no way to return to an initial "perfect" state, except through rare back-mutation to the original condition. Recombination of genes, through sex, allows some of the progeny to inherit all the best alleles. They may then survive, through natural selection, while their less perfect sibs are weeded out. This is not DNA repair, in the more casual bacterial sense, although it accomplishes the same result through its eugenic program.

The parallel with human thought seems to me clear here. Combining ideas and throwing out those that are flawed, or which seem not to work in practice, allow the most useful conglomerate of ideas to survive. I am not here insisting on a "natural selection" of ideas (see below, "Selection Arenas"), I am merely underlining that deriving information from a variety of sources can actually allow for more stability than being dependent on just one source of information, which may be damaged in transmission. This is true whether the information is genetic, or the tribal legends, or a computer disk that brushes a magnet.

Variability as a Benefit of Sex and Intelligence

We are more used to thinking of variability than stability as a benefit of sex and intelligence. It seems obvious that species that can take advantage of diverse niches should survive, but it is not so obvious that the offspring of a parent that has succeeded in surviving to reproduce would do well to differ from that parent. The two major theories as to why sex would actually benefit such offspring are called the "Tangled Bank" (Ghiselin 1974, 1987; Bell 1982) and the "Red Queen" (Van Valen 1973).

The "Tangled Bank" addresses the variability of the environment. The seeds of a plant, falling variously in sunny and shaded spots, on fertile or barren ground, are entering a lottery of chances in life. The asexual parent, in effect, buys many tickets with the same number, whereas the sexual parent a collection of many different numbers (Williams 1975). The "better" strategy will depend on the distribution of patches in the environment. If many of the patches are identical, it would be a better strategy to choose the one most-frequent winning number. If the offspring can move and seek out favorable habitat, it may again be best to concentrate on one optimal ticket number. If there are few offspring in varying patches, it may still be best to stick with one ticket, so that at least some will have the optimal chance. However, if enough offspring land in each differing patch so that one of the variable numbers is likely to be superbly well adapted, it may be best to opt for the many different tickets.

These are the familiar arguments for the advantages of learning over instinct, or vice versa. In a highly predictable environment, instinct may well be more efficient, allowing the organism to proceed with a minimum of lost time or energy or disastrous mistakes. If environmental factors such as potential food supply are highly variable, it may pay to learn the appropriate information by trial and error, or, much better, from others.

One further facet is the avoidance of sibling competition. If several seeds fall near each other, but each exploits a slightly different composition, for instance, of soil minerals, there may be more joint survival than if they were an identical clone. This argument presumably varies case by case, since successful clonal organisms can also form dense stands or colonies that tolerate and often benefit each other. The argument probably arose, in fact, from considering the learned diversification of human and higher primate siblings, who often seem to choose different styles of life (Goodall 1986).

The second set of considerations was called the "Red Queen" by Van Valen (1973), after the Red Queen's remark in *Alice Through the Looking Glass*, "it takes all the running *you* can do to keep in the same place" (Carroll 1872). That is, there is not just a variable environment, but a changing environment, as other species evolve.

Some of these species are competitors, predators, and food. Plants evolve defensive thorns and secondary compounds. Herbivores evolve ways to chew, to peel the thorns, and to detoxify the poisons. Carnivores evolve communal ambushes, whereas prey evolve communal vigilance. And, humans evolve the intelligence to remember foraging patterns through places and seasons and to make weapons for defense, and tools, baskets, and cooking pots for maintenance.

Humans' chief competitors are, of course, each other. "Machiavellian intelligence"—that is, deceit, manipulation and cooperation in the face of each other and their environment—were perhaps the crucial contexts for the evolution of primate and human minds (Byrne and Whiten 1988).

Hamilton et al. (1981) pointed out that an even stronger pressure than competition for rapid diversification, and hence sexual reproduction, is the pressure of parasites. Bacteria such as *Escherichia coli* have three generations an hour. Even parasitic worms have many generations during the life cycle of a single higher vertebrate host. The vertebrate parent can be sure that the parasites have been adapting to itself and its generation: its offspring will fare better if they are immunologically different from the parent.

This is one selective pressure that seems to have had little to do with intelligence, up to the last two or three generations. We are now clearly waging a war of medical science and social concern against newly resistant malarias and entirely new diseases such as AIDS. Before this, all societies have had health traditions, such as the appropriate places for defecation. It is possible that early learning involved the use of medicinal plants in all the world's cultures (now known even in chimpanzees; Wrangham and Goodall [1989]). However, these were conceived as defenses against constant menaces. Hamilton et al.'s vision of rapid parasite evolution answered by genetic recombination in each sexual generation does not have an obvious mental parallel until recognition of a medical race against ever-changing epidemics in modern times.

Selection Arenas

In Stearns' (1987b) conception, "selection arenas" are life stages in which a parent "decides" which of its offspring will receive further investment, or in which siblings "decide" by early contest which will survive, without further damaging competition. My picture is my family's apple tree, which yearly shed scores of small green windfalls for each apple that ripened. (cf. the botanical literature on overproduction of zygotes, listed in Stearns [1987b]). Another version is the 80% of fertilized human ova that do not become live infants: the large majority shed before implantation, without the mother knowing she may be pregnant (Diamond 1987). Perhaps 50% of spontaneous abortions result from chromosomal abnormalities in the fetus, only about 15% from maternal malfunction (Lauritsen, in Diamond [1987]).

Stearns (1987b) pointed out that the costs of sexual reproduction are minimized, and benefits maximized, if parents can easily shed the less fit offspring. Sex as a way to avoid Muller's Ratchet, in particular, depends on recombination shifting the better alleles at random to a few offspring, while other gametes and fertilized eggs get most of the deleterious ones. If the less favored ova are not culled by environmental selection, but before the parents even incur the cost of developing them as embryos, there are far fewer costs to recombination of previously well-coordinated gene complexes.

Here, we reach a parallel to foresight and to creative intelligence. To the extent that a person deliberates different courses of action and then chooses one behavior to act out, that person is constructing a "selection arena" for ideas. Griffin (1984) argued a similar case for the evolution of consciousness in animals. One of the most plausible adaptive advantages for thought is precisely this ability to consider behaviors economically as mental models, without the time, energy, and danger of trying out all of them. Without arguing the question of whether we actually have free will, this is what we *experience* as free will. It is not surprising that it is difficult to describe even Stearns' biological selection arenas without using the word "decision."

Stages of Evolution of Sex and Intelligence

One of the attempted comparisons between intelligence and genetic evolution is Dawkins' (1976) concept of "memes." Memes would be elements of thought, as genes are genetic elements. Successful memes would be those that spread and survive, like adaptive genes. There are fundamental problems with the analogy of meme and gene. One is that we have no single particulate level of ideas. It is as though we were trying to do genetics while confusing genes, gene complexes, and chromosomes—a pre-Mendelian situation. This situation raises the problem of blending inheritance, which could undermine Darwin's entire conception of natural selection of transmitted characteristics.

The analogy between sex and intelligence does not have this problem. Both processes are ways of acquiring information from others, which may then be transmitted. If ideas actually blend like miscible liquids, rather than reshuffling like genes, it is hard to picture natural selection acting at the level of competing ideas. It could still favor individuals with greater intelligence, who could deal with the passing flux of ideas.

It could be argued, however, that there are particulate or digital aspects to our social intelligence. Language itself is made up of phonemes that seem to be innately recognized as discrete entities. Those not used in surrounding adult language are dropped after a critical phase in the second year of life (Werker 1989). Words are far less discretely bounded than phonemes, but it is impossible to imagine language without words or other symbols of limited meanings. On a much larger scale, we impose limits on concepts, reifying notions that are actually vague or complex. What about "The U.S. Government," "Feminism," or "Gaia" as nouns that we think actually have a meaning and that are as simple to say as "my son" or "my cat?" Humphrey (1976) and I (Jolly 1966, 1988) have argued elsewhere that this reification derives from the mind of a social primate, which depends on being aware of not only the individual identity of objects but also the identity and desires of other individuals.

The power of the digital computer has added a new dimension to our transmission of information. Here, it is easy to see the hierarchy of mutually exclusive decisions that builds from the simplest 0-or-1 bit to elaborate programs. There are even "fuzzy" computer programs and recognitions, which deal with the imperfect information of the real world but are still built on a digital base. We understand and build computers: they seem rather more like the working of the genome than our own, imperfectly understood intelligence.

It seems likely to me that as intelligence progresses, with worldwide computer links of the now-fashionable information highway, we will see increasing formalization of the system. In this sense, we may progress from the "hit-or-miss" communications that characterized early evolving sex and bacterial interchange through conjugation and through viral transfer of genes, toward something that is more rigidly codified.

This is not to say that we will evolve a two-parent system of intelligence, with half of the ideas from each! One of the strengths of intelligence is precisely its ability to draw information from many "parents." However, the codification of intelligence to be transmitted seems a likely trend of the future—the beginnings of the trend that today we call formal education.

From that point, you can move to science fiction. What about academics and their laboratories, libraries, and data banks as our society's germ line of intelligence, separated from the doers and decision makers by a generation's time-lag of ideas? What about two or more intellectual sexes—not necessarily drawn along present sexual lines? (My candidate for the "females" would be computer programmers, who control the machinery of elaboration of ideas and become a scarce resource for the rest of the intellectual community.) What about our species' present symbiosis with crop plants and animals (and perhaps eventually self-replicating computers), where we control their genetics, in part, and their opportunities to reproduce, that is, somewhat like the early bacterial community that became the cell. Is the superorganism of human society and its symbionts as viable as the cellular and multicellular organisms that reproduce by sex?

The answers depend on how far the particular attributes of the system of exchanging information through sex may be generalized to any system of information exchange. This chapter is a very modest beginning. Two things are clear, at least. The first is that there have been only two such innovations in the whole history of life. Each has transformed the species that discovered it, and the biosphere around them. The second is that in keeping with their evolutionary importance, both sex and thinking are great fun.

20 Life History, Females, and Evolution: A Commentary

Beverly McLeod

When I was a graduate student in cultural anthropology during the 1970s, the "man the hunter" theory still was in full swing. Human society developed, it was theorized, out of bonds among a group of males cooperating to kill large prey. The social division of labor by gender seen in modern society was patterned by our prehistoric ancestors: men went out to capture the bacon, while women stayed home to cook it and look after the kids. Even the origins of human language were ascribed to the male hunting band's necessity for verbal communication through tall savanna grass. The implications were clear; human nature could best be understood by searching for the antecedents of human social, sexual, and political behavior in the anatomy and behavior of male primates.

Even though the pre-eminent role of "man the hunter" has been challenged, we still look to biological anthropologists including primatologists and paleontologists (and now, geneticists) to discover the essence of human nature. But rather than focusing on the influence of "nature" in explaining human behavior, my training in anthropology and social psychology, and the extracurricular atmosphere of the feminist movement, encouraged me to give more weight to the effect of "nurture." Cultural anthropologists looked less at where *Homo sapiens* came from (biological evolution) than at where the species had been going (cultural "evolution"). Feminists asserted that biology is *not* destiny, and ethnographies of the world's cultures demonstrated the malleability of gender roles.

Looking at human behavior through a cultural lens has been a mind-opening experience. Never again could I assume that the behavior of people in my own culture was "natural." Reading the material in this book, in which humans are viewed through an evolutionary lens, has been a similarly mind-opening experience. Like an adoptee who suddenly discovers her biological parents, I have become aware of a hitherto hidden dimension to my life as a human female—my roots as a female mammal and primate.

Focusing on females can tell us much about our continuity with other species and about the distinctiveness of *Homo sapiens.* Viewing the lives of females across species as well as across cultures suggests the following ideas and questions about evolution, females, and life history.

Evolution

The Human Female as a Series of Evolutionary Layers

Consider five women on a sunny beach. One is lying on her stomach, sunbathing. The second is sitting in a beach chair, nursing her infant. The third is sharing a picnic with her youngster. The fourth is jogging along the water's edge. And the fifth is reading a book on child development, or the health benefits of exercise, and discussing it with her companion. What do they illustrate about human evolution?

The sunbather reveals her reptilian ancestry in her body shape, a torso with four apendages. The nursing mother, far from being the hallmark of humankind, recalls our shared ancestry with all mammals. Our primate history is demonstrated by the mother feeding her child; young primates cannot survive by instinct alone. The fourth woman shows a distinctively human characteristic, upright posture and gait. Had she been walking and carrying a baby, she would have illustrated what is not only "most human" but also "most humanly female." The fifth woman illustrates the human capacity for combining information in an intelligent fashion (as discussed by Jolly, [chap. 19]), reflecting, planning, and making sense of the world in a manner that affects behavior.

More Animallike Humans and More Humanlike Animals

Besides language, two of the features I consider, as a social scientist, most characteristic of humans as a species are our social emotions and intimate relationships, but they may have deep roots in our evolutionary past. Studies of nonhuman primates indicate the importance of kinship; for monkeys and apes, as well as humans, relationships with kin—especially relationships between mothers and their offspring—form the basic fabric of society (Smuts chap 5; Pavelka chap. 7). For primates, the personality and behavior of individuals makes a difference to other individuals. For example, although we may be able to understand the behavior of ants by viewing each as interchangeable with any other, understanding human behavior requires that we study individuals over the course of their lives (Morbeck chaps. 1, 9).

The social relationships and emotions of monkeys and apes seem very humanlike. We can understand intuitively and identify with the grief of a mother chimp whose baby dies and the depression suffered by an orphaned chimp because we may experience these in our own lives. For the primates, especially the apes, we may be able to put ourselves more readily in the place of the organism. We can appreciate the survival value of strong emotional ties and recognize the potential for costs as well as benefits to our attachments to others (Smuts chap. 5).

FEMALES

Lack of Knowledge about the Human Female Body

It is amazing to think that debate still rages about the evolutionary significance of the anatomical features of human females. Given that humans are fatter than other creatures, and women fatter than men (and given that American culture is obsessed with the fat of breasts and hips), it is remarkable that scientific investigation into adipose tissue is so recent (Pond chap. 11; McFarland chap. 12). Once again, our evolutionary legacy is indicated by the similar patterns of fat deposits among all mammals. But beyond that, many questions remain. Do enlarged breasts serve as a sexual marker or an extra repository for maternal fat? Are fatter animals reproductively fitter? Is there a "threshold" level of fatness required for reproductive capability that varies by diet and workload (Vitzthum chap. 18)? How much flexibility has evolved in this interaction?

New Perspectives on Human Gender Differences

The physical differences that seem so apparent between living men and women are not as easy to discern in the fossil or nonhuman record. Comparing male and female bodies and behavior across primate species yields no clearly definable core of maleness or femaleness but rather a mosaic of similarities and differences (Morbeck chaps. 1, 9; Zihlman chaps. 8, 13).

One might argue that the female body better exhibits the tremendous flexibility of reproductive options made possible during evolution than does the male body. Female bodies and lives—much more than those of males—reflect the tremendous changes in the process of reproduction, from reptiles that complete the reproductive process by laying eggs, to mammals that gestate young inside their bodies and suckle them after birth, to primates that carry infants continuously and have long-lasting social and emotional ties with them, to humans who exaggerate and lengthen the primate pattern. Human females have a higher percentage of body fat and a wider pelvis than males (Zihlman chap. 8, 13), allowing humans to produce and nurture bigger-brained offspring.

Despite these dramatic changes, there is continuity and striking similarity between apes and humans in the pattern of female reproduction and raising of young, so much so that Jane Goodall is said to have remarked that she learned how to be a good mother to her own infant son by imitating chimpanzee mothers.

A Slow and Steady Reproductive Strategy

Reproduction is so different for males and females that it does not make sense to view them through an undifferentiated construct of reproductive success. In many species, males are either "studs" or "duds"; each male has sexual access to many or very few females. They can spread their genes widely or hardly at all. Thus, it makes sense to focus on the strategies they use, to find out what makes some males players while others sit on the sidelines and never get a chance in the reproductive game. Some researchers have focused on strategies of male-male competition; others ask whether this is the only way to play the game and focus on males' friendly relations with females (Smuts chap. 5; Pavelka chap. 7).

But it does not make sense to look at females from the same perspective. The motto for elephant seal males, it would seem, is "mate early, mate often" (Reiter chap. 4), but this strategy is successful only for males. For females, there are costs associated with early motherhood, even in a species like elephant seals with only a brief period of maternal involvement with their young. For primate females also, there is a trade-off between high fertility and high survivorship of infants (Fedigan chap. 2). For many mammals, especially monkeys and apes, longevity is the key; females with a long life span started to reproduce later, but had twice as many surviving infants, and a much higher rate of infant survivorship, than shorter-lived females (Fedigan chap. 2).

For males, mating can be the only contribution to reproduction. But for females, mating is only the first step in a long process. Focusing on females leads us to look at rearing rather than mating. Reproductive success may be measured in number of offspring for males, but the difference in the number of offspring produced among females may be slight in the first generation. If quantity counts for males, for females, quality may be crucial. Only by nurturing each offspring sufficiently can a female ensure that her genes will be multiplied in the next generation. Only by integrating her social life with her biological function can a female reproduce successfully.

The Mother-Infant Duo as a Biological and Social System

We are accustomed to thinking about a pregnant female and her fetus as an inseparable biological system. But, as several chapters in this book indicate, the biological and social lives of mothers and infants are intertwined after birth as well. The infant's very life depends on the quality of mothering it receives, but the mother is also affected by the infant. A woman's present fertility may depend directly on her infant's suckling (Vitzthum chap. 18), and the ability of human females to mobililize calcium from bone to nourish infants may compromise their health in the postreproductive years (Galloway chap. 10).

A common thread linking the research on females across species and cultures is the complexity of female lives. Women in modern societies may feel torn between their occupational and familial roles. But even before the supermom image, female mammals have had to do double duty caring for youngsters while ensuring their own survival.

The burdens of motherhood are often greater in species that share a longer portion of evolutionary history with *Homo sapiens.* Infant seals and sea lions require suckling, as do infant monkeys and humans, but they do not require carrying (Zihlman chap. 8, 13). Infant monkeys assist their mobile moms by hanging on and learning to feed themselves. Human mothers compromise their mobility and agility by having to carry an infant with little help—and sometimes active resistance—from the infant itself. And in most societies, mothers must supply their offspring with food for many years.

Some bread-earning activities dovetail nicely with child care, whereas others conflict with it. Women whose economic role consists of gathering, planting, harvesting, preparing, or marketing food close to home can often care for their children while they work. In contrast, other tasks preclude bringing children, especially as they become somewhat more independent or the terrain becomes more treacherous (Panter-Brick chap. 17). I see this in my own life where reading, thinking, writing, and teaching cannot be done while simultaneously tending a child.

In societies without artificial birth control, women may have many children, but at a slow pace. Older children can assist in both economic activities and childcare. The balancing act is different in societies such as the United States, where people tend to have fewer children, with a shorter interbirth interval. Some women in these societies adopt an alternating pattern, doing childcare exclusively for several years before returning to economic activities. Childbearing in these societies tends to be confined to a short segment of a woman's life, rather than continuing throughout her reproductive years. Some careers require complete devotion to the job for several critical years that often coincide with a woman's prime childbearing years, creating a conflict between economic tasks and child rearing.

Life History

Maternal Experience and Support

Economic independence is achieved at different points in the lives of the young in various species and cultures. Nonhuman animals usually achieve survival capability before they reach reproductive maturity. Humans in some societies are capable of economic survival around the time of sexual maturity, whereas young adults in other societies are able to reproduce long before they are capable of economic survival in their society.

It takes time for females to achieve both survival and reproductive capability. Even if young females can support themselves before they begin to reproduce, they may not be able to support an infant. The pups of the youngest elephant seal mothers are at the greatest risk of dying in infancy (Reiter chap. 4). One suspects the same may be true for humans and other mammals.

The support of other females apparently enhances a female monkey's reproductive success (Smuts chap. 5). It is logical to expect the same effect among humans. Women who have the support of kin may find it easier to bear and raise more children. Those children may be more likely to survive and reproduce if they receive care from female kin if their mother becomes incapacitated or dies.

A Complex Interplay among Biological and Sociocultural Influences

Because human female life history is a complex interplay among biological and sociocultural influences, there are no invariable or unidirectional influences between biology and social factors. When the social environment changes, so may biology, and vice versa. The degree of sexual dimorphism in humans may reflect social and economic change and/or the division of labor by gender, but not in a simple or straightforward manner (Borgognini Tarli and Repetto chap. 14). Economic change and cultural contact may influence the reproductive choices of women, which in turn affect the survivability of their children (Draper chap. 16). The economic system of the Tamang people, requiring mothers to perform seasonal agricultural work far from home in conditions dangerous for toddlers, poses greater risks for infants born at a certain time of year and for children of a certain age (Panter-Brick chap. 17).

Conditions during an earlier phase of life affect body function at a later phase. Many of these conditions are considered pathological, resulting in such questions as, "What causes, and how can we treat, reproductive malfunctions and osteoporosis?" Instead, Vitzthum (chap. 18) and Galloway (chap. 10) inquired, "What could be the evolutionary significance of reproductive malfunctions and osteoporosis?" Vitzthum speculates that the functioning of the female reproductive system reflects the environmental conditions a woman experienced during her adolescence, while Galloway theorizes that the functioning of the female skeletal system reflects the hormonal conditions a woman experienced during her reproductive years.

Varying Experiences of Infants and Young across Species and Cultures

This book has described the varieties of female life history. But focusing on females leads inevitably to focusing on the young. Perhaps there is also much to be gained by examining the lives of infants across species and cultures, comparing the survival and well-being of the young under different types of maternal care.

For example, what accounts for the similar infant survival rate for seal lion and elephant seal pups, despite the very different maternal care they receive (Ono chap. 3)? What conclusions can we draw by noting that both elephant seal pups (Reiter chap. 4) and Tamang toddlers (Panter-Brick chap. 17) undergo abrupt transitions between intense association with their mothers and survival for at least some portion of their time on their own. In contrast, sea lion pups (Ono chap. 3) and Kami toddlers (Panter-Brick chap. 17) have a more consistent level of contact with their mothers but at very different levels of intensity. How does childhood differ for young children in farming communities, who begin learning adult economic tasks from an early age, compared to those in forager communities (Morelli chap. 15) and modern industrial societies? Is the transition from childhood to adulthood easier for children in societies in which adults sit on the floor rather than on chairs, or in which adults as well as children regularly take an afternoon nap?

Humans: Meaning-Making Creatures

Fedigan remarked that primatologists study the individual lives of their subjects in the context of preassigned life-history features, whereas social scientists study individuals with life histories marked by features significant to that person. How could it be otherwise, for humans are meaning-making creatures?

It is ironic that all human societies attribute the greatest meaning to the life-history features that mark inevitable stages among all animals—birth, sexual maturity, reproduction, and death—and to the differences between the sexes. For example, the birth of a human infant may be cause for celebration or anguish, depending on the society and family into which it is born, and on its sex.

Although male and female infants apparently receive identical care from other mammal and primate mothers (Ono chap. 3; Reiter chap. 4; Hiraiwa-Hasegawa chap. 6), in humans, the background of gendered adult roles influences the lives of infants and young children. Society drapes a cultural overlay on biological sex differences, and children are expected to begin conforming to their gender role at a young age (Morelli chap. 15).

Lack of Knowledge about Human Life Histories

Although multigenerational studies among some monkey and ape families have been conducted (Pavelka chap. 7), comparable studies of human families are more difficult to compile. We may have more long-term observational data across generations on nonhuman primates—and many other animals—than on humans. Thus, it is tempting to read human truths into the behavior of other animals.

One hallmark of *Homo sapiens* is variability, even in as basic a feature as reproduction. Whereas female monkeys have a narrow range for age at first reproduction and number of offspring, human females may begin reproducing before they reach their teens or as late as three decades later, and produce zero, one, or over a dozen children.

Conclusion

This book has focused on the lives of females across species and cultures, illuminating diversity as well as commonality. The differences between female elephant seals—whose major maternal contribution is converting their body fat into a temporary super rich milk diet for their pups—and chimpanzees—whose tenderness toward infants and even apparent grief at the death of kin are well known—are dramatic. Yet both species share the mammalian female pattern with humans.

Because maternity can easily be determined, looking at the behavior of individual nonhuman primates can tell us much about human female and maternal behavior. Perhaps primatology offers a richer source of self-knowledge to women than to men. Against the standard of reproductive success, it is becoming clear what a good mother is. It is much less clear what a good father is, evolutionarily speaking. Humans are, in a sense, negotiating unknown territory in expecting intense relationships between fathers and offspring, although we still are building on primate friendships between males and infants.

The comparative and evolutionary approach taken in this book—which looks at whole animals and whole lives at all levels from physiological features to social behavior (Morbeck chaps. 1, 9)—illustrates the potential consequences of individual behavior for evolution and the consequences of evolution on the flexibility of individual behavior.

Literature Cited

Aiello, Leslie, and Christopher Dean. 1990. *An Introduction to Human Evolutionary Anatomy.* San Diego: Academic Press.

Ainsworth, Mary D. Salter, M. C. Blehar, Everett Waters, and S. Wall. 1978. *Patterns of Attachment.* Hillsdale, NJ: Erlbaum.

Akre, J. 1989. "Infant Feeding: The Physiological Basis." *Bulletin of the World Health Organization* 67 (suppl.).

Alberts, S. 1987. "Parental Care in Captive Siamangs (*Hylobates syndactylus*)." *Zoo Biology* 6: 401–406.

Albright, J. A., and R. A. Brand, eds. 1987. *The Scientific Basis of Orthopaedics.* Norwalk, CT: Appleton and Lange.

Alexander, Richard D. 1987. *The Biology of Moral Systems.* New York: Aldine De Gruyter.

Alexander, Richard D., J. L. Hoogland, R. D. Howard, K. M. Noonan, and P. W. Sherman. 1979. "Sexual Dimorphism and Breeding in Pinnipeds, Ungulates, Primates and Humans." In *Evolutionary Biology and Human Social Behavior: An Anthropological Perspective*, ed. N. A. Chagnon and W. Irons, pp. 402–435. North Scituate, MA: Duxbury Press.

Allen, Timothy F. H., and Thomas W. Hoekstra. 1992. *Toward a Unified Ecology.* New York: Columbia University Press.

Altmann, Jeanne. 1974. "Observational Study of Behavior: Sampling Methods." *Behaviour* 49: 227–267.

———. 1980. *Baboon Mothers and Infants.* Cambridge, MA: Harvard University Press.

———. 1983. "The Costs of Reproduction in Baboons." In *Behavioral Energetics: The Cost of Survival in Vertebrates*, ed. W. P. Aspey and S. I. Lustick, pp. 67–88. Columbus: Ohio State University Press.

———. 1986a. "Adolescent Pregnancies in Non-Human Primates: An Ecological and Developmental Perspective." In *School-Age Pregnancy and Parenthood, Biosocial Dimensions*, ed. J. B. Lancaster and B. A. Hamburg, pp. 247–262. New York: Aldine DeGruyter.

———. 1986b. "Parent-Offspring Interactions in Anthropoid Primates: An Evolutionary Perspective." In *Evolution of Animal Behavior: Paleontological and Field Approaches*, ed. Matthew H. Nitecki and Jennifer A. Kitchell, pp. 161–178. New York: Oxford University Press.

Altmann, Jeanne, and S. Alberts. 1987. "Body Mass and Growth Rates in a Wild Primate Population." *Oecologia* 72:15–20.

Altmann, Jeanne, and P. Muruthi. 1988. "Differences in Daily Life between Semiprovisioned and Wild Feeding Baboons." *American Journal of Primatology* 15:213–221.

Altmann, Jeanne, and A. Samuels. 1992. "Costs of Maternal Care: Infant-Carrying in Baboons." *Behavioral Ecology and Sociobiology* 29:391–398.

Altmann, Jeanne, Stuart A. Altmann, Glenn Hausfater, and Sue Ann McCuskey. 1977. "Life History of Yellow Baboons: Physical Development, Reproductive Parameters, and Infant Mortality." *Primates* 18(2):315–330.

Altmann, Jeanne, Glen Hausfater, and Stuart A. Altmann. 1985. "Demography of Amboseli Baboons, 1963–1983." *American Journal of Primatology* 8:113–125.

———. 1988. "Determinants of Reproductive Success in Savannah Baboons, *Papio cynocephalus.*" In *Reproductive Success: Studies of Individual Variation in Contrasting Breeding Systems*, ed. T. H. Clutton-Brock, pp. 403–418. Chicago: University of Chicago Press.

Altmann, Stuart A. 1991. "Diets of Yearling Female Primates (*Papio cynocephalus*) Predict Lifetime Fitness." *Proceedings of the National Academy of Sciences, U.S.A.* 88:420–423.

Altmann, Stuart A., and Jeanne Altmann. 1979. "Demographic Constraints on Behavior and Social Organization." In *Primate Ecology and Human Origins: Ecological Influences and Social Organization*, ed. I. S. Bernstein and E. O. Smith, pp. 47–63. New York: Garland STMP Press.

Andersen, A. N., and V. Schioler. 1982. "Influence of Breast-feeding Pattern on Pituitary-Ovarian Axis of Women in an Industrialized Community."

American Journal Obstetrics and Gynecology 143: 673.

Anemone, Robert L., Elizabeth S. Watts, and Daris R. Swindler. 1991. "Dental Development of Known-Age Chimpanzees, *Pan troglodytes* (Primates, Pongidae)." *American Journal of Physical Anthropology* 86:229–241.

Aramburu, C. E., and A. Ponce. 1983. *Familia y Trabajo en el Peru Rural.* Lima: Instituto Andino de Estudios en Poblacion y Desarollo.

Archer, J., and B. Lloyd. 1982. *Sex and Gender.* Harmondsworth, Middlesex: Penguin Books Ltd.

———. 1985. *Sex and Gender.* Cambridge: Cambridge University Press.

Arend, R., F. L. Gove, and L. A. Sroufe. 1979. "Continuity of Individual Adaptation from Infancy to Kindergarten: A Predictive Study of Ego-Resiliency and Curiosity in Pre-Schoolers." *Child Development* 50:950–959.

Armbrecht, H. J. 1989. "Changes in Intestinal Calcium Absorption and Vitamin D Metabolism with Age." In *Mineral Homeostasis in the Elderly*, ed. C. W. Bales. *Current Topics in Nutrition and Disease.* Vol. 21, pp. 127–140. New York: Alan R. Liss.

Armelagos, George J., and Dennis P. Van Gerven. 1980. "Sexual Dimorphism and Human Evolution: An Overview." *Journal of Human Evolution* 9:437–446.

Arntz, W., W. G. Pearcy, and F. Trillmich. 1991. "Biological Consequences of the 1982–83 El Niño in the Eastern Pacific." In *Pinnipeds and El Niño Responses to Environmental Stress*, ed. F. Trillmich and K. A. Ono, pp. 22–42. Heidelberg: Springer-Verlag.

Ashwell, M., S. A. McCall, T. J. Cole, and A. K. Dixon. 1986. "Fat Distribution and Its Metabolic Complications: Interpretations." In *Human Body Composition and Fat Distribution*, ed. N. G. Norgan, pp. 227–242. Wageningen, The Netherlands: Euro Nut.

Asquith, Pamela J. 1989. "Provisioning and the Study of Free-Ranging Primates: History, Effect, and Prospects." *Yearbook of Physical Anthropology* 32:129–158.

———. 1991. "Primate Research Groups in Japan: Orientations and East-West Differences." In *The Monkeys of Arashiyama: Thirty-Five Years of Research in Japan and the West*, ed. Linda M. Fedigan and Pamela J. Asquith, pp. 81–98. Albany: State University of New York Press.

Austad, Steven N., and Kathleen E. Fischer. 1992. "Primate Longevity: Its Place in the Mammalian Scheme." *American Journal of Primatology* 28: 251–261.

Aziz, Philippe. 1977. *La Civilisation Etrusque.* Genève: Ferni.

Bailey, R. C., and I. DeVore. 1989. "Research on the Efe and Lese Populations of the Ituri Forest, Zaire." *American Journal of Physical Anthropology* 78:459–471.

Bailey, S. M. 1982. "Absolute and Relative Sex Differences in Body Composition." In *Sexual Dimorphism in Homo sapiens*, ed. Roberta Hall, pp. 363–390. New York: Praeger.

Bailey, S. M., and V. L. Katch. 1981. "The Effect of Body Size on Sexual Dimorphism in Fatness, Volume and Muscularity." *Human Biology* 53:337–349.

Baker, P. T. 1988. "Part IV. Human Adaptability." In *Human Biology: An Introduction to Human Evolution, Variation, Growth, and Adaptability*, ed. G. A. Harrison, J. M. Tanner, D. R. Pilbeam, and P. T. Baker, pp. 437–547. Oxford: Oxford University Press.

Baker, P. T., and M. A. Little, eds. 1976. *Man in the Andes.* Stroudsburg, PA: Dowden, Hutchinson and Ross.

Barlet, J.-P. 1985. "Prolactin and calcium metabolism in pregnant ewes." *Journal of Endocrinology* 107:171–175.

Barry, H., III, M. K. Bacon, and I. L. Child. 1957. "A Cross Cultural Survey of Some Sex Differences in Socialization." *Journal of Abnormal and Social Psychology* 55:327–332.

Barry, H., I. L. Child, and M. K. Bacon. 1959. "Relation of Child Training to Subsistence Economy." *American Anthropologist* 61:51–63.

Bartholomew, G. A. 1970. "A Model for the Evolution of Pinniped Polygyny." *Evolution* 24:546–549.

Bartoloni, G. 1988. "A Few Comments on the Social Position of Woman in the Proto-historic Coastal Areas of Western Italy Made on the Basis of a Study of Funerary Goods." *Rivista di Antropologia* 66 (suppl.):317–336.

Bates, Marston. 1950. *The Nature of Natural History.* New York: Charles Scribner's Sons.

———. 1961. "Ecology and Evolution." In *Evolution After Darwin.* Vol. 1. *The Evolution of Life: Its Origin, History and Future*, ed. S. Tax, pp. 547–568. Chicago: University of Chicago Press.

Bell, G. 1982. *The Masterpiece of Nature—The Evolution and Genetics of Sexuality.* Berkeley: University of California Press.

Bell, G., and V. Koufopanov. 1986. "The Cost of Reproduction." In *Oxford Surveys in Evolutionary Biology*, ed. R. Dawkins and M. Ridley, pp. 83–131. Oxford: Oxford University Press.

Ben-David, M., and M. L'Hermite. 1976. "Prolactin and Menopause." In *Consensus on Menopause Research*, ed. P. A. van Keep, R. B. Greenblatt, and M. Albeaux-Fernet, pp. 17–23. Baltimore: University Park Press.

Ben Shaul, D. M. 1962. "The Composition of the Milk of Wild Animals." *International Zoo Yearbook* 4:333–342.

Bennett, A. F. 1989. "Integrated Studies of Locomotor Performance." In *Complex Organismal Functions: Integration and Evolution in Vertebrates*, ed. D. B. Wake and G. Roth, pp. 191–202. New York: John Wiley & Sons.

Bennett, K. A. 1981. "On the Expression of Sex Dimorphism." *American Journal of Physical Anthropology* 56:59–61.

Bentley, G. R. 1985. "Hunter-Gatherer Energetics and Fertility: A Reassessment of the !Kung San." *Human Ecology* 13:79–109.

———. 1994. "Ranging Hormones: Do Hormonal Contraceptives Ignore Human Biological Variation and Evolution?" *Annals of the New York Academy of Sciences* 709:201–203.

Bercovitch, F. B. 1987. "Female Weight and Reproductive Condition in a Population of Olive Baboons (*Papio anubis*)." *American Journal of Primatology* 12:189–195.

———. 1989. "Body Size, Sperm Competition, and Determinants of Reproductive Success in Male Savanna Baboons." *Evolution* 43:1507–1521.

Berman, Carol M. 1982. "The Ontogeny of Social Relationships with Group Companions among Free-Ranging Infant Rhesus Monkeys. 1: Social Networks and Differentiation." *Animal Behaviour* 30:149–162.

Berman, M. L., K. Hanson, and I. L. Hellman. 1972. "Effect of Breast-Feeding on Postpartum Menstruation, Ovulation, and Pregnancy in Alaskan Eskimos." *American Journal Obstetrics and Gynecology* 114:524.

Bernstein, Irwin S. 1991. "The Correlation between Kinship and Behaviour in Non-Human Primates." In *Kin Recognition*, ed. P. G. Hepper, pp. 6–29. Cambridge: Cambridge University Press.

Bernstein, Irwin S., and L. E. Williams. 1986. "The Study of Social Organization." In *Comparative Primate Biology*. Vol. 2A. *Behavior, Conservation and Ecology*, ed. G. Mitchell and J. Erwin, pp. 195–213. New York: Alan R. Liss.

Betzig, Laura, Monique Borgerhoff Mulder, and Paul Turke, eds. 1988. *Human Reproductive Behaviour: A Darwinian Perspective*. Cambridge: Cambridge University Press.

Biesele, M., M. Guenther, R. Hitchcock, R. Lee, and J. MacGregor. 1989. "Hunters, Clients, and Squatters: The Contemporary Socioeconomic Status of Botswana Basarwa." *African Studies Monographs* 9:109–151.

Bietti Sestieri, Anna-Maria. 1985. "Roma e il Lazio Antico Agli Inizi dell'Eta del Ferro." In *Roma e il Lazio dell'Eta della Pietra alla Formazione della Citta*, ed. A. P. Anzidel, A. M. Bietti Sestieri, and A. de Santis, pp. 149–175. Roma: Quasar.

———. 1986. "A Che Serve Scavare una Necropoli. Viaggio Attraverso le Trasformazioni di una Comunita Laziale e della sua Struttura Sociale dal IX al VII Secolo a.C." *Sapere* 52(1):21–30.

Bietti Sestieri, Anna-Maria, A. De Santis, and Loredana Savadei. 1988. "The Relevance of Anthropological Data for the Identification of Uncommon Social Roles: The Case of the Iron Age Cemetery of Osteria dell'Osa (Rome)." *Rivista di Antropologia* 66 (suppl.):349–380.

Bikle, D. D., E. Jee, B. P. Halloran, and J. G. Haddad. 1984. "Free 1,25 Vitamin D Levels in Serum from Normal Subjects, Pregnant Subjects and Subjects with Liver Disease." *Journal of Clinical Investigations* 74:1966–1971.

Billewicz, W. Z., H. M. Fellowes, and C. A. Hytten. 1976. "Comments on the Critical Metabolic Mass and the Age of Menarche." *Annals of Human Biology* 3:51.

Björntorp, P. 1987. "Adipose Tissue Distribution and Morbidity." *Recent Advances in Obesity Research* 5:60–65.

Björntorp, P., U. Smith, and P. Lonnroth, eds. 1988. "Health Implications of Regional Obesity." *Acta Medica Scandinavia* 723 (suppl.).

Blackburn, M. W., and D. H. Calloway. 1979. "Energy Expenditure and Consumption of Mature, Pregnant and Lactating Women." *Journal of American Dietary Association* 69:29–37.

Blurton Jones, N. G. 1989. "The Costs of Children and the Adaptive Scheduling of Births: Towards a Sociobiological Perspective on Demography." In *The Sociobiology of Sexual and Reproductive Strategies*, ed. E. Voland, C. Vogel, and A. Rasa, pp. 265–282. London: Chapman and Hall.

Blurton Jones, N. G., K. Hawkes, and J. O'Connell. 1989. "Measuring and Modelling Costs of Children in Two Foraging Societies." In *Comparative Socioecology: The Behavioral Ecology of Humans and Other Mammals*, ed. V. Standen and R. Foley, pp. 367–390. London: Blackwell Scientific Publications.

Bock, Walter J. 1980. "The Definition and Recognition of Biological Adaptation." *American Zoologist* 20:217–227.

———. 1989. "Organisms as Functional Machines: A Connectivity Explanation." *American Zoologist* 29:1119–1132.

Bock, Walter, J., and Gerd von Wahlert. 1965. "Adaptation and the Form-Function Complex." *Evolution* 19:269–299.

Boesch, Christophe. 1991a. "The Effects of Leopard Predation on Grouping Patterns in Forest Chimpanzees." *Behaviour* 117:220–242.

———. 1991b. "Teaching among Wild Chimpanzees." *Animal Behaviour* 41:530–532.

Boesch, Christophe, and H. Boesch. 1981. "Sex Differences in the Use of Natural Hammers by Wild Chimpanzees: A Preliminary Report." *Journal of Human Evolution* 10:585–593.

———. 1984. "Possible Causes of Sex Differences in the Use of Natural Hammers by Wild Chimpanzees." *Journal of Human Evolution* 13:415–440.

———. 1989. "Hunting Behavior of Wild Chimpanzees in the Tai National Park." *American Journal of Physical Anthropology* 78:547–573.

———. 1990. "Tool Use and Tool Making in Wild Chimpanzees." *Folia Primatologica* 54:86–99.

Boinski, S. 1988. "Sex Differences in the Foraging Behavior of Squirrel Monkeys in a Seasonal Habitat." *Behavioral Ecology and Sociobiology* 23:177–186.

Bolton, C., R. Bolton, L. Gross, A. Koehl, C. Michelson, R. L. Munroe, and R. H. Munroe. 1976. Pastoralism and Personality: An Andean Replication. *Ethos* 4:463–481.

Boness, D. J., O. T. Oftedal, and K. A. Ono. 1991. The Effect of El Niño on Pup Development in the California Sea Lion (*Zalophus californianus*). I. Early Postnatal Growth. In *Pinnipeds and El Niño: Responses to Environmental Stress*, ed. F. Trillmich and K. A. Ono, pp. 173–179. Berlin: Springer-Verlag.

Bongaarts, John. 1980. "Does Malnutrition Affect Fertility? A Summary of Evidence." *Science* 208: 564–569.

Bongaarts, John, and Robert C. Potter. 1983. *Fertility, Biology and Behavior: An Analysis of the Proxi*mate Determinants. San Diego: Academic Press.

Bonte, M., and H. van Balen. 1969. "Prolonged Lactation and Family Spacing in Rwanda." *Journal of Biosocial Science* 1:97.

Bonte, M., E. Akingeneye, M. Gashakamba, E. Mbarutso, and M. Nolens. 1974. "Influence of the Socio-Economic Level on the Conception Rate During Lactation." *International Journal of Fertility* 19:97.

Borchert, C. M. 1985. "A Critique of Neo-Darwinism and Its Implications for the Evolution of Language." Ph.D. diss., University of California, Santa Cruz.

Borchert, C. M., and A. L. Zihlman. 1990. "The Ontogeny and Phylogeny of Symbolizing." In *The Life of Symbols*, ed. M. LeC. Foster and L. J. Botscharow, pp. 15–44. Boulder, CO: Westview Press.

Borgognini Tarli, Silvana M., and Elisabetta Marini. 1990. *Annotated Bibliography on Sexual Dimorphism in Primates.* Pisa: E.T.S.

Borgognini Tarli, Silvana M., and Melchiorre Masali. 1983. "Dimorfismo Sessuale e Ruoli. Un Modello Microevolutivo (Egiziani Predinastici e Dinastici)." *Antropologia Contemporanea* 6:179–192.

Borgognini Tarli, Silvana M., and Francesca Mazzotta. 1986. "Ethnogenese der Italiker aus der Sicht der Anthropologie. Physical Anthropology of Italy from the Bronze Age to the Barbaric Age." In *Ethnogenese der Europäischen Völker*, ed. W. Bernhard and A. Kandler Palsson, pp. 147–172. Stuttgart: G. Fischer.

Borgognini Tarli, Silvana M., and Elena Repetto. 1986a. "Methodological Considerations on the Study of Sexual Dimorphism in Past Human Populations." *Human Evolution* 1:51-66.

———. 1986b. "Skeletal Indicators of Subsistence Patterns and Activity Regime in the Mesolithic Sample from Grotta dell'Uzzo (Trapani, Sicily): A Case Study." *Human Evolution* 1:331–352.

———. 1989. "Sexual Dimorphism in Europe during the Mesolithic Transition: Methodological and Applied Aspects." In *Hominidae*, ed. G. Giacobini, pp. 459–466. Milano: Jaca Book.

———. 1997. "Sex Differences in Human Populations: Change through Time." In *The Evolving Female: A Life-History Perspective*, ed. M. E. Morbeck, A. Galloway, and A. L. Zihlman, chap. 14. Princeton, NJ: Princeton University Press.

Borgognini Tarli, Silvana M., Melchiorre Masali, and Antonella Aimar. 1987. "Dimorfismo Sessuale Nell'antico Egitto: Popolazioni Predinastiche e Dinastiche a Confronto." *Antropologia Contemporanea* 10:29–41.

Borkan, G. A., and A. H. Norris. 1977. "Fat Redistribution and the Changing Body Dimensions of the Adult Male." *Human Biology* 49:495–514.

Bouchard, C. 1987. "Genetics of Body Fat, Energy Expenditure and Adipose Tissue Metabolism." *Recent Advances in Obesity Research* 5:6–25.

———. 1990. "Variation in Body Fat: The Contribution of the Genotype." In *Obesity: Towards a Molecular Approach*, ed. G. A. Bray, D. Ricquier, and B. M. Spiegelman, pp. 17–28. New York: Wiley/Liss.

Bouchard, C., and F. E. Johnston, eds. 1988. *Fat Distribution during Growth and Later Health Outcomes.* New York: Alan R. Liss.

Bouillon, R. 1983. "Vitamin D Metabolites in Human Pregnancy." In *Perinatal Calcium and Phosphorus Metabolism*, ed. M. F. Holick, T. K. Gray, and C. S. Anast, pp. 291–300. New York: Elsevier.

Boutin, Stan. 1990. "Food-Supplementation Experiments with Terrestrial Vertebrates: Patterns, Problems, and the Future." *Canadian Journal of Zoology* 68:203–220.

Bowlby, John. 1969. *Attachment and Loss: I. Attachment.* International Psychological Library 79. London: Hogarth.

———. 1973. *Attachment and Loss.* Vol. 2. *Separation: Anxiety and Anger.* London: Hogarth Press.

———. 1982. *Attachment and Loss.* Vol. 1. *Attachment.* 2d ed. London: Hogarth Press.

Box, Hilary O. 1975. "A Social Developmental Study of Young Monkeys (*Callithrix jacchus*) within a Captive Family Group." *Primates* 16: 419–435.

———, ed. 1991. *Primate Responses to Environmental Change.* New York: Chapman and Hall.

Boyce, Mark S., ed. 1988. *Evolution of Life Histories of Mammals: Theory and Pattern.* New Haven, CT: Yale University Press.

Boyd, R., and P. J. Richerson. 1985. *Culture and the Evolutionary Process.* Chicago: University of Chicago Press.

Boyde, Alan. 1990. "Developmental Interpretations of Dental Microstructure." In *Primate Life History and Evolution*, ed. C. J. DeRousseau, pp. 229–267. New York: Wiley-Liss.

Boyden, T. W., R. W. Pamenter, P. Stanforth, T. Rotkis, and J. H. Wilmore. 1983. "Sex Steroids and Endurance Running in Women." *Fertility and Sterility* 39:629.

Brace, C. Loring. 1973. "Sexual Dimorphism in Human Evolution." *Yearbook of Physical Anthropology* 16:31-49.

Brace, C. Loring, and Alan S. Ryan. 1980. "Sexual Dimorphism and Human Tooth Size Differences." *Journal of Human Evolution* 9:417–435.

Bray, G. A. 1988. "Role of Fat Distribution During Growth and Its Relationship to Health." *American Journal of Clinical Nutrition* 47:551–552.

Bremerman, H. J. 1979. "Theory of Spontaneous Cell Fusion. Sexuality in Cell Populations as an Evolutionary Stable Strategy. Applications to Immunology and Cancer." *Journal of Theoretical Biology* 76:311–334.

———. 1987. "The Adaptive Significance of Sexuality." In *The Evolution of Sex and Its Consequences*, ed. S. C. Stearns, pp. 135–161. Basel: Birkhauser Verlag.

Breslau, N. A., and J. E. Zerwekh. 1986. "Relationship of Estrogen and Pregnancy to Calcium Homeostasis in Pseudohypoparathyroidism." *Journal of Clinical Endocrinology and Metabolism* 62: 45–51.

Brewer, M. M., M. R. Bates, and L. P. Vannoy. 1989. "Postpartum Changes in Maternal Weight and Body Fat Depots in Lactating vs. Nonlactating Women." *American Journal of Clinical Nutrition* 49:259–265.

Brighton, S. W. 1988. "The Prevalence of Rheumatoid Arthritis in a Rural African Population." *Journal of Rheumatology* 15:405–408.

Brizzee, K. R., and W. P. Dunlap. 1986. "Growth." In *Comparative Primate Biology.* Vol. 3. *Reproduction and Development*, ed. W. R. Dukelow and J. Erwin, pp. 363–413. New York: Alan R. Liss.

Brockelman, W. Y. and D. Schilling. 1984. "Inheritance of Stereotyped Gibbon Calls." *Nature* 312:634–636.

Bromage, T. 1992. "The Ontogeny of *Pan troglodytes* Craniofacial Architectural Relationships and Implications for Early Hominids." *Journal of Human Evolution* 231:235–251.

Bromage, T., and M. C. Dean 1985. "Re-Evaluations of the Age at Death of Immature Fossil Hominids." *Nature* 317:525–527 .

Brommage, R., and H. F. DeLuca. 1985. "Regulation of Bone Mineral Loss During Lactation." *American Journal of Physiology* 248:E182–187.

Bronson, F. H. 1989. *Mammals and Reproductive Biology.* Chicago: University of Chicago Press.

Brooks, Daniel, and Deborah A. McLennan. 1991. *Phylogeny, Ecology, and Behavior: A Research Program in Comparative Biology.* Chicago: University of Chicago Press.

Brown, J. K. 1973. "The Subsistence Activities of Women and the Socialization of Children." *Ethos* 1:413–423.

Brown, P. J., and M. Konner. 1987. "An Anthropological Perspective on Obesity." *Annals of the New York Academy of Sciences* 499:29–46.

Bruner, J., and H. Haste, ed. 1987. *Making Sense: The Child's Construction of the World.* New York: Methuen.

Bruton, Michael N., ed. 1989a. *Alternative Life-History Styles of Animals.* Boston: Kluwer Academic.

———. 1989b. "The Ecological Significance of Alternative Life-History Styles." In *Alternative Life-History Styles of Animals*, ed. M. N. Bruton, pp. 503–533. Boston: Kluwer Academic.

Bucht, E., M. Telenius-Berg, G. Lundell, and H.-E. Sjoberg. 1986. "Immunoextracted Calcitonin in Milk and Plasma for Total Thyroidectomized Women: Evidence of Monomeric Calcitonin in Plasma during Pregnancy and Lactation." *Acta Endocrinologica, Copenhagen* 113:529–535.

Burns, T. P. 1992. "Adaptedness, Evolution, and a Hierarchical Concept of Fitness." *Journal of Theoretical Biology* 154:219–237.

Burton, Frances D. 1972. "An Analysis of the Muscular Limitation on Opposability in Seven Species of Cercopithecinae." *American Journal of Physical Anthropology* 36:169–188.

Butte, N. F., and C. Garza. 1986. "Anthropometry in the Appraisal of Lactation Performance among Well-Nourished women." In *Human Lactation 2: Maternal and Environmental Factors*, ed. M. Hamosh and A. S. Goldman, pp. 61–67. New York: Plenum Press.

Byrne, R., and A. Whiten. 1988. *Machiavellian Intelligence.* Oxford: Oxford University Press.

Calder, William A., III. 1984. *Size, Function and Life History.* Cambridge: Harvard University Press.

Cameron, J. L. 1989a. "Influence of Nutrition on the Hypothalamic-Pituitary-Gonadal Axis in Primates." In *The Menstrual Cycle and Its Disorders: Influence of Nutrition, Exercise and Neurotransmitters*, ed. K. M. Pirke, W. Wuttke, and V. Schweiger, pp. 66–78. Berlin: Springer-Verlag.

———. 1989b. "Nutritional and Metabolic Determinants of GNRH Secretion in Primate Species." In *Control of the Onset of Puberty III*, ed. H. A. Delemarre-van de Waal et al., pp. 275–284. New York: Elsevier Science Publishers.

Campaigne, B. N., V. L. Katch, P. Freedson, and S. Sady. 1979. "Measurement of Breast Volume in Females: Description of a Reliable Method." *Annals of Human Biology* 6:363–367.

Campbell, O. 1987. "A Prospective Study of Breast-Feeding Patterns and the Resumption of Ovarian Activity Postpartum." Ph.D. Diss., Johns Hopkins University.

Caniggia, A., R. Nuti, M. Galli, F. Lore, V. Turchetti, and G. A. Righi. 1984. "The Hormonal Form of Vitamin D in the Pathophysiology and Therapy of Postmenopausal Osteoporosis." *Journal of Endocrinological Investigation* 7:373–378.

Cann, C. E. 1989. "Pregnancy and Lactation Cause Reversible Trabecular Bone Loss in Humans." *Journal of Bone and Mineral Research* 4:53–84.

Cant, J.G.H. 1981. "Hypothesis for the Evolution of Human Breasts and Buttocks." *American Naturalist* 117:199–204.

———. 1987a. "Positional Behavior of Female Bornean Orangutans (*Pongo pygmaeus*)." *American Journal of Primatology* 12:71–90.

———. 1987b. "Effects of Sexual Dimorphism in Body Size on Feeding Postural Behavior of Sumatran Orangutans (*Pongo pygmaeus*)." *American Journal of Physical Anthropology* 74:143–148.

Carey, J. W. 1988. "Health, Social Support, and Social Networks in a Rural Andean Community of Southern Peru." Ph.D. diss., University of Massachusetts, Amherst.

Carpenter, C. J., and A. Huston-Stein. 1980. "Activity Structure and Sex-Typed Behavior in Preschool Children." *Child Development* 51:862–872.

Carroll, Lewis. 1872. *Through the Looking Glass, and What Alice Found There.* London: Macmillan.

Carter, D. B., and L. A. McCloskey. 1983. "Peers and the Maintenance of Sex-Typed Behavior: The Development of Children's Conceptions of Cross-Gender Behavior in Their Peers." *Social Cognition* 2:294–314.

Carter, Dennis R., Marcy Wong, and Tracy E. Orr. 1991. "Musculoskeletal Ontogeny, Phylogeny, and Functional Adaptation." *Journal of Biomechanics* 24 (suppl.): 3–16.

Cartmill, M. 1974. "Rethinking Primate Origins." *Science* 184:436–443.

———. 1992. " New Views on Primate Origins." *Evolutionary Anthropology* 1:105–111.

Casimir, M. J. 1975. "Feeding Ecology and Nutrition of an Eastern Gorilla Group in the Mt. Kahuzi Region *(Republique du Zaire)*." *Folia Primatologica* 24:81–136.

Cateni, Gabriele. n.d. *Etruschi: Scene di Vita Quotidiana*. Pisa: Pacini.

Cattet, M. 1990. "Predicting Nutritional Condition in Black Bears and Polar Bears on the Basis of Morphological and Physiological Measurements." *Canadian Journal of Zoology* 68:32–39.

Caughley, G. 1970. "Eruptions of Ungulate Populations, with Emphasis on the European Thar in New Zealand." *Ecology* 51:53–72.

Cavalli-Sforza, L. L., and W. F. Bodmer 1971. *The Genetics of Human Populations*. San Francisco: W. H. Freeman.

Chakraborty R., and P. P. Majumder. 1982. "On Bennett's Measure of Sex Dimorphism." *American Journal of Physical Anthropology* 59:295–298.

Chambers, T. J. 1982. "Osteoblasts Release Osteoclasts from Calcitonin-Induced Quiescence." *Journal of Cell Science* 57:247–260.

Charlesworth, Brian. 1973. "Selection in Populations with Overlapping Generations. V. Natural Selection and Life Histories." *American Naturalist* 107:303–311.

———. 1980. *Evolution of Age-Structured Populations*. New York: Cambridge University Press.

———. 1990. "Natural Selection and Life-History Patterns." In *Genetic Effects on Aging II*, ed. D. E. Harrison, pp. 21–40. Caldwell, NJ: Telford Press.

Charnov, Eric L. 1991. "Evolution of Life History Variation among Female Mammals." *Proceedings of the National Academy of Sciences U.S.A.* 88: 1134–1137.

———. 1993. *Life History Variants. Some Explorations of Symmetry in Evolutionary Ecology*. New York: Oxford University Press.

Charnov, Eric L., and David Berrigan. 1993. "Why Do Female Primates Have Such Long Lifespans and So Few Babies? Or Life in the Slow Lane." *Evolutionary Anthropology* 2:191–194.

Chavez, A., and C. Martinez. 1973. "Nutrition and Development of Infants from Poor Rural Areas. III. Maternal Nutrition and Its Consequences on Fertility." *Nutritional Reports International* 7: 1–8.

Cheney, Dorothy L. 1978. "Interactions of Immature Male and Female Baboons with Adult Females." *Animal Behaviour* 26:389–408.

———. 1987. "Interactions and Relationships between Groups." In *Primate Societies*, ed. B. B. Smuts, D. L. Cheney, R. M. Seyfarth, R. W. Wrangham, and T. T. Struhsaker, pp. 267–281. Chicago: University of Chicago Press.

Cheney, Dorothy L., and R. M. Seyfarth. 1990. *How Monkeys See the World*. Chicago: University of Chicago Press.

Cheney, Dorothy L., Phyllis C. Lee, and Robert M. Seyfarth. 1981. "Behavioral Correlates of Nonrandom Mortality among Free-Ranging Female Vervet Monkeys." *Behavioral Ecology and Sociobiology* 5:262–278.

Cheney, Dorothy L., Robert M. Seyfarth, Sandy J. Andelman, and Phyllis C. Lee. 1988. "Reproductive Success in Vervet Monkeys." In *Reproductive Success: Studies of Individual Variation in Contrasting Breeding Systems*, ed. T. H. Clutton-Brock, pp. 384–402. Chicago: University of Chicago Press.

Chepko-Sade, B. Diane, and Zuleyma Tang Halpin. 1987. *Mammalian Dispersal Patterns: The Effects of Social Structure on Population Genetics*. Chicago: University of Chicago Press.

Chevalier-Skolnikoff, S., B. M. F. Galdikas, and A. Z. Skolnikoff. 1982. "The Adaptive Significance of Higher Intelligence in Wild Orangutans: A Preliminary Report." *Journal of Human Evolution* 11:639–652 .

Chisholm, James S. 1993. "Death, Hope and Sex: Life-History Theory and the Development of Reproductive Strategies." *Current Anthropology* 34: 1–24.

Chism, Janice, Thelma Rowell, and Dana Olson. 1984. "Life History Patterns of Female Patas Monkeys." In *Female Primates: Studies by Women Primatologists*, ed. M. F. Small, pp. 175–190. New York: Alan R. Liss.

Chivers, David J. 1974. *The Siamang in Malaya. Contributions to Primatology*, Vol. 4. Karger: Basel.

———. 1977. "The Feeding Behavior of Siamang (*Symphalangus syndactylus*)." In *Primate Ecology: Studies of Feeding and Ranging Behavior of Lemurs, Monkeys and Apes*, ed. T. H. Clutton-Brock, pp. 355–382. New York: Academic Press.

Christiansen, C., B. J. Riis, and P. Rodbro. 1987. "Prediction of Rapid Bone Loss in Postmenopausal Women." *Lancet* 8542:1105–1108.

Christiansen, C., P. Rodro, and B. Heinild. 1976. "Unchanged Total Body Calcium in Normal Human Pregnancy." *Acta Obstetrica et Gynecologica Scandinavica* 55:141–143.

Christiansen, C., B. J. Riis, L. Nilas, P. Rodbro, and L. Deftos. 1985. "Uncoupling of Bone Formation and Resorption by Combined Oestrogen and Progestagen Therapy in Postmenopausal Osteoporosis." *Lancet* 2:800–801.

Clark, Anne B. 1991. "Individual Variation in Responsiveness to Environmental Change." In *Primate Responses to Environmental Change*, ed. H. O. Box, pp. 91–110. New York: Chapman and Hall.

Clark, J. D. and J. W. K. Harris. 1985. "Fire and Its Roles in Early Hominid Lifeways." *African Archaeological Review* 3:3–27.

Clarke, Margaret R., and Kenneth E. Glander. 1984. "Reproductive Success in a Group of Free-Ranging Howling Monkeys." In *Female Primates: Studies by Women Primatologists*, ed. M. F. Small, pp. 111–126. New York: Alan R. Liss.

Clarys, J. P., A. D. Martin, and D. T. Drinkwater. 1984. "Gross Tissue Weights in the Human Body by Cadaver Dissection." *Human Biology*, 56:459–473.

Clutton-Brock, Timothy H. 1977. "Some Aspects of Intraspecific Variation in Feeding and Ranging Behaviour in Primates." In *Primate Ecology: Studies of Feeding and Ranging Behaviour in Lemurs, Monkeys and Apes*, ed. T. H. Clutton-Brock, pp. 539–556. London: Academic Press.

———, ed. 1988. *Reproductive Success: Studies of Individual Variation in Contrasting Breeding Systems.* Chicago: University of Chicago Press.

———. 1991. *The Evolution of Parental Care.* Princeton, NJ: Princeton University Press.

Clutton-Brock, T. H., F. E. Guiness, and S. D. Albon. 1982. *Red Deer: Behavior and Ecology of Two Sexes.* Chicago: University of Chicago Press.

Clutton-Brock, T. H., Paul Harvey, and B. Rudder. 1977. "Sexual Dimorphism, Socionomic Sex Ratio, and Body Weight in Primates." *Nature* 269:797–799.

Coe, C. L., A. C. Connolly, H. C. Kraemer, and S. Levine. 1973. "Reproductive Development and Behavior of Captive Female Chimpanzees." *Primates* 20:571–582 .

Coe, Christopher L., Sandra G. Weiner, Leon T. Rosenberg, and Seymour Levine. 1985. "Endocrine and Immune Responses to Separation and Maternal Loss in Nonhuman Primates." In *The Psychobiology of Attachment and Separation*, ed. M. Reite and T. Field, pp. 163–199. New York: Academic Press.

Coe, Christopher L., Leon T. Rosenberg, and Seymour Levine. 1988. "Immunological Consequences of Psychological Disturbance and Maternal Loss in Infancy." In *Advances in Infancy Research.* Vol. 5., ed. C. Rovee-Collier et al., pp. 98–136. Norwood, NJ: Ablex.

Coehlo, Anthony M. 1974. "Socio-Bioenergetics and Sexual Dimorphism in Primates." *Primates* 15:263–369.

———. 1985. "Baboon Dimorphism: Growth in Weight, Length and Adiposity from Birth to 8 Years of Age." In *Nonhuman Primate Models for Human Growth and Development*, ed. E. Watts, pp. 125–153. New York: Alan R. Liss.

———. 1986. "Time and Energy Budgets." In *Comparative Primate Biology.* Vol. 2. *Behavior, Conservation and Ecology*, ed. G. Mitchell and J. Erwin, pp. 141–168. New York: Alan R. Liss.

Coehlo, Anthony M., D. M. Glassman, and C. A. Bramblett. 1984. "The Relation of Adiposity and Body Size to Chronological Age in Olive Baboons." *Growth* 48:445–454.

Cole, Lamont C. 1954. "The Population Consequences of Life History Phenomena." *Quarterly Review of Biology* 29 (2):103–137.

Cole, M. 1985. "The Zone of Proximal Development: Where Culture and Cognition Create Each Other." In *Culture, Comrnunication, and Cognition: Vygotskian Perspectives*, ed. J. V. Wertsch, pp. 146–161. Cambridge: Cambridge University Press.

Collins, D. A. and W. C. McGrew. 1988. "Habitats of Three Groups of Chimpanzees (*Pan troglodytes)* in Western Tanzania Compared." *Journal of Human Evolution* 17:553–574.

Colussi, G., M. Surian, M. E. DeFerrari, G. Rombola, and L. Minetti. 1987. "The Changes in Plasma Diffusible Levels and Renal Tubular Handling of Magnesium during Pregnancy: A Longitudinal Study." *Bone and Mineral* 2:311–319.

Condry, J., and S. Condry. 1976. "Sex Differences: A Study of the Eye of the Beholder." *Child Development* 47:812–819.

Cone, C. A. 1979. "Personality and Subsistence: Is the Child the Parent to the Person?" *Ethnology* 18:291–301.

Conforti, A., Y. Biale, M. Levi, S. Shany, R. Shainkin-Kestenbaum, and G. M. Berlyne. 1980. "Changes in Serum Calcitonin, Parathyroid Hormone and 25-Hydroxycholecalciferol Levels in

Pregnancy and in Labor." *Mineral and Electrolyte Metabolism* 3:323–328.

Conroy, Glenn C., and C. James Mahoney. 1991. "Mixed Longitudinal Study of Dental Emergence in the Chimpanzee, *Pan troglodytes* (Primates, Pongidae)." *American Journal of Physical Anthropology* 86:243–254.

Constandse-Westermann, Trinette S., and Raymond R. Newell. 1982. "Mesolithic Trauma: Demographical and Chronological Trends in Western Europe." In *Proceedings of the Fourth European Meeting of the Paleopathology Association*, ed. G. T. Haneveld and W. R. K. Perizonius, pp. 70–76. Middelburg Antwerpen.

———. 1988. "Patterns of Extraterritorial Ornament Dispersion: An Approach to the Measurement of Mesolithic Exogamy." *Rivista di Antropologia* 66 (suppl.):75–126.

Conton, L. 1985. "Social, Economic and Ecological Parameters of Infant Feeding in Usino, Papua New Guinea." *Ecology of Food and Nutrition* 16:39–54.

Coolidge, H. 1933. "*Pan paniscus*: Pygmy Chimpanzee from South of the Congo River." *American Journal of Physical Anthropology* 18:1–57.

Coppinger, Raymond P., and Charles Kay Smith. 1990. "A Model for Understanding the Evolution of Mammalian Behavior." *Current Mammalogy*, ed. H. H. Genoway, pp. 335–374. New York: Plenum Press.

Cords, M. 1986. "Interspecific and Intraspecific Variation in Diet of Two Forest Guenons, *Cercopithecus ascanius* and *Cercopithecus mitus*." *Journal of Animal Ecology* 55:811–827.

Corghi, E., S. Ortolani, M. L. Bianchi, P. Favini, P. Vigo, and E. E. Polli. 1984. "Basal Plasma Levels of Calcitonin and Bone Mineral Mass in Normal and Uremic Women, Effect of Menopause." *Biochemical Pharmacology* 38:263–265.

Correnti, Venerando. 1960. "Sulla Valutazione dei Caratteri Morfologici. Ricerche su Giovani Sportivi." *Rivista di Antopologia* 67:58–88.

Corruccini, Robert S. 1972. "Allometry Correction in Taximetrics." *Systematic Zoology* 21:375–383.

Costa, D. P., B. J. Le Boeuf, A. C. Huntley, and C. L. Ortiz. 1986. "The Energetics of Lactation in the Northern Elephant Seal, *Mirounga angustirostris*." *Journal of Zoology* 209:21–33.

Costa, D. P., F. Trillmich, and J. P. Croxall. 1988. "Intraspecific Allometry of Neonatal Size in the Antarctic Fur Seal *(Arctocephalus gazella)*." *Behavioral Ecology and Sociobiology* 22:361–364.

Costa, D. P., G. A. Antonelis, and R. L. DeLong. 1991. "Effects of El Niño on the Foraging Energetics of the California Sea Lion." In *Pinnipeds and El Niño Responses to Environmental Stress*, ed. F. Trillmich and K. A. Ono, pp. 156–165. Heidelberg: Springer-Verlag.

Cox, C. R., and B. J. Le Boeuf. 1977. "Female Incitation of Male Competition: A Mechanism in Sexual Selection." *American Naturalist* 111:317–335.

Cramer, D. L. 1977. "Craniofacial Morphology of *Pan paniscus*: A Morphometric and Evolutionary Appraisal." *Contributions to Primatology*. Vol. 10. Basel: Karger.

Cramer, D. L., and A. L. Zihlman. 1978. "Sexual Dimorphism in the Pygmy Chimpanzee, *Pan paniscus*." In *Recent Advances in Primatology*. Vol. 3. *Evolution*, ed. D. J. Chivers and K. A. Joysey, pp. 487–490. London: Academic Press.

Crompton, A. W., and F. A. Jenkins. 1973. "Mammals from Reptiles. A Review of Mammalian Origins." *Annual Review of Earth and Planetary Science* 1:131–155.

Crompton, A. W., and P. Parker. 1978. "Evolution of the Mammalian Masticatory Apparatus." *Annual Review of Earth and Planetary Science* 1: 131–155.

Cronin, J. E., and V. M. Sarich. 1976. "Molecular Evidence for Dual Origin of Mangabeys among Old World Monkeys." *Nature* 260:700–702.

Cronin J. E., V. M. Sarich, and O. Ryder. 1984. "Molecular Evolution and Speciation in the Lesser Apes." In *The Lesser Apes*, ed. H. Preuschoft; D. J. Chivers, W. Brockelman, and N. Creel, pp. 467–485. Edinburgh: Edinburgh University Press.

Cumming, D. C., G. D. Wheeler, and V. J. Harber. 1994. "Physical Activity, Nutrition, and Reproduction." *Annals of the New York Academy of Sciences* 709:287–298.

Cummings, S. R. 1987. "Epidemiology of Osteoporotic Fractures: Selected Topics." *Osteoporosis: Current Concepts*, ed. A. F. Roche and J. D. Gussler, pp. 3–8. Columbus, OH: Ross Laboratories.

Cushard, W. G., M. A. Creditor, J. M. Canterbury, and E. Reiss. 1972. "Physiologic Hyperparathyroidism in Pregnancy." *Journal of Clinical Endocrinology* 34:767–771.

Cutler, R. G. 1976. "Evolution in Longevity in Primates." *Journal of Human Evolution* 5:169–202.

———. 1981. "Life-Span Extension." In *Aging: Bi-*

ology and Behavior, ed. J. L. McGaugh and S. B. Kiesler, pp. 31–76. New York: Academic Press.

Dahlberg, F. 1981. *Woman the Gatherer.* New Haven, CT: Yale University Press.

Darwin, Charles. 1859. *On the Origin of Species by Means of Natural Selection or the Preservation of Favored Races in the Struggle for Life.* London: Murray.

———. 1861. "On the Two Forms, or Dimorphic Condition, in the Species of *Primula*, and on Their Remarkable Sexual Relations." In *The Collected Papers of Charles Darwin*, ed. P. H. Barrett, pp. 45–63. Chicago: University of Chicago Press, 1977.

———. 1871 *Descent of Man and Selection in Relation to Sex.* New York: Modern Library.

———. 1872. *The Expression of the Emotions in Man and Animals.* London: D. Appleton.

Datta, S. B., and G. Beauchamp. 1991. "Effects of Group Demography on Dominance Relationships among Female Primates. I. Mother-Daughter and Sister-Sister Relations." *American Naturalist* 138:201–226.

DaVanzo, J., W. P. Butz, and J. P. Habicht. 1983. "How Biological and Behavioral Influences on Mortality in Malaysia Vary during the First Year of Life." *Population Studies* 37:381.

Davis, D. 1961. "Origin of the Mammalian Feeding Mechanism." *American Zoologist* 1:229–234.

Davis, K., and J. Blake. 1956. "Social Structure and Fertility: An Analytical Framework." *Economic Development and Cultural Change* 4:211–235.

Dawkins, Richard. 1976. *The Selfish Gene.* Oxford: Oxford University Press.

———. 1986. *The Blind Watchmaker. Why the Evidence of Evolution Reveals a Universe without Design.* New York: W.W. Norton.

Dean, M. Christopher. 1989. "The Developing Dentition and Tooth Structure in Hominoids." *Folia Primatologica* 53:160–176.

Delgado, H., A. Lechtig, R. Martorell, E. Brineman, and R. E. Klein. 1978. "Nutrition, Lactation, and Postpartum Amenorrhea." *American Journal of Clinical Nutrition* 31:322–337.

Deloache, J. S., J. Cassidy, and C. J. Carpenter. 1987. "The Three Bears Are All Boys: Mothers' Gender Labeling of Neutral Picture Book Characters." *Sex Roles* 17:163–178.

DeLong, R. L., G. A. Antonelis, C. W. Oliver, B. S. Stewart, M. C. Lowry, and P. K. Yochem. 1991. "Effects of the 1982–83 El Niño on Several Population Parameters and Diet of California Sea Lions on the California Channel Islands." In *Pinnipeds and El Niño Responses to Environmental Stress*, ed. F. Trillmich and K. A. Ono, pp. 166–172. Heidelberg: Springer-Verlag.

De Robertis, Eddy M., Guillermo Oliver, and Christopher V. E. Wright. 1990. "Homeobox Genes and the Vertebrate Body Plan." *Scientific American* 263:46–52.

DeRousseau, C. Jean. 1985. "Aging in the Musculoskeletal System of Rhesus Monkeys: III. Bone Loss." *American Journal of Physical Anthropology* 68:157–168.

———, ed. 1990. *Primate Life History and Evolution.* New York: Wiley-Liss.

Deuster, P. A., S. B. Kyle, P. B. Moser, R. A. Vigersky, A. Singh, and E. B. Schoomaker. 1986. "Nutritional Intakes and Status of Highly Trained Amenorrheic and Eumenorrheic Women Runners." *Fertility and Sterility* 46:636.

Deutsch, C., M. P. Haley, and B. J. Le Boeuf. 1990. "Reproductive Effort of Male Northern Elephant Seals: Estimates from Mass Loss." *Canadian Journal of Zoology* 68:2580–2593.

Deutsch, C., D. E. Crocker, D. P. Costa, and B. J. Le Boeuf. 1994. "Sex and Age-Related Variation in Reproductive Effort of Northern Elephant Seals." In *Elephant Seals, Population Ecology, Behavior and Physiology*, ed. B. J. Le Boeuf and R. M. Laws, pp. 169–210. Berkeley: University of California Press.

DeVore, I., ed. 1965. *Primate Behavior.* New York: Holt Rinehart.

de Waal, Frans. 1982. *Chimpanzee Politics: Power and Sex among Apes.* New York: Harper and Row.

———. 1984. "Sex Differences in the Formation of Coalitions among Chimpanzees." *Ethology and Sociobiology* 5:239–255.

———. 1987. "The Integration of Dominance and Social Bonding in Primates." *Quarterly Review of Biology* 61:459–479.

———. 1989. *Peacemaking in Primates.* Cambridge: Cambridge University Press.

———. 1992. "Intentional Deception in Primates." *Evolutionary Anthropology* 1:86–92.

Diamond, J. M. 1987. "Causes of Death before Birth." *Nature* 329:487–488.

Diaz, S. 1989. "Determinants of Lactational Amenorrhea." *International Journal of Gynaecological Obstetrics* (suppl.) 1:83–89.

Dickemann, M. 1979. "The Reproductive Structure of Stratified Societies: A Preliminary Model." In *Evolutionary Biology and Human Social Organi-*

zation: An Anthropological Perspective, ed. N. A. Chagnon and W. Irons, pp. 331–367. North Scituate, MA: Duxbury Press.

Dixson, Alan F. 1981. *The Natural History of the Gorilla*. London: Weidenfeld and Nicolson.

Dobbing, J., ed. 1985. *Maternal Nutrition and Lactational Infertility*. Nestle Nutrition Workshop Series, Vol. 9. New York: Raven Press.

Dobzhansky, Theodosius. 1937. *Genetics and the Origin of Species*. New York: Columbia University Press.

Dollard, John. 1935. *Criteria for the Life History, with Analysis of Six Notable Documents*. New York: P. Smith.

Doran, Diane. 1989. "Chimpanzee and Pygmy Chimpanzee Positional Behavior: The Influence of Environment, Body Size, Morphology and Ontogeny on Locomotion and Posture." Ph.D. diss., State University of New York, Stony Brook.

Drake, T. S., R. A. Kaplan, and T. A. Lewis. 1979. "The Physiologic Hyperparathyroidism of Pregnancy: Is It Primary or Secondary?" *Obstetrics and Gynecology* 53:746–749.

Draper, P. 1973. "Crowding among Hunter Gatherers: The !Kung Bushmen." *Science* 182:301–303.

———. 1975a. "Cultural Pressure on Sex Differences." *American Ethnologist* 2:602–616.

———. 1975b. "!Kung Women: Contrasts in Sex Role Egalitarianism of Foraging and Sedentary Contexts." In *Toward an Anthropology of Women*, ed. Rayna Reiter, pp. 77–109. New York: Monthly Review Press.

———. 1976. "Social and Economic Constraints on !Kung Childhood." In *Kalahari Hunter Gatherers*, ed. R. B. Lee and I. DeVore, pp. 200–217. Cambridge, Mass.: Harvard University Press.

———. 1978. "The Learning Environment for Aggression and Anti-Social Behavior among the !Kung." In *Learning Non-Aggression*, ed. Ashley Montagu, pp. 31–53. New York: Oxford University Press.

———. 1991. "Room to Maneuver: !Kung Women Cope with Men." In *Sanctions and Sanctuary: Cultural Perspectives on the Beating of Wives*, ed. Jacquelyn Campbell, Dorothy Counts, and Judith K. Brown, pp. 43–61. Boulder, CO: Westview Press.

———. 1997. "Institutional, Evolutionary, and Demographic Contexts of Gender Roles: A Case Study of !Kung Bushmen." In *The Evolving Female: A Life-History Perspective*, ed. M. E. Morbeck, A. Galloway, and A. L. Zihlman, chap. 16. Princeton, NJ: Princeton University Press.

Draper, P., and A. Buchanan. 1992. "If you Have a Child You Have a Life: Demographic and Cultural Perspectives on Fathering in Old Age in !Kung Society." In *Father-Child Relations: Cultural and Biosocial Contexts*, ed. Barry Hewlett, pp. 132–152. New York: Aldine de Gruyter.

Draper, P., and E. Cashdan. 1988. "Technological Change and Child Behavior among the !Kung." *Ethnology* 27:339–365.

Draper, P., and Henry Harpending. 1982. "Father Absence and Reproductive Strategy: An Evolutionary Perspective." *Journal of Anthropological Research* 38:255–273.

Draper, P., and M. Kranichfeld. 1990. "Coming in from the Bush: The Significance of Household and Village Organization in Economic Change." *Human Ecology* 18(4): 363–384.

———. In press. "Temporal Trends in Hypergynous Out-Mating among !Kung Women and in Survival of Their Children."

Drickamer, L. C. 1974. "A Ten-Year Summary of Reproductive Data for Free-Ranging *Macaca mulatta*." *Folia Primatologica* 21:60–80.

Dukelow, W. Richard, and Kyle B. Dukelow. 1989. "Reproductive and Endocrinological Measures of Stress and Nonstress in Nonhuman Primates." *American Journal of Primatology* (suppl.) 1:17–24.

Dunbar, R. I. M. 1977. "Feeding Ecology of Gelada Baboons: A Preliminary Report." In *Primate Ecology*, ed. T. H. Clutton-Brock, pp. 251–273. New York: Academic Press.

———. 1980. "Demographic and Life History Variables of a Population of Gelada Baboons (*Theropithecus gelada*)." *Journal of Animal Ecology* 49:485–506.

———. 1987. "Demography and Reproduction." In *Primate Societies*, B. B. Smuts, D. L. Cheney, R. M. Seyfarth, R. W. Wrangham, and T. T. Struhsaker, pp. 240–249. Chicago: University of Chicago Press.

———. 1988. *Primate Social Systems*. New York: Comstock; London: Croom Helm.

Dunbar, R. I. M., and P. Dunbar. 1988. "Maternal Time Budgets of Gelada Baboons." *Animal Behaviour* 36:970–980.

Durant, Will. 1956. *The Story of Civilization: Our Oriental Heritage*. Italian edition. Milano: Mondadori.

———. 1957. *The Story of Civilization: Caesar and Christ*. Italian edition. Milano: Mondadori.

Durnin, J. V. G. A. 1978. "Indirect Calorimetry in Man: A Critique of Practical Problems." *Proceedings of the Nutrition Society* 37:5–12.

Durnin, J. V. G. A., and S. Drummond. 1988. "The Role of Working Women in a Rural Environment When Nutrition Is Marginally Adequate: Problems of Assessment." In *Capacity for Work in the Tropics*, ed. K. J. Collins and D. F. Roberts, pp. 77–83. Cambridge: Cambridge University Press.

Dyer, K. F. 1982. *Catching Up the Men. Women in Sport.* London: Junction Books Ltd.

Ebenman, Bo, and Lennart Persson. 1988. *Size-Structured Populations.* Berlin: Springer-Verlag.

Eisenberg, E. 1969. "Effects of Estrogen on Calcium Metabolism in Man." In *Metabolic Effects of Gonadal Hormones and Contraceptive Steroids*, ed. H. A. Salhanick, D. M. Kipnis, and R. L. Vande Wiele and others, pp. 503–517. New York: Plenum Press.

Eisenberg, J. F. 1981. *The Mammalian Radiations, Analysis of Trends in Evolution, Adaptation and Behavior.* Chicago: University of Chicago Press.

Eldredge, Niles. 1985. *Unfinished Synthesis: Biological Hierarchies and Modern Evolutionary Thought.* New York: Oxford University Press.

Eldredge, Niles, and Stanley N. Salthe. 1984. "Hierarchy and Evolution." In *Oxford Surveys in Evolutionary Biology*, Vol. 1, ed. R. Dawkins and M. Ridley, pp. 184–208. Oxford: Oxford University Press.

Ellefson, J. 1968. "Territorial Behavior in the Common White-Handed Gibbon *Hylobates lar* Linn." In *Primates: Studies in Adaptation and Variability*, ed. P. C. Jay, pp. 180–199. New York: Holt, Rinehart and Winston.

———. 1974. "A Natural History of White-Handed Gibbons in the Malayan Peninsula." *Gibbon and Siamang.* Vol. 3., ed. D. M. Rumbaugh, pp. 1–136. Basel: Karger.

Ellis, R. A., and W. Montagna. 1962. "The Skin of the Primates. VI. The Skin of the Gorilla (*Gorilla gorilla*)." *American Journal of Physical Anthropology* 20:79–94.

Ellison, P. T. 1982. "Skeletal Growth, Fatness and Menarcheal Age: A Comparison of Two Hypotheses." *Human Biology* 54:269.

———. 1989. "Human Ovarian Function and Reproductive Ecology: New Hypotheses." *American Journal of Physical Anthropology* 78:217.

———. 1990. "Human Ovarian Function and Reproductive Ecology: New Hypotheses." *American Anthropologist* 92:933–952.

———. 1994. "Salivary Steroids and Natural Variation in Human Ovarian Function." *Annals of the New York Academy of Sciences* 709:287–298.

Ellison, P. T., and C. Lager. 1986. "Moderate Recreational Running Is Associated with Lowered Salivary Progesterone Profiles in Women." *American Journal of Obstetrics and Gynecology* 154: 1000–1003.

Ellison, P. T., N. R. Peacock, and C. Lager. 1989. "Ecology and Ovarian Function among Lese Women of the Ituri Forest, Zaire." *American Journal of Physical Anthropology* 78:519–526.

Ellison, P. T., S. F. Lipson, M. T. O'Rourke, G. R. Bentley, A. M. Harrigan, C. Panter-Brick, and V. J. Vitzthum. 1993a. "Letter: Population Variation in Ovarian Function." *Lancet* 342:433–434.

Ellison, P. T., C. Panter-Brick, S. F. Lipson, and M. T. O'Rourke. 1993b. "The Ecological Context of Human Ovarian Function." *Human Reproduction* 8:2248–2258.

Ember, C. 1973. "Female Task Assignment and the Social Behavior of Boys." *Ethos* 1:424–439.

Emlen, S. T., and L. W. Oring. 1977. "Ecology, Sexual Selection, and the Evolution of Mating Systems." *Science* 197:215–223.

Endler, J. A. 1986. *Natural Selection in the Wild.* Princeton, NJ: Princeton University Press.

Enomoto, T. 1974. "The Sexual Behavior of Japanese Macaques." *Primates* 20:563–570.

Erchak, G. M. 1976. "The Nonsocial Behavior of Young Kpelle Children and the Acquisition of Sex Roles." *Journal of Cross-Cultural Psychology* 7: 223–234.

Ereshefsky, Marc. ed. 1992. *The Units of Evolution: Essays on the Nature of Species.* Boston: MIT Press.

Escobar, G. 1976. "Social and Political Structure of Nuñoa." In *Man in the Andes*, ed. P. T. Baker and M. A. Little, pp. 60–84. Stroudsburg, PA: Dowden, Hutchinson, and Ross.

Eslami, S. S. 1988. "Predictors of Ovulation in Breast-Feeding Women in Manila, The Philippines." Ph.D. diss., Johns Hopkins University.

Estioko-Griffin, A. 1985. "Women Hunters: The Implications for Pleistocene Prehistory and Contemporary Ethnography." In *Women in Asia and the Pacific*, ed. M. J. Goodman, pp. 61–81. Honolulu: University of Hawaii Press.

Estioko-Griffin, A., and P. Bion Griffin. 1981. "Woman the Hunter: The Agta." In *Woman the Gatherer*, ed. F. Dahlberg, pp. 121–151. New Haven, CT: Yale University Press.

Eveleth, P. B. 1975. "Differences between Ethnic

Groups in Sex Dimorphism of Adult Height." *Annals of Human Biology* 2:35–39.

Fa, John E., and Donald G. Lindburg, eds. *In press. Evolutionary Behavior of the Macaques.* Cambridge: Cambridge University Press.

Fagen, R. 1981. *Animal Play Behaviour.* Oxford: Oxford University Press.

Fairbanks, Lynn A., and M. T. McGuire. 1985. "Relationships of Vervet Monkeys with Sons and Daughters from One through Three Years of Age." *Animal Behavior* 33:40–50.

Falk, Dean. 1990. "Brain Evolution in *Homo*: The 'Radiator' Theory." *Behavioral and Brain Sciences* 13:333–381 .

———. 1992. *Braindance: New Discoveries about Human Origins and Brain Evolution.* New York: Henry Holt.

Fay, M. J., M. Agnagna, J. Moore, and R. Oko. 1989. "Gorillas *(Gorilla gorilla gorilla)* in the Likouala Swamp Forests of North Central Congo: Preliminary Data on Populations and Ecology." *International Journal of Primatology* 10:477–486.

Fedigan, L. M. 1992. *Primate Paradigms.* Montreal: Eden Press, 1982. Reprint, Chicago: University of Chicago Press.

———. 1997. "Changing Views of Female Life Histories." In *The Evolving Female: A Life-History Perspective*, ed. M. E. Morbeck, A. Galloway, and A. L. Zihlman, chap. 2. Princeton, NJ: Princeton University Press.

Fedigan, L. M., and Pamela J. Asquith, eds. 1991. *The Monkeys of Arashiyama: Thirty-Five Years of Research in Japan and the West.* Albany, NY: State University of New York Press.

Fedigan, L. M., and M. J. Baxter. 1984. "Sex Differences and Social Organizations in Free-Ranging Spider Monkeys (*Ateles geoffroy*)." *Primates* 25: 279–294.

Fedigan, L. M., and Mary S. M. Pavelka. 1994. "The Physical Anthropology of Menopause." In *Strength in Diversity: A Reader in Physical Anthropology*, ed. A. Herring and L. Chan, pp. 103–126. Toronto: Canadian Scholars Press.

Fedigan, L. M., and Lisa M. Rose. 1995. "Interbirth Interval Variation in Three Sympatric Species of Neotropical Monkey." *American Journal of Primatology* 37(1):9–24.

Fedigan, L. M., L. Fedigan, S. Gouzoules, H. Gouzoules and N. Koyama. 1986. "Lifetime Reproductive Success in Female Japanese Macaques." *Folia Primatologica* 47:143–157.

Fedigan, Laurence. 1991. "Life Span and Reproduction in Japanese Macaque Females." In *The Monkeys of Arashiyama: Thirty-Five Years of Research in Japan and the West*, ed. Linda M. Fedigan and Pamela J. Asquith, pp. 140–154. Albany: State University of New York Press.

Feldkamp, S. D., R. L. DeLong, and G. A. Antonelis. 1991. "Effects of the El Niño 1983 on the Foraging Patterns of California Sea Lions *(Zalophus californianus)* near San Miguel Island, California." In *Pinnipeds and El Niño Responses to Environmental Stress*, ed. F. Trillmich and K. A. Ono, pp. 146–155. Heidelberg: Springer-Verlag.

Ferro-Luzzi, A. 1988. "Marginal Energy Malnutrition: Some Speculations on Primary Energy Sparing Mechanisms." In *Capacity for Work in the Tropics*, ed. K. J. Collins and D. F. Roberts, pp. 141–164. Cambridge: Cambridge University Press.

Ferroni, M. A. 1980. "Urban Bias of Peruvian Food Policy." Ph.D. diss., Cornell University, Ithaca, NY.

Finch, Caleb E. 1990. *Longevity, Senescence, and the Genome.* Chicago: University of Chicago Press.

Fink, A. E., G. Fink, H. Wilson, J. Bennle, S. Carroll, and H. Dick. 1992. "Lactation, Nutrition and Fertility and the Secretion of Prolactin and Gonadotrophins in Mopan Mayan Women." *Journal of Biosocial Science* 24:35–52.

Finney, D. J. 1971. *Statistical Method in Biological Assay.* 2d ed. London: Griffin.

Fivush, R. 1989. "Exploring Sex Differences in the Emotional Content of Mother-Child Conversations About the Past." *Sex Roles* 20:675–691.

Flowerdew, J. R. 1987. *Mammals: Their Reproductive Biology and Population Ecology.* London: Edward Arnold.

Fooden J., and R. J. Izor. 1983. "Growth Curves, Dental Emergence Norms and Supplementary Morphological Observations in Known-Age Captive Orangutans." *American Journal of Primatology* 5:285–301.

Forbes, G. B. 1987. *Human Body Composition: Growth, Aging, Nutrition and Activity.* New York: Springer-Verlag.

Ford, C. S., and F. A. Beach. 1951. *Patterns of Sexual Behavior.* New York: Ace Books.

Ford, K., S. L. Huffman, A. K. M. A. Chowdhury, S. Becker, H. Allen, and J. Menken. 1989. "Birth-Interval Dynamics in Rural Bangladesh and Maternal Weight." *Demography* 26:425–237.

Forsum, E., A. Sadurskis, and J. Wager. 1988. "Resting Metabolic Rate and Body Composition of Healthy Swedish Women during Pregnancy." *American Journal of Clinical Nutrition* 47:942–947.

Fossey, D. 1972. "Vocalizations of the Mountain Gorilla (*Gorilla gorilla beringei*)." *Animal Behaviour* 20:36–53.

———. 1979. "Development of the Mountain Gorilla (*Gorilla gorilla beringei*): The First Thirty-Six Months." In *The Great Apes*, ed. D. Hamburg and E. R. McCown, pp. 139–184. Menlo Park, CA: Benjamin/Cummings.

———. 1983. *Gorillas in the Mist*. Boston: Houghton Mifflin.

Franceschini, R., P. L. Venturini, A. Cataldi, T. Barreca, N. Ragni, and E. Rolandi. 1989. "Plasma Beta-Endorphin Concentrations during Suckling in Lactating Women." *British Journal of Obstetrics and Gynaecology* 96:711–713.

Francis, J., and C. Heath. 1991. "The Effects of El Niño on the Frequency and Sex Ratio of Suckling Yearlings in the California Sea Lion." In *Pinnipeds and El Niño Responses to Environmental Stress*, ed. F. Trillmich and K. A. Ono, pp. 193–201. Heidelberg: Springer-Verlag.

Frayer, David W. 1977. "Dental Sexual Dimorphism in the European Upper Palaeolithic and Mesolithic." *Journal of Dental Research* 56:871.

———. 1980. "Sexual Dimorphism and Cultural Evolution in the Late Pleistocene and Holocene of Europe." *Journal of Human Evolution* 9:399–415.

———. 1981. "Body Size, Weapon Use and Natural Selection in the European Upper Palaeolithic and Mesolithic." *American Anthropologist* 83:57–73.

Frayer, David W., and Milford H. Wolpoff. 1985. "Sexual Dimorphism." *Annual Review of Anthropology* 14:429–473.

Frisch, J. E. 1973. "The Hylobatid Dentition." *Gibbon and Siamang*. Vol. 2, pp. 55–95. Basel: Karger.

Frisch, R. E. 1978a. "Nutrition, Fatness and Fertility: The Effect of Food Intake on Reproductive Ability." In *Nutrition and Human Reproduction*, ed. W. H. Mosley, pp. 91–122. New York: Plenum Press.

———. 1978b. "Population, Food Intake, and Fertility." *Science* 199:22.

———. 1976. "Critical Metabolic Mass and the Age at Menarche." *Annals of Human Biology* 3:489.

———. 1988. "Fatness and Fertility." *Scientific American* 258:70–77.

———. 1990a. "Body Fat, Menarche, Fitness, and Fertility." In *Adipose Tissue and Reproduction. Progress in Reproductive Biology and Medicine*. Vol. 14. ed. R. Frisch, pp. 1–26. Basel: Karger.

———. 1990b. "Body Weight, Body Fat and Ovulation." *Trends in Endocrinology and Metabolism* 2:181–187.

Frisch, R. E., and J. W. McArthur. 1974. "Menstrual Cycles: Fatness as a Determinant of Minimum Weight for Height Necessary for their Maintenance or Onset." *Science* 185:949–951.

Frisch, R. E., and R. Revelle. 1970. "Height and Weight at Menarche and a Hypothesis of Critical Body Weight and Adolescent Events." *Science* 169:397–399.

———. 1971. "Height and Weight at Menarche and a Hypothesis of Menarche." *Archives of Disease in Childhood* 46:695–701.

Frisch, R. E., A. von Gotz-Weibergen, and J. W. McArthur. 1981. "Delayed Menarche and Amenorrhea of College Athletes in Relation to Age of Onset of Training." *Journal of the American Medical Association* 246:1559–1563.

Frost, Harold M. 1987. "The Mechanostat: A Proposed Pathogenic Mechanism of Osteoporosis and the Bone Mass Effects of Mechanical and Nonmechanical Agents." *Bone Mineral* 2:73–85.

———. 1991. "A New Direction for Osteoporosis Research: A Review and Proposal." *Bone* 12:429–437.

Fujisawa, Y., K. Kida, and H. Matsuda. 1976. "Role of Change in Vitamin D Metabolism with Age in Calcium and Phosphorus Metabolism in Normal Human Subjects." *Journal of Clinical Endocrinology and Metabolism* 59:719–726.

Fukuda, Fumio. 1988. "Influence of Artificial Food Supply on Population Parameters and Dispersal in the Hakone T Troop of Japanese Macaques." *Primates* 29:477–492.

Furuichi, T. 1989. "Social Interactions and the Life History of Female *Pan paniscus* in Wamba, Zaire." *International Journal of Primatology* 10:173–197.

Futuyma, Douglas J. 1986. *Evolutionary Biology*. 2d ed. Sunderland, MA: Sinauer Associates.

Gadgil, Madhav, and William H. Bossert. 1970. "Life Historical Consequences of Natural Selection." *American Naturalist* 104 (935):1–24.

Galdikas, B. 1979. "Orangutan Adaptation at Tanjung Puting Reserve: Mating and Ecology." In

The Great Apes, ed. D. A. Hamburg and E. R. McCown, pp. 195–233. Menlo Park, CA.: Benjamin/Cummings.

———. 1981. "Orangutan Reproduction in the Wild." In *Reproductive Biology of Great Apes*, ed. C. Graham, pp. 281–300. New York: Academic Press.

———. 1983. "The Orangutan Long Call and Snag Crashing at Tanjung Puting Reserve." *Primates* 24:371–384.

———. 1984. "Adult Female Sociality among Wild Orangutans at Tanjung Puting Reserve." In *Female Primates: Studies by Women Primatologists*, ed. M. Small, pp. 217–235. New York: Alan R. Liss.

———. 1985a. "Orangutan Sociality at Tanjung Puting." *American Journal of Primatology* 9:101–119.

———. 1985b. "Subadult Male Orangutan Sociality and Reproductive Behavior at Tanjung Puting." *American Journal of Primatology* 8:87–99.

———. 1988. "Orangutan Diet, Range, and Activity at Tanjung Puting, Central Borneo." *International Journal of Primatology* 9:1–35.

Galdikas, B., and P. Vasey. 1992. "Why Are Orangutans So Smart? Ecological and Social Hypotheses." In *Social Processes and Mental Abilities in Non-Human Primates*, ed. F. D. Burton, pp. 183–224. Lewiston, NY: Edwin Mellen Press.

Galdikas, B., and J. Wood. 1990. "Birth Spacing Patterns in Humans and Apes." *American Journal of Physical Anthropology* 83:185–191.

Galloway, Alison. 1988. "Long Term Effects of Reproductive History on Bone Mineral Content in Women." Ph.D. diss., University of Arizona, Tucson.

———. 1997. "The Cost of Reproduction and the Evolution of Postmenopausal Osteoporosis." In *The Evolving Female: A Life-History Perspective*, ed. M. E. Morbeck, A. Galloway, and A. L. Zihlman, chap. 10. Princeton, NJ: Princeton University Press.

———. *In press.* "Determination of Parity from the Maternal Skeleton: An Appraisal." *Rivista di Anthropologie.*

Gallup, G. G. 1982. "Permanent Breast Enlargement: A Sociological Analysis." *Journal of Human* Evolution 11:597–601.

Garel, J.-M. 1987. "Hormonal Control of Calcium Metabolism during the Reproductive Cycle in Mammals." *Physiological Reviews* 67:1–66.

Garn, Stanley M., J. M. Nagy, and S. T. Sandusky. 1972. "Differential Sex Dimorphism in Bone Diameters of Subjects of European and African Ancestry." *American Journal of Physical Anthropology* 37:127–130.

Garn, Stanley M., T. V. Sullivan, and V. M. Hawthorne. 1987. "Differential Rates of Fat Change Relative to Weight Change at Different Body Sites." *International Journal of Obesity* 11:519–525.

Gautier-Hion, Anne. 1980. "Seasonal Variations of Diet Related to Species and Sex in a Community of *Cercopithecus* Monkeys." *Journal of Animal Ecology* 49:237–269.

Gautier-Hion, Anne, and Jean-Paul Gautier. 1985. "Sexual Dimorphism, Social Units, and Ecology among Sympatric Forest Guenons." In *Human Sexual Dimorphism*, ed. J. Ghesquire, R. D. Martin, and F. Newcombe, pp. 61–77. London: Taylor and Francis.

Gautier-Hion, Anne, François Bouliere, Jean-Pierre Gautier, and Jonathan Kingdon, eds. 1988. *A Primate Radiation: Evolutionary Biology of the African Guenons.* Cambridge: Cambridge University Press.

Gavan, James A. 1953. "Growth and Development of the Chimpanzee: A Longitudinal and Comparative Study." *Human Biology* 25:93–143.

———. 1971. "Longitudinal, Postnatal Growth in Chimpanzees." In *The Chimpanzee.* Vol. 4, ed. G. H. Bourne, pp. 46–102. Basel: Karger.

Gavan, James, and Daris R. Swindler. 1966. "Growth rates and phylogeny in primates." *American Journal of Physical Anthropology* 24:181–190.

Geissmann, T. 1991. "Reassessment of Age at Sexual Maturity in Gibbons (*Hylobates spp.).*" *American Journal of Primatology* 23:11–22.

Gentry, R. L. 1981. "Northern Fur Seal *Callorhinus ursinus* (Linnaeus, 1758)." In *Handbook of Marine Mammals.* Vol. 1. *The Walrus, Sea Lions, Fur Seals and Sea Otter*, ed. S. H. Ridgway and R. J. Harrison, pp. 143–160. San Diego: Academic Press.

Gentry, R. L., and G. L. Kooyman, eds. 1986. *Fur Seals: Maternal Strategies on Land and at Sea.* Princeton, NJ: Princeton University Press.

Gertner, J. M., D. R. Coustan, A. S. Kliger, L. E. Mallette, N. Ravin, and A. E. Broadus. 1986. "Pregnancy as State of Physiologic Absorptive Hypercalciuria." *American Journal of Medicine* 81:451–456.

Ghiselin, Michael T. 1974. *The Economy of Nature and the Evolution of Sex*. Berkeley: University of California Press.

———. 1987. "Species Concepts, Individuality, and Objectivity." *Biology and Philosophy* 2:127–143.

Gittins, S. P. 1980. "Territorial Behavior in the Agile Gibbon." *International Journal of Primatology* 1:381–399.

Glander, Kenneth E. 1980. "Reproduction and Population Growth in Free-Ranging Mantled Howler Monkeys." *American Journal of Physical Anthropology* 53:25–36.

Glasier, A., and A. S. McNeilly. 1990. "Physiology of Lactation." *Bailliere's Clinical Endocrinology and Metabolism* 4:379–395.

Glasier, A., A. S. McNeilly, and D. T. Baird. 1986. "Induction of Ovarian Activity by Pulsatile Infusion of LHRH in Women with Lactational Amenorrhoea." *Clinical Endocrinology* 24:243–252.

Godfrey, Laurie, M. Sutherland, D. Boy, and N. Gomberg. 1991. "Scaling of Limb Joint Surface Areas in Anthropoid Primates and Other Mammals." *Journal of Zoology, London* 223:603–625.

Goldizen, Ann W. 1987. "Tamarins and Marmosets: Communal Care of Offspring." In *Primate Societies*, ed. B. B. Smuts, D. L. Cheney, R. M. Seyfarth, R. W. Wrangham, and T. T. Struhsaker, pp. 34–43. Chicago: University of Chicago Press.

Gomendio, Montzerato. 1989. "Suckling Behavior and Fertility in Rhesus Macaques (*Macaca mulatta*)." *Journal of Zoology* 217:449–467.

Goodall, A. 1977. "Feeding and Ranging Behaviour of a Mountain Gorilla Group (*Gorilla gorilla beringei*) in the Tshibinda-Kahuzi Region (Zaire)." In *Primate Ecology*, ed. T. H. Clutton-Brock, pp. 450–479. London: Academic Press.

Goodall, Jane. 1983. "Population Dynamics During a 15-Year Period in One Community of Free-Living Chimpanzees in the Gombe National Park." *Zeitschrift für Tierpsychologie* 61:1–60.

———. 1986. *The Chimpanzees of Gombe: Patterns of Behavior*. Cambridge, MA: Harvard University Press.

———. 1990. *Through a Window: My Thirty Years with the Chimpanzees of Gombe*. Boston: Houghton Mifflin.

Goodman, Alan H., and Jerome C. Rose. 1990. "Assessment of Systemic Physiological Perturbations from Dental Enamel Hypoplasias and Associated Histological Structures." *Yearbook of Physical Anthropology* 33:59–110.

Goodman, Alan H., John Lallo, George J. Armelagos, and Jerome C. Rose. 1984. "Health Changes at Dickson Mounds, Illinois (A.D. 950–1300)." In *Paleopathology at the Origins of Agriculture*, ed. M. N. Cohen and G. J. Armelagos, pp. 271–305. London: Academic Press.

Goodman, Alan H., R. Brooke Thomas, Alan C. Swedlund, and George J. Armelagos. 1988. "Biocultural Perspectives on Stress in Prehistoric, Historical, and Contemporary Population Research." *Yearbook of Physical Anthropology* 31:169–202.

Goodman, M., P. Griffin, A. Estioko-Griffin, and J. Grove. 1985. "The Compatibility of Hunting and Mothering among the Agta Hunter-Gatherers of the Philippines." *Sex Roles* 12:1199–1209.

Gould, E. 1983. "Mechanisms of Mammalian Auditory Communication." In *Advances in the Study of Mammalian Behavior*, Special Publication No. 7, ed. J. F. Eisenberg and D. G. Kleiman, pp. 265–342. Shippensburg, PA: American Society of Mammalogists.

Gould, Stephen Jay. 1977. *Ontogeny and Phylogeny*. Cambridge, MA: Belknap Press of Harvard University Press.

———. 1980. "Caring Groups and Selfish Genes." In *The Panda's Thumb: More Reflections in Natural History*, pp. 85–92. New York: W. W. Norton.

———. 1990. *The Individual in Darwin's World*. The Second Edinburgh Medal Address. Edinburgh: Edinburgh University Press.

Gouzoules, Harold. 1980. "A Description of Genealogical Rank Changes in a Troop of Japanese Monkeys (*Macaca fuscata*)." *Primates* 21:262–267.

Gouzoules, Harold, Sarah Gouzoules, and Linda Fedigan. 1982. "Behavioral Dominance and Reproductive Success in Female Japanese Monkeys (*Macaca fuscata*)." *Animal Behaviour* 30:1138–1150.

Gouzoules, Sarah, and Harold Gouzoules. 1987. "Kinship." In *Primate Societies*, ed. B. B. Smuts, D. L. Cheney, R. M. Seyfarth, R. W. Wrangham, and T. T. Struhsaker, pp 299–305. Chicago: University of Chicago Press.

Grand, T. I 1972. "A Mechanical Interpretation of Terminal Branch Feeding." *Journal of Mammalogy* 53:198–201.

———. 1977a. "Body Weight: Its Relation to Tissue Composition, Segment Distribution, and Motor Function. I. Interspecific Comparisons." *American Journal of Physical Anthropology* 47: 211–239.

———. 1977b. "Body Weight: Its Relation to Tissue Composition, Segment Distribution and Motor Function. II. Development of *Macaca mulatta*." *American Journal of Physical Anthropology* 47:241–248.

———. 1983. "The Anatomy of Growth and Its Relation to Locomotor Capacity in *Macaca*." In *Advances in the Study of Mammalian Behavior*, Special Publication No. 7, ed. J. F. Eisenberg and D. G. Kleiman, pp. 5–23. Shippensburg, PA: American Society of Mammalogists.

———. 1990. "The Functional Anatomy of Body Mass." In *Body Size in Mammalian Paleobiology*, ed. J. Damuth and B. J. MacFadden, pp. 39–47. Cambridge: Cambridge University Press.

Gray, J. Patrick, and Linda D. Wolfe. 1980. "Height and Sexual Dimorphism in Stature among Human Societies." *American Journal of Physical Anthropology* 53:441–456.

Gray, R. H. 1981. "Birth Intervals, Postpartum Sexual Abstinence, and Child Health." In *Child-Spacing in Tropical Africa: Traditions and Change*, ed. H. J. Page and R. Lesthaeghe. New York: Academic Press.

Gray, R. H., O. M. Campbell, R. Apelo, et al. 1990. "Risk of Ovulation during Lactation." *Lancet* 335:25–29.

Gray, S. J. 1993. "Comparisons of the Effects of Breast-Feeding Practices on Birth-Spacing in Three Societies: Nomadic Turkana, Gainj, and Quechua." *Journal of Biosocial Science* 26:69–90.

Gray, T. K. 1983. "Vitamin D and Human Pregnancy." In *Perinatal Calcium and Phosphorus Metabolism*, ed. M. F. Holick, T. K. Gray, and C. S. Anast, pp. 281–290. New York: Elsevier.

Greenberg, Gary, and Ethel Tobach, eds. 1988. *Evolution of Social Behavior and Integrative Levels.* The T. C. Schneirla Conference Series, Vol. 3. Hillsdale, NJ: Lawrence Erlbaum.

Greenfield, H., and J. Clark. 1975. "Energy Compensation in Childbearing in Young Lufa Women." In *Papua New Guinea Medical Journal Proceedings Tenth Annual Symposium.* Port Moresby: Papua New Guinea Medical Society.

Greer, F. R., R. C. Tsang, and R. S. Levin. 1982a. "Increasing Serum Calcium and Magnesium Concentrations in Breast-Fed Infants: Longitudinal Studies of Minerals in Human Milk and in Sera of Nursing Mothers and Their Infants." *Journal of Pediatrics* 100:59–64.

Greer, F. R., R. C. Tsang, J. E. Searcy, R. S. Levin, and J. J. Steichen. 1982b. "Mineral Homeostasis during Lactation—Relationship to Serum 1,25-dihydroxyvitamin D, 25-hydroxyvitamin D, Parathyroid Hormone and Calcitonin." *American Journal of Clinical Nutrition* 36:431–437.

Grene, Marjorie. 1987. "Hierarchies in Biology." *American Scientist* 75:504–510.

———. 1988. "Hierarchies and Behavior." In *Evolution of Social Behavior and Integrative Levels*, eds. G. Greenberg and E. Tobach, pp. 3–17. Hillsdale, NJ: Lawrence Erlbaum.

Griffin, D. W. 1984. *Animal Thinking.* Cambridge, MA: Harvard University Press.

Gross, B. A. and C. J. Eastman. 1983. "Effect of Breast-Feeding Status on Prolactin Secretion and Resumption of Menstruation." *Medical Journal of Australia* 1:313.

Grossman, K. E., and K. Grossman. 1990. "The Wider Concept of Attachment in Cross-Cultural Research." *Human Development* 33:31–47.

Groves, C. P. 1986. "Systematics of the Great Ape." In *Comparative Primate Biology.* Vol. 1. *Systematics, Evolution and Anatomy*, ed. D. R. Swindler and J. Erwin, pp. 187–217. New York: Alan R. Liss.

Gubhaju, B. B. 1985. "Regional and Socio-Economic Differentials in Infant and Child Mortality in Rural Nepal." *Contributions to Nepalese Studies* 13:33–44.

———. 1986. "Effect of Birth Spacing on Infant and Child Mortality in Rural Nepal." *Journal of Biosocial Science* 18:435–447.

Habicht, J.-P., J. DaVanzo, W. P. Butz, and L. Meyers. 1985. "The Contraceptive Role of Breast-feeding." *Population Studies* 39:213.

Hagelberg, Erika, Ian C. Gray, and Alec J. Jeffreys. 1991. "Identification of the Skeletal Remains of a Murder Victim by DNA Analysis." *Nature* 352: 427–429.

Hager, L. D. 1991. "The Evidence for Sex Differences in the Hominid Fossil Record." In *The Archaeology of Gender*, ed. D. Walde and N. Willows, pp. 46–49. Calgary: Archaeological Association of the University of Calgary.

Hagerty, M. A., B. J. Howle, S. Tan, and T. D. Shultz. 1988. "Effect of Low- and High-Fat Intakes on the Hormonal Milieu of Premenopausal Women." *American Journal of Clinical Nutrition* 47:653–659.

Haimoff, E. H. 1984. "Acoustic and Organizational Features of Gibbon Songs." In *The Lesser Apes*, ed. H. Preuschoft, D. J. Chivens, W. Brockelman, and N. Creel. pp. 333–353. Edinburgh: Edinburgh University Press .

Hall, Brian K. 1988. "The Embryonic Development of Bone." *American Scientist* 76:174–181.

Hall, K. K. L., and Irven DeVore. 1965. "Baboon Social Behavior." In *Primate Behavior: Field Studies of Monkeys and Apes*, ed. I. DeVore, pp. 53–110. New York: Holt, Rinehart and Winston.

Halliday, T. R. 1980. *Sexual Strategy.* Oxford: Oxford University Press.

Hamilton, Margaret E. 1982. "Sexual Dimorphism in Skeletal Samples." In *Sexual Dimorphism: A Question of Size*, ed. R. L. Hall, pp. 107–163. New York: Praeger.

Hamilton, W. D. 1966. "The Moulding of Senescence by Natural Selection." *Journal of Theoretical Biology* 12:122–145.

Hamilton, W. D., P. A. Henderson, and N. Moran. 1981. "Fluctuation of Environment and Coevolved Antagonist Polymorphism as Factors in the Maintenance of Sex." In *Natural Selection and Social Behavior*, ed. R. D. Alexander and D. W. Tinkle, pp. 363–381. New York: Chiron Press.

Harcourt, Alexander H. 1979. "Social Relationships between Adult Male and Female Mountain Gorillas in the Wild." *Animal Behaviour* 27:325–342.

———. 1987. "Dominance and Fertility among Female Primates." *Journal of Zoology (London)* 217: 449–467.

Harcourt, Alexander H., and K. J. Stewart. 1984. "Gorillas' Time Feeding: Aspects of Methodology, Body Size, Competition and Diet." *African Journal of Ecology* 22:207–215.

Harcourt, Alexander H., D. Fossey and K. J. Stewart. 1980. "Reproduction by Wild Gorillas and Some Comparisons with the Chimpanzee." *Journal of Reproduction and Fertility* (suppl.) 28:59–70.

Harcourt, Alexander H., Kelly J. Stewart, and Dian Fossey. 1981. "Gorilla Reproduction in the Wild." In *Reproductive Biology of Great Apes*, ed. C. Graham, pp. 265–279. New York: Academic Press.

Harlow, Harry F. 1958. "The Nature of Love." *American Psychologist* 13:673–685.

Harrison, G. A., ed. 1983. *Energy and Effort.* London: Taylor and Francis.

———. 1985. "Anthropometric Differences and Their Clinical Correlates." *Recent Advances in Obesity Research* 4:144–149.

Hart, C. W. M., and A. R. Piling. 1960. *The Tiwi of North Australia.* New York: Holt Rinehart and Winston.

Hartup, W. W. 1979. "The Social Worlds of Childhood." *American Psychologist* 34:944–950.

Harvey, Paul H. 1986. "Energetic Costs of Reproduction." *Nature* 321:648–649.

———. 1990. "Life-History Variation: Size and Mortality Patterns." Primate Life History and Evolution, ed. C. Jean DeRousseau, pp. 81–88. New York: Wiley-Liss.

Harvey, Paul H., and Timothy Clutton-Brock. 1985. "Life History Variation in Primates." *Evolution* 39:559–581.

Harvey, Paul H., and Andy Purvis. 1991. "Comparative Methods for Explaining Adaptations." *Nature* 351:619–624.

Harvey, Paul H., R. D. Martin, and T. H. Clutton-Brock. 1987. "Life Histories in Comparative Perspective." In *Primate Societies*, ed. B. B. Smuts, D. L. Cheney, R. M. Seyfarth, R. W. Wrangham, and T. T. Struhsaker, pp. 181–196. Chicago: University of Chicago Press.

Harvey, Paul H., D. E. L. Promislow, and A. F. Read. 1989a. "Causes and Correlates of Life History Differences among Mammals." In *Comparative Socioecology: The Behavioural Ecology of Humans and Other Mammals*, ed. V. Standen and R. A. Foley, pp. 305–318. Oxford: Blackwell Scientific Publications.

Harvey, Paul H., Andrew F. Read, and Daniel E. L. Promislow. 1989b. "Life History Variation in Placental Mammals: Unifying the Data with Theory." In *Oxford Surveys in Evolutionary Biology.* Vol. 6, ed. P. H. Harvey and L. Partridge, pp. 13–31. Oxford: Oxford University Press.

Harvey, Paul H., Linda Partridge, and T. R. E. Southwood. 1991. *The Evolution of Reproductive Strategies.* Cambridge: Cambridge University Press.

Hasegawa, M. 1992. "Evolution of Hominoids as Inferred from DNA Sequences." In *Topics in Primatology.* Vol. 1. *Human Origins*, ed. T. Nishida, pp. 347–357. Tokyo: Tokyo University Press.

Hasegawa, Toshikazu. 1990. "Sex Differences in Ranging Patterns." In *The Chimpanzees of the Mahale Mountains*, ed. T. Nishida, pp. 99–114. Tokyo: University of Tokyo Press.

Hasegawa, Toshikazu, and Mariko Hiraiwa-Hasegawa. 1990. "Sperm Competition and Mating Behavior." In *The Chimpanzees of Mahale: Sexual and Life History Strategies*, ed. T. Nishida, pp. 115–132. Tokyo: University of Tokyo Press.

Hatfield, H. S., and L. G. C. Pugh. 1951. "Thermal

Conductivity of Human Fat and Muscle." *Nature, London* 168:918–919.

Hayaki, Hitoshige. 1985. "Social Play of Juvenile and Adolescent Chimpanzees in the Mahale Mountains National Park, Tanzania." *Primates* 26:343–360.

———. 1988. "Association Partners of Young Chimpanzees in the Mahale Mountains National Park, Tanzania." *Primates* 29:147–161.

Hayslip, C. C., T. A. Klein, H. L. Wray, and W. E. Duncan. 1989. "The Effects of Lactation on Bone Mineral Content in Healthy Postpartum Women." *Obstetrics and Gynecology* 73:588–592.

Heaney, R. P. 1965. "A Unified Concept of Osteoporosis." *American Journal of Medicine* 39:877–880.

Heaney, R. P., and R. R. Recker. 1986. "Distribution of Calcium Absorption in Middle-Aged Women." *American Journal of Clinical Nutrition* 43:299–305.

Heaney, R. P., and T. G. Skillman. 1971. "Calcium Metabolism in Normal Human Pregnancy." *Journal of Clinical Endocrinology* 33:661–670.

Heaney, R. P., R. R. Recker, and P. D. Saville. 1977. "Calcium Balance and Calcium Requirements in Middle-Aged Women." *American Journal of Clinical Nutrition* 30:1603–1611.

Heath, C. B., K. A. Ono, D. J. Boness, and J. M. Francis. 1991. The Influence of El Niño on Female Attendance Patterns in California Sea Lion. In *Pinnipeds and El Niño: Responses to Environmental Stress*, ed. F. Trillmich and K. A. Ono, pp. 138–145. Berlin: Springer-Verlag.

Hendrix, L. 1985. "Economy and Child Training Re-Examined." *Ethos* 13:246–261.

Hewlett, B. 1988. "Sexual Selection and Paternal Investment among Aka Pygmies." In *Human Reproductive Behavior: A Darwinian Perspective*, ed. L. Betzig, M. Borgerhoff Mulder, and P. Turke, pp. 263–276. Cambridge: Cambridge University Press.

———. 1989. "Multiple Caretaking among African Pygmies." *American Anthropologist* 91:186–191.

Hiernaux, Jean. 1968. "Variabilité du Dimorphisme Sexuel de la Stature en Afrique Subsaharienne et en Europe." In *Anthropologie und Humangenetik*, ed. T. Bielicki, pp. 42–50. Stuttgart: Fischer.

Hill, Kim. 1993. "Life History Theory and Evolutionary Anthropology." *Evolutionary Anthropology* 2(3):78–88.

Hill, Kim, and H. Kaplan. 1988. "Trade-offs in Male and Female Reproductive Strategies among the Ache." In *Human Reproductive Behaviour*, ed. L. Betzig, M. Borgerhoff Mulder, and P. Turke, pp. 277–305. Cambridge: Cambridge University Press.

Hill, P., L. Garbaczewski, N. Haley, and E. L. Wynder. 1984. "Diet and Follicular Development." *American Journal of Clinical Nutrition* 39:771–777.

Hillman, L., S. Sateesha, M. Haussler, E. Slatopolsky, and J. Haddad. 1981. "Control of Mineral Homeostasis During Lactation: Interrelationships of 25-hydroxyvitamin D, 24,25-dihydroxyvitamin D, 1,25-dihydroxyvitamin D, Parathyroid Hormone, Calcitonin, Prolactin, and Estradiol." *American Journal of Obstetrics and Gynecology* 139:471–476.

Hillman, Laura S. 1990. "Nutritional Factors Affecting Mineral Homeostasis and Mineralization in the Term and Preterm Infant." In *Nutrition and Bone Development*, ed. D. J. Simmons, pp. 55–92. Oxford: Oxford University Press.

Hiraiwa, Mariko. 1981. "Maternal and Alloparental Care in a Troop of Free-Ranging Japanese Monkeys." *Primates* 22:309–329.

Hiraiwa-Hasegawa, Mariko. 1989. "Sex Differences in the Behavioral Development of Chimpanzees at Mahale." In *Understanding Chimpanzees*, ed. P. Heltne and L. A. Marquardt, pp. 104–115. Cambridge, MA: Harvard University Press.

———. 1990. "Maternal Investment before Weaning." In *The Chimpanzees of Mahale: Sexual and Life History Strategies*, ed. T. Nishida, pp. 257–266. Tokyo: University of Tokyo Press.

———. 1997. "Development of Sex Differences in Nonhuman Primates." In *The Evolving Female: A Life-History Perspective*, ed. M. E. Morbeck, A. Galloway, and A. L. Zihlman, chap. 6. Princeton, NJ: Princeton University Press.

Ho, Mae-Wan. 1984. "Environment and Heredity in Development and Evolution." In *Beyond Neo-Darwinism: An Introduction to the New Evolutionary Paradigm*, eds. M.-W. Ho. and P. T. Saunders, pp. 267–289. New York: Academic Press.

———. 1988. "Genetic Fitness and Natural Selection: Myth and Metaphor." In *Evolution of Social Behavior and Integrative Levels*, eds. G. Greenberg and E. Tobach, pp. 85–111. Hillsdale, NJ: Lawrence Erlbaum.

Hodnett, Dean W., Hector F. DeLuca, and Neal A. Jorgensen. 1992. "Bone Mineral Loss during Lactation Occurs in Absence of Parathyroid Tissue." *American Journal of Physiology* 262:E230–233.

Hordon, L. D., and M. Peacock. 1987. "Vitamin D Metabolism in Women with Femoral Neck Fracture." *Bone and Mineral* 2:413–426.

Horn, Henry S. 1978. "Optimal Tactics of Reproduction and Life-History." In *Behavioural Ecology: An Evolutionary Approach*, ed. J. R. Krebs and N. B. Davies, pp. 411–429. Sunderland, MA: Sinauer.

Horsman, A., M. Simpson, P. A. Kirby, and B. E. C. Nordin. 1977. "Non-linear Loss in Oophorectomized Women." *British Journal of Radiology* 50:504–507.

Howell, N. 1979. *Demography of the Dobe !Kung*. New York: Academic Press.

Howie, P. W., and A. S. McNeilly. 1982. "Effect of Breast-Feeding Patterns on Human Birth Intervals." *Journal of Reproductive Fertility* 65:545–557.

Howie, P. W., A. S. McNeilly, M. J. Houston, A. Cook, and H. Boyle. 1982. "Fertility after Childbirth: Infant Feeding Patterns, Basal PRL Levels and Post-Partum Ovulation." *Clinical Endocrinology* 17:315.

Howlett, T. A., S. Tomlin, L. Ngahfoong, L. H. Rees, B. A. Bullen, G. S. Skirinar, and J. W. McArthur. 1984. "Release of Beta Endorphin and Met-Enkephalin During Exercise in Normal Women: Response to Training." *British Medical Journal* 288:1950.

Hrdy, Sarah B. 1976. "Care and Exploitations of Nonhuman Primate Infants by Conspecifics Other than the Mother." In *Advances in the Study of Behavior*. Vol. 6, ed. J. Rosenblatt, R. A. Hinde, E. Show, and C. Beer, pp. 101–151. New York: Academic Press.

Hron-Stewart, K. M. 1988. "Gender Differences in Mothers' Strategies for Helping Toddlers Solve Problems." Paper presented at the Biennial International Conference on Infancy Studies, Washington, D.C., April, 1988.

Huffman, M. 1987. "Consort Intrusion and Female Mate Choice in Japanese Macaques." *Ethology* 75:221–234.

Hughes, R. E., and E. Jones. 1985. "Intake of Dietary Fibre and the Age of Menarche." *Annals of Human Biology* 12:325–332.

Humphrey, N. K. 1976. "The Social Functions of Intellect." In *Growing Points in Ethology*, ed. P. P. G. Bateson and R. A. Hinde, pp. 303–317. Cambridge: Cambridge University Press.

Huq, N., J. C. King, B. P. Halloran, P. Buckendahl, D. Cooke, and F. M. Costa. 1988. "Calcium Metabolism in Pregnant and Lactating Women—A Longitudinal Study." *Federation of American Societies for Experimental Biology Journal* 2: A645.

Hurley, D. L., et al. 1986. "Does Estrogen Treatment in Postmenopausal Women Affect Calcitonin Secretion? [abstract]." Paper Presented at the Eighth Annual Meeting, American Society for Bone and Mineral Research, Anaheim, June 21–24, 1986.

Hurme, V. O., and G. van Wagenen. 1961. "Basic Data on the Emergence of Permanent Teeth in the Rhesus Monkey (*Macaca mulatta*)." *Proceedings of American Philosophical Society* 105:105–140.

Hurov, Jack R. 1991. "Rethinking Primate Locomotion: What Can We Learn from Development?" *Journal of Motor Behavior* 23:211–218.

Hurtado, M. A., K. Hawkes, K. Hill, and H. Kaplan. 1985. "Female Subsistence Strategies among Ache Hunter-Gatherers of Eastern Paraguay." *Human Ecology* 13:1–28.

Huss-Ashmore, R. 1980. "Fat and Fertility: Demographic Implications of Differential Fat Storage." *Yearbook of Physical Anthropology* 23:65–91.

Huxley, Julian S. 1932. *Problems of Relative Growth*. London: Methuen.

Hyde, Arthur Sidney. 1937. "The Life History of Henslow's Sparrow, *Passerherbulus henslowi* (Audubon)." *Miscellaneous Publications of the Museum of Zoology, University of Michigan* 41:1–71.

Idani, G. 1991. "Social Relationships between Immigrant and Resident Bonobo (*Pan paniscus*) Females at Wamba." *Folia Primatologica* 57:3–95.

Ingold, Tim. 1989. "An Anthropologist Looks at Biology." *Man* (N.S.) 25:208–229.

Inhobe, H. 1992. "Observations on Meat-Eating Behavior of Wild Bonobos (*Pan paniscus*) at Wamba, Republic of Zaire." *Primates* 33:247–250.

İşcan, Mehmet Y., ed. 1989. *Age Markers in the Human Skeleton*. Springfield, IL: Charles C. Thomas.

İşcan, Mehmet Y., and Kenneth A. R. Kennedy, eds. 1989. *Reconstruction of Life from the Skeleton*. New York: Alan R. Liss.

Itani, J. 1983. "Sociological Studies of Japanese Monkeys." *Recent Progress of Natural Sciences in Japan* 8:89–94.

Iverson, S. 1988. "Composition, Intake and Gastric Digestion of Milk Lipids in Pinnipeds." Ph.D. diss., University of Maryland, College Park.

Iverson, S., O. T. Oftedal, and D. J. Boness. 1991. "The Effect of El Niño on Pup Development in the California Sea Lion (*Zalophus californianus*). II. Milk Intake." In *Pinnipeds and El Niño: Responses to Environmental Stress*, ed. F. Trillmich and K. A. Ono, pp. 180–184. Berlin: Springer-Verlag.

Iwamoto, Toshitaka. 1987. "Feeding Strategies of Primates in Relation to Social Status." In *Animal Societies: Theories and Facts*, ed. Y. Ito, J. L. Brown, and J. Kikkawa, pp. 243–252. Tokyo: Japan Scientific Societies Press.

———. 1988. "Food and Energetics of Provisioned Wild Japanese Macaques (*Macaca fuscata*)." In *Ecology and Behavior of Food-Enhanced Primate Groups*, ed. John E. Fa and Charles H. Southwick, pp. 79–94. New York: Alan R. Liss.

Janson, Charles H., and Carel P. Van Shaik. 1993. "Ecological Risk Aversion in Juvenile Primates: Slow and Steady Wins the Race." In *Juvenile Primates: Life History, Development, and Behavior*, ed. M. E. Pereira and L. A. Fairbanks, pp. 57–74. New York: Oxford University Press.

Jayo, J. M., C. A. Shiveley, J. R. Kaplan, and S. B. Manuck. 1993. "Effects of Exercise and Stress on Body Fat Distribution in Male Cynomolgus Monkeys." *International Journal of Obesity* 17:597–604.

Jelliffe, E. F. P. 1976. "Maternal Nutrition and Lactation." In *Breast-Feeding and the Mother*. CIBA Foundation Symposium 45. Elsevier, Amsterdam.

Jimenez, M. H., and N. Newton. 1979. "Activity and Work during Pregnancy and the Postpartum Period: A Cross-Cultural Study of 202 Societies." *American Journal of Obstetrics and Gynecology* 135:171–176.

Jochim, M. A. 1981. *Strategies for Survival—Cultural Behavior in an Ecological Context*. Academic Press: New York.

Johanson, D. C. 1974. "Some Metric Aspects of the Permanent and Deciduous Dentition of the Pygmy Chimpanzee (*Pan paniscus*)." *American Journal of Physical Anthropology* 41:39–48.

Johnston, C. C., S. L. Hui, R. M. Witt, R. Appledorn, R. S. Baker, and C. Longcope. 1985. "Early Menopausal Changes in Bone Mass and Sex Steroids." *Journal of Clinical Endocrinology and Metabolism* 615:905–911.

Johnston, F. E., A. F. Roche, L. M. Schell, and H. N. Wettenhall. 1975. "Critical Weight at Menarche: Critique of a Hypothesis." *American Journal of Diseases of Childhood* 129:19.

Johnston, F. E., W. S. Laughlin, A. B. Harper, and A. E. Ensroth. 1982. "Physical Growth of St. Lawrence Island Eskimos: Body Size, Proportion and Composition." *American Journal of Physical Anthropology* 58:397–402.

Johnston, F. E., S. Cohen, and A. Beller. 1985. "Body Composition and Temperature Regulation in Newborns." *Journal of Human Evolution* 14: 341–345.

Jolly, Alison. 1966. "Lemur Social Behavior and Primate Intelligence." *Science* 153:501–506.

———. 1984. "The Puzzle of Female Feeding Priority." In *Female Primates: Studies by Women Primatologists*, ed. M. Small, pp. 197–215. New York: Alan R. Liss.

———. 1988. "The Evolution of Purpose." In *Machiavellian Intelligence*, ed. R. Byrne and A. Whiten, Oxford: Oxford University Press.

———. 1997. "Social Intelligence and Sexual Reproduction: Evolutionary Strategies." In *The Evolving Female: A Life-History Perspective*, ed. M. E. Morbeck, A. Galloway, and A. L. Zihlman, chap. 19. Princeton, NJ: Princeton University Press.

Kano, T. 1980. "Social Behavior of Wild Pygmy Chimpanzees (*Pan paniscus*) of Wamba: A Preliminary Report." *Journal of Human Evolution* 9:243–260.

———. 1982. "The Social Group of Pygmy Chimpanzees (*Pan paniscus*) of Wamba." *Primates* 23:171–188.

———. 1992. *The Last Ape: Pygmy Chimpanzee Behavior and Ecology*, Stanford, CA: Stanford University Press.

Kappeler, M. 1984. "Vocal Bouts and Territorial Maintenance in the Moloch Gibbon." In *The Lesser Apes*, ed. H. Preuschoft et al., pp. 376–389. Edinburgh: Edinburgh University Press.

Kardiner, Abram. 1945. *The Psychological Frontiers of Society*. New York: Columbia University Press.

Karra, M. V., A. Kirksey, O. Galal, M. S. Bassily, G. G. Harrison, and N. W. Jerome. 1988. "Zinc, Calcium, Magnesium Concentrations in Milk From American and Egyptian Women throughout the First 6 Months of Lactation." *American Journal of Clinical Nutrition* 47:642–648.

Kawai, Masao. 1958. "On the System of Social Rank in a Natural Group of Japanese Monkeys." *Primates* 1:11–48.

Kawamura, Shunzo. 1958. "Matriarchal Social Order in the Minoo-B Group: A Study on the Rank System of Japanese Macaques." *Primates* 1:149–156.

Kemnitz, J. W., R. W. Goy, T. J. Flitsch, J. J. Lohmiller, and J. A. Robinson. 1989. Obesity in Male and Female Rhesus Monkeys: Fat Distribution, Glucoregulation, and Serum Androgen Levels. *Journal of Clinical Endocrinology and Metabolism* 69:287–293.

Kennedy, Kenneth A. R. 1989. Skeletal markers of occupation stress. In *Reconstruction of Life from the Skeleton*, eds. M. Y. İşçan and K. A. R. Kennedy, pp. 129–160. New York: Alan R. Liss.

Kent, G. N., R. I. Price, D. H. Gutteridge, M. Smith, J. R. Allen, C. I. Bhagat, M. P. Barnes, C. J. Hickling, R. W. Retallack, S. G. Wilson, R. D. Devlin, C. Davies, and A. St. John. 1990. "Human Lactation: Forearm Trabecular Bone Loss, Increased Bone Turnover, and Renal Conservation of Calcium and Inorganic Phosphate with Recovery of Bone Mass Following Weaning." *Journal of Bone and Mineral Research* 5:361–369.

Kent, J. C., R. I. Price, D. H. Gutteridge, K. J. Rosman, M. Smith, J. R. Allen, C. J. Hickling, and S. L. Blakeman. 1991. "The Efficiency of Intestinal Calcium Absorption is Increased in Late Pregnancy but not in Established Lactation." *Calcified Tissue International* 48:293–295.

Kerley, Ellis R. 1966. "Skeletal Age Changes in the Chimpanzees." *Tulane Studies in Zoology* 13:71–80.

King, Barbara J. 1991. "Social Information Transfer in Monkeys, Apes and Hominids." *Yearbook of Physical Anthropology* 34:97–115.

King, C. E. 1982. "The Evolution of Life Span." In *Evolution and Genetics of Life Histories*, ed. H. Dingle and J. P. Hegman, pp. 121–138. New York: Springer-Verlag.

King, J. C., B. P. Halloran, N. Huq, T. Diamond, and P. E. Buckendahl. 1992. "Calcium Metabolism during Pregnancy and Lactation." In *Mechanisms Regulating Lactation and Infant Nutritient Utilization*, ed. M. F. Picciano and B. Lonnerdal, pp. 129–146. New York: Wiley-Liss.

Kingdon, Jonathan. 1993. *Self-Made Man: Human Evolution from Eden to Extinction?* New York: John Wiley.

Kinzey, W. 1984. "The Dentition of the Pygmy Chimpanzee, *Pan paniscus*." In *The Pygmy Chimpanzee*, ed. R. Susman, pp. 65–88. New York: Plenum Press.

Kirkwood, T. B. L. 1985. "Comparative and Evolutionary Aspects of Longevity." In *Handbook of the Biology of Aging*, ed. C. E. Finch and E. L. Schneider, pp. 27–44. New York: Van Nostrand Reinhold.

Kleerekoper, M., and S. M. Krane, eds. 1990. *Clinical Disorders of Bone and Mineral Metabolism*. New York: Mary Ann Liebow.

Klopfer, P. H., and M. S. Klopfer. 1968. "Maternal 'Imprinting' in Goats: Fostering of Alien Young." *Zeitschrift fur Tierpsychologie* 25:862–866.

Kluckhohn, Clyde. 1945. "The Personal Document in Anthropological Science." In *The Use of Personal Documents in History, Anthropology, and Sociology*, ed. L. Gottschalk et al., pp. 78–173. New York: Social Science Research Council.

Knauer, M. J. 1984. "Breast-Feeding Patterns and Postpartum Fertility in Urban Canadian Women." Ph.D. diss. University of Toronto.

Kohn, Alan J. 1989. "Natural History and the Necessity of the Organism." *American Zoologist* 29: 1095–1103.

Konner, M., and C. Worthman. 1980. "Nursing Frequency, Gonadal Function, and Birth Spacing Among !Kung Hunter-Gatherers." *Science* 207: 788–791.

Koppert, G. J. A. 1988. "Alimentation et Culture chez les Tamang, les Ghale et les Kami du Nepal." Thèse de 3ème cycle, Faculté de Droit et de Science Politique, Aix-Marseille.

Kovarik, J., W. Woloszczuk, W. Linkesch, and R. Pavelka. 1980. "Calcitonin in Pregnancy." *Lancet* 1:199–200.

Koyama, Naoki. 1970. "Changes in Dominance Rank and Division of a Wild Japanese Monkey Troop in Arashiyama." *Primates* 11:335–390.

Koyama, Naoki, Kohshi Norikoshi, and Tetsuzo Mano. 1975. "Population Dynamics of Japanese Monkeys at Arashiyama." In *Contemporary Primatology*, ed. S. Kondo, M. Kawai, and A. Ehara, pp. 411–417. Basel: Karger.

Koyama, Naoki, Kohshi Norikoshi, Tetsuzo Mano, and Yukio Takahata. 1980. "Population Changes of Japanese Monkeys at Arashiyama." In *Demographic Study on the Society of Wild Japanese Monkeys*, ed. Y. Sugiyama, pp. 19–33. Kyoto: Primate Research Institute, Kyoto University.

Kranichfeld, Marion. 1991. "Cultural Transition and Reproduction: The Dobe Area !Kung San." Ph.D. diss., Pennsylvania State University, University Park.

Kranichfeld, Marion, and P. Draper. 1990. "Bio Demography of Cultural Transition: Mating Competition, Work Roles and Reproductive Success among !Kung Men." Paper presented at the Second Annual Meeting of the Human Behavior and Evolution Society. August, University of California, Los Angeles.

Krogman, Wilton M. 1969. "Growth Changes in Skull, Face, Jaws, and Teeth of the Chimpanzee." In *The Chimpanzee*. Vol. 1, ed. G. H. Bourne, pp. 104–164. Basel: Karger.

Krogman, Wilton M., and Mehmet Y. ;a.Işçan 1986. *The Human Skeleton in Forensic Medicine*. 2d ed. Springfield, IL: Charles C. Thomas.

Krotkiewski, M., P. Björntorp, L. Sjöström, and U. Smith. 1983. "Impact of Obesity on Metabolism in Men and Women: Importance of Regional Adipose Tissue Distribution." *Journal of Clinical Investigation* 72:1150–1162.

Kuhn, C. M., S. R. Butler, and S. M. Schanberg. 1978. "Selective Depression of Serum Growth Hormone during Maternal Deprivation in Rat Pups." *Science* 199:445–447.

Kumar, R., W. R. Cohen, P. Silva, and F. H. Epstein. 1979. "Elevated 1,25-dihydroxyvitamin D Plasma Levels in Normal Human Pregnancy and Lactation." *Journal of Clinical Investigation* 63: 342–344.

Kummer, Hans. 1968. *Social Organization of Hamadryas Baboons*. Chicago: University of Chicago Press.

Kurland, Jeffry A. 1977. "Kin Selection in the Japanese Monkey." *Contributions to Primatology*, Vol. 12. Basel: Karger.

Kuroda, S. 1980. "Social Behavior of the Pygmy Chimpanzees." *Primates* 21:181–197.

———. 1984. "Interactions over Food among Pygmy Chimpanzees." In *The Pygmy Chimpanzee*, ed. R. L. Susman, pp. 301–324. New York: Plenum Press.

———. 1989. "Development Retardation and Behavioral Characteristics of Pygmy Chimpanzees." In *Understanding Chimpanzees*, ed. P. G. Heltne and L. A. Marquardt, pp. 184–193. Cambridge, MA: Harvard University Press.

Laber-Laird, K., C. A. Shively, M. Karstaedt, and B. C. Bullock. 1991. "Assessment of Abdominal Fat Deposition in Female Cynomolgus Monkeys." *International Journal of Obesity* 15:213–220.

Lager, C., and P. T. Ellison. 1990. "Effect of Moderate Weight Loss on Ovarian Function Assessed by Salivary Progesterone Measurements." *American Journal of Human Biology* 2:303–312.

Lamb, M. J. 1977. *The Biology of Aging*. London: Blackie Press.

Lancaster, Jane B. 1972. "Play-Mothering: the Relations between Juvenile Females and Young Infants among Free-Ranging Vervets (*Cercopithecus aethiops*)." *Folia Primatologica* 15:161–182.

———. 1985. "Evolutionary Perspectives on Sex Differences in the Higher Primates." In *Gender and the Life Course*, ed. Alice S. Rossi, pp. 3–28. New York: Aldine.

Langness, L. L. 1965. *The Life History in Anthropological Science*. New York: Holt, Rinehart and Winston.

Langness, L. L., and Gelya Frank. 1981. *Lives: An Anthropological Approach to Biography*. Novato, CA: Chandler and Sharp.

Lanska, D. J., M. J. Lanska, A. J. Hartz, and A. A. Rimm. 1985. "Factors Influencing the Anatomic Location of Fat Tissue in 52,935 Women." *International Journal of Obesity* 9:29–30.

Lanyon, L. E., and C. T. Rubin. 1985. "Functional Adaptation in Skeletal Structures." In *Functional Vertebrate Morphology*, ed. M. Hildebrand, D. M. Bramble, K. F. Liem, and D. B. Wake, pp. 1–25. Cambridge, MA: Harvard University Press.

Laporte, L., and A. L. Zihlman 1983. "Plates, Climates, and Hominoid Evolution." *South African Journal of Science* 79:96–110.

Larsen, Clark Spencer. 1981. "Skeletal and Dental Adaptations to the Shift to Agriculture on the Georgia Coast." *Current Anthropology* 22:422–423.

———. 1984. "Health and Disease in Prehistoric Georgia: The Transition to Agriculture." In *Paleopathology at the Origins of Agriculture*, ed. M. N. Cohen and G. J. Armelagos, pp. 367–392. London: Academic Press.

Laudenslager, Mark L., and Martin L. Reite. 1984. "Losses and Separations: Immunological Consequences and Health Implications." In *Review of Personality and Social Psychology*. Vol. 5. *Emotions, Relationships, and Health*, ed. P. Shaver, pp. 285–312. Beverly Hills, CA: Sage.

Lauder, George V. 1990. "Functional Morphology and Systematics: Studying Functional Patterns in an Historical Context." *Annual Review of Ecological Systematics* 21:317–340.

Lauder, George V., and K. F. Liem. 1989. "The Role of Historical Factors in the Evolution of Complex Organismal Functions." In *Complex Organismal Functions: Integration and Evolution in Vertebrates*, eds. D. B. Wake and G. Roth, pp. 63–78. New York: John Wiley & Sons.

Lawrence, M., J. Singh, F. Lawrence, and R. G. Whitehead. 1985. "The Energy Cost of Common Daily Activities in African Women: Increased Expenditure in Pregnancy?" *American Journal of Clinical Nutrition* 42:753–763.

Leakey, M. D., and R. L. Hay. 1979. "Pliocene Footprints in the Laetolil Beds at Laetoli, Northern Tanzania." *Nature* 278:317–323.

Leatherman, T. L., J. S. Juerssen, L. Markowitz, and R. B. Thomas. 1986. "Illness and Political Economy—The Andean Dialectic." *Cultural Survival Quarterly* 10(3):19–21.

Le Boeuf, B. J. 1974. "Male-Male Competition and Reproductive Success in Elephant Seals." *American Zoologist* 14:163–176.

———. 1991. "Pinniped Mating Systems: On Land, Ice, and in the Water." In *The Behavior of Pinnipeds*, ed. D. Renouf. London: Chapman and Hall.

———. 1994. "Variation in the Diving Patterns of Northern Elephant Seals with Age, Mass, Sex, and Reproductive Condition." In *Elephant Seals, Population Ecology, Behavior, and Physiology*, ed. B. J. Le Boeuf and R. M. Laws, pp. 237–252. Berkeley: University of California Press.

Le Boeuf, B. J., and K.T. Briggs. 1977. "The Cost of Living in a Seal Harem." *Mammalia* 41:167–195.

Le Boeuf, B. J., and S. Mesnick. 1991. "Sexual Behavior of Male Northern Elephant Seals: Lethal Injuries to Adult Females." *Behaviour* 116:143–162.

Le Boeuf, B. J.. and J. Reiter. 1988. "Lifetime Reproductive Success in Northern Elephant Seals." In *Reproductive Success: Studies of Individual Variation in Contrasting Breeding Systems*, ed. T. H. Clutton-Brock, pp. 344–362. Chicago: University of Chicago Press.

———. 1991. "Biological Effects Associated with El Niño Southern Oscillation, 1982–83, on Northern Elephant Seals Breeding at Año Nuevo, California." In *Pinnipeds and El Niño, Responses to Environmental Stress*, ed. F. Trillmich and K. Ono, pp. 206–218. Berkeley: University of California Press.

Le Boeuf, B. J., D. P. Costa, A. C. Huntley, and S. D. Feldkamp. 1988. "Continuous, Deep Diving in Female Northern Elephant Seals, *Mirounga angustirostris*." *Canadian Journal of Zoology* 66: 446–458.

Le Boeuf, B. J., R. Condit, and J. Reiter. 1989. "Parental Investment and the Secondary Sex Ratio in Northern Elephant Seals." *Behavioral Ecology and Sociobiology* 25:109–117.

Le Boeuf, B. J., R. J. Whiting, and R. F. Gantt. 1972. "Perinatal Behavior in Northern Elephant Seal Females and Their Young." *Behaviour* 43: 121–156.

Lee, Phyllis C. 1987. "Nutrition, Fertility and Maternal Investment in Primates." *Journal of Zoology* 213:409–422.

———. 1991. "Adaptations to Environmental Change: An Evolutionary Perspective." In *Primate Responses to Environmental Change*, ed. H. O. Box, pp. 39–56. New York: Chapman and Hall.

Lee, Phyllis C., and Cynthia J. Moss. 1986. "Early Maternal Investment in Male and Female African Elephant Calves." *Behavioral Ecology and Sociobiology* 18:353–361.

Lee, Phyllis C., P. Majluf, and I. J. Gordon. 1991. "Growth, Weaning and Maternal Investment from a Comparative Perspective." *Journal of Zoology, London* 225:99–114.

Lee, R. B. 1968. "What Hunters Do for a Living, or, How to Make Out on Scarce Resources." In *Man the Hunter*, ed. R. B. Lee and I. DeVore, pp. 30–48. Chicago: Aldine.

———. 1969. "!Kung Bushman Subsistence: An Input-Output Analysis." In *Environment and Cultural Behavior*, ed. A. P. Vayda, pp. 47–79. New York: Natural History Press.

———. 1972. "Population Growth and the Beginning of Sedentary Life among the !Kung Bushmen." In *Population Growth: Anthropological Implications*, ed. B. Spooner, pp. 329–342. Cambridge, MA: MIT Press.

———. 1979. *The !Kung San: Men, Women and Work in a Foraging Society*. Cambridge: Cambridge University Press.

———. 1984. *The Dobe !Kung*. New York: Holt, Rinehart and Winston.

Lee, R. B., and I. DeVore. 1968. *Man the Hunter*. Chicago: Aldine.

———, eds. 1976. *Kalahari Hunter Gatherers*. Cambridge, MA: Harvard University Press.

LeGros Clark, W. E. 1959. *The Antecedents of Man*. Edinburgh: Edinburgh University Press.

Leighton, D. R. 1987. "Gibbons: Territoriality and Monogamy." In *Primate Societies*, ed. B. Smuts D. L. Cheney, R. M. Seyfarth, R. W. Wrangham, and T. T. Struhsaker, pp. 135–145. Chicago: University of Chicago Press.

Leonard, W. R. 1987. "Nutritional Adaptation and

Dietary Change in the Southern Peruvian Andes." Ph.D. diss., University of Michigan, Ann Arbor.

Leslie, J. 1988. "Women's Work and Child Nutrition in the Third World." *World Development* 16:1341–1362.

Leslie, P. W., and P. H. Fry. 1989. "Extreme Seasonality of Births among Nomadic Turkana Pastoralists." *American Journal of Physical Anthropology* 79:103–116.

Lessells, C. M. 1991. "The Evolution of Life Histories." In *Behavioral Ecology: An Evolutionary Approach.* London: Blackwell Scientific.

Levine, N. E. 1987. "Differential Child Care in Three Tibetan Communities: Beyond Son Preference." *Population and Development Review* 13: 281–304.

Lewis, D. S., H. A. Bertrand, E. J. Masoro, H. C. McGill, K. D. Carey, and C. A. McMahon. 1983. "Preweaning Nutrition and Fat Development in Baboons." *Journal of Nutrition* 113:2253–2259.

Lewis, D. S., H. A. Bertrand, E. J. Masoro, H. C. McGill Jr., H. D. Carey, and C. A. McMahon. 1984. "Effect of Interaction of Gender and Energy Intake on Lean Body Mass and Fat Mass Gain in Infant Baboons." *Journal of Nutrition* 114:2021–2026.

Lewis, P., B. Rafferty, M. Shelley, and C. J. Robinson. 1971. "A Suggested Physiological Role of Calcitonin: The Protection of the Skeleton during Pregnancy and Lactation." *Journal of Endocrinology* 49:9–10.

Lewis, W. M., Jr. 1987. "The Cost of Sex." In *The Evolution of Sex and its Consequences*, ed. S. C. Stearns, pp. 33–58. Basel: Birkhauser Verlag.

Lewontin, R. C. 1965. "Selection for Colonizing Ability." In *The Genetics of Colonizing Species*, ed. H. G. Baker and G. L. Stebbins, pp. 79–94. New York: Academic Press.

Liem, Karel F. 1990. "Key Evolutionary Innovations, Differential Diversity, and Symecomorphosis." In *Evolutionary Innovations*, ed. M. H. Nitecki, pp. 147–170. Chicago: University of Chicago Press.

Lindsay, R. 1987. "Estrogen Therapy in the Prevention and Management of Osteoporosis." *American Journal of Obstetrics and Gynecology* 156: 1347–1351.

Lindstedt, S. L., and M. S. Boyce. 1985. "Seasonality, Fasting Endurance, and Body Size in Mammals." *American Naturalist* 125:873–878.

Lindstedt, Stan D., and Steve D. Swain. 1988. "Body Size as a Constraint of Design and Function." In *Evolution of Life Histories of Mammals: Theory and Pattern*, ed. M. S. Boyce, pp. 93–105. New Haven, CT: Yale University Press.

Lloyd, B. 1987. "Social Representations of Gender." In *Making Sense: The Child's Construction of the World*, ed. J. Bruner and H. Haste, pp. 147–162. New York: Methuen.

Lluch-Belda, D. 1969. *El lobo marino de California Zalophus californianus californianus* (Lesson 1828) Allen 1880. In *Observaciones Sobre Su Ecologia y Explotacion*, ed. D. Lluch-Belda, L. Allen, and S. G. Losocki. Mexico City: Instituto Mexicano de Recursos Naturales Renovables.

Łomnicki, Adam. 1988. *Population Ecology of Individuals.* Princeton, NJ: Princeton University Press.

Lopez-Contreras, M. E., N. Farid-Coupal, M. Landaeta de Jimenez, and G. Laxague. 1983. "Sex Dimorphism of Height in Two Venezuelan Populations." In *Human Growth and Development*, ed. J. Borms, R. Sand, C. Susanne, and M. Hebbelinck, pp. 277–281. New York: Plenum Press.

Lorenz, K. 1970. *Studies in Animal and Human Behavior.* Translated by R. D. Martin. London: Methuen.

Lovejoy, C. O. 1981. "The Origin of Man." *Science* 211:341–350.

———. 1988. "The Evolution of Human Walking." *Scientific American* 259:82–89.

———. 1993. "Modeling Human Origins: Are We Sexy Because We're Smart, or Smart Because We're Sexy?" In *The Origin and Evolution of Humans and Humanness*, ed. T. Rasmussen, pp. 1–28. Boston: Jones and Bartlett.

Lovell, Nancy C. 1990a. *Patterns of Injury and Illness in Great Apes: A Skeletal Analysis.* Washington, D.C.: Smithsonian Institution Press.

———. 1990b. "Skeletal and Dental Pathology of Free-Ranging Mountain Gorillas." *American Journal of Physical Anthropology* 81:399–412.

Low, B. S. 1993. "Ecological Demography: A Synthetic Focus in Evolutionary Anthropology." *Evolutionary Anthropology* 1:177–187.

Loy, James. 1988. "Effects of Supplementary Feeding on Maturation and Fertility in Primate Groups." In *Ecology and Behavior of Food-Enhanced Primate Groups*, eds. John E. Fa and Charles H. Southwick, pp. 153–166. New York: Alan R. Liss.

Lucas, P. W. 1981. "An Analysis of Canine Size and Jaw Shape in Some Old and New World Nonhuman Primates." *Journal of Zoology* 195:437–448.

Lucas, P. W., and R. T. Corlett. 1991. "Quantitative Aspects of the Relationship between Dentition and Diets." In *Feeding and the Texture of Food*, eds. J. F. V. Vincent and P. J. Lillford, pp. 93–121. Cambridge: Cambridge University Press.

Luerssen, J. S., and L. B. Markowitz. 1986. "To Market, to Market: Monetization and Vulnerability in a Highland Peruvian Town." Abstract of a Paper Presented at the Eighty-Fifth Annual Meetings of the American Anthropological Association, p. 239.

Lukacs, John R. 1989. "Dental Paleopathology: Methods for Reconstructing Dietary Patterns." In *Reconstruction of Life from the Skeleton*, ed. M. Y. İsçan and K. A. R. Kennedy, pp. 261–286. New York: Alan R. Liss.

Lund, B., and A. Selnes 1979. "Plasma 1,25-dihydroxyvitamin D Levels in Pregnancy and Lactation." *Acta Endocrinologica* 92:330–335.

Lunn, P. G. 1985. "Maternal Nutrition and Lactational Infertility: The Baby in the Driving Seat." In *Maternal Nutrition and Lactational Infertility*, ed. J. Dobbing, pp. 41–64. New York: Raven Press.

———. 1992. "Breast-Feeding Patterns, Maternal Milk Output, and Lactational Infecundity." *Journal of Biosocial Science* 24:317–324.

Lustig, R., R. J. Hershcopf, and H. L. Bradlow. 1990. "The Effects of Body Weight and Diet on Estrogen Metabolism and Estrogen-Dependent Disease." In *Adipose Tissue and Reproduction. Progress in Reproductive Biology and Medicine*. Vol. 14, ed. R. Frisch, pp. 107–124. Basel: Karger.

MacArthur, R. H., and E. O. Wilson. 1967. *The Theory of Island Biogeography*. Princeton, NJ: Princeton University Press.

Macchiarelli, Roberto, Loretana Salvadei, and Paola Catalano. 1988. "Biocultural Changes and Continuity throughout the First Millennium B.C. in Central Italy: Anthropological Evidence and Perspectives." *Rivista di Antropologia* 66(suppl.): 249–272.

MacKinnon, J. 1974. "The Behaviour and Ecology of Wild Orang-utans (*Pongo pygmaeus*)." *Animal Behaviour* 22:3–74.

MacLean, P. D. 1985. "Brain Evolution Relating to Family, Play, and the Separation Call." *Archives of General Psychiatry* 42:405–417.

———. 1990. *The Triune Brain in Evolution*. New York: Plenum Press.

MacMahon, James A., Donald L. Phillips, James V. Robinson, and David J. Schimpf. 1978. "Levels of Biological Organization: An Organism-Centered Approach." *BioScience* 28:700–704.

Macy, S. K. 1982. "Mother-Pup Interactions in the Northern Fur Seal." Ph.D. diss., University of Washington, Seattle.

Main, Mary, and Nancy Kaplan. 1985. "Security in Infancy, Childhood, and Adulthood: A Move to the Level of Representation." *Monographs of the Society for Research in Child Development* 50:66–104.

Mandelbaum, David H. 1973. "The Study of Life History: Gandhi." *Current Anthropology* 14:177–206.

Mann, Alan, Michelle Lampl, and Janet Monge. 1990. "Patterns of Ontogeny in Human Evolution: Evidence from Dental Development." *Yearbook of Physical Anthropology* 33:111–150.

Maple, T., and M. A. Hoff. 1982. *Gorilla Behavior*. New York: Van Nostrand Reinhold.

Margulis, L. 1981. *Symbiosis in Cell Evolution*. W. H. Freeman, San Francisco.

Margulis, L., and D. Sagan. 1986. *Origins of Sex*. New Haven, CT: Yale University Press.

Marler, P. 1976. "Social Organization, Communication, and Graded Signals: The Chimpanzee and Gorilla." In *Growing Points in Ethology*, ed. P. P. G. Bateson and R. A Hinde. pp. 239–280. Cambridge: Cambridge University Press.

Marshall, J. T., and E. R. Marshall. 1976. "Gibbons and their Territorial Songs." *Science* 193:235–237.

Marshall, J. T., and J. Sujardito. 1986. "Gibbon Systematics." In *Comparative Primate Biology*. Vol. 1. *Systematics, Evolution and Anatomy*, ed. D. R. Swindler and J. Erwin, pp. 137–185. New York: Alan R. Liss.

Marshall, L. 1976. *The !Kung of Nyae Nyae*. Cambridge, MA: Harvard University Press.

Marshall, W. A., and J. M. Tanner. 1986. "Puberty." In *Human Growth: A Comprehensive Treatise*. Vol. 2, ed. F. Falkner and J. M. Tanner, pp. 171–210. London and New York: Plenum Press.

Martill, David M. 1991. "Bones as Stones: The Contribution of Vertebrate Remains to the Lithologic Record." In *The Processes of Fossilization*, ed. S. K. Donovan, pp. 270–292. London: Belhaven Press.

Martin, R. B., and David B. Burr. 1989. *Structure, Function, and Adaptation of Compact Bone*. New York: Raven Press.

Martin, R. D. 1990. *Primate Origins and Evolution:*

A Phylogenetic Reconstruction. Princeton, NJ: Princeton University Press.

———. 1992. "Female Cycles in Relation to Paternity in Primate Societies." In *Paternity in Primates: Genetic Tests and Theories*, ed. R. D. Martin, A. F. Dixson, and E. J. Wickings, pp. 238–274. Basel: Karger.

Masali, Melchiorre, and Silvana M. Borgognini Tarli. 1985. "Dimorfismo Sessuale e Ruoli Sociali: Ipotesi Microevolutive su Un'antica Popolazione Egiziana." In *Un Ponte Fra Paradigmi: Filogenesi ed Epigenesi del Comportamento Sociale Umano*, ed. A. Milanaccio, pp. 271–292. Milano: Angeli.

Masui, K., A. Nishimura, H. Ohsawa, and Y. Sugiyama. 1973. "Population Study of Japanese Monkeys at Takasakiyama I." *Journal of the Anthropological Society of Nippon* 81:236–248.

Masui, K., Y. Sugiyama, A. Nishimura, and H. Ohsawa. 1975. "The Life Table of Japanese Monkeys at Takasakiyama." In *Contemporary Primatology*, ed. S. Kondo, M. Kawai, and A. Ehara, pp. 401–406. Basel: Karger.

Mattacks, C. A., and C. M. Pond. 1988. "Site-Specific and Sex Differences in the Rates of Fatty Acid/Triacylglycerol Substrate Cycling in Adipose Tissue and Muscle of Sedentary and Exercised Dwarf Hamsters (*Phodopus sungorus*)." *International Journal of Obesity* 12:585–597.

Mattacks, C. A., D. Sadler, and C. M. Pond. 1987. "The Effects of Exercise on the Activities of Hexokinase and Phosphofructokinase in Superficial, Intra-abdominal and Intermuscular Adipose Tissue of Guinea-Pigs." *Comparative Biochemistry and Physiology* 87B:533–542.

Maynard Smith, J. 1971. "The Origin and Maintenance of Sex." In *Group Selection*, ed. G. C. Williams, pp. 163–175. Chicago: Aldine-Atherton.

———. 1978. *The Evolution of Sex*. Cambridge: Cambridge University Press.

Mayr, Ernst. 1982. *The Growth of Biological Thought. Diversity, Evolution, and Inheritance*. Cambridge, MA: Harvard University Press.

———. 1988. *Toward a New Philosophy of Biology, Observations of an Evolutionist*. Cambridge, MA: Harvard University Press.

McCance, F. R. S. and E. M. Widdowson. 1951. "A Method of Breaking Down the Body Weights of Living Persons into Terms of Extracellular Fluid, Cell Mass, and Fat, and Some Applications of It to Physiology and Medicine." *Proceedings of the Royal Society of London, Series B* 138:115–130.

McCown, E. R. 1982. "Sex Differences: the Female as the Baseline for Species Description." In *Sexual Dimorphism in Homo sapiens*, ed. R. L. Hall, pp. 37–83. New York: Praeger.

McFalls, Joseph A., and Marguerite Harvey McFalls. 1984. *Disease and Fertility*. San Diego: Academic Press.

McFarland, R. 1992. "Body Composition and Reproduction in Female Pigtail Macaques." Ph.D. diss., University of Washington, Seattle.

———. 1997. "Primate Females: Fat or Fit?" In *The Evolving Female: A Life-History Perspective*, ed. M. E. Morbeck, A. Galloway, and A. L. Zihlman, chap. 12. Princeton, NJ: Princeton University Press.

McGrew, William C. 1979. "Evolutionary Implications of Sex Differences in Chimpanzee Predation and Tool Use." In *The Great Apes*, ed. D. A. Hamburg and E. R. McCown, pp. 440–463. Menlo Park, CA: Benjamin/Cummings.

———. 1981. "The Female Chimpanzee as a Human Evolution Prototype." In *Woman the Gatherer*, ed. F. Dahlberg, pp. 35–73. New Haven, CT: Yale University Press.

———. 1992. *Chimpanzee Material Culture: Implications for Human Evolution*. Cambridge: Cambridge University Press.

McGrew, William C., P. J. Baldwin, and C. E. G. Tutin. 1981. "Chimpanzees in a Hot, Dry and Open Habitat: Mt. Assirik, Senegal, West Africa." *Journal of Human Evolution* 10:227–244.

McKenna, James. 1990. "Evolution and Sudden Infant Death Syndrome (SIDS). Part I: Infant Responsivity to Parental Contact." *Human Nature* 1:145–177.

McKinney, Michael L., 1988. *Heterochrony in Evolution: A Multidisciplinary Approach*. New York: Plenum Press.

———. 1991. "Completeness of the Fossil Record: An Overview." In *The Processes of Fossilization*, ed. S. K. Donovan, pp. 66–83. London: Belhaven Press.

McKinney, Michael L., and Kenneth J. McNamara. 1991. *Heterochrony: The Evolution of Ontogeny*. New York: Plenum Press.

McLean, Paul D. 1985. "Brain Evolution Relating to Family, Play, and the Separation Call." *Archives of General Psychiatry* 42:405–417.

McLeod, Beverly. 1997. "Life History, Females, and Evolution: A Commentary." In *The Evolving Female: A Life-History Perspective*, ed. M. E. Morbeck, A. Galloway, and A. L. Zihlman, chap. 20. Princeton, NJ: Princeton University Press.

McNeilly, A. S. 1993. "Lactational Amenorrhea." *Endocrinology and Metabolism Clinics of North America* 22:59–73.

McNeilly, A. S., A. F. Glasier, J. Jonassen, and P. W. Howie 1982. "Evidence for Direct Inhibition of Ovarian Function by Prolactin." *Journal of Reproductive Fertility* 65:559.

McNeilly, A. S., A. Glasier, and P. W. Howie 1985. "Endocrine Control of Lactational Infertility." In *Maternal Nutrition and Lactational Infertility*, ed. J. Dobbing, pp. 1–24. New York: Raven Press.

McNeilly, A. S., C. C. K. Tay, and A. Glasier. 1994. "Physiological Mechanisms Underlying Lactational Amenorrhea." *Annals of the New York Academy of Sciences* 709:145–155.

Mead, M. 1935. *Sex and Temperament in Three Primitive Societies.* New York: William Morrow.

———. 1949. *Male and Female, a Study of the Sexes in a Changing World.* New York: William Morrow.

Meder, A. 1986. "Physical and Activity Changes Associated with Pregnancy in Captive Lowland Gorillas (*Gorilla gorilla gorilla*)." *American Journal of Primatology* 11:111–116.

———. 1990. "Sex Differences in the Behavior of Immature Captive Lowland Gorillas." *Primates* 31:51–64.

Melnick, Don J. and Guy A. Hoelzer. 1993. "What Is mtDNA Good for in the Study of Primate Evolution?" *Evolutionary Anthropology* 2:2–10.

Melton, D. A. 1991. "Pattern Formation during Animal Development." *Science.* 252:234–241.

Melton, L. J., and S. R. Cummings. 1987. "Heterogeneity of Age-Related Fractures: Implications for Epidemiology." *Bone and Mineral* 2:321–331.

Menken, J., J. Trussell, and S. Watkins. 1981. "The Nutrition-Fertility Link: An Evaluation of the Evidence." *Journal of Interdisciplinary History* 9: 425–441.

Merbs, Charles F. 1989. "Trauma." In *Reconstruction of Life from the Skeleton*, ed. M. Y. İşcan and K. A. R. Kennedy, pp. 161–189. New York: Alan R. Liss.

Meszéna, G., and E. Pásztor. 1990. "Population Regulation and Optimal Life-History Strategies." In *Organizational Constraints on the Dynamics of Evolution*, eds. J. Maynard Smith and G. Vida, pp. 321–331. New York: Manchester University Press.

Michejda, Maria. 1980. "Growth Standards in the Skeletal Age of Rhesus Monkey (*Macaca mulatta*), Chimpanzee (*Pan troglodytes*), and Man." *Developmental Biology Standard* 45:45–50.

Milhaud, G., M. Benezech-Lefevre, and M. S. Moukhtar. 1978. "Deficiency of Calcitonin in Age-Related Osteoporosis." *Biomedicine* 29:272–276.

Miller, S. C., B. P. Halloran, H. F. DeLuca, and W. S. S. Jee. 1982. "Role of Vitamin D in Maternal Skeletal Changes During Pregnancy and Lactation: A Histomorphometric Study." *Calcified Tissue International* 34:245–252.

Miller, S. C., J. G. Shupe, E. H. Redd, M. A. Miller, and T. H. Omura. 1986. "Changes in Bone Mineral and Bone Formation Rates During Pregnancy and Lactation in Rats." *Bone* 7:283–287.

Mishkin, B. 1946. "The Contemporary Quechua." In *Handbook of South American Indians.* Vol. 2, ed. J. H. Steward. Washington, D.C.: Bureau of American Ethnology.

Missakian, Elisabeth A. 1974. "Mother-Offspring Grooming Relations in Rhesus Monkeys." *Archives in Sexual Behavior* 3:135–141.

Mitani, J. 1985. "Sexual Selection and Adult Male Orangutan Long Calls." *Animal Behaviour* 33: 272–283.

———. 1992. "Singing Behavior of Male Gibbons: Field Observations and Experiments." In *Topics in Primatology.* Vol. 1. *Human Origins*, ed. T. Nishida, pp. 199–210. Tokyo: Tokyo University Press.

Miyamoto, M. M., J. L. Slightom, and M. Goodman. 1988. "Phylogenetic Relations of Humans and African Apes from DNA Sequences in the Globin Region." *Science* 238:369–373.

Moerman, M. L. 1982. "Growth of the Birth Canal in Adolescent Girls." *American Journal of Obstetrics and Gynecology* 143:528–532.

Molleson, T. L. 1986. "Skeletal Age and Palaeodemography." In *The Biology of Human Aging*, ed. A. H. Bittles and K. J. Collins, pp. 95–118. Cambridge: Cambridge University Press.

Morbeck, Mary Ellen. 1975. "*Dryopithecus africanus* Forelimb." *Journal of Human Evolution* 4: 39–46.

———. 1979. "Forelimb Use and Positional Adaptation in *Colobus guereza*: Integration of Behavioral, Ecological, and Anatomical Data." In *Environment, Behavior, and Morphology: Dynamic Interactions in Primates*, ed. M. E. Morbeck, N. Preuschoft, and N. Gomberg, pp. 95–117. New York: Gustav Fischer.

———. 1991a. "Biology, Behavior, and Evolution." (Review of *Comparative Primate Biology*, volumes 1, 2A, 3.) *Reviews in Anthropology* 20: 113–123.

———. 1991b. "Bones, Gender, and Life History." In *The Archaeology of Gender*, ed. D. Walde and Noreen D. Willows, pp. 39–45. Calgary: Archaeological Association of the University of Calgary.

———. 1994. "Object Manipulation, Gestures, Posture and Locomotion." In *Hominid Culture in Primate Perspective*, ed. D. Quiatt and J. Itani, pp. 117–135. Denver: University of Colorado Press.

———. 1997a. "Life History, The Individual, and Evolution." In *The Evolving Female: A Life-History Perspective*, ed. M. E. Morbeck, A. Galloway, and A. L. Zihlman, chap. 1. Princeton, NJ: Princeton University Press.

———. 1997b. "Reading Life History in Teeth, Bones, and Fossils." In *The Evolving Female: A Life-History Perspective*, ed. M. E. Morbeck, A. Galloway, and A. L. Zihlman, chap. 9. Princeton, NJ: Princeton University Press.

Morbeck, Mary Ellen, and A. L. Zihlman. 1988. "Body Composition and Limb Proportions in Orangutans." In *Orang-utan Biology*, ed. J. Schwartz, pp. 285–297. New York: Oxford University Press.

———. 1989. "Body Size and Proportions in Chimpanzees with Special Reference to *Pan troglodytes schweinfurthii* from Gombe National Park, Tanzania." *Primates* 30:369–382.

Morbeck, Mary Ellen, A. L. Zihlman, R. Sumner, and A. Galloway. 1991. "Poliomyelitis and Skeletal Asymmetry in Gombe Chimpanzees." *Primates* 32:77–91.

Morbeck, Mary Ellen, A. Galloway and A. L. Zihlman. 1992. "Gombe Chimpanzee Sex Differences in the Pelvis and Observations of Pubic and Preauricular Areas." *Primates* 33:129–132.

Morbeck, Mary Ellen, A. Galloway, K. M. Mowbray, and A. L. Zihlman. 1994. "Skeletal Asymmetry and Hand Preference during Termite Fishing by Gombe Chimpanzees." *Primates* 35: 99–103.

Morelli, G. A. 1987. "A Comparative Study of Efe (Pygmy) and Lese One-, Two-, and Three-Year-Olds of the Ituri Forest of Northeastern Zaire: The Influence of Subsistence-Related Variables, and Children's Age and Gender on Social-Emotional Development." Ph.D. diss., University of Massachusetts, Amherst.

———. 1997. "Growing Up Female in a Farmer and Forager Community." In *The Evolving Female: A Life-History Perspective*, ed. M. E. Morbeck, A. Galloway, and A. L. Zihlman, chap. 15. Princeton, NJ: Princeton University Press.

Morgan, E. 1982. *The Aquatic Ape: A Theory of Human Evolution*. London: Souvenir Press.

Mori, Akio. 1975. "Signals Found in the Grooming Interactions of Wild Japanese Monkeys of the Koshima Troop." *Primates* 20:371–398.

———. 1979. "Analysis of Population Changes by Measurement of Body Weight in the Koshima Troop of Japanese Monkeys." *Primates* 20:371–397.

———. 1984. "An Ethological Study of Pygmy Chimpanzees in Wamba, Zaire: A Comparison with Chimpanzees." *Primates* 25:255–278.

Morimoto, S. M., T. Tsudi, Y. O. Okada, T. Onishi, and Y. Kumahara. 1980. "The Effect of Oestrogens on Human Calcitonin Secretion after Calcium Infusion in Elderly Female Subjects." *Clinical Endocrinology* 13:135–143.

Morris, D. 1967. *The Naked Ape: A Zoologist's Study of the Human Animal*. London: J. Cape.

Moser, P. R., R. D. Reynolds, S. Acharya, M. R. Howard, and M. B. Andon. 1988. "Calcium and Magnesium Dietary Intakes and Plasma and Milk Concentrations of Nepalese Lactating Women." *American Journal Clinical Nutrition* 47:735–739.

Muller, H. J. 1964. "The Relation of Recombination to Mutational Advance." *Mutation Research* 1:2–9.

Munroe, R. H., H. S. Shimmin, and R. L. Munroe. 1984. "Gender Understanding and Sex Role Preference in Four Cultures." *Developmental Psychology* 20:673–682.

Napier, J. R., and P. Napier. 1967. *Handbook of Living Primates*. London: Academic Press.

Nei, M. 1987. *Molecular Evolutionary Genetics*. New York: Columbia University Press.

Nestle Foundation. 1982. *Annual Report*. Lausanne, Switzerland.

Newell, Raymond R. 1984. "On the Mesolithic Contribution to the Social Evolution of Western European Society." In *European Social Evolution: Archaeological Perspectives*, ed. J. Bintliff, pp. 69–82. Bradford, MA: University of Bradford Press.

Newell, Raymond R., and Trinette S. Constandse-Westermann. 1986. "Testing an Ethnographic Analogue of Mesolithic Social Structure and the Archaeological Resolution of Mesolithic Ethnic Groups and Breeding Populations." *Proceedings of the Koninklijke Nederlandse Akademie van Wetenschappen*, Series B 89(3):243–310.

Nicolson, Nancy A. 1987. "Infant, Mothers, and Other Females." In *Primate Societies*, ed. B. B. Smuts, D. L. Cheney, R. M. Seyfarth, R. W. Wrangham, and T. T. Struhsaker, pp. 330–342. Chicago: University of Chicago Press.

Nishida, Toshisada. 1987. "Local Traditions and Cultural Transmission." In *Primate Societies*, ed. B. B. Smuts, D. L. Cheney, R. M. Seyfarth, R. W. Wrangham, and T. T. Struhsaker, pp. 462–474. Chicago: University of Chicago Press.

———. 1989. "Social Interactions between Resident and Immigrant Female Chimpanzees." In *Understanding Chimpanzees*, ed. P. G. Heltne and L. A. Marquardt, pp. 68–89. Cambridge, MA: Harvard University Press.

———, ed. 1990. *The Chimpanzees of the Mahale Mountains. Sexual and Life History Strategies.* Tokyo: University of Tokyo Press.

Nishida, Toshisada, and Mariko Hiraiwa-Hasegawa. 1987. "Chimpanzees and Bonobos: Cooperative Relationships among Males." In *Primate Societies*, ed. B. B. Smuts D. L. Cheney, R. M. Seyfarth, R. W. Wrangham, and T. T. Struhsaker, pp. 165–177. Chicago: University of Chicago Press.

Nishida, Toshisada, Hiroyuki Takasaki, and Yukio Takahata. 1990. "Demography and Reproductive Profiles." In *The Chimpanzees of the Mahale Mountains: Sexual and Life History Strategies*, ed. Toshisada Nishida, pp. 63–97. Tokyo: University of Tokyo Press.

Nissen, H. W., and A. H. Riesen. 1964. "The Eruption of the Permanent Dentition of Chimpanzee." *American Journal of Physical Anthropology* 22:285–294.

Noel, G. L., H. K. Suh, and A. G. Frantz. 1974. "Prolactin Release During Nursing and Breast Stimulation in Postpartum and Nonpostpartum Subjects." *Journal of Clinical Endocrinology and Metabolism* 38:413.

Nordin, B. E. C., A. Horsman, R. Brook, and D. A. Williams. 1976. "The Relationship between Oestrogen Status and Bone Loss in Postmenopausal Women." *Clinical Endocrinology* 5:353s–161s.

Nordin, B. E. C., A. Horsman, D. H Marshall, M. Simpson, and G. M. Waterhouse. 1979. "Calcium Requirement and Calcium Therapy." *Clinical Orthopaedics and Related Research* 140:216–239.

Nordin, B. E. C., M. B. Baker, A. Horsman, and M. Peacock. 1985. "A Prospective Trial of the Effect of Vitamin D Supplementation on Metacarpal Bone Loss in Elderly Women." *American Journal of Clinical Nutrition* 42:470–474.

Norman, A. W., and G. Litwack. 1987. *Hormones.* San Diego: Academic Press.

Novacek, Michael J. 1993. "Reflections on Higher Mammalian Phylogenetics." *Journal of Mammalian Evolution* 1:3–30.

Oftedal, Olav T. 1984. "Milk Composition, Milk Yield, and Energy Output at Peak Lactation: A Comparative Review." *Symposium of the Zoological Society, London* 51:33–85.

Oftedal, Olav T., and S. Iverson. 1987. "Hydrogen Isotope Methodology for Measurement of Milk Intake and Energetics of Growth in Suckling Pinnipeds." In *Approaches to Marine Mammal Energetics*, ed. A. C. Huntley, D. P. Costa, G. A. Worthy, and M. A. Castallini, pp. 67–96. Special Publ. No. 1. Lawrence, Kans.: Society for Marine Mammology.

Oftedal, Olav T., D. Boness, and R. Tedman. 1987a. "The Behavior, Physiology, and Anatomy of Lactation in Pinnipedia." In *Current Mammalogy.* Vol. 1, ed. H. H. Genoways, pp. 175–245. New York: Plenum Press.

Oftedal, Olav T., S. Iverson, and D. Boness. 1987b. "Milk And Energy Intakes of Suckling California Sea Lion Pups (*Zalophus californianus*) in Relation to Sex, Growth, and Predicted Maintenance Requirements." *Physiological Zoology* 60: 560–575.

O'Neill, R. V., D. L. DeAngelis, J. B. Wade, T. F. H. Allen. 1986. *A Hierarchical Concept of Ecosystems.* Princeton, NJ: Princeton University Press.

Ono, Kathryn A. 1991. "Introductory Remarks and the Natural History of the California Sea Lion." In *Pinnipeds and El Niño Responses to Environmental Stress*, ed. F. Trillmich and K. A. Ono, pp. 110–111. Heidelberg: Springer-Verlag.

———. 1997. "Sea Lions, Life History, and Reproduction." In *The Evolving Female: A Life-History Perspective*, ed. M. E. Morbeck, A. Galloway, and A. L. Zihlman, chap. 3. Princeton, NJ: Princeton University Press.

Ono, K. A., and D. J. Boness. 1996. "Sexual Dimorphism in Sea Lion Pups: Differential Maternal Investment, or Sex-Specific Differences in Energy Allocation?" *Behaviour.*

Ono, K. A., D. J. Boness, and O. T. Oftedal. 1987. "The Effect of a Natural Environmental Disturbance on Maternal Investment and Pup Behavior in the California Sea Lion." *Behavioral Ecology and Sociobiology* 21:109–118.

Orlove, B. S. 1987. "Stability and Change in Highland Andean Dietary Patterns." In *Food and Evolution: Toward a Theory of Human Food Habits*, ed. M. Harris and E. B. Ross, pp. 481–515. Philadelphia: Temple University Press.

Ortiz, C. L., D. P. Costa, and B. J. Le Boeuf. 1978. "Water and Energy Flux in Elephant Seal Pups Fasting Under Natural Conditions." *Physiological Zoology* 51:166–178.

Ortner, Donald J., and Arthur C. Aufderheide, eds. 1991. *Human Paleopathology, Current Syntheses, and Future Options.* Washington, D.C.: Smithsonian Institution Press.

Orzack, Steven Hecht, and Elliott Sober. 1994. "Optimality Models and the Test of Adaptationism." *American Naturalist* 143:361–380.

Oxnard, C. E., S. S. Lieberman, and B. R. Gelvin. 1985. "Sexual Dimorphism in Dental Dimensions of Higher Primates." *American Journal of Primatology* 8:127–152.

Pääbo, Svante. 1993. "Ancient DNA." *Scientific American* 269:86–92.

Pahuja, P. N., and H. F. DeLuca. 1981. "Stimulation of Intestinal Calcium Transport and Bone Calcium Mobilization by Prolactin in Vitamin D Deficient Rats." *Science* 214:1038–1049.

Painter, M. 1983. "The Political Economy of Food Production in Peru." *Studies in Comparative International Development* 18:34–52.

Pallottino, Massimo. 1984. *Etruscologia.* VII. Italian edition. Milano: Hoepli. [1978. *The Etruscans.* English. London: Penguin Books.]

Palombit, R. A. 1995. "Longitudinal Patterns of Reproduction in Wild Female Siamang *Hylobates syndactylus* and White-Handed Gibbons *Hylobates lar*." *International Journal of Primatology* 16(5): 73–76.

Panter-Brick, Catherine. 1989. "Motherhood and Subsistence Work—the Tamang of Rural Nepal." *Human Ecology* 17:205–228.

———. 1991. "Lactation, Birth Spacing, and Maternal Workloads among Two Castes in Rural Nepal." *Journal of Biosocial Science* 23:137–154.

———. 1992a. "The Energy Cost of Common Tasks in Rural Nepal: Levels of Energy Expenditure Compatible with Sustained Physical Activity." *European Journal of Applied Physiology and Occupational Physiology* 64:477–484.

———. 1992b. "Women's Working Behaviour and Maternal-Child Health in Rural Nepal." In *Physical Activity and Health*, ed. N. Norgan, pp. 190–206. Cambridge: Cambridge University Press.

———. 1993a. "Seasonality and Levels of Energy Expenditure during Pregnancy and Lactation for Rural Nepali Women." *American Journal of Clinical Nutrition* 57:620–628.

———. 1993b. "Seasonal Organisation of Work Patterns." In *Seasonality and Human Ecology*, ed. S. J. Ulijaszek and S. Strickland, pp. 220–234. Cambridge: Cambridge University Press.

———. 1997. "Women's Work and Energetics: A Case Study from Nepal." In *The Evolving Female: A Life-History Perspective*, ed. M. E. Morbeck, A. Galloway, and A. L. Zihlman, chap. 17. Princeton, NJ: Princeton University Press.

Panter-Brick, Catherine, D. S. Lotstein, and P. T. Ellison. 1993. "Seasonality of Reproductive Function and Weight Loss in Rural Nepali Women." *Human Reproduction* 8:684–690.

Parfitt, A. M., B. Chir, J. C. Gallagher, R. P. Heaney, C. C. Johnston, R. Neer, and G. D. Whedon. 1982. "Vitamin D and Bone Health in the Elderly." *American Journal of Clinical Nutrition* 36:1014–1031.

Parfitt, D. B., K. R. Church, and J. L. Cameron. 1991. "Restoration of Pulsatile LH Secretion after Fasting in Rhesus Monkeys (*Macaca mulatta*): Dependence on Size of the Refeed Meal." *Endocrinology* 129:749–56.

Parker, G. A., and J. Maynard Smith. 1990. "Optimality Theory in Evolutionary Biology." *Nature* 348:27–33.

Partridge, Linda, and Paul H. Harvey. 1988. "The Ecological Context of Life History Evolution." *Science* 241:1449–1455.

Pavelka, Mary S. McDonald. 1997. "The Social Life of Female Japanese Monkeys." In *The Evolving Female: A Life-History Perspective*, ed. M. E. Morbeck, A. Galloway, and A. L. Zihlman, chap. 7. Princeton, NJ: Princeton University Press.

Pavelka, Mary S. McDonald, and Linda Marie Fedigan. 1991. "Menopause: A Comparative Life History Perspective." *Yearbook of Physical Anthropology* 34:13–38.

Peacock, Nadine R. 1985. "Time Allocation, Work, and Fertility among Efe Pygmy Women of Northeast Zaire." Ph.D. diss., Harvard University, Cambridge, MA.

———. 1990. "Comparative and Cross-Cultural Approaches to the Study of Human Female Reproductive Failure." In *Primate Life History and Evolution*, ed. C. J. DeRousseau, pp. 195–220. New York: Wiley-Liss.

Peacock, Nadine R. 1991. "Rethinking the Sexual Division of Labor: Reproduction and Work Effort among the Efe." In *Gender at the Crossroads of Knowledge: Feminist Anthropology in the Postmodern Era*, ed. M. di Leonardo, pp. 333–360. Berkeley: University of California Press.

Pedersen, A. B., M. J. Bartholemew, L. A. Dolence, L. P Aljadir, K. L. Netteburg, and T. Lloyd. 1991. "Menstrual Differences Due to Vegetarian and Non-Vegetarian Diets." *American Journal of Clinical Nutrition* 53:879–885.

Pereira, Michael E. 1993. "Juvenility in Animals." In *Juvenile Primates: Life History, Development, and Behavior*, ed. M. E. Pereira and L. A. Fairbanks, pp. 17–27. New York: Oxford University Press.

Pereira, Michael E., and Jeanne Altmann. 1985. "Development of Social Behavior in Free-Living Nonhuman Primates." In *Nonhuman Primate Models for Human Growth and Development*, ed. E. S. Watts, pp. 217–309. New York: Alan R. Liss.

Pereira, Michael E., and Lynn A. Fairbanks, eds. 1993. *Juvenile Primates: Life History, Development, and Behavior.* New York: Oxford University Press.

Pereira, Michael E., and C. M. Pond. 1995. "Organization of White Adipose Tissue in Lemuridae." *American Journal of Primatology* 35:(1):1–13.

Perez, A., P. Vela, R. Potter, and G. S. Masnick. 1971. "Timing and Sequence of Resuming Ovulation and Menstruation after Childbirth." *Population Studies* 25:491.

Perez Cano, R., M. J. Montoya, R. Moruno, A. Vazquez, F. Galan, and M. Garrido. 1989. "Calcitonin Reserve in Healthy Women and Patients with Postmenopausal Osteoporosis." *Calcified Tissue International* 45:203–208.

Peroni, Renato. 1969. "Per Uno Studio dell'Economia di Scambio in Italia nel Quadro dell'Ambiente Culturale dei Secoli Intorno al 100 a.C." *La Parola del Passato* 24:134–160.

Perzigian, Anthony J., Patricia A. Tench, and Donna J. Braun. 1984. "Prehistoric Health in the Ohio River Valley." In *Paleopathology at the Origins of Agriculture*, ed. M. N. Cohen and G. J. Armelagos, pp. 347–366. London: Academic Press.

Peters, Robert H. 1983. *The Ecological Implications of Body Size.* Cambridge: Cambridge University Press.

Pianka, E. R. 1970. "On '*r*' and '*K*' Selection." *American Naturalist* 104:592–597.

Pierret-Risoud, B., and J.-P. Risoud. 1985. "Dynamique de Système Agraire et Developpement—Le Cas du Village de Salme au Nepal." Thèse de Docteur-Ingenieur en Agro-economie, ENSA, Montpellier.

Pike, J. W., J. B. Parker, M. R. Haussler, A. Boass, and S. U. Toverud. 1979. "Dynamic Changes in Circulating 1,25-Dihydroxyvitamin D during Reproduction in Rats." *Science* 204:1427–1429.

Piperno, Dolores R. 1988. *Phytolith Analysis: An Archaeological and Geological Perspective.* San Diego: Academic Press.

Pirke, K. M., U. Schweiger, W. Lemmel, J. C. Krieg, and M. Berger. 1985. "The Influence of Dieting on the Menstrual Cycle of Healthy Young Women." *Journal of Clinical Endocrinology and Metabolism* 60:1174–1179.

Pirke, K. M., U. Schweiger, T. Strowitzki, R. J. Tuschi, G. Laessle, A. Broocks, B. Huber, and R. Middendorf. 1989. "Dieting Causes Menstrual Irregularities in Normal Weight Young Women through Impairment of Episodic Lutenizing Hormone Secretion." *Fertility and Sterility* 51:263–268.

Pitkin, R. M., W. A. Reynolds, G. A. Williams, and G. K. Hargis. 1979. "Calcium Metabolism in Normal Pregnancy: a Longitudinal Study." *American Journal of Obstetrics and Gynecology* 133: 781–790.

Pollock, J. I. 1979. "Female Dominance in *Indri indri.*" *Folia Primatologia* 31:143–164.

Polson, D. W., M. Sagle, H. D. Mason, J. Adams, H. S. Jacobs, and S. Franks. 1986. "Ovulation and Normal Luteal Function during LHRH Treatment of Women with Hyperprolactinaemic Amenorrhoea." *Clinical Endocrinology* 24:531–537.

Pond, Caroline M. 1977. "The Significance of Lactation in the Evolution of Mammals." *Evolution* 31:177–199.

———. 1978. "Morphological Aspects and the Ecological and Mechanical Consequences of Fat Deposition in Wild Vertebrates." *Annual Review of Ecology and Systematics* 9:519–570.

———. 1984. "Physiological and Ecological Importance of Energy Storage in Evolution of Lactation: Evidence for a Common Pattern of Anatomical Organization of Adipose Tissue in Mammals." *Symposium, Zoological Society London* 51: 1–32.

———. 1986. "The Natural History of Adipocytes." *Science Progress, Oxford* 70:45–71.

———. 1987. "Some Conceptual and Comparative Aspects of Body Composition Measurements." In *Techniques in the Behavioral and Neural Sciences.* Vol. 1, *Feeding and Drinking*, ed. F. M. Toates and N. E. Rowland, pp. 499–529. Amsterdam: Elsevier Science.

———. 1991. "Adipose Tissue in Human Evolution." In *The Aquatic Ape: Fact or Fiction?* ed. M. Roede, J. Wind, J. M. Patrick, and V. Reynolds, pp. 193–220. London: Souvenir Press.

———. 1992a. "The Structure and Function of Adipose Tissue in Humans, with Comments on the Evolutionary Origin and Physiological Consequences of Sex Differences." *Collegium Antropologicum Zagreb* 16:135–143.

———. 1992b. "An Evolutionary and Functional View of Mammalian Adipose Tissue." *Proceedings of the Nutrition Society* 51:367–377.

———. 1994. "The Structure and Organization of Adipose Tissue in Naturally Obese Nonhibernating Mammals." In *Obesity in Europe '93*: Proceedings of the Fifth European Congress of Obesity, ed. H. Ditschuneit, F. A. Gries, H. Hauner, V. Schusdziarra, and J. G. Wechsler, pp. 395–402. London: J. Libbey.

———. 1997. "The Biological Origins of Adipose Tissue in Humans." In *The Evolving Female: A Life-History Perspective*, ed. M. E. Morbeck, A. Galloway, and A. L. Zihlman, chap. 11. Princeton, NJ: Princeton University Press.

Pond, Caroline M., and C.A. Mattacks. 1985a. "Anatomical Organization of Mammalian Adipose Tissue." *Fortschritte der Zoologie* 30:485–489.

———. 1985b. "Body Mass and Natural Diet as Determinants of the Number and Volume of Adipocytes in Eutherian Mammals." *Journal of Morphology* 185:183–193.

———. 1987a. "Comparative Aspects of Hexokinase and Phosphofructokinase Activity in Intermuscular Adipose Tissue." *Comparative Biochemistry and Physiology* 87B:543–551.

———. 1987b. "The Anatomy of Adipose Tissue in Captive *Macaca* Monkeys and Its Implications for Human Biology." *Folia Primatologia* 48:164–185.

———. 1988. "The Distribution, Cellular Structure, and Metabolism of Adipose Tissue in the Fin Whale *Balaenoptera physalus." Canadian Journal of Zoology* 66:534–537.

———. 1989. "Biochemical Correlates of the Structural Allometry and Site-Specific Properties of Mammalian Adipose Tissue." *Comparative Biochemistry and Physiology* 92A:455–463.

Pond, Caroline M., and M. A. Ramsay. 1992. "Allometry of the Distribution of Adipose Tissue in Carnivora." *Canadian Journal of Zoology* 70:342–347.

Pond, Caroline M., C. A. Mattacks, R. H. Colby, and M. A. Ramsay. 1992a. "The Anatomy, Chemical Composition and Metabolism of Adipose Tissue in Wild Polar Bears (*Ursus maritimus*)." *Canadian Journal of Zoology* 70:326–341.

Pond, Caroline M., C. A. Mattacks, and D. Sadler. 1992b. "The Effects of Exercise and Feeding on the Activity of Lipoprotein Lipase in Nine Different Adipose Depots of Guinea-Pigs." *International Journal of Biochemistry* 24:1825–1831.

Pond, Caroline M., C. A. Mattacks, P. C. Calder, and J. Evans. 1993. "Site-Specific Properties of Human Adipose Depots Homologous to those of Other Mammals." *Comparative Biochemistry and Physiology* 104A:819–824.

Pond, Caroline M., C. A. Mattacks, R. H. Colby, and N. J. Tyler. 1993. "The Anatomy, Chemical Composition, and Maximum Glycolytic Capacity of Adipose Tissue in Wild Svalbard Reindeer (*Rangifer tarandus platyrhynchus*) in Winter." *Journal of Zoology, London* 229:17–40.

Pond, Caroline M., C. A. Mattacks, and M. A. Ramsay. 1994. "The Anatomy and Chemical Composition of Adipose Tissue in Wild Wolverines (*Gulo gulo*) in northern Canada." *Journal of Zoology, London* 232:603–616.

Pond, Caroline M., C. A. Mattacks, and P. Prestrud. 1995. "Variability in the Distribution and Composition of Adipose Tissue in Wild Arctic Foxes (*Alopex lagopus*) on Svalbard." *Journal of Zoology, London* 236:593–610.

Popkin, B. M., R. E. Bilsborrow, and J. S. Akin. 1982. "Breast-Feeding Patterns in Low-Income Countries." *Science* 218:1088.

Popp, Joseph L. 1983. "Ecological Determinism in the Life Histories of Baboons." *Primates* 24:198–210.

Portman, A. 1990. *A Zoologist Looks at Humankind.* New York: Columbia University Press.

Prentice, A., and A. Prentice. 1988. "Reproduction against the Odds." *New Scientist* 118:42–46.

Prentice, A., A. M. Prentice, and R. G. Whitehead. 1981. "Breast-Milk Concentration of Rural African Women. 2. Long-Term Variations within a Community." *British Journal of Nutrition* 45: 495–503.

Prentice, A. M., and R. G. Whitehead. 1987. The Energetics of Human Reproduction. *Symposium of the Zoological Society of London* 57:275–304.

Prentice, A. M., A. Paul, A. Prentice, A. E. Black, T. J. Cole, and R. G. Whitehead. 1986. Cross-Cultural Differences in Lactational Performance. In *Human Lactation 2. Maternal and Environmental Factors*, ed. M. Hamosh and A. S. Goldman, pp. 13–44. New York: Plenum Press.

Preuschoft, H., D. Chivers, W. Brockelman, and N. Creel. 1984. *The Lesser Apes: Evolutionary and Behavioral Biology*. Edinburgh: Edinburgh University Press.

Prior, J. C. 1985. "Luteal Phase Defects and Anovulation: Adaptive Alterations Occurring with Conditioning Exercise." *Seminars in Reproductive Endocrinology* 3:27.

Promislow, D. E. L., and P. H. Harvey. 1990. "Living Fast and Dying Young: A Comparative Analysis of Life-History Variation among Mammals." *Journal of Zoology, London* 220:417–437.

Pusey, Anne E. 1983. "Mother-Offspring Relationships in Chimpanzees after Weaning." *Animal Behaviour* 31:363–377.

———. 1990. "Behavioural Changes in Adolescence in Chimpanzees." *Behaviour* 115:203–246.

Pusey, Anne E., and Craig Packer. 1987. "Dispersal and Philopatry." In *Primate Societies*, ed. B. B. Smuts, D. L. Cheney, R. M. Seyfarth, R. W. Wrangham, and T. T. Struhsaker, pp. 250–266. Chicago: University of Chicago Press.

Quandt, S. A. 1985. "Biological and Behavioral Predictors of Exclusive Breastfeeding Duration." *Medical Anthropology* 9:139–151.

———. 1986. "Patterns of Variation in Breastfeeding Behaviors." *Social Science and Medicine* 23: 445–453.

———. 1987. "Maternal Recall Accuracy for Dates of Infant Feeding Transitions." *Human Organization* 46:152.

Radin, Paul, ed. 1926. *Crashing Thunder: The Autobiography of an American Indian*. New York: Appleton.

Raemakers, J. J., and P. M. Raemakers. 1985. "Field Playback of Loud Calls to Gibbons (*Hylobates lar*): Territorial, Sex-Specific, and Species-Specific Responses." *Animal Behaviour* 33:481–493.

Rajalakshmi, R. 1980. "Gestation and Lactation Performance in Relation to the Plane of Maternal Nutrition." In *Maternal Nutrition during Pregnancy and Lactation*, ed. H. Aebi and R. G. Whitehead, pp. 184–202. Bern: Hans Huber.

Ralls, Katharine. 1976. "Mammals in Which Females Are Larger than Males." *Quarterly Review of Biology* 51:245–276.

———. 1977. "Sexual Dimorphism in Mammals: Avian Models and Unanswered Questions." *American Naturalist* 11:917–938.

Ramsay, M. A., C. A. Mattacks, and C. M. Pond. 1992. "Seasonal and Sex Differences in the Structure and Chemical Composition of Adipose Tissue in Wild Polar Bears (*Ursus maritimus*)." *Journal of Zoology, London* 228:533–544.

Rasmussen, P. 1986. "Effect of Calcium Deprivation of Rat Dams on Fetuses and Newborn Offspring." *Calcified Tissue International* 38:289–292.

Rawlins, Richard G. 1975. Age changes in the pubic symphysis of *Macaca mulatta*. *American Journal of Physical Anthropology* 42:477–488.

Rawlins, Richard G., and Matt J. Kessler, eds. 1986a. *The Cayo Santiago Macaques: History, Behavior, and Biology*. Albany: State University of New York Press.

———. 1986b. "Demography of the Free-Ranging Cayo Santiago Macaques (1976–1983)." In *The Cayo Santiago Macaques: History, Behavior, and Biology*, ed. Richard G. Rawlins and Matt J. Kessler, pp. 47–72. Albany: State University of New York Press.

Rebuffé-Scrive, M. 1987. "Regional Adipose Tissue Metabolism in Women during and after Reproductive Life and in Men." *Recent Advances in Obesity Research* 5:82–91.

Rebuffé-Serive, M., L. Enk, N. Crona, P. Lonnroth, L. Abrahamsson, U. Smith, and P. Björntorp. 1985. Fat Cell Metabolism in Different Regions in Women: Effect of Menstrual Cycle, Pregnancy, and Lactation. *Journal of Clinical Investigation* 75:1973–1976.

Reeves, J. 1979. "Estimating Fatness." *Science* 204: 881.

Reite, Martin, and Deborah Snyder. 1982. "Physiology of Maternal Separation in a Bonnet Macaque Infant." *American Journal of Primatology* 2:115–120.

Reite, Martin, C. Seiler, and R. Short. 1978. "Loss of Your Mother is More than Loss of a Mother." *American Journal of Psychiatry* 135:370–371.

Reiter, J. 1984. "Studies of Female Competition and Reproductive Success in Northern Elephant Seals." Ph.D. diss., University of California, Santa Cruz.

———. 1997. "Life History and Reproductive Success of Female Northern Elephant Seals." In *The*

Evolving Female: A Life-History Perspective, ed. M. E. Morbeck, A. Galloway, and A. L. Zihlman, chap. 4. Princeton, NJ: Princeton University Press.

Reiter, J., K. J. Panken, and B. J. Le Boeuf. 1981. "Female Competition and Reproductive Success in Northern Elephant Seals." *Animal Behavior* 29:670–687.

Reiter, J., Nell L. Stinson, and Burney J. Le Boeuf. 1978. "Northern Elephant Seal Development: The Transition from Weaning to Nutritional Independence." *Behavioral Ecology and Sociobiology* 3:337–367.

Reiter, J., and B. J. Le Boeuf. 1991. "Life History Consequences of Variation in Age at Primiparity in Northern Elephant Seals." *Behavioral Ecology and Sociobiology* 28:153–160.

Reitz, R. E., T. A. Duane, J. R. Woods, and R. L. Weinstein. 1977. "Calcium, Magnesium, Phosphorus, and Parathyroid Hormone Interrelationships in Pregnancy and Newborn Infants." *Obstetrics and Gynecology* 50:701–705.

Rhine, Ramon J., Guy W. Norton, and B. J. Westlund. 1984. "The Waning of Dependence in Infant Free-Ranging Yellow Baboons (*Papio cynocephalus*) of Mikumi National Park." *American Journal of Primatology* 7:213–228.

Richard, Alison F. 1985. *Primates in Nature.* San Francisco: Freeman.

Richard, Alison F., Pothin Rakotomanga, and Marion Schwartz. 1991. "Demography of *Propithecus verreauxi* at Beza Mahafaly, Madagascar: Sex Ratio, Survival, and Fertility, 1984–1988." *American Journal of Physical Anthropology* 84:307–322.

Richelson, L. S., H. W. Wahner, L. J. Melton III, and B. L. Riggs. 1984. "Relative Contributions of Aging and Estrogen Deficiency to Postmenopausal Bone Loss." *New England Journal of Medicine* 311(20):1273–1275.

Riedman, M. L., and C. L. Ortiz 1979. "Changes in Milk Composition during Lactation in the Northern Elephant Seal." *Physiological Zoology* 52:240–249.

Riggs, B. L. 1991. "Overview of Osteoporosis." *Western Journal of Medicine* 154:63–77.

Riggs, B. L., and L. J Melton. 1983. "Evidence for Two Distinct Syndromes of Involutional Osteoporosis." *American Journal of Medicine* 756:899–901.

———. 1986. "Involutional Osteoporosis." *New England Journal of Medicine* 31426:1676–1686.

———, eds. 1988. *Osteoporosis.* New York: Raven Press.

Riggs, B. L., and K. I. Nelson. 1985. "Effect of Long-Term Treatment with Calcitriol on Calcium Absorption and Mineral Metabolism in Postmenopausal Osteoporosis." *Journal of Clinical Endocrinology and Metabolism* 61:457–461.

Riis, B., and D. Christiansen. 1984. "Does Calcium Potentiate the Bone-Preserving Effect of Oestrogen Treatment in Early Post-Menopausal Women by a Change in Vitamin D Metabolism?" *Maturitas* 6:65–70.

Riis, B., P. Rodbro, and C. Christiansen. 1986. "The Role of Serum Concentrations of Sex Steroids and Bone Turnover in the Development and Occurrence of Postmenopausal Osteoporosis." *Calcified Tissue International* 38:318–322.

Riopelle, A. J., and P. A. Hale. 1975. "Nutritional and Environmental Factors Affecting Gestation Length in Rhesus Monkeys." *American Journal of Clinical Nutrition* 28:1170–1176.

Riopelle, A. J., P. A. Hale, and E. S. Watts. 1976. "Protein Deprivation in Primates. VII. Determinants of Size and Skeletal Maturity at Birth in Rhesus Monkeys." *Human Biology* 48:203–222.

Roberts, S. B., and W. A. Coward. 1985. "Lactational Performance in Relation Energy Intake in the Baboon." *American Journal of Clinical Nutrition* 41:1270–1276.

Roberts, S. B., A. A. Paul, T. J. Cole, and R. G. Whitehead. 1982. "Seasonal Changes in Activity, Birth Weight, and Lactational Performance in Rural Gambian Women." *Transactions of the Royal Society of Tropical Medicine and Hygiene* 76:668–678.

Robinson, C. J., E. Spanos, M. F. James, J. W. Pike, M.-R. Haussler, A. M. Makeen, C. J. Hillyard, and I. MacIntyre. 1982. "Role of Prolactin in Vitamin D Metabolism and Calcium Absorption during Lactation in the Rat." *Journal of Endocrinology* 94:443–453.

Robinson, John G. 1988a. "Group Size in Wedge-Capped Capuchin Monkeys *Cebus olivaceus* and the Reproductive Success of Males and Females." *Behavioral Ecology and Sociobiology* 23:187–197.

———. 1988b. "Demography and Group Structure in Wedge-Capped Capuchin Monkeys, *Cebus olivaceus.*" *Behaviour* 104:202–232.

Robyn, C., S. Meuris, and P. Hennart. 1985. "Endocrine Control of Lactational Infertility." In *Maternal Nutrition and Lactational Infertility*, ed. J. Dobbing, pp. 25–40. New York: Raven Press.

Rodman, Peter S. 1977. "Feeding behavior of orangutans of the Kutai Nature Reserve, East Kalimantan." In *Primate Ecology*, ed. T. H. Clutton-Brock, pp. 383–413. London: Academic Press.

———. 1984. "Foraging and Social Systems of Orangutans and Chimpanzees." In *Adaptations for Foraging in Nonhuman Primates*, ed. P. S. Rodman and J. G. H. Cant, pp. 134–160. Columbia University Press, New York.

———. 1988. "Diversity and Consistency in Ecology and Behavior." In *Orangutan Biology*, ed. J. H. Schwartz, pp. 31–51. New York: Oxford University Press.

Rodman, Peter S. and John C. Mitani. 1987. "Orangutans: Sexual Dimorphism in a Solitary Species." In *Primate Societies*, ed. B. B. Smuts, D. L. Cheney, R. M. Seyfarth, R. W. Wrangham, and T. T. Struhsaker, pp. 146–154. Chicago: University of Chicago Press.

Rodseth, Lars, Richard W. Wrangham, Alisa M. Harrigan, and Barbara B. Smuts. 1991. "The Human Community as a Primate Society." *Current Anthropology* 32:221–249.

Roff, Derek A. 1992. *The Evolution of Life Histories. Theory and Analysis.* New York: Chapman and Hall.

Rogoff, B. 1990. *Apprenticeship in Thinking: Cognitive Development in Social Context.* New York: Oxford University Press.

Rogoff, B., and G. A. Morelli. 1989. "Perspectives on Children's Development from Cultural Psychology. Children and Their Development: Knowledge Base Research Agenda and Social Policy Application." *American Psychologist* (special issue) 44:343–348.

Rogoff, B., M. J. Sellers, S. Piorrata, N. Fox, and S. H. White. 1975. "Age of Assignment of Roles and Responsibilities to Children: A Cross-Cultural Survey." *Human Development* 18:353–369.

Rosaldo, M. Z., and L. Lamphere, eds. 1974. *Woman, Culture, and Society.* Stanford, CA: Stanford University Press.

Rose, F. G. G. 1960. *Classification of Kin, Age Structure, and Marriage amongst the Groote Eylandt Aborigines: A Study in Method and a Theory of Australian Kinship.* London: Pergamon Press.

Rose, Jerome C., Barbara A. Burnett, Mark S. Nassaney, and Michael W. Blaeuer. 1984. "Paleopathology and the Origins of Maize Agriculture in the Lower Mississippi Valley and Caddoan Culture Areas." In *Paleopathology at the Origins of Agriculture*, ed. M. N. Cohen and G. J. Armelagos, pp. 393–424. London: Academic Press.

Rose, M. R. 1983. "Theories of Life History Evolution." *American Zoologist* 23:15–23.

Rosetta, Lyliane. 1993. "Female Reproductive Dysfunction and Intense Physical Training. In *Oxford Reviews of Reproductive Biology*, ed. S. R. Milligan, pp. 113–141. Oxford: Oxford University Press.

Ross, Caroline. 1991. "Life History Patterns of New World Monkeys." *International Journal of Primatology* 12:481–501.

———. 1992. "Life History Patterns and Ecology of Macaque Species." *Primates* 32:207–215.

Ross, Philip E. 1992. "Eloquent Remains." *Scientific American* 266:114–125.

Rothbart, M. K, and M. Rothbart. 1976. "Birth Order, Sex of Child, and Maternal Help-Giving." *Sex Roles* 2:39–46.

Rowe, Timothy. 1988. "Definition, Diagnosis, and Origin of Mammalia." *Journal of Vertebrate Paleontology* 8:241–264.

Rowell, Therma E., and Jane Chism. 1986. "The Ontogeny of Sex Differences in the Behavior of Patas Monkeys." *International Journal of Primatology* 7:83–107.

Rubenstein, Daniel I. 1993. "On the Evolution of Juvenile Life-Styles in Mammals." In *Juvenile Primates: Life History, Development, and Behavior*, ed. M. E. Pereira and L. A. Fairbanks, pp. 38–56. New York: Oxford University Press.

Ruegsegger, P., M. A. Dombacher, E. Ruegsegger, J. A. Fischer, and M. Anliker. 1984. "Bone Loss in Premenopausal and Postmenopausal Women." *Journal of Bone and Joint Surgery* 66-A7:1015–1023.

Ruff, Christopher B. 1989. "New Approaches to Structural Evolution of Limb Bones in Primates." *Folia Primatologica* 53:142–159.

Ruff, Christopher B., and J. A. Runestad. 1992. "Primate Limb Bone Structural Adaptations." *Annual Review of Anthropology* 21:407–433.

Rutenberg, G. W., A. M. Coehlo Jr., D. S. Lewis, K. D. Carey, and H. C. McGill Jr. 1987. "Body Composition in Baboons: Evaluating a Morphometric Method." *American Journal of Primatology* 12:275–285.

Sade, Donald Stone. 1980. "Population Biology of Free-Ranging Rhesus Monkeys on Cayo Santiago, Puerto Rico." In *Biosocial Mechanisms of Popula-*

tion Regulation, ed. M. N. Cohen, R. S. Malpass, and H. G. Klein, pp. 171–187. New Haven, CT: Yale University Press.

———. 1990. "Intrapopulation Variation in Life-History Parameters." In *Primate Life History and Evolution*, ed. C. Jean DeRousseau, pp. 181–194. New York: Wiley-Liss.

Sade, Donald Stone, K. Cushing, P. Cushing, J. Dunard, A. Figueroa, J. Kaplan, C. Lauer, D. Rhodes, and J. Schneider. 1976. "Population Dynamics in Relation to Social Structure on Cayo Santiago." *Yearbook of Physical Anthropology* 20: 253–262.

Sadurskis, A., N. Kabir, J. Wager, and E. Forsum. 1989. "Energy Metabolism, Body Composition, and Milk Production in Healthy Swedish Women during Lactation." *American Journal of Clinical Nutrition* 48:44–49.

Salmon, E. T. 1967. *Samnium and the Samnites.* London: Cambridge University Press.

Salthe, Stanley N. 1985. *Evolving Hierarchical Systems: Their Structure and Representation.* New York: Columbia University Press.

Samaan, N. A., G. D. Anderson, and M. Adam-Mayne. 1975. "Immunoreactive Calcitonin in the Mother, Neonate, Child, and Adult." *American Journal of Obstetrics and Gynecology* 121:622–625.

Saphier, P. W., T. C. B. Stamp, C. R. Kelsey, and N. Loveridge. 1987. "PTH Bioactivity in Osteoporosis." *Bone and Mineral* 3:75–83.

Sapolsky, Robert M. 1990. "Stress in the Wild." *Scientific American* 262:116–123.

Sarich, V. M., and J. Cronin. 1976. "Molecular Systematics of the Primates." In *Molecular Anthropology*, ed. M. Goodman and R. Tashian, pp. 141–170. New York: Plenum Press.

Saul, F. P., and J. M. Saul. 1989. Osteobiography: A Maya Example. In *Reconstruction of Life from the Skeleton*, eds. M. Y. İşcan and K. A. R. Kennedy, pp. 287–302. New York: Alan R. Liss.

Saunders, Shelley R., and M. Anne Katzenberg. 1992. *Skeletal Biology of Past Peoples: Research Methods.* New York: John Wiley.

Sauther, M. L., and L. T. Nash. 1987. Effect of Reproductive State and Body Size on Food Consumption in Captive *Galago senegalensis braccatus. American Journal of Physical Anthropology* 73:81–88.

Schaefer, O. 1977. "Are Eskimos More or Less Obese than Other Canadians? A Comparison of Skinfold Thickness and Ponderal Index in Canadian Eskimos." *American Journal of Clinical Nutrition* 30:1623–1628.

Schallenberger, E., D. W. Richardson, and E. Knobil. 1981. "Role of Prolactin in the Lactational Amenorrhea of the Rhesus Monkey (*Macaca mulatta*)." *Biological Reproduction* 25:370.

Schaller, G. 1963. *The Mountain Gorilla.* Chicago: University of Chicago Press.

Schick, Kathy D., and Nicholas Toth. 1993. *Making Silent Stones Speak: Human Evolution and the Dawn of Technology.* New York: Simon & Schuster.

Schlegel, A. 1989. "Gender Issues and Cross-Cultural Research." *Behavior Science Research* 23: 265–280.

Schmid, P., and Z. Stratil. 1986. "Growth Changes, Variations, and Sexual Dimorphism of the Gorilla Skull." In *Primate Evolution*, ed. J. Else, pp. 239–247. Cambridge: Cambridge University Press.

Schneck, Daniel J. 1990. *Engineering Principles of Physiologic Function.* New York: New York University Press.

Schultz, Adolph H. 1933. "Chimpanzee Fetuses." *American Journal of Physical Anthropology* 18:61–79.

———. 1935. "Eruption and Decay of the Permanent Teeth in Primates." *American Journal of Physical Anthropology* 19:489–581.

———. 1937. "Fetal Growth and Development of the Rhesus Monkey." *Contributions to Embryology* 26:73–97.

———. 1940. "Growth and Development of the Chimpanzee." *Contributions to Embryology* 128: 1–63.

———. 1941. "Growth and Development of the Orangutan." *Contributions to Embryology* 29:57–110.

———. 1944. "Age Changes and Variability in Gibbons." *American Journal of Physical Anthropology* 2:1–129.

———. 1956. "Postembryonic Age Changes." *Primatologia—Handbuch der Primatenkunde.* 1: 887–964.

———. 1960. "Age Changes in Primates and Their Modification in Man." In *Human Growth*, ed. J. M. Tanner, pp. 1–20. New York: Pergamon Press.

———. 1962. "Metric Age Changes and Sex Differences in Primate Skulls." *Zeitschrift für Morphologie und Anthropologie* 52:239–253.

———. 1968. "The Recent Hominoid Primates." In *Perspectives on Human Evolution.* Vol. 1, ed.

S. L. Washburn and P. C. Jay, pp. 122–195. New York: Holt, Rinehart and Winston.

Schulz, Alolph H. 1969a. *The Life of Primates.* London: Weidenfeld and Nicolson.

———. 1969b. "The Skeleton of Chimpanzees." In *The Chimpanzee. Anatomy, Behavior, and Diseases of Chimpanzees.* Vol. 1, ed. G. H. Bourne, pp. 50–103. Basel: Karger.

———. 1973. "The Skeleton of the Hylobatidae and Other Observations on Their Morphology." *Gibbon and Siamang* 2:1–54. Basel: Karger.

Schurmann, C. L., and J. A. R. A. M. van Hooff. 1986. "Reproductive Strategies of the Orangutan: New Data and a Reconsideration of Existing Sociosexual Models." *International Journal of Primatology* 7:265–287.

Schwarcz, Henry P., and Margaret J. Schoeninger. 1991. "Stable Isotope Analyses in Human Nutritional Ecology." *Yearbook of Physical Anthropology* 34:283–321.

Schwartz, Jeffrey H., and Herbert L. Langdon. 1991. "Innervation of the Human Upper Primary Dentition: Implications for Understanding Tooth Initiation and Rethinking Growth and Eruption Patterns." *American Journal of Physical Anthropology* 86:273–286.

Schweiger, U., R. Laessle, H. Pfister, C. Hoehl, M. Schwingen Schloegel, M. Schweiger, and K. M. Pirke. 1987. "Diet-Induced Menstrual Irregularities: Effects of Age and Weight Loss." *Fertility and Sterility* 48:746.

Segal, K. R., B. Lutin, E. Presta, J. Wang, and T. E. Van Itallie. 1985. Estimation of Human Body Composition by Electrical Impedance Methods: A Comparative Study. *Journal of Applied Physiology* 58:1565–1571.

Seger, J., and H. J. Brockman. 1987. "What Is Bet-Hedging?" In *Oxford Surveys of Evolutionary Biology.* Vol. 4: ed. P. H. Harvey and L. Partridge, pp. 182–211. Oxford: Oxford University Press.

Seton, Ernest Thompson. 1909. *Life-Histories of Northern Mammals.* New York: C. Scribner.

Seyfarth, Robert M. 1978. "Social Relationships among Adult Male and Female Baboons. 2: Behavior throughout the Female Reproductive Cycle." *Behaviour* 64:227–247.

Shangold, M. M., and H. S. Levine. 1982. "The Effect of Marathon Training upon Menstrual Function." *American Journal of Obstetrics and Gynecology* 143:862.

Shattock, S. G. 1909. On Normal Tumour-Like Formations of Fat in Man and the Lower Animals. *Proceedings of the Royal Society of Medicine and Pathology* 2:207–270.

Shea, Brian T. 1988. "Heterochrony in Primates." In *Heterochrony in Evolution: A Multidisciplinary Approach*, ed. M. L. McKinney, pp. 237–266. New York: Plenum Press.

———. 1989. "Heterochrony in Human Evolution: The Case for Neoteny Reconsidered." *Yearbook of Physical Anthropology* 32:69–101.

———. 1990. "Dynamic Morphology: Growth, Life History, and Ecology in Primate Evolution." In *Primate Life History and Evolution*, ed. C. J. DeRousseau, pp. 325–352. New York: Wiley-Liss.

———. 1992. "Developmental Perspective on Size Change and Allometry in Evolution." *Evolutionary Anthropology* 4:124–134.

Shenolikar, I. S. 1970. "Absorption of Dietary Calcium in Pregnancy." *American Journal of Clinical Nutrition* 23:63–67.

Shigehara, Nobuo. 1980. "Epiphyseal Union, Tooth Eruption, and Sexual Maturation in the Common Tree Shrew, with Reference to its Systematic Problem." *Primates* 2:1–19.

Shively, C. A., T. B. Clarkson, L. C. Miller, and K. W. Weingard. 1987. "Body Fat Distribution as a Risk Factor for Coronary Artery Atherosclerosis in Female Cynomolgus Monkeys." *Arteriosclerosis* 7:226–231.

Short, R. V. 1984. "Breast-Feeding." *Scientific American* 250:35.

Shostak, M. 1976. *Nisa: The Life and Words of a !Kung Woman.* Cambridge, MA: Harvard University Press.

———. 1981. *Nisa: The Life and Words of a !Kung Woman.* New York: Vintage.

———. 1990. *Nisa: The Life and Words of a !Kung Woman.* New ed. London: Earthscan.

Sibley, C. G., and J. E. Ahlquist. 1984. "The Phylogeny of the Hominoid Primates, as Indicated by DNA-DNA Hybridization." *Journal of Molecular Evolution* 20:2–15.

Siebert, Joseph R., and Daris R. Swindler. 1991. "Perinatal Dental Development in the Chimpanzee (*Pan troglodytes*)." *American Journal of Physical Anthropology* 86:287–294.

Sigg, H., A. Stolba, J.-J. Abegglen, and V. Dasser. 1982. "Life History of Hamadryas Baboons: Physical Development, Infant Mortality, Reproductive Parameters, and Family Relationships." *Primates* 23:473–487.

Silberberg, M., and R. Silberberg. 1976. "Steroid Hormones and Bone." In *The Biochemistry and*

Physiology of Bone. Vol. 4. 2d ed. ed. G. H. Bourne, pp. 401–484. New York: Academic Press.

Silk, Joan B. 1986. "Eating for Two: Behavioral and Environmental Correlates of Gestational Length among Free-Ranging Baboons *Papio cynocephalus*." *International Journal of Primatology* 7:583–602.

Silk, Joan B., Amy Samuels, and Peter Rodman. 1981. "The Influence of Kinship, Rank, and Sex on Affiliation and Aggression between Adult Female and Immature Bonnet Macaques (*Macaca radiata*)." *Behaviour* 78:111–177.

Sillen, A. 1992. "Sr/Ca of *Australopithecus robustus* and associated fauna from Swartkrans." *American Journal of Physical Anthropology* (suppl.) 14:151.

Sillen, A., and M. Kavanagh. 1982. "Strontium and Paleodietary Research: A Review." *Yearbook of Physical Anthropology* 25:67–90.

Sillen, A., and P. Smith. 1984. "Weaning Patterns Are Reflected in Strontium-Calcium Ratios of Juvenile Skeletons." *Journal of Archaeological Science* 11:237–245.

Simmons, David Jason, ed. 1990. *Nutrition and Bone Development*. New York: Oxford University Press.

Simpson, E. R., and C. R. Mendelson. 1990. "The Role of Adipose Tissue in Estrogen Biosynthesis In the Human." In *Adipose Tissue and Reproduction: Progress in Reproductive Biology and Medicine* Vol. 14, ed. R. Frisch, pp. 85–106. Basel: Karger.

Simpson, George Gaylord. 1941. "The Role of the Individual in Evolution." *Journal of the Washington Academy of Sciences* 31:1–20.

———. 1944. *Tempo and Mode of Evolution*. New Haven, CT: Yale University Press.

———. 1949. *The Meaning of Evolution*. New Haven, CT: Yale University Press.

Simpson, S. W., C. O. Lovejoy, and R. S. Meindl. 1990. "Hominoid Dental Maturation." *Journal of Human Evolution* 19:285–297.

———. 1991. "Relative Dental Development in Hominoids and Its Failure to Predict Somatic Growth Velocity." *American Journal of Physical Anthropology* 86:113–120.

———. 1992. "Further Evidence on Relative Dental Maturation and Somatic Development Rate in Hominoids." *American Journal of Physical Anthropology* 87:29–38.

Sinclair, David. 1989. *Human Growth After Birth*. 5th ed. Oxford: Oxford University Press.

Sjöström, L. 1981. "Can the Relapsing Patient Be Identified?" *Recent Advances in Obesity Research* 3:85–93.

———. 1988. "Measurement of Fat Distribution." *Current Topics in Nutritional Disease* 18:43–61.

Sjöström, L., and P. Björntorp. 1974. "Body Composition and Adipose Tissue Cellularity in Human Obesity." *Acta Medica Scandinavica* 195:201–211.

Smadga, J. 1986. "Géodynamique des Milieux d'un Versant de Mousson en Moyenne Montagne Himalayenne—Le Versant de Salme, Nepal Central." Thèse de Doctorat en Géographie, Université de Paris I.

Small, Meredith F., and D. J. Smith. 1984. "Sex Differences in Maternal Investment by *Macaca mulatta*." *Behavioral Ecology and Sociobiology* 14: 313–314.

Smith, B. Holly. 1986. "Dental Development in *Australopithecus* and Early *Homo*." *Nature* 323: 327–330.

———. 1989. "Dental Development as a Measure of Life History in Primates." *Evolution* 43:683–688.

———. 1991. "Dental Development and the Evolution of Life History in Hominidae." *American Journal of Physical Anthropology* 86:157–174.

———. 1992. "Life History and the Evolution of Human Maturation." *Evolutionary Anthropology* 1:134–142.

———. 1994. "Patterns of Dental Development in *Homo, Australopithecus, Pan* and *Gorilla*." *American Journal of Physical Anthropology* 94:307–325.

Smith, C., and B. Lloyd. 1978. "Maternal Behavior and Perceived Sex of Infant: Revisited." *Child Development* 49:1263–1265.

Smith, C. C. 1977. "Feeding Behaviour and Social Organization in Howling Monkeys." In *Primate Ecology: Studies of Feeding and Ranging Behavior in Lemurs, Monkeys and Apes*, ed. T. H. Clutton-Brock, pp. 97–106. London: Academic Press.

Smith, E. A. 1979. "Human Adaptation and Energetic Efficiency." *Human Ecology* 7:53–74.

———. 1983. "Anthropological Applications of Optimal Foraging Theory: A Critical Review." *Current Anthropology* 24:625–640.

Smith, E. L., and C. Gilligan. 1989. "Mechanical Forces and Bone." In *Bone and Mineral Research*. Vol. 6, ed. W. A. Peck, pp. 139–173. Amsterdam: Elsevier Science.

Smith, K. K. 1992. "The Evolution of the Mammalian Pharynx." *Zoological Journal of the Linnean Society* 104:313–349.

Smith, R. H. 1991. "Genetic and Phenotypic Aspects of Life-History Evolution in Animals." In *Advances in Ecological Research*. Vol. 21, ed. M. Begon, A. H. Fitter, and A. MacFadyen, pp. 63–120. London: Academic Press.

Smuts, Barbara B. 1983. "Special Relationships between Adult Male and Female Olive Baboons: Selective Advantages." In *Primate Social Relationships: An Integrated Approach*, ed. R. A. Hinde, pp. 262–266. Oxford: Blackwell.

———. 1985. *Sex and Friendship in Baboons*. Hawthorne, New York: Aldine de Gruyter.

———. 1987. "Gender, Aggression and Influence." In *Primate Societies*, ed. B. B. Smuts, D. L. Cheney, R. M. Seyfarth, R. W. Wrangham, and T. T. Struhsaker, pp. 400–412. Chicago: University of Chicago Press.

———. 1997. "Social Relationships and Life History of Primates." In *The Evolving Female: A Life-History Perspective*, ed. M. E. Morbeck, A. Galloway, and A. L. Zihlman, chap. 5. Princeton, NJ: Princeton University Press.

Smuts, Barbara B., and Nancy A. Nicolson. 1989. "Reproduction in Wild Female Olive Baboons." *American Journal of Primatology* 19:229–246.

Smuts, Barbara B., and John M. Watanabe. 1990. Social relationships and ritualized greetings in adult male baboons (*Papio cynocephalus anubis*). *International Journal of Primatology* 11:147–172.

Smuts, Barbara B., Dorothy Cheney, R. M. Seyfarth, R. W. Wrangham, and T. T. Struhsaker, eds. 1987. *Primate Societies*. Chicago: University of Chicago Press.

Sober, E., ed. 1984. *Conceptual Issues in Evolutionary Biology: An Anthology*. Cambridge, MA: MIT Press.

Sokoll, L. J., and B. Dawson-Hughes. 1989. "Effect of Menopause and Aging on Serum Total and Ionized Calcium and Protein Concentrations." *Calcified Tissue International* 44:181–185.

Solway, J. S., and R. B. Lee. 1990. "Foragers, Genuine or Spurious? Situating the Kalahari San in History." *Current Anthropology* 31:109–122

Southwood, T. R. E. 1988. "Tactics, Strategies, and Templets." *Oikos* 52:3–18.

Sowers, M. F., G. Corton, B. Shapiro, M. L. Jannausch, M. Crutchfield, M. L. Smith, J. F. Randolph, and B. Hollis. 1993. "Changes in Bone Density with Lactation." *Journal of the American Medical Association* 269:3130–3135.

Spanos, E., D. W. Colston, I. M. Evans, L. S. Galante, S. J. Macauley, and I. MacIntyre. 1976. "Effect of Prolactin on Vitamin D Metabolism." *Molecular and Cellular Endocrinology* 5:163–167.

Specker, B. L., R. C. Tsang, and M. L. Ho. 1991. "Changes in Calcium Homeostasis over the First Year Postpartum: Effect of Lactation and Weaning." *Obstetrics and Gynecology* 78:56–62.

Sprague, David S. 1992. "Life History and Male Intertroop Mobility among Japanese Macaques (*Macaca fuscata*)." *International Journal of Primatology* 13:437–53.

Sroufe, L. Alan. 1984. "The Organization of Emotional Development." In *Approaches to Emotion*, ed. K. R. Scherer and Paul Ekman, pp. 109–157. Hillsdale, NJ: Erlbaum.

Stallings, J., C. Panter-Brick, and C. M. Worthman. 1994. "Prolactin Levels in Nursing Tamang and Kami Women: Effects of Nursing Practices on Lactational Amenorrhea." *American Journal of Physical Anthropology* (suppl.) 18:185–186.

Stearns, Stephen C. 1976. "Life-History Tactics: A Review of the Ideas." *Quarterly Review of Biology* 51:3–45.

———. 1977. "The Evolution of Life History Traits: A Critique of the Theory and a Review of the Data." *Annual Review of Ecology and Systematics* 8:145–171.

———. 1986. "Natural Selection and Fitness, Adaptation, and Constraint." In *Patterns and Processes in the History of Life*, ed. D. M. Raup and D. Jablonski, pp. 23–44. Berlin: Springer-Verlag.

———. 1987a. "Why Sex Evolved and the Difference It Makes." In *The Evolution of Sex and Its Consequences*, ed. S. C. Stearns, pp. 15–32. Basel: Birkhauser Verlag.

———. 1987b. "The Selection-Arena Hypothesis." In *The Evolution of Sex and Its Consequences*, ed. S. C. Stearns, pp 337–349. Basel: Birkhauser Verlag.

———. 1992. *The Evolution of Life Histories*. New York: Oxford University Press.

Stearns, Stephen C., and Jacob C. Koella. 1986. "The Evolution of Phenotypic Plasticity in Life-History Traits: Predictions of Reaction Norms for Age and Size at Maturity." *Evolution* 40:893–913.

Stevenson, J. C., C. J. Hilyard, and I. MacIntyre. 1979. "A Physiological Role for Calcitonin: Protection of the Maternal Skeleton." *Lancet* 2:769–770.

Stewart, Kelly J. 1981. Social Development of Wild Mountain Gorillas. Ph.D. diss., University of Cambridge.

Stewart, Kelly J., and Alexander H. Harcourt. 1987. "Gorillas: Variation in Female Relationships." In *Primate Societies*, ed. B. B. Smuts, D. L. Cheney, R. M. Seyfarth, R. W. Wrangham, and T. T. Struhsaker, pp. 155–164. Chicago: University of Chicago Press.

Stini, William A. 1969. "Nutritional Stress and Growth: Sex Differences in Adaptive Response." *American Journal of Physical Anthropology* 31: 417–426.

———. 1982. "Sexual Dimorphism and Nutrient Reserves." In *Sexual Dimorphism in Homo sapiens: A Question of Size?* ed. R. L. Hall, pp. 391–419. New York: Praeger.

———. 1990. "'Osteoporosis': Etiologies, Prevention and Treatment." *Yearbook of Physical Anthropology* 33:151–194.

Stini, William A., C. W. Weber, S. R. Kemberling, and L. A. Vaughan. 1980. "Lean Tissue Growth and Disease Susceptibility in Bottle-Fed versus Breast-Fed Infants." In *Social and Biological Predictors of Nutritional Status, Physical Growth, and Neurological Development*, ed. L. S. Greene and F. E. Johnston, pp. 81–106. New York: Academic Press.

Streier, Karen, F. D. C. Mendes, J. Rimoli, and A. O. Rimoli. 1993. "Demography and Social Structure in One Group of Muriquis (*Brachyteles arachnoides*)." *International Journal of Primatology* 14:513–526.

Stringer, Christopher, M. Christopher Dean, and Robert D. Martin. 1990. "A Comparative Study of Cranial and Dental Development within a Recent British Sample and among Neandertals." In *Primate Life History and Evolution*, ed. C. J. DeRousseau, pp. 115–152. New York: Wiley-Liss.

Strum, Shirley C. 1987. *Almost Human: A Journey into the World of Baboons*. New York: Random House.

Strum, Shirley C., and David Western. 1982. "Variations in Fecundity with Age and Environment in Olive Baboons." *American Journal of Primatology* 3:61–76.

Stuart-Macadam, Patty. 1992. "Porotic Hyperostosis: A New Perspective." *American Journal of Physical Anthropology* 87:39–47.

Suchey, J. M., D. V. Wiseley, D. F. Green, and T. T. Noguchi. 1979. "Analysis of Dorsal Pitting in the Os Pubis in an Extensive Sample of Modern American Females." *American Journal of Physical Anthropology* 51:517–540.

Sugiyama, Yukimaru. 1976. "Life History of Male Japanese Monkeys." In *Advances in the Study of Behavior*, ed. J. S. Rosenblatt, R. A. Hinde, E. Shaw, and C. Beer, pp. 255–284. New York: Academic Press.

Sugiyama, Yukimaru, and Hideyuki Ohsawa. 1982. "Population Dynamics of Japanese Monkeys with Special Reference to the Effect of Artificial Feeding." *Folia Primatologica* 39:238–269.

Sujardito, J. 1982. "Locomotor Behavior of the Sumatran Orangutan (*Pongo pygmaeu abelii*) at Ketambe, Gunung Leuser National Park." *Malayan Nature Journal* 35:57–64.

Sugiyama, Yukimaru, and J. A. R. A. M. van Hooff. 1986. "Age-Sex Class Differences in the Postural Behavior of the Sumatran Orangutan in the Gunung Leuser National Park, Indonesia." *Folia Primatologica* 47:14–25.

Sumner, Dale Richman, Jr., Mary Ellen Morbeck, and John J. Lobick. 1989. "Apparent Age-Related Bone Loss among Adult Female, Gombe Chimpanzees." *American Journal of Physical Anthropology* 79:225–234.

Sumner, E. L. 1936. "A Life History Study of the California Quail." *California Fish and Game* 21: 167–256.

Susman, R. L. 1984. *The Pygmy Chimpanzee*. New York: Plenum Press.

Sussman, Robert W. 1991a. "Demography and Social Organization of Free-Ranging *Lemur catta* in the Beza Mahafaly Reserve, Madagascar." *American Journal of Physical Anthropology* 84:43–58.

———. 1991b. "Primate Origins and the Evolution of Angiosperms." *American Journal of Primatology* 23:209–223.

Swartz, Sharon M. 1989. "The Functional Morphology of Weight Bearing: Limb Joint Surface Area Allometry in Anthropoid Primates." *Journal of Zoology, London* 218:441–460.

Swindler, Daris. 1985. "Nonhuman Primate Dental Development and Its Relationship to Human Dental Development." In *Nonhuman Primate Models for Human Growth and Development*, ed. E. S. Watts, pp. 67–94. New York: Alan R. Liss.

Symington, Margaret McFarland. 1987. "Sex Ratio and Maternal Rank in Wild Spider Monkeys: When Daughters Disperse." *Behavioral Ecology and Sociobiology* 20:421–425.

Symons, Donald. 1978. "The Question of Functions: Dominance and Play." In *Social Play in Primates*, ed. E. O. Smith, pp. 193–230. New York: Academic Press.

Szalay, F. S., and R. K. Costello. 1991. "Evolution of Permanent Estrus Displays in Hominids." *Journal of Human Evolution* 20:439–464.

Tague, R. C. 1988. "Bone Resorption of the Pubis and Preauricular Area in Humans and Nonhuman Mammals." *American Journal of Physical Anthropology* 76:251–267.

———. 1990. "Morphology of the Pubis and Preauricular Area in Relation to Parity and Age at Death in *Macaca mulatta*," *American Journal of Physical Anthropology* 82:517–525.

Tague, R. C., and C. O. Lovejoy. 1986. "The Obstetric Pelvis of A.L. 288-1 (Lucy)." *Journal of Human Evolution* 15:237–255.

Takahata, Y. 1980. "The Reproductive Biology of a Free-Ranging Troop of Japanese Monkeys." *Primates* 21:303–329.

Takahata, Y., T. Hasegawa, and T. Nishida 1984. "Chimpanzee Predation in the Mahale Mountains from August 1979 to May 1982." *International Journal of Primatology* 5:213–233.

Tanaka, I. 1992. "Three Phases of Lactation in Free-Ranging Japanese Macaques." *Animal Behaviour* 44:129–139.

Tanner, J. M. 1962. *Growth at Adolescence.* Oxford: Blackwell.

———. 1978. *Foetus into Man: Physical Growth from Conception to Maturity.* Cambridge, MA: Harvard University Press.

———. 1990. *Foetus into Man: Physical Growth from Conception to Maturity*, 2d ed. Cambridge, MA: Harvard University Press.

Tanner, J. M., and R.H. Whitehouse. 1975. "Revised Standards for Triceps and Subscapular Skinfolds in British Children." *Archives of Diseases in Childhood* 50:142–145.

Tanner, J. M., M. E. Wilson, and C. G. Rudman. 1990. "Pubertal Growth Spurt in the Female Rhesus Monkey: Relation to Menarche and Skeletal Maturation." *American Journal of Human Biology* 2:101–106.

Tanner, N. T., and A. L. Zihlman 1976a. "Women in Evolution. Part 1. Innovation and Selection in Human Origins." *Signs: Journal of Women in Culture and Society* 1(3):585–608.

Tanner, N. T., and A. L. Zihlman. 1976b. "Discussion Paper. The Evolution of Human Communication: What Can Primates Tell Us?" *New York Academy of Science Annals* 280:467–480.

Tay, C. C. K., A. F. Glasier, and A. S. McNeilly. 1992. "The 24 h Pattern of Pulsatile LH, FSH, and Prolactin Release during the First 8 Weeks of Lactational Amenorrhoea in Breast-feeding Women." *Human Reproduction* 7:951–958.

Tenaza, R. R. 1976. "Songs, Choruses and Countersinging of Kloss's Gibbons (*Hylobates klossii*) in Siberut Island, Indonesia." *Zeitschrift Tierpsychologie* 40:37–52.

Thieme, Frederick P., and William J. Schull. 1957. "Sex Determination from the Skeleton." *Human Biology* 29:242–273.

Thomas, R. B. 1976. "Energy Flow at High Altitude." In *Man in the Andes*, ed. P. T. Baker and M. Little, pp. 379–404. Stroudsburg, PA: Dowden, Hutchison and Ross.

Thomas, R. B., and B. Winterhalder. 1976. "Physical and Biotic Environment of Southern Highland Peru." In *Man in the Andes*, ed. P. T. Baker and M. A. Little, pp. 21–59. Stroudsburg, PA: Dowden, Hutchinson, and Ross.

Thomas, R. D. K., and W.-E. Reif. 1993. "The Skeleton Space: A Finite Set of Organic Designs." *Evolution* 47:341–360.

Thomas, V. G. 1990. "Control of Reproduction in Animal Species with High and Low Body Fat Reserves." In *Adipose Tissue and Reproduction: Progress in Reproductive Biology and Medicine*, Vol. 14, ed. R. Frisch, pp. 85–106. Basel: Karger.

Thomas, William I., and Florian Znaniecki. 1918–1920. *The Polish Peasant in Europe and America.* 5 vols. Boston: Richard G. Badger.

Thomas, Yan. 1990. "La Divisione dei Sessi nel Diritto Romano." In *Storia delle Donne: L'Antichita*, ed. G. Duby and M. Perrot, pp. 103–176. Bari: Laterza.

Thompsen, S. D., K. A. Ono, O. T. Oftedal, and D. J. Boness. 1987. "Thermoregulation and Resting Metabolic Rate of California Sea Lion *(Zalophus californianus)* Pups." *Physiological Zoology* 60:730–736.

Thompson-Handler, N., R. K. Malenky, and N. Badrian 1984. "Sexual Behavior of *Pan paniscus* under Natural Conditions in the Lomako Forest, Equateur, Zaire." In *The Pygmy Chimpanzee*, ed. R. L. Susman, pp. 347–368. New York: Plenum Press.

Thomson, A. M., and F. Hytten. 1977. "Physiologic Basis of Nutritional Needs during Pregnancy and Lactation." In *Nutritional Impacts on Women throughout Life with Emphasis on Reproduction*, ed. K. S. Moghissi and T. N. Evans, pp. 10–22. New York: Harper and Row.

Thomson, Keith Stewart. 1988. *Morphogenesis and Evolution.* New York: Oxford University Press.

———. 1992. "Macroevolution: The Morphological Problem." *American Zoologist* 32:106–112.

Thorne, Alan G., and Milford Wolpoff. 1992. "The Multiregional Evolution of Humans." *Scientific American* 266:76–83.

Tiegs, R. D., J. J. Body, H. W. Wahner, J. Barta, B. L. Riggs, and H. Heath III. 1985. "Calcitonin Secretion in Postmenopausal Osteoporosis." *New England Journal of Medicine* 312:1097–1100.

Toner, M. M., M. N. Sanka, M. E. Foley, and K. B. Pandolf. 1986. "Effects of Body Mass and Morphology on Thermal Response in Water." *Journal of Applied Physiology* 60:521–525.

Torun, B., J. McGuire, and R. D. Mendoza. 1982. "Energy Costs of Activities and Tasks of Women from a Rural Region of Guatemala." *Nutrition Research* 2:127–136.

Toverud, S. U., and A. Boass. 1979. "Hormonal Control of Calcium Metabolism in Lactation." *Vitamins and Hormones*, 37:303–347.

Toverud, S. U., C. Harper, and P. L. Munson. 1976. "Calcium Metabolism during Lactation Enhanced Effects of Thyrocalcitonin." *Endocrinology* 99:371–378.

———. 1978. "Calcium Metabolism during Lactation: Enhanced Blood Levels of Calcitonin." *Endocrinology* 103:472–479.

Tracer, D. P. 1991. "Fertility-Related Changes in Maternal Body Compostion among the Au of Papua New Guinea." *American Journal of Physical Anthropology* 85:395–405.

Tremollieres, F. A., D. D. Stong, D. J. Baylink, and S. Mohan. 1992. "Progesterone and Promegestone Stimulate Human Bone Cell Proliferation and Insulin-Like Growth Factor-2 Production." *Acta Endocrinologica* 126:329–337.

Trevathan, W. R. 1987. *Human Birth: An Evolutionary Perspective.* New York: Aldine de Gruyter.

Trillmich, Fritz. 1986. "Maternal Investment and Sex-allocation in the Galapagos Fur Seal, *Arctocephalus galapagoensis.*" *Behavioral Ecology and Sociobiology* 19:157–164.

———. 1990. "The Behavioral Ecology of Maternal Effort in Fur Seals and Sea Lions." *Behaviour* 114:3–20.

Trillmich, Fritz, and T. Dellinger. 1991. "The Effects of El Niño on Galapagos Pinnipeds." In *Pinnipeds and El Niño Responses to Environmental Stress*, ed. F. Trillmich and K. A. Ono, pp. 66–74. Heidelberg: Springer-Verlag.

Trivers, R. C. 1972. "Parental Investment and Sexual Selection." In *Sexual Selection and the Descent of Man: 1871–1971*, ed. Bernard Campbell, pp. 136–179. Chicago: Aldine.

Trivers, R. C., and D. E. Willard. 1973. "Natural Selection of Parental Ability to Vary the Sex Ratio of Offspring." *Science* 179:90–92.

Tronick, E., G. A. Morelli, and S. Winn. 1987. "Multiple Caretaking of Efe (Pygmy) Infants." *American Anthropologist* 89:96–106.

———. 1989. "The Caretaker-Child Strategic Model: Efe and Aka Child Rearing as Exemplars of the Multiple Factors Affecting Child Rearing—A Reply to Hewlett." *American Anthropologist* 91:192–194.

Trussel, J. 1978. "Menarche and Fatness: Reexamination of the Critical Body Composition Hypothesis." *Science* 200:1506–1509.

———. 1980. "Statistical Flaws in Evidence for the Frisch Hypothesis That Fatness Triggers Menarche." *Human Biology* 52:711.

Turner, T. R., P. L. Whitten, C. J. Jolly, and J. G. Else. 1987. "Pregnancy Outcome in Free-Ranging Vervet Monkeys (*Cercopithecus aethlops*)." *American Journal of Primatology* 12:197–203.

Tutin, C. E. G. 1994. "Reproductive success story. Variability among chimpanzees and comparisons with gorillas." In *Chimpanzee Cultures*, ed. R. W. Wrangham, W. C. McGrew, F. B. M. de Waal, P. G. Heltne, and L. A. Marquardt, pp. 181–193. Cambridge, MA: Harvard University Press.

Tutin, C. E. G., and K. Benirschke. 1991. "Possible Osteomyelitis of Skull Causes Death of a Wild Lowland Gorilla in the Lopé Reserve, Gabon." *Journal of Medical Primatology* 2:357–360.

Tutin, C. E. G., and M. Fernandez. 1985. "Foods Consumed by Sympatric Populations of *Gorilla g. gorilla* and *Pan t. troglodytes* in Gabon." *International Journal of Primatology* 6:27–43.

———. 1988. "Foods Consumed by Sympatric Populations of *Gorilla g. gorilla* and *Pan t. troglodytes* in Gabon: Some Preliminary Data." *International Journal of Primatology* 6:27–43.

———. 1992. "Insect-Eating by Sympatric Lowland Gorillas (*G. g. gorilla*) and Chimpanzees (*Pan troglodytes*) in the Lopé Reserve, Gabon." *American Journal of Primatology* 28:29–40.

Tutin, C. E. G., M. Fernandez, M. E. Rogers, E. A. Williamson, and W. C. McGrew. 1991. "Foraging Profiles of Sympatric Lowland Gorillas and Chimpanzees in the Lopé Reserve, Gabon." *Philosophical Transactions of the Royal Society of London B* 354:179–186.

Tutin, C. E. G., M. Fernandez, M. E. Rogers, and E. A. Williamson. 1992. "A Preliminary Analysis of the Social Structure of Lowland Gorillas in the Lopé Reserve, Gabon." In *Topics in Primatology.* Vol. 2, *Behavior, Ecology and Conservation*, ed. N. Itoigawa, Y. Sugiyama, G. P. Sackett, and R. K. R. Thompson, pp. 245–253. Tokyo: University of Tokyo Press.

Tuttle, R., and D. Watts. 1985. "The Positional Behavior and Adaptive Complexes of *Pan gorilla.*" In *Primate Morphophysiology, Locomotor Analyses, and Human Bipedalism*, ed. S. Kondo, pp. 261–288. Tokyo: Tokyo University Press.

Tyson, J. E. 1977. "Neuroendocrine Control of Lactational Infertility." *Journal of Biosocial Science* (suppl.) 4:23–40.

Ueda, S., F. Matsuda, and T. Honjo. 1988. "Multiple Recombinational Events in Primate Immunoglobin Epsilon and Alpha Genes Suggest Closer Relationships of Humans to Chimpanzees than to Gorillas." *Journal of Molecular Evolution* 27:77–83.

Uehara, Shigeo. 1986. "Sex and Group Differences in Feeding on Animals by Wild Chimpanzees in the Mahale Mountains National Park, Tanzania." *Primates* 27:1–13.

———. 1990. "Utilization Patterns of a Marsh Grassland within the Tropical Rain Forest by the Bonobos (*Pan paniscus*) of Yalosidi, Republic of Zaire." *Primates* 31:311–322.

Uehara, Shigeo, and T. Nishida. 1987. "Body Weights of Wild Chimpanzees (*Pan troglodytes schweinfurthii*) of the Mahale Mountains National Park, Tanzania." *American Journal of Physical Anthropology* 72:315–321.

Uehara, Shigeo, Toshisada Nishida, Miya Hamai, Toshikazu Hasegawa, Hitoshige Hayaki, Michael A. Huffman, Kenji Kawanaka, Satoshi Kobayashi, John C. Mitani, Yukio Takahata, Hiroyuki Takasaki, and Takahiro Tsukahara. 1992. "Characteristics of Predation by the Chimpanzees in the Mahale Mountains National Park, Tanzania." In *Topics in Primatology* Vol. 1, *Human Origins*, ed. T. Nishida, W. C. McGrew, P. Marler, M. Pickford, and F. B. M. de Waal, pp. 143–158. Tokyo: Tokyo University Press.

Valenzuela, G. J., L. A. Muson, N. M. Tarbaux, and J. R. Farley. 1987. "Time-Dependent Changes in Bone, Placental, Intestinal and Hepatic Alkaline Phosphatase Activities in Serum during Human Pregnancy." *Clinical Chemistry* 33:1801–1806.

Valsiner, J. 1988. "Ontogeny of Construction of Culture within Socially Organized Environmental Settings." In *Child Development within Culturally Structured Environments: Social Co-Construction and Environmental Guidance in Development.* Vol. 2, ed. J. Valsiner, pp. 283–297. Norwood, NJ: Ablex.

van den Berghe, P. L. 1979. *Human Family Systems.* New York: Elsevier Science.

Van Esterik, P., and T. Elliott. 1986. "Infant Feeding Style in Urban Kenya." *Ecology of Food and Nutrition* 18:183.

van Ginnekan, J. K. 1977. "Fertility Regulation during Human Lactation: The Chance of Conception during Lactation." *Journal of Biosocial Science* (suppl.) 4:41–54.

Van Itallie, T., and H. Kissileff. 1990. "Human Obesity: A Problem in Body Energy Economics." In *Handbook of Behavioral Neurobiology.* Vol. 10. *Neurobiology of Food and Fluid Intake*, ed. E. Stricker, pp. 207–240. New York: Plenum Press.

Van Valen, L. 1973. "A New Evolutionary Law." *Evolutionary Theory* 1:1–30.

van Wagenen, G., and H. R. Catchpole. 1956. "Physical Growth of the Rhesus Monkey (*Macaca mulatta*)." *American Journal of Physical Anthropology* 14:245–273.

Vaughn, B. E., A. Weickgenant, and C. B. Kopp. 1981. "Socialization Supporting Mastery and Achievement: Evidence for Gender-Related Differences in Behavior of Mothers toward Their Very Young Children." Unpublished manuscript, University of California, Los Angeles.

Via, Sara, and Russell Lande. 1985. "Genotype-Environment Interaction and the Evolution of Phenotypic Plasticity." *Evolution* 39:505–522.

Vitzthum, V. J. 1988. "Variation in Infant Feeding Practices in an Andean Community." In *Multidisciplinary Studies in Andean Anthropology.* Michigan Discussions in Anthropology 8:137–156.

———. 1989. "Nursing Behavior and Its Relation to Duration of Post-Partum Amenorrhoea in an Andean Community." *Journal of Biosocial Science* 21:145–160.

———. 1990. "An Adaptational Model of Ovarian Function." Population Studies Center, Research Report no. 90–200. Ann Arbor: University of Michigan.

———. 1992a. "Infant Nutrition and the Consequences of an Expanding Market Economy." *Ecology of Food and Nutrition* 28:45–63.

———. 1992b. "Lack of Concordance between

Maternal Recall and Observational Data on Breast-Feeding." *American Journal of Physical Anthropology* (suppl.) 14:168.

———. 1992c. "The Physiological Basis of Developmental Calibration." In Abstracts of the Ninety-First Annual Meeting of the American Anthropological Association, p. 334.

———. 1994a. "Causes and Consequences of Heterogeneity in Infant Feeding Practices among Indigenous Andean Women." *Annals of the New York Academy of Sciences* 709:221–224.

———. 1994b. "Suckling Patterns: Lack of Concordance between Maternal Recall and Observational Data." *American Journal of Human Biology* 6:551–562.

———. 1994c. "The Comparative Study of Breast-Feeding Structure and Its Relation to Human Reproductive Ecology." *Yearbook of Physical Anthropology* 37:307–349.

———. 1995. "Resolution of a Paradox: The Evolution of a Flexibly Responsive Reproductive System." *Human Biology. In press.*

———. 1997. "Flexibility and Paradox: The Nature of Adaptation in Human Reproduction." In *The Evolving Female: A Life-History Perspective*, ed. M. E. Morbeck, A. Galloway, and A. L. Zihlman, chap. 18. Princeton, NJ: Princeton University Press.

Vitzthum, J., and S. Smith. 1989. "Evaluation of Data on Menstrual Status and Activity: An Evolutionary Perspective." *American Journal of Physical Anthropology* 78:318–319.

Vitzthum, V. J., P. T. Ellison, and S. Sukalich. 1994. "Salivary Progesterone Profiles of Indigenous Andean Women." *American Journal of Physical Anthropology.* (suppl.) 18:201–202.

Vogel, Steven. 1988. *Life's Devices: The Physical World of Animals and Plants.* Princeton, NJ: Princeton University Press.

Vrba, Elisabeth S. 1989. "Levels of Selection and Sorting with Special Reference to the Species Level." In *Oxford Surveys in Evolutionary Biology*, ed. P. H. Harvey and L. Partridge, pp. 111–168. Oxford: Oxford University Press.

———. 1990. "Life History in Relation to Life's Hierarchy." In *Primate Life History and Evolution*, ed. C. J. DeRousseau, pp. 37–46. New York: Wiley-Liss.

Vrba, Elisabeth S., and Niles Eldredge. 1984. "Individuals, Hierarchies, and Processes: Towards a More Complete Evolutionary Theory." *Paleobiology* 10:146–171.

Vrba, Elisabeth S., and Stephen J. Gould. 1986. "The Hierarchical Expansion of Sorting and Selection: Sorting and Selection Cannot Be Equated." *Paleobiology* 12:217–228.

Vygotsky, L. S. 1978. *Mind in Society: The Development of Higher Psychological Processes.* Cambridge, MA: Harvard University Press.

Wake, D. B., and A. Larson. 1987. "Multidimensional Analysis of an Evolving Lineage." *Science* 238:42–48.

Wake, D. B., and G. Roth, eds. 1989a. *Complex Organismal Functions: Integration and Evolution in Vertebrates.* New York: John Wiley & Sons.

———. 1989b. "Introduction." In *Complex Organismal Functions: Integration and Evolution in Vertebrates*, ed. D. B. Wake and G. Roth, pp. 1–5. New York: John Wiley & Sons.

Wake, Marvalee H. 1990. "The Evolution of Integration of Biological Systems: An Evolutionary Perspective through Studies on Cells, Tissues, and Organs." *American Zoologist* 30:897–906.

Walike, B. C., C. J. Goodner, D. J. Koerker, E. W. Chiceckel, and L. W. Kalnasky. 1977. "Assessment of Obesity in Pigtailed Monkeys." *Journal of Medical Primatology* 6:151–162.

Walker, A. 1991. "The Origin of the Genus *Homo.*" In *Evolution of Life*, ed. S. Osawa and T. Honjo, pp. 379–389. Tokyo: Springer-Verlag.

Walker, A., and Richard Leakey, eds. 1993. T*he Nariokotome Homo erectus Skeleton.* Cambridge, MA: Harvard University Press.

Walker, A., and M. Teaford. 1989. "Inferences from Quantitative Analysis of Dental Microwear." *Folia Primatologica* 53:177–189.

Wallach, S. 1987. "Calcitonin Therapy in Osteoporosis." In *Osteoporosis: Current Concepts*, ed. A. F. Roche and J. D. Gussler, pp. 72–78. Columbus, OH: Ross Laboratories.

Walters, Jeffrey R. 1987. "Transition to Adulthood." In *Primate Societies*, ed. B. B. Smuts, D. L. Cheney, R. M. Seyfarth, R. W. Wrangham, and T. T. Struhsaker, pp. 358–369. Chicago: University of Chicago Press.

Walters, Jeffrey R., and Robert M. Seyfarth. 1987. "Conflict and Cooperation." In *Primate Societies*, ed. B. B. Smuts, D. L. Cheney, R. M. Seyfarth, R. W. Wrangham, and T. T. Struhsaker, pp. 306–317. Chicago: University of Chicago Press.

Wardlaw, G. M., and A. M. Pike. 1986. "The Effect of Lactation on Peak Adult Shaft and Ultra-Distal Forearm Bone Mass in Women." *American Journal of Clinical Nutrition* 44:283–286.

Waser, Peter. 1977. "Feeding, Ranging and Group Size in the Mangabey (*Cercocebus albigena*)." In *Primate Ecology: Studies of Feeding and Ranging Behavior in Lemurs, Monkeys and Apes*, ed. T. H. Clutton-Brock, pp. 183–222. New York: Academic Press.

Washburn, S. L. 1951. "The Analysis of Primate Evolution with Particular Reference to the Origin of Man." *Cold Spring Harbor Symposia on Quantitative Biology* 15: 67–78.

———. 1968. "The Study of Human Evolution." *Condon Lectures.* Eugene: Oregon State System of Higher Education. pp. 1–48.

———. 1981. "Longevity in Primates." In *Aging: Biology and Behavior*, ed. J. L. McGaugh and S. B. Kiesler, pp. 11–29. New York: Academic Press.

Wasser, S. K., and D. Y. Isenberg. 1986. "Reproductive Failure among Women: Pathology or Adaptation?" *Journal of Psychosomatic Obstetrics and Gynecolology* 5:153–175.

Watanabe, Kunio. 1979. "Alliance Formation in a Free-Ranging Troop of Japanese Macaques." *Primates* 20:459–474.

Watanabe, Kunio, A. Mori, and M. Kawai. 1992. "Characteristic Features of the Reproduction of Koshima Monkeys, *Macaca fuscata fuscata*: A Summary of Thirty-Four Years of Observation." *Primates* 33:1–32.

Watson, Lawrence C., and Maria-Barbara Watson-Franke. 1985. *Interpreting Life Histories: An Anthropological Inquiry.* New Brunswick, NJ: Rutgers University Press.

Watts, David P. 1988. "Environmental Influences on Mountain Gorilla Time Budgets." *American Journal of Primatology* 15:195–211.

———. 1989. "Infanticide in Mountain Gorillas: New Cases and a Reconsideration of the Evidence." Ethology 81:1–18.

———. 1990a. "Mountain Gorilla Life Histories, Reproductive Competition, and Sociosexual Behavior and Some Implications for Captive Husbandry." *Zoo Biology* 9:185–200.

———. 1990b. "Ecology of Gorillas and Its Relation to Female Transfer in Mountain Gorillas." *International Journal of Primatology* 11:21–45.

———. 1991a. "Mountain Gorilla Reproduction and Sexual Behavior." *American Journal of Primatology* 24:211–225.

———. 1991b. "Strategies of Habitat Use by Mountain Gorillas" *Folia Primatologica* 56:1–16.

Watts, David P., and Anne Pusey. 1993. "Behavior of Juvenile and Adolescent Great Apes." In *Juvenile Primates: Life History, Development, and Behavior*, ed. M. E. Pereira and L. A. Fairbanks, pp. 148–167. New York: Oxford University Press.

Watts, Elizabeth S. 1985a. "Adolescent Growth and Development of Monkeys, Apes, and Humans." In *Nonhuman Primate Models for Human Growth and Development*, ed. E. S. Watts, pp. 41–65. New York: Alan R. Liss.

———, ed. 1985b. *Nonhuman Primate Models for Human Growth and Development.* New York: Alan R. Liss.

———. 1986a. "Evolution of the Human Growth Curve." In *Human Growth: A Comprehensive Treatment.* Vol. 1, *Development Biology and Prenatal Growth*, ed. F. Falkner and J. M. Tanner, pp. 153–166. New York: Plenum Press.

———. 1986b. "Skeletal Development." In *Comparative Primate Biology.* Vol. 3. *Reproduction and Development*, eds. W. R. Dukelow and J. Erwin, pp. 415–439. New York: Alan R. Liss.

———. 1990. "Evolutionary Trends in Primate Growth and Development." In *Primate Life History and Evolution*, ed. C. J. DeRousseau, pp. 89–104. New York: Wiley-Liss.

Watts, Elizabeth S., and James A. Gavan. 1982. "Postnatal Adolescent Spurt." *Human Biology* 54: 53–70.

Weaver, David S. 1986. "Forensic Aspects of Fetal and Neonatal Skeletons." In *Forensic Osteology: Advances in the Identification of Human Remains*, ed. K. J. Reichs, pp. 90–100. Springfield, IL: Charles C. Thomas.

Weiss, Kenneth M. 1990. "Duplication with Variation: Metameric Logic in Evolution from Genes to Morphology." *Yearbook of Physical Anthropology* 33:1–23.

Weitzman, N., B. Birns, and R. Friend. 1985. "Traditional and Nontraditional Mothers' Communication with Their Daughters and Sons." *Child Development* 56:894–898.

Welch, M. R. 1978. "Linking Subsistence Economy and Socialization Practices." *Journal of Social Psychology* 105:315–316.

Wenger, M. 1983. "Gender Role Socialization in an East African Community: Social Interaction between 2- to 3-Year-Olds and Older Children in Social Ecological Perspective." Ph.D. diss., Harvard University, Cambridge, MA.

Wenlock, R. W. 1977. "Birth Spacing and Prolonged Lactation in Rural Zambia." *Journal of Biosocial Science* 9:481.

Werker, J. F. 1989. "Becoming a Native Listener." *American Scientist* 77:54–59.

Wertsch, J. V. 1985. *Vygotsky and the Social Formation of Mind.* Cambridge, MA: Harvard University Press.

Western, David. 1979. "Size, Life History, and Ecology in Mammals." *African Journal of Ecology* 17:185–204.

———. 1983. "Production, Reproduction and Size of Mammals." *Oecologia* 59:269–271.

Western, David, and James Ssemakula. 1982. "Life History Patterns in Birds and Mammals and Their Evolutionary Interpretation." *Oecologia* 54:281–290.

White, F. J., and R. W. Wrangham. 1988. "Feeding Competition and Patch Size in the Chimpanzee Species *Pan paniscus* and *Pan troglodytes.*" *Behaviour* 105:148–164.

Whitehead, M., G. Lane, O. Young, S. Campbell, G. Abeyasekera, C. J. Hillyard, I. MacIntyre, K. G. Phang, and J. C. Stevenson. 1981. "Interrelations of Calcium-Regulating Hormones during Normal Pregnancy." *British Medical Journal* 283:10–12.

Whiting, B. B., ed. 1965. *Six Cultures.* Cambridge, MA: Harvard University Press.

Whiting, B. B., and C. P. Edwards. 1973. "A Cross-Cultural Analysis of Sex Differences in the Behavior of Children Aged 3 through 11." *Journal of Social Psychology* 91:171–188.

———. 1988. *Children of Different Worlds: The Formation of Social Behavior.* Cambridge, MA: Harvard University Press.

Whiting, B. B., and J. W. M. Whiting. 1975. *Children of Six Cultures: A Psycho-Cultural Analysis.* Cambridge, MA: Harvard University Press.

Whiting, J. W. M., and B. Whiting. 1973. "Altruistic and Egoistic Behavior in Six Cultures." In *Cultural Illness and Health: Essays in Human Adaption*, ed. L. Nader and T. W. Maretzkl, pp. 56–66. Washington, D.C.: American Anthropological Association.

Whitsett, J. A., M. Ho, R. C. Tsand, E. J. Norman, and K. G. Adams. 1981. "Synthesis of 1,25-Dihydroxyvitamin D3 by Human Placenta *in Vitro.*" *Journal of Clinical Endocrinology Metabolism* 53: 484–488.

Whitten, P. L. 1983. "Diet and Dominance among Female Vervet Monkeys (*Cercopithecus aethiops*)." *American Journal of Primatology* 5:139–159.

Widdowson, E. M. 1976. "Changes in the Body and Its Organs during Lactation: Nutritional Implications." In *Breast-Feeding and the Mother.* Ciba Foundation Symposium #45. pp. 103–113. Amsterdam: Elsevier.

Wieland, P., J. A. Fischer, U. Trechsel, H.-R. Roth, K. Vetter, H. Schneider, and A. Huch. 1980. "Perinatal Parathyroid Hormone, Vitamin D Metabolites, and Calcitonin in Man." *Journal of Physiology* 239:385–390.

Wiesner, T. S., and R. Gallimore. 1977. "My Brother's Keeper: Child and Sibling Caretaking." *Current Anthropology* 18:169–190.

Wiessner, P. 1982. "Risk, Reciprocity and Social Influences on !Kung San Economics." In *Politics and History in Band Societies*, ed. E. Leacock and R. B. Lee, pp. 61–84. Cambridge: Cambridge University Press.

Wilkie, D. 1988. "Hunters and Farmers of the African Forest." In *People of the Tropical Rain Forest*, ed. J. S. Denslow and C. Padoch, pp. 111–126. Berkeley: University of California Press.

Williams, George C. 1966. "Natural Selection, the Costs of Reproduction, and a Refinement of Lack's Principle." *American Naturalist* 100:687–690.

———. 1975. *Sex and Evolution.* Princeton, N.J: Princeton University Press.

———. 1985. "A Defense of Reductionism in Evolutionary Biology." In *Oxford Surveys in Evolutionary Biology.* Vol. 2, ed. R. Dawkins and M. Ridley, pp. 1–27. Oxford: Oxford University Press.

———. 1992. *Natural Selection: Domains, Levels, and Challenges.* New York: Oxford University Press.

Williams, T. M. 1990. "Heat Transfer in Elephants: Thermal Partitioning Based on Skin Temperature Profiles." *Journal of Zoology, London* 222:235–245.

Williamson, E. A., C. E. G. Tutin, M. E. Rogers, and M. Fernandez. 1990. "Composition of the Diet of Lowland Gorillas at Lopé in Gabon." *American Journal of Primatology* 21:265–277.

Willner, L. A., and Robert D. Martin. 1985. "Some Basic Principles of Mammalian Sexual Dimorphism." In *Human Sexual Dimorphism*, ed. J. Ghesquire, R. D. Martin, and F. Newcombe, pp. 1–42. London: Taylor and Francis.

Willoughby, D. P. 1978. *All about Gorillas.* South Brunswick, NJ, and New York: A. S. Barnes.

Wilmsen, E. 1982. "Exchange, Interaction, and Settlement in North-Western Botswana: Past and Present Perspectives." In *Settlement in Botswana*, ed. R. Hitchcock and M. Smith, pp. 98–109. Johannesburg: Heinemann.

Wilmsen, E. 1988. "The Antecedents of Contemporary Pastoralism in Western Ngamiland." *Botswana Notes and Records* 20:29–39.

Wilson, Allan C., and Rebecca L. Cann. 1992. "The Recent African Genesis of Humans." *Scientific American* 266:68–73.

Wilson, E. O. 1975. *Sociobiology.* Cambridge, MA: Harvard University Press.

Winkler, L. A. 1987. "Sexual Dimorphism in the Cranium of Infant and Juvenile Orangutans." *Folia Primatologica* 49:117–126.

Winkler, P., H. Loch, and C. Vogel. 1984. "Life History of Hanuman Langurs (*Presbytis entellus*): Reproductive Parameters, Infant Mortality, and Troop Development." *Folia Primatologica* 43:1–23.

Winn, S., G. A. Morelli, and E. Z. Tronick. 1989. "The Infant and the Group: A Look at Efe Care-Taking Practices." In *The Cultural Context of Infancy*, ed. J. K. Nugent, B. M. Lester, and T. B. Brazelton, pp. 87–109. Norwood, NJ: Ablex.

Winterhalder, B. 1980. "Environmental Analysis in Human Evolution and Adaptation Research." *Human Ecology* 8:134–170.

Winterhalder, B., and E. A. Smith. 1981. *Hunter-Gatherer Foraging Strategies.* Chicago: University Press.

Wiske, P. S., S. Epstein, N. H. Bell, S. F. Queener, J. Edmondson, and C. C. Johnston. 1979. "Increases in Immunoreactive Parathyroid Hormone with Age." *New England Journal of Medicine* 300:1419–1421.

Wolfe, Linda D. 1984. "Female Rank and Reproductive Success among Arashiyama B Japanese Macaques (*Macaca fuscata*)." *International Journal of Primatology* 5:133–143.

Wolfe, Linda D., and J. Patrick Gray. 1982a. "Subsistence Practices and Human Sexual Dimorphism of Stature." *Journal of Human Evolution* 11:575–580.

———. 1982b. "A Cross-Cultural Investigation into the Sexual Dimorphism of Stature." In *Sexual Dimorphism in Homo sapiens: A Question of Size?* ed. R. L. Hall, pp. 197–230. New York: Praeger.

Wolff, J. O. 1988. "Maternal Investment and Sex Ratio Adjustment in American Bison Calves." *Behavioral Ecology and Sociobiology* 23:127–133.

Wolfheim, Janet H. 1977. "Sex Differences in Behavior in a Group of Captive Juvenile Talapoin Monkeys (*Miopithecus talapoin*)." *Behaviour* 63: 110–128.

Wolpert, Lewis. 1991. *The Triumph of the Embryo.* New York: Oxford University Press.

Wolpoff, Milford H. 1975. "Sexual Dimorphism in the Australopithecines." In *Paleoanthropology, Morphology and Paleoecology*, ed. R. H. Tuttle, pp. 245–284. The Hague: Mouton.

———. 1976. "Some Aspects of the Evolution of Early Hominid Sexual Dimorphism." *Current Anthropology* 17:579–606.

Wood, Bernard A. 1985. "Sexual Dimorphism in the Fossil Record." In *Human Sexual Dimorphism*, ed. J. Ghesquiere, R. D. Martin, and F. Newcombe, pp. 105–123. London: Taylor and Francis.

Wood, E. J., and P. T. Bladon. 1985. *The Human Skin.* London: E. Arnold.

Wood, James W. 1990. "Fertility in Anthropological Populations." *Annual Review of Anthropology* 19:211–242.

Wood, James W., D. Lai, P. L. Johnson, K. L. Campbell, and I. A. Maslar. 1985. "Lactation and Birth Spacing in Highland New Guinea." *Journal of Biosocial Science* (suppl.) 9:159.

Wood, James W., George R. Milner, Henry C. Harpending, and Kenneth M. Weiss. 1992. "The Osteological Paradox." *Current Anthropology* 33: 343–370 (with comments).

Worthington-Roberts, B. S., and S. Rodwell Williams. 1989. *Nutrition in Pregnancy and Lactation.* St Louis: Times Mirror/Mosby.

Worthman, C. M. 1990. "Socioendocrinology: Key to a Fundamental Synergy." In *Socioendocrinology of Primate Reproduction*, ed. T. E. Ziegler and F. B. Bercovitch, pp. 187–212. New York: Wiley-Liss.

———. 1993. "Biocultural Interactions in Human Development." In *Juvenile Primates: Life History, Development, and Behavior*, ed. M. E. Pereira and L. A. Fairbanks, pp. 339–358. New York: Oxford University Press.

Worthman, C. M., C. L. Jenkins, J. F. Stallings, and D. Lai. 1993. "Attenuation of Nursing-Related Ovarian Suppression and High Fertility in Well-Nourished, Intensively Breast-Feeding Amele Women of Lowland Papua New Guinea." *Journal of Biosocial Science* 25:425–443.

Wrangham, Richard W. 1989. "On the Evolution of Ape Social Systems." *Social Science Information* 18:334–368.

Wrangham,Richard W., and J. Goodall. 1989.

"Chimpanzee Use of Medicinal Leaves." In *Understanding Chimpanzees*, ed. P. Heltne and L. Marquardt, pp. 22–37. Cambridge, MA: Harvard University Press.

Wrangham, Richard W., and B. Smuts. 1980. "Sex Differences in the Behavioural Ecology of Chimpanzees in the Gombe National Park, Tanzania." *Journal of Reproduction and Fertility* (suppl.) 28:13–31.

Wrangham, Richard W., A. P. Clark, and G. Isabirye-Basuta. 1992. "Female Social Relationships and Social Organization of Kibale chimpanzees." In *Topics in Primatology*. Vol. 1, *Human Origins*, ed. T. Nishida et al., pp. 81–98. Tokyo: Tokyo University Press.

Wright, P. 1984. "Biparental Care in *Aotus trivirgatus* and *Callicebus moloch*." In *Female Primates: Studies by Women Primatologists*, ed. M. Small, pp. 59–75. New York: Alan R. Liss.

Yamagiwa, J. 1987. "Male Life History and the Social Structure of Wild Mountain Gorillas (*Gorilla gorilla beringei*)." In *Evolution and Coadaptation in Biotic Communities*, ed. S. Kawano, J. H. Connell, and T. Hidaka, pp. 31–51. Tokyo: University of Tokyo Press .

Yellen, J. E. 1990. "The Transformation of the Kalahari !Kung." *Scientific American* 262:96–104.

Zhang, D.-Y., and G. Wang. 1994. "Evolutionary Stable Reproductive Strategies in Sexual Organisms: An Integrated Approach to Life-History Evolution and Sex Allocation." *American Naturalist* 144:65–75.

Ziegler, Toni E., and Fred B. Bercovitch, eds. 1989. *Socioendocrinology of Primate Reproduction*. New York: Wiley-Liss.

Zihlman, A. L. 1976. "Sexual Dimorphism and Its Behavioral Implications in Early Hominoids." In *Neuvième Congres, Union Internationale des Sciences prehistoriques et Protohistoriques: "Les Plus Anciens Hominides,"* ed. P. V. Tobias and Y. Coppens, pp. 268–293. Paris: CNRS.

———. 1981. "Women as Shapers of the Human Adaptation." In *Woman the Gatherer*, ed. F. Dahlberg, pp. 75–120. New Haven, CT: Yale University Press.

———. 1982. "Sexual Dimorphism in *Homo erectus*." In *L'Homo erectus et la Place de l'Homme Tautavel parmi les Hominides: Premier Congres International de Paleontologie Humaine*. Vol. 2, pp. 947–970. Paris: CNRS.

———. 1985. "*Australopithecus afarensis*: Two Sexes or Two Species?" In *Hominid Evolution: Past, Present and Future*, ed. P. V. Tobias, pp. 213–220. New York: Alan R. Liss.

———. 1992. "Locomotion as a Life History Character: The Contribution of Anatomy." *Journal of Human Evolution* 22:315–325.

———. 1993. "Sex Differences and Gender Hierarchies among Primates: An Evolutionary Perspective." In *Sex and Gender Hierarchies*, ed. B. Miller, pp. 32–56. Cambridge: University Press.

———. 1997a. "Natural History of Apes: Life-History Features in Females and Males." In *The Evolving Female: A Life-History Perspective*, ed. M. E. Morbeck, A. Galloway, and A. L. Zihlman, chap. 8. Princeton, NJ: Princeton University Press.

———. 1997b. "Women's Bodies, Women's Lives: An Evolutionary Perspective." In *The Evolving Female: A Life-History Perspective*, ed. M. E. Morbeck, A. Galloway, and A. L. Zihlman, chap. 13. Princeton, NJ: Princeton University Press.

Zihlman, A. L., and L. Brunker. 1979. "Hominid Bipedalism: Then and Now." *Yearbook of Physical Anthropology* 22:132–162.

Zihlman, A. L., and B. Cohn. 1988. "The Adaptive Response of Human Skin to the Savanna." *Human Evolution* 3:397–409.

Zihlman, A. L., and D. L. Cramer. 1978. "Skeletal Differences between Pygmy (*Pan paniscus*) and Common Chimpanzees (*Pan troglodytes*)." *Folia Primatologica* 29:86–94.

Zihlman, A. L., and N. Tanner. 1978. "Gathering and the Hominid Adaptation." In *Female Hierarchies*, ed. L. Tiger and H. T. Fowler, pp. 163–194. Chicago: Beresford Book Service.

Zihlman, A. L., M. E. Morbeck, and J. Goodall. 1990. "Skeletal Biology and Individual Life History of Gombe Chimpanzees." *Journal of Zoology, London* 221:37–6l.

Zuckerman, S. 1932. *The Social Life of Monkeys and Apes*. London: Kegan Paul Trench Trubner.

Index